D1571261

VOLUME FIVE HUNDRED AND THREE

METHODS IN ENZYMOLOGY

Protein Engineering for Therapeutics, Part B

METHODS IN ENZYMOLOGY

VOLUME FIVE HUNDRED AND THREE

METHODS IN ENZYMOLOGY

Protein Engineering for Therapeutics, Part B

EDITED BY

K. DANE WITTRUP

Departments of Chemical Engineering & Biological Engineering
Koch Institute for Integrative Cancer Research
Massachusetts Institute of Technology
Cambridge
Massachusetts
USA

GREGORY L. VERDINE

Departments of Chemistry and Chemical Biology
Stem Cell and Regenerative Biology, and
Molecular and Cellular Biology, Harvard University, and
Program in Cancer Chemical Biology
Dana - Farber Cancer Institute
Boston
Massachusetts
USA

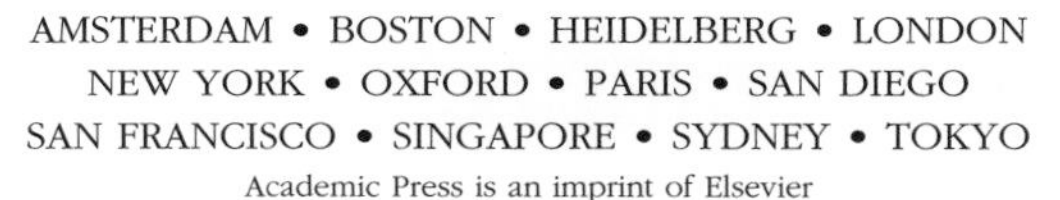
AMSTERDAM • BOSTON • HEIDELBERG • LONDON
NEW YORK • OXFORD • PARIS • SAN DIEGO
SAN FRANCISCO • SINGAPORE • SYDNEY • TOKYO
Academic Press is an imprint of Elsevier

Academic Press is an imprint of Elsevier
525 B Street, Suite 1900, San Diego, CA 92101-4495, USA
225 Wyman Street, Waltham, MA 02451, USA
32 Jamestown Road, London NW1 7BY, UK

First edition 2012

For information on all Academic Press publications
visit our website at elsevierdirect.com

ISBN: 978-0-12-396962-0
ISSN: 0076-6879

Printed and bound in United States of America
12 13 14 10 9 8 7 6 5 4 3 2 1

Contents

Contributors ix
Preface xiii
Volumes in series xv

Section IV. Peptides 1

1. Stapled Peptides for Intracellular Drug Targets 3

Gregory L. Verdine and Gerard J. Hilinski

1. Introduction 4
2. All-Hydrocarbon Stapled α-Helical Peptides 6
3. The Design of Stapled Peptides 9
4. Stapled Peptide Synthesis 12
5. Olefin Metathesis 16
6. N-terminal and Internal Modifications 16
7. Cleavage of Stapled Peptides from the Solid Support 19
8. Stapled Peptide Purification 20
9. Biophysical Characterization 20
10. Cell Permeability 23
11. *In Vitro* Target Interaction and Activity Assays 26
12. *In Vivo* Efficacy 27
13. Strategies for Stapled Peptide Optimization 28
14. Summary 30

Acknowledgments 30
References 30

2. Mapping of Vascular ZIP Codes by Phage Display 35

Tambet Teesalu, Kazuki N. Sugahara, and Erkki Ruoslahti

1. Introduction 36
2. Methods 41
3. Concluding Remarks and Perspectives 52

Acknowledgments 53
References 53

3. Engineering Cyclic Peptide Toxins 57
Richard J. Clark and David J. Craik

1. Introduction 58
2. Peptide Design 59
3. Peptide Synthesis 60
4. Structural Analysis 66
5. Stability Assays 68
6. Special Considerations for Cyclic Peptides 69
Acknowledgments 71
References 71

4. Peptide Discovery Using Bacterial Display and Flow Cytometry 75
Jennifer A. Getz, Tobias D. Schoep, and Patrick S. Daugherty

1. Introduction 76
2. Protocols 79
3. Conclusions 94
Acknowledgments 95
References 95

Section V. Scaffolds 99

5. Designed Ankyrin Repeat Proteins (DARPins): From Research to Therapy 101
Rastislav Tamaskovic, Manuel Simon, Nikolas Stefan, Martin Schwill, and Andreas Plückthun

1. Introduction 102
2. Applications of DARPins 107
3. Protocols for DARPins in Biomedical Applications 112
Acknowledgment 129
References 129

6. Target-Binding Proteins Based on the 10th Human Fibronectin Type III Domain (^{10}Fn3) 135
Shohei Koide, Akiko Koide, and Daša Lipovšek

1. Introduction 136
2. Library Design 137
3. Choice of Selection Platform 139
4. Phage Display, mRNA Display, and Yeast-Surface Display of ^{10}Fn3-Based Libraries 141
5. Production 151

6. Conclusion 153
Acknowledgments 154
References 154

7. Anticalins: Small Engineered Binding Proteins Based on the Lipocalin Scaffold 157

Michaela Gebauer and Arne Skerra

1. Introduction 158
2. Cloning and Expression of Lipocalins and Anticalins in *E. coli* 166
3. Construction of a Genetic Anticalin Library 170
4. Preparation and Selection of a Phage Display Library for Anticalins 172
5. Preparation and Selection of a Bacterial Surface Display Library for Anticalins 176
6. Colony Screening for Anticalins with Specific Target-Binding Activity 177
7. Screening for Anticalins with Specific Target-Binding Activity Using Microtiter Plate Expression in *E. coli* 179
8. Measuring Target Affinity of Anticalins in an ELISA 181
9. Measuring Target Affinity of Anticalins via Surface Plasmon Resonance 182
10. Application of Anticalins in Biochemical Research and Drug Development 184
References 186

8. T Cell Receptor Engineering 189

Jennifer D. Stone, Adam S. Chervin, David H. Aggen, and David M. Kranz

1. Introduction 190
2. Stability and Affinity Engineering of T Cell Receptors by Yeast Surface Display 192
3. Affinity Engineering and Selection of T Cell Receptors by T Cell Display 200
4. Expression, Purification, and Applications of Soluble scTv Proteins 207
5. Recipes for Media and Buffers 215
Acknowledgments 218
References 219

9. Engineering Knottins as Novel Binding Agents 223

Sarah J. Moore and Jennifer R. Cochran

1. Introduction 224
2. Knottins as Scaffolds for Engineering Molecular Recognition 225
3. Engineering Knottins by Yeast Surface Display 227
4. Knottin Library Construction 230
5. Screening Yeast-Displayed Knottin Libraries 234
6. Knottin production by chemical synthesis or recombinant expression 237
7. Cell binding assays 243

8. Summary 247
Acknowledgments 247
References 247

Section VI. Pharmacokinetics **253**

10. Practical Theoretic Guidance for the Design of Tumor-Targeting Agents **255**
K. Dane Wittrup, Greg M. Thurber, Michael M. Schmidt, and John J. Rhoden

1. Introduction 256
2. What Molecular Size Is Best for Tumor Uptake? 256
3. Will Targeting Increase Nanoparticle Accumulation in a Tumor? 259
4. How Does Affinity Affect Biodistribution? 261
5. What Dose Is Necessary in Order to Overcome the "Binding Site Barrier"? 263
6. Conclusions 264
References 265

11. Reengineering Biopharmaceuticals for Targeted Delivery Across the Blood–Brain Barrier **269**
William M. Pardridge and Ruben J. Boado

1. Introduction 270
2. Blood–Brain Barrier Receptor-Mediated Transport and Molecular Trojan Horses 271
3. Reengineering Recombinant Proteins for Targeted Brain Delivery 274
4. Genetic Engineering of Expression Plasmid DNA Encoding IgG Fusion Proteins 278
5. Pharmacokinetics and Brain Uptake of IgG Fusion Proteins 279
6. CNS Pharmacological Effects of IgG Fusion Proteins 284
7. Immune Response Against IgG Fusion Proteins 287
8. Summary 288
References 289

12. Engineering and Identifying Supercharged Proteins for Macromolecule Delivery into Mammalian Cells **293**
David B. Thompson, James J. Cronican, and David R. Liu

1. Introduction 294
2. Methods 301
3. Conclusion 317
References 318

Author Index *321*
Subject Index *341*

Contributors

David H. Aggen

Department of Biochemistry, University of Illinois, Urbana, Illinois, USA

Ruben J. Boado

Department of Medicine, UCLA, Los Angeles, California, USA; ArmaGen Technologies, Inc., Santa Monica, California, USA

Adam S. Chervin

Department of Biochemistry, University of Illinois, Urbana, Illinois, USA

Richard J. Clark

School of Biomedical Sciences, The University of Queensland, Brisbane, Queensland, Australia

Jennifer R. Cochran

Department of Bioengineering, Cancer Institute, and Bio-X Program, Stanford University, Stanford, California, USA

David J. Craik

Institute for Molecular Bioscience, The University of Queensland, Brisbane, Queensland, Australia

James J. Cronican

Howard Hughes Medical Institute, Department of Chemistry and Chemical Biology, Harvard University, Cambridge, Massachusetts, USA

Patrick S. Daugherty

Department of Chemical Engineering, Institute for Collaborative Biotechnologies, University of California, Santa Barbara, California, USA

Michaela Gebauer

Lehrstuhl für Biologische Chemie, Technische Universität München, Freising-Weihenstephan, Germany

Jennifer A. Getz

Department of Chemical Engineering, Institute for Collaborative Biotechnologies, University of California, Santa Barbara, California, USA

Gerard J. Hilinski

Department of Stem Cell and Regenerative Biology, Harvard University, Cambridge, Massachusetts, USA

Akiko Koide
Department of Biochemistry and Molecular Biology, The University of Chicago, Chicago, Illinois, USA

Shohei Koide
Department of Biochemistry and Molecular Biology, The University of Chicago, Chicago, Illinois, USA

David M. Kranz
Department of Biochemistry, University of Illinois, Urbana, Illinois, USA

Daša Lipovšek
Department of Protein Design, Adnexus, A Bristol-Myers Squibb R&D, Company Waltham, Massachusetts, USA

David R. Liu
Howard Hughes Medical Institute, Department of Chemistry and Chemical Biology, Harvard University, Cambridge, Massachusetts, USA

Sarah J. Moore
Department of Bioengineering, Cancer Institute, and Bio-X Program, Stanford University, Stanford, California, USA

William M. Pardridge
Department of Medicine, UCLA, Los Angeles, California, USA

Andreas Plückthun
Department of Biochemistry, University of Zurich, Winterthurerstrasse, Zurich, Switzerland

John J. Rhoden
Department of Chemical Engineering, Massachusetts Institute of Technology, Cambridge, Massachusetts, USA

Erkki Ruoslahti
Center for Nanomedicine, Sanford-Burnham Medical Research Institute at UCSB, Santa Barbara, California, USA; Cancer Center, Sanford-Burnham Medical Research Institute, La Jolla, California, USA

Michael M. Schmidt
Department of Biological Engineering, Massachusetts Institute of Technology, Cambridge, Massachusetts, USA

Tobias D. Schoep
Department of Chemical Engineering, Institute for Collaborative Biotechnologies, University of California, Santa Barbara, California, USA

Martin Schwill
Department of Biochemistry, University of Zurich, Winterthurerstrasse, Zurich, Switzerland

Manuel Simon
Department of Biochemistry, University of Zurich, Winterthurerstrasse, Zurich, Switzerland

Arne Skerra
Lehrstuhl für Biologische Chemie, Technische Universität München, Freising-Weihenstephan, Germany

Nikolas Stefan
Department of Biochemistry, University of Zurich, Winterthurerstrasse, Zurich, Switzerland

Jennifer D. Stone
Department of Biochemistry, University of Illinois, Urbana, Illinois, USA

Kazuki N. Sugahara
Cancer Center, Sanford-Burnham Medical Research Institute, La Jolla, California, USA

Rastislav Tamaskovic
Department of Biochemistry, University of Zurich, Winterthurerstrasse, Zurich, Switzerland

Tambet Teesalu
Center for Nanomedicine, Sanford-Burnham Medical Research Institute at UCSB, Santa Barbara, California, USA; Cancer Center, Sanford-Burnham Medical Research Institute, La Jolla, California, USA

David B. Thompson
Howard Hughes Medical Institute, Department of Chemistry and Chemical Biology, Harvard University, Cambridge, Massachusetts, USA

Greg M. Thurber
Department of Chemical Engineering, Massachusetts Institute of Technology, Cambridge, Massachusetts, USA

Gregory L. Verdine
Departments of Chemistry and Chemical Biology, Stem Cell and Regenerative Biology, and Molecular and Cellular Biology, Harvard University, and Program in Cancer Chemical Biology, Dana - Farber Cancer Institute, Boston, Massachusetts, USA

K. Dane Wittrup
Departments of Chemical Engineering & Biological Engineering, Koch Institute for Integrative Cancer Research, Massachusetts Institute of Technology, Cambridge, Massachusetts, USA

Preface

These two volumes of *Methods in Enzymology* cover engineering approaches to the development of protein biopharmaceuticals, which represent a significant and rapidly growing proportion of drug sales. Particular advantages of proteins as drugs relative to small organic molecules include high affinity and specificity afforded by a larger molecular recognition surface and much lower probability of off-target toxicities due to metabolic byproducts. The primary disadvantage to date has been the pharmacokinetic inaccessibility of intracellular drug targets to proteins and peptides, although vigorous efforts at overcoming this limitation are beginning to bear fruit.

The protein biopharmaceutical field was born with the advent of recombinant DNA expression systems for natural human protein agonists such as insulin, human growth hormone, erythropoietin, and granulocyte colony-stimulating factor. Humanization of mouse antibodies opened the playing field to novel molecules that specifically bound to and blocked receptors and ligands important in a variety of diseases. In vitro directed evolution technologies have enabled further exploration of nonnative structures and topologies for target antagonism or delivery of therapeutic payloads. It should be pointed out that to date protein engineering has made minimal progress in engineering the agonists which were the first protein biopharmaceuticals. Such engineered agonists could enable more subtle redirection of innate homeostatic regulatory pathways than the relatively crude tools of antibody antagonism or parenteral oversupply of naturally occurring protein and peptide agonists.

We provide here only an overview representing the wide spectrum of approaches in this field. The first section covers some aspects of antibodies, by far the dominant class of protein biopharmaceuticals at the present time. The second section provides examples where protein targeting is exploited to deliver a payload conjugated to the proteins. In the third section of the first volume, examples of engineered therapeutic enzymes are provided. In the first section of the second volume, peptides are considered, enabling chemical synthesis and more facile intracellular delivery, given their smaller size. The second section describes a number of leading efforts to engineer molecular recognition onto a scaffold other than that of an antibody.

In the third and final section, pharmacokinetics of protein drugs is discussed with respect to delivery to tumors, intracellular targets, and across the blood–brain barrier.

We particularly thank the authors of these contributions for their thorough and clear exposition of the state of the art in their respective specializations.

K. Dane Wittrup and Gregory L. Verdine

Methods in Enzymology

Volume I. Preparation and Assay of Enzymes
Edited by Sidney P. Colowick and Nathan O. Kaplan

Volume II. Preparation and Assay of Enzymes
Edited by Sidney P. Colowick and Nathan O. Kaplan

Volume III. Preparation and Assay of Substrates
Edited by Sidney P. Colowick and Nathan O. Kaplan

Volume IV. Special Techniques for the Enzymologist
Edited by Sidney P. Colowick and Nathan O. Kaplan

Volume V. Preparation and Assay of Enzymes
Edited by Sidney P. Colowick and Nathan O. Kaplan

Volume VI. Preparation and Assay of Enzymes *(Continued)*
Preparation and Assay of Substrates
Special Techniques
Edited by Sidney P. Colowick and Nathan O. Kaplan

Volume VII. Cumulative Subject Index
Edited by Sidney P. Colowick and Nathan O. Kaplan

Volume VIII. Complex Carbohydrates
Edited by Elizabeth F. Neufeld and Victor Ginsburg

Volume IX. Carbohydrate Metabolism
Edited by Willis A. Wood

Volume X. Oxidation and Phosphorylation
Edited by Ronald W. Estabrook and Maynard E. Pullman

Volume XI. Enzyme Structure
Edited by C. H. W. Hirs

Volume XII. Nucleic Acids (Parts A and B)
Edited by Lawrence Grossman and Kivie Moldave

Volume XIII. Citric Acid Cycle
Edited by J. M. Lowenstein

Volume XIV. Lipids
Edited by J. M. Lowenstein

Volume XV. Steroids and Terpenoids
Edited by Raymond B. Clayton

VOLUME XVI. Fast Reactions
Edited by KENNETH KUSTIN

VOLUME XVII. Metabolism of Amino Acids and Amines (Parts A and B)
Edited by HERBERT TABOR AND CELIA WHITE TABOR

VOLUME XVIII. Vitamins and Coenzymes (Parts A, B, and C)
Edited by DONALD B. MCCORMICK AND LEMUEL D. WRIGHT

VOLUME XIX. Proteolytic Enzymes
Edited by GERTRUDE E. PERLMANN AND LASZLO LORAND

VOLUME XX. Nucleic Acids and Protein Synthesis (Part C)
Edited by KIVIE MOLDAVE AND LAWRENCE GROSSMAN

VOLUME XXI. Nucleic Acids (Part D)
Edited by LAWRENCE GROSSMAN AND KIVIE MOLDAVE

VOLUME XXII. Enzyme Purification and Related Techniques
Edited by WILLIAM B. JAKOBY

VOLUME XXIII. Photosynthesis (Part A)
Edited by ANTHONY SAN PIETRO

VOLUME XXIV. Photosynthesis and Nitrogen Fixation (Part B)
Edited by ANTHONY SAN PIETRO

VOLUME XXV. Enzyme Structure (Part B)
Edited by C. H. W. HIRS AND SERGE N. TIMASHEFF

VOLUME XXVI. Enzyme Structure (Part C)
Edited by C. H. W. HIRS AND SERGE N. TIMASHEFF

VOLUME XXVII. Enzyme Structure (Part D)
Edited by C. H. W. HIRS AND SERGE N. TIMASHEFF

VOLUME XXVIII. Complex Carbohydrates (Part B)
Edited by VICTOR GINSBURG

VOLUME XXIX. Nucleic Acids and Protein Synthesis (Part E)
Edited by LAWRENCE GROSSMAN AND KIVIE MOLDAVE

VOLUME XXX. Nucleic Acids and Protein Synthesis (Part F)
Edited by KIVIE MOLDAVE AND LAWRENCE GROSSMAN

VOLUME XXXI. Biomembranes (Part A)
Edited by SIDNEY FLEISCHER AND LESTER PACKER

VOLUME XXXII. Biomembranes (Part B)
Edited by SIDNEY FLEISCHER AND LESTER PACKER

VOLUME XXXIII. Cumulative Subject Index Volumes I–XXX
Edited by MARTHA G. DENNIS AND EDWARD A. DENNIS

VOLUME XXXIV. Affinity Techniques (Enzyme Purification: Part B)
Edited by WILLIAM B. JAKOBY AND MEIR WILCHEK

Volume XXXV. Lipids (Part B)
Edited by John M. Lowenstein

Volume XXXVI. Hormone Action (Part A: Steroid Hormones)
Edited by Bert W. O'Malley and Joel G. Hardman

Volume XXXVII. Hormone Action (Part B: Peptide Hormones)
Edited by Bert W. O'Malley and Joel G. Hardman

Volume XXXVIII. Hormone Action (Part C: Cyclic Nucleotides)
Edited by Joel G. Hardman and Bert W. O'Malley

Volume XXXIX. Hormone Action (Part D: Isolated Cells, Tissues, and Organ Systems)
Edited by Joel G. Hardman and Bert W. O'Malley

Volume XL. Hormone Action (Part E: Nuclear Structure and Function)
Edited by Bert W. O'Malley and Joel G. Hardman

Volume XLI. Carbohydrate Metabolism (Part B)
Edited by W. A. Wood

Volume XLII. Carbohydrate Metabolism (Part C)
Edited by W. A. Wood

Volume XLIII. Antibiotics
Edited by John H. Hash

Volume XLIV. Immobilized Enzymes
Edited by Klaus Mosbach

Volume XLV. Proteolytic Enzymes (Part B)
Edited by Laszlo Lorand

Volume XLVI. Affinity Labeling
Edited by William B. Jakoby and Meir Wilchek

Volume XLVII. Enzyme Structure (Part E)
Edited by C. H. W. Hirs and Serge N. Timasheff

Volume XLVIII. Enzyme Structure (Part F)
Edited by C. H. W. Hirs and Serge N. Timasheff

Volume XLIX. Enzyme Structure (Part G)
Edited by C. H. W. Hirs and Serge N. Timasheff

Volume L. Complex Carbohydrates (Part C)
Edited by Victor Ginsburg

Volume LI. Purine and Pyrimidine Nucleotide Metabolism
Edited by Patricia A. Hoffee and Mary Ellen Jones

Volume LII. Biomembranes (Part C: Biological Oxidations)
Edited by Sidney Fleischer and Lester Packer

VOLUME LIII. Biomembranes (Part D: Biological Oxidations)
Edited by SIDNEY FLEISCHER AND LESTER PACKER

VOLUME LIV. Biomembranes (Part E: Biological Oxidations)
Edited by SIDNEY FLEISCHER AND LESTER PACKER

VOLUME LV. Biomembranes (Part F: Bioenergetics)
Edited by SIDNEY FLEISCHER AND LESTER PACKER

VOLUME LVI. Biomembranes (Part G: Bioenergetics)
Edited by SIDNEY FLEISCHER AND LESTER PACKER

VOLUME LVII. Bioluminescence and Chemiluminescence
Edited by MARLENE A. DELUCA

VOLUME LVIII. Cell Culture
Edited by WILLIAM B. JAKOBY AND IRA PASTAN

VOLUME LIX. Nucleic Acids and Protein Synthesis (Part G)
Edited by KIVIE MOLDAVE AND LAWRENCE GROSSMAN

VOLUME LX. Nucleic Acids and Protein Synthesis (Part H)
Edited by KIVIE MOLDAVE AND LAWRENCE GROSSMAN

VOLUME 61. Enzyme Structure (Part H)
Edited by C. H. W. HIRS AND SERGE N. TIMASHEFF

VOLUME 62. Vitamins and Coenzymes (Part D)
Edited by DONALD B. MCCORMICK AND LEMUEL D. WRIGHT

VOLUME 63. Enzyme Kinetics and Mechanism (Part A: Initial Rate and Inhibitor Methods)
Edited by DANIEL L. PURICH

VOLUME 64. Enzyme Kinetics and Mechanism (Part B: Isotopic Probes and Complex Enzyme Systems)
Edited by DANIEL L. PURICH

VOLUME 65. Nucleic Acids (Part I)
Edited by LAWRENCE GROSSMAN AND KIVIE MOLDAVE

VOLUME 66. Vitamins and Coenzymes (Part E)
Edited by DONALD B. MCCORMICK AND LEMUEL D. WRIGHT

VOLUME 67. Vitamins and Coenzymes (Part F)
Edited by DONALD B. MCCORMICK AND LEMUEL D. WRIGHT

VOLUME 68. Recombinant DNA
Edited by RAY WU

VOLUME 69. Photosynthesis and Nitrogen Fixation (Part C)
Edited by ANTHONY SAN PIETRO

VOLUME 70. Immunochemical Techniques (Part A)
Edited by HELEN VAN VUNAKIS AND JOHN J. LANGONE

VOLUME 71. Lipids (Part C)
Edited by JOHN M. LOWENSTEIN

VOLUME 72. Lipids (Part D)
Edited by JOHN M. LOWENSTEIN

VOLUME 73. Immunochemical Techniques (Part B)
Edited by JOHN J. LANGONE AND HELEN VAN VUNAKIS

VOLUME 74. Immunochemical Techniques (Part C)
Edited by JOHN J. LANGONE AND HELEN VAN VUNAKIS

VOLUME 75. Cumulative Subject Index Volumes XXXI, XXXII, XXXIV–LX
Edited by EDWARD A. DENNIS AND MARTHA G. DENNIS

VOLUME 76. Hemoglobins
Edited by ERALDO ANTONINI, LUIGI ROSSI-BERNARDI, AND EMILIA CHIANCONE

VOLUME 77. Detoxication and Drug Metabolism
Edited by WILLIAM B. JAKOBY

VOLUME 78. Interferons (Part A)
Edited by SIDNEY PESTKA

VOLUME 79. Interferons (Part B)
Edited by SIDNEY PESTKA

VOLUME 80. Proteolytic Enzymes (Part C)
Edited by LASZLO LORAND

VOLUME 81. Biomembranes (Part H: Visual Pigments and Purple Membranes, I)
Edited by LESTER PACKER

VOLUME 82. Structural and Contractile Proteins (Part A: Extracellular Matrix)
Edited by LEON W. CUNNINGHAM AND DIXIE W. FREDERIKSEN

VOLUME 83. Complex Carbohydrates (Part D)
Edited by VICTOR GINSBURG

VOLUME 84. Immunochemical Techniques (Part D: Selected Immunoassays)
Edited by JOHN J. LANGONE AND HELEN VAN VUNAKIS

VOLUME 85. Structural and Contractile Proteins (Part B: The Contractile Apparatus and the Cytoskeleton)
Edited by DIXIE W. FREDERIKSEN AND LEON W. CUNNINGHAM

VOLUME 86. Prostaglandins and Arachidonate Metabolites
Edited by WILLIAM E. M. LANDS AND WILLIAM L. SMITH

VOLUME 87. Enzyme Kinetics and Mechanism (Part C: Intermediates, Stereo-chemistry, and Rate Studies)
Edited by DANIEL L. PURICH

VOLUME 88. Biomembranes (Part I: Visual Pigments and Purple Membranes, II)
Edited by LESTER PACKER

VOLUME 89. Carbohydrate Metabolism (Part D)
Edited by WILLIS A. WOOD

VOLUME 90. Carbohydrate Metabolism (Part E)
Edited by WILLIS A. WOOD

VOLUME 91. Enzyme Structure (Part I)
Edited by C. H. W. HIRS AND SERGE N. TIMASHEFF

VOLUME 92. Immunochemical Techniques (Part E: Monoclonal Antibodies and General Immunoassay Methods)
Edited by JOHN J. LANGONE AND HELEN VAN VUNAKIS

VOLUME 93. Immunochemical Techniques (Part F: Conventional Antibodies, Fc Receptors, and Cytotoxicity)
Edited by JOHN J. LANGONE AND HELEN VAN VUNAKIS

VOLUME 94. Polyamines
Edited by HERBERT TABOR AND CELIA WHITE TABOR

VOLUME 95. Cumulative Subject Index Volumes 61–74, 76–80
Edited by EDWARD A. DENNIS AND MARTHA G. DENNIS

VOLUME 96. Biomembranes [Part J: Membrane Biogenesis: Assembly and Targeting (General Methods; Eukaryotes)]
Edited by SIDNEY FLEISCHER AND BECCA FLEISCHER

VOLUME 97. Biomembranes [Part K: Membrane Biogenesis: Assembly and Targeting (Prokaryotes, Mitochondria, and Chloroplasts)]
Edited by SIDNEY FLEISCHER AND BECCA FLEISCHER

VOLUME 98. Biomembranes (Part L: Membrane Biogenesis: Processing and Recycling)
Edited by SIDNEY FLEISCHER AND BECCA FLEISCHER

VOLUME 99. Hormone Action (Part F: Protein Kinases)
Edited by JACKIE D. CORBIN AND JOEL G. HARDMAN

VOLUME 100. Recombinant DNA (Part B)
Edited by RAY WU, LAWRENCE GROSSMAN, AND KIVIE MOLDAVE

VOLUME 101. Recombinant DNA (Part C)
Edited by RAY WU, LAWRENCE GROSSMAN, AND KIVIE MOLDAVE

VOLUME 102. Hormone Action (Part G: Calmodulin and Calcium-Binding Proteins)
Edited by ANTHONY R. MEANS AND BERT W. O'MALLEY

VOLUME 103. Hormone Action (Part H: Neuroendocrine Peptides)
Edited by P. MICHAEL CONN

VOLUME 104. Enzyme Purification and Related Techniques (Part C)
Edited by WILLIAM B. JAKOBY

VOLUME 105. Oxygen Radicals in Biological Systems
Edited by LESTER PACKER

VOLUME 106. Posttranslational Modifications (Part A)
Edited by FINN WOLD AND KIVIE MOLDAVE

VOLUME 107. Posttranslational Modifications (Part B)
Edited by FINN WOLD AND KIVIE MOLDAVE

VOLUME 108. Immunochemical Techniques (Part G: Separation and Characterization of Lymphoid Cells)
Edited by GIOVANNI DI SABATO, JOHN J. LANGONE, AND HELEN VAN VUNAKIS

VOLUME 109. Hormone Action (Part I: Peptide Hormones)
Edited by LUTZ BIRNBAUMER AND BERT W. O'MALLEY

VOLUME 110. Steroids and Isoprenoids (Part A)
Edited by JOHN H. LAW AND HANS C. RILLING

VOLUME 111. Steroids and Isoprenoids (Part B)
Edited by JOHN H. LAW AND HANS C. RILLING

VOLUME 112. Drug and Enzyme Targeting (Part A)
Edited by KENNETH J. WIDDER AND RALPH GREEN

VOLUME 113. Glutamate, Glutamine, Glutathione, and Related Compounds
Edited by ALTON MEISTER

VOLUME 114. Diffraction Methods for Biological Macromolecules (Part A)
Edited by HAROLD W. WYCKOFF, C. H. W. HIRS, AND SERGE N. TIMASHEFF

VOLUME 115. Diffraction Methods for Biological Macromolecules (Part B)
Edited by HAROLD W. WYCKOFF, C. H. W. HIRS, AND SERGE N. TIMASHEFF

VOLUME 116. Immunochemical Techniques
(Part H: Effectors and Mediators of Lymphoid Cell Functions)
Edited by GIOVANNI DI SABATO, JOHN J. LANGONE, AND HELEN VAN VUNAKIS

VOLUME 117. Enzyme Structure (Part J)
Edited by C. H. W. HIRS AND SERGE N. TIMASHEFF

VOLUME 118. Plant Molecular Biology
Edited by ARTHUR WEISSBACH AND HERBERT WEISSBACH

VOLUME 119. Interferons (Part C)
Edited by SIDNEY PESTKA

VOLUME 120. Cumulative Subject Index Volumes 81–94, 96–101

VOLUME 121. Immunochemical Techniques (Part I: Hybridoma Technology and Monoclonal Antibodies)
Edited by JOHN J. LANGONE AND HELEN VAN VUNAKIS

VOLUME 122. Vitamins and Coenzymes (Part G)
Edited by FRANK CHYTIL AND DONALD B. MCCORMICK

VOLUME 123. Vitamins and Coenzymes (Part H)
Edited by FRANK CHYTIL AND DONALD B. MCCORMICK

VOLUME 124. Hormone Action (Part J: Neuroendocrine Peptides)
Edited by P. MICHAEL CONN

VOLUME 125. Biomembranes (Part M: Transport in Bacteria, Mitochondria, and Chloroplasts: General Approaches and Transport Systems)
Edited by SIDNEY FLEISCHER AND BECCA FLEISCHER

VOLUME 126. Biomembranes (Part N: Transport in Bacteria, Mitochondria, and Chloroplasts: Protonmotive Force)
Edited by SIDNEY FLEISCHER AND BECCA FLEISCHER

VOLUME 127. Biomembranes (Part O: Protons and Water: Structure and Translocation)
Edited by LESTER PACKER

VOLUME 128. Plasma Lipoproteins (Part A: Preparation, Structure, and Molecular Biology)
Edited by JERE P. SEGREST AND JOHN J. ALBERS

VOLUME 129. Plasma Lipoproteins (Part B: Characterization, Cell Biology, and Metabolism)
Edited by JOHN J. ALBERS AND JERE P. SEGREST

VOLUME 130. Enzyme Structure (Part K)
Edited by C. H. W. HIRS AND SERGE N. TIMASHEFF

VOLUME 131. Enzyme Structure (Part L)
Edited by C. H. W. HIRS AND SERGE N. TIMASHEFF

VOLUME 132. Immunochemical Techniques (Part J: Phagocytosis and Cell-Mediated Cytotoxicity)
Edited by GIOVANNI DI SABATO AND JOHANNES EVERSE

VOLUME 133. Bioluminescence and Chemiluminescence (Part B)
Edited by MARLENE DELUCA AND WILLIAM D. MCELROY

VOLUME 134. Structural and Contractile Proteins (Part C: The Contractile Apparatus and the Cytoskeleton)
Edited by RICHARD B. VALLEE

VOLUME 135. Immobilized Enzymes and Cells (Part B)
Edited by KLAUS MOSBACH

VOLUME 136. Immobilized Enzymes and Cells (Part C)
Edited by KLAUS MOSBACH

VOLUME 137. Immobilized Enzymes and Cells (Part D)
Edited by KLAUS MOSBACH

VOLUME 138. Complex Carbohydrates (Part E)
Edited by VICTOR GINSBURG

VOLUME 139. Cellular Regulators (Part A: Calcium- and Calmodulin-Binding Proteins)
Edited by ANTHONY R. MEANS AND P. MICHAEL CONN

VOLUME 140. Cumulative Subject Index Volumes 102–119, 121–134

VOLUME 141. Cellular Regulators (Part B: Calcium and Lipids)
Edited by P. MICHAEL CONN AND ANTHONY R. MEANS

VOLUME 142. Metabolism of Aromatic Amino Acids and Amines
Edited by SEYMOUR KAUFMAN

VOLUME 143. Sulfur and Sulfur Amino Acids
Edited by WILLIAM B. JAKOBY AND OWEN GRIFFITH

VOLUME 144. Structural and Contractile Proteins (Part D: Extracellular Matrix)
Edited by LEON W. CUNNINGHAM

VOLUME 145. Structural and Contractile Proteins (Part E: Extracellular Matrix)
Edited by LEON W. CUNNINGHAM

VOLUME 146. Peptide Growth Factors (Part A)
Edited by DAVID BARNES AND DAVID A. SIRBASKU

VOLUME 147. Peptide Growth Factors (Part B)
Edited by DAVID BARNES AND DAVID A. SIRBASKU

VOLUME 148. Plant Cell Membranes
Edited by LESTER PACKER AND ROLAND DOUCE

VOLUME 149. Drug and Enzyme Targeting (Part B)
Edited by RALPH GREEN AND KENNETH J. WIDDER

VOLUME 150. Immunochemical Techniques (Part K: *In Vitro* Models of B and T Cell Functions and Lymphoid Cell Receptors)
Edited by GIOVANNI DI SABATO

VOLUME 151. Molecular Genetics of Mammalian Cells
Edited by MICHAEL M. GOTTESMAN

VOLUME 152. Guide to Molecular Cloning Techniques
Edited by SHELBY L. BERGER AND ALAN R. KIMMEL

VOLUME 153. Recombinant DNA (Part D)
Edited by RAY WU AND LAWRENCE GROSSMAN

VOLUME 154. Recombinant DNA (Part E)
Edited by RAY WU AND LAWRENCE GROSSMAN

VOLUME 155. Recombinant DNA (Part F)
Edited by RAY WU

VOLUME 156. Biomembranes (Part P: ATP-Driven Pumps and Related Transport: The Na, K-Pump)
Edited by SIDNEY FLEISCHER AND BECCA FLEISCHER

Volume 157. Biomembranes (Part Q: ATP-Driven Pumps and Related Transport: Calcium, Proton, and Potassium Pumps)
Edited by Sidney Fleischer and Becca Fleischer

Volume 158. Metalloproteins (Part A)
Edited by James F. Riordan and Bert L. Vallee

Volume 159. Initiation and Termination of Cyclic Nucleotide Action
Edited by Jackie D. Corbin and Roger A. Johnson

Volume 160. Biomass (Part A: Cellulose and Hemicellulose)
Edited by Willis A. Wood and Scott T. Kellogg

Volume 161. Biomass (Part B: Lignin, Pectin, and Chitin)
Edited by Willis A. Wood and Scott T. Kellogg

Volume 162. Immunochemical Techniques (Part L: Chemotaxis and Inflammation)
Edited by Giovanni Di Sabato

Volume 163. Immunochemical Techniques (Part M: Chemotaxis and Inflammation)
Edited by Giovanni Di Sabato

Volume 164. Ribosomes
Edited by Harry F. Noller, Jr., and Kivie Moldave

Volume 165. Microbial Toxins: Tools for Enzymology
Edited by Sidney Harshman

Volume 166. Branched-Chain Amino Acids
Edited by Robert Harris and John R. Sokatch

Volume 167. Cyanobacteria
Edited by Lester Packer and Alexander N. Glazer

Volume 168. Hormone Action (Part K: Neuroendocrine Peptides)
Edited by P. Michael Conn

Volume 169. Platelets: Receptors, Adhesion, Secretion (Part A)
Edited by Jacek Hawiger

Volume 170. Nucleosomes
Edited by Paul M. Wassarman and Roger D. Kornberg

Volume 171. Biomembranes (Part R: Transport Theory: Cells and Model Membranes)
Edited by Sidney Fleischer and Becca Fleischer

Volume 172. Biomembranes (Part S: Transport: Membrane Isolation and Characterization)
Edited by Sidney Fleischer and Becca Fleischer

Volume 173. Biomembranes [Part T: Cellular and Subcellular Transport: Eukaryotic (Nonepithelial) Cells]
Edited by Sidney Fleischer and Becca Fleischer

Volume 174. Biomembranes [Part U: Cellular and Subcellular Transport: Eukaryotic (Nonepithelial) Cells]
Edited by Sidney Fleischer and Becca Fleischer

Volume 175. Cumulative Subject Index Volumes 135–139, 141–167

Volume 176. Nuclear Magnetic Resonance (Part A: Spectral Techniques and Dynamics)
Edited by Norman J. Oppenheimer and Thomas L. James

Volume 177. Nuclear Magnetic Resonance (Part B: Structure and Mechanism)
Edited by Norman J. Oppenheimer and Thomas L. James

Volume 178. Antibodies, Antigens, and Molecular Mimicry
Edited by John J. Langone

Volume 179. Complex Carbohydrates (Part F)
Edited by Victor Ginsburg

Volume 180. RNA Processing (Part A: General Methods)
Edited by James E. Dahlberg and John N. Abelson

Volume 181. RNA Processing (Part B: Specific Methods)
Edited by James E. Dahlberg and John N. Abelson

Volume 182. Guide to Protein Purification
Edited by Murray P. Deutscher

Volume 183. Molecular Evolution: Computer Analysis of Protein and Nucleic Acid Sequences
Edited by Russell F. Doolittle

Volume 184. Avidin-Biotin Technology
Edited by Meir Wilchek and Edward A. Bayer

Volume 185. Gene Expression Technology
Edited by David V. Goeddel

Volume 186. Oxygen Radicals in Biological Systems (Part B: Oxygen Radicals and Antioxidants)
Edited by Lester Packer and Alexander N. Glazer

Volume 187. Arachidonate Related Lipid Mediators
Edited by Robert C. Murphy and Frank A. Fitzpatrick

Volume 188. Hydrocarbons and Methylotrophy
Edited by Mary E. Lidstrom

Volume 189. Retinoids (Part A: Molecular and Metabolic Aspects)
Edited by Lester Packer

Volume 190. Retinoids (Part B: Cell Differentiation and Clinical Applications)
Edited by Lester Packer

Volume 191. Biomembranes (Part V: Cellular and Subcellular Transport: Epithelial Cells)
Edited by Sidney Fleischer and Becca Fleischer

Volume 192. Biomembranes (Part W: Cellular and Subcellular Transport: Epithelial Cells)
Edited by Sidney Fleischer and Becca Fleischer

Volume 193. Mass Spectrometry
Edited by James A. McCloskey

Volume 194. Guide to Yeast Genetics and Molecular Biology
Edited by Christine Guthrie and Gerald R. Fink

Volume 195. Adenylyl Cyclase, G Proteins, and Guanylyl Cyclase
Edited by Roger A. Johnson and Jackie D. Corbin

Volume 196. Molecular Motors and the Cytoskeleton
Edited by Richard B. Vallee

Volume 197. Phospholipases
Edited by Edward A. Dennis

Volume 198. Peptide Growth Factors (Part C)
Edited by David Barnes, J. P. Mather, and Gordon H. Sato

Volume 199. Cumulative Subject Index Volumes 168–174, 176–194

Volume 200. Protein Phosphorylation (Part A: Protein Kinases: Assays, Purification, Antibodies, Functional Analysis, Cloning, and Expression)
Edited by Tony Hunter and Bartholomew M. Sefton

Volume 201. Protein Phosphorylation (Part B: Analysis of Protein Phosphorylation, Protein Kinase Inhibitors, and Protein Phosphatases)
Edited by Tony Hunter and Bartholomew M. Sefton

Volume 202. Molecular Design and Modeling: Concepts and Applications (Part A: Proteins, Peptides, and Enzymes)
Edited by John J. Langone

Volume 203. Molecular Design and Modeling: Concepts and Applications (Part B: Antibodies and Antigens, Nucleic Acids, Polysaccharides, and Drugs)
Edited by John J. Langone

Volume 204. Bacterial Genetic Systems
Edited by Jeffrey H. Miller

Volume 205. Metallobiochemistry (Part B: Metallothionein and Related Molecules)
Edited by James F. Riordan and Bert L. Vallee

VOLUME 206. Cytochrome P450
Edited by MICHAEL R. WATERMAN AND ERIC F. JOHNSON

VOLUME 207. Ion Channels
Edited by BERNARDO RUDY AND LINDA E. IVERSON

VOLUME 208. Protein–DNA Interactions
Edited by ROBERT T. SAUER

VOLUME 209. Phospholipid Biosynthesis
Edited by EDWARD A. DENNIS AND DENNIS E. VANCE

VOLUME 210. Numerical Computer Methods
Edited by LUDWIG BRAND AND MICHAEL L. JOHNSON

VOLUME 211. DNA Structures (Part A: Synthesis and Physical Analysis of DNA)
Edited by DAVID M. J. LILLEY AND JAMES E. DAHLBERG

VOLUME 212. DNA Structures (Part B: Chemical and Electrophoretic Analysis of DNA)
Edited by DAVID M. J. LILLEY AND JAMES E. DAHLBERG

VOLUME 213. Carotenoids (Part A: Chemistry, Separation, Quantitation, and Antioxidation)
Edited by LESTER PACKER

VOLUME 214. Carotenoids (Part B: Metabolism, Genetics, and Biosynthesis)
Edited by LESTER PACKER

VOLUME 215. Platelets: Receptors, Adhesion, Secretion (Part B)
Edited by JACEK J. HAWIGER

VOLUME 216. Recombinant DNA (Part G)
Edited by RAY WU

VOLUME 217. Recombinant DNA (Part H)
Edited by RAY WU

VOLUME 218. Recombinant DNA (Part I)
Edited by RAY WU

VOLUME 219. Reconstitution of Intracellular Transport
Edited by JAMES E. ROTHMAN

VOLUME 220. Membrane Fusion Techniques (Part A)
Edited by NEJAT DÜZGÜNEŞ

VOLUME 221. Membrane Fusion Techniques (Part B)
Edited by NEJAT DÜZGÜNEŞ

VOLUME 222. Proteolytic Enzymes in Coagulation, Fibrinolysis, and Complement Activation (Part A: Mammalian Blood Coagulation Factors and Inhibitors)
Edited by LASZLO LORAND AND KENNETH G. MANN

VOLUME 223. Proteolytic Enzymes in Coagulation, Fibrinolysis, and Complement Activation (Part B: Complement Activation, Fibrinolysis, and Nonmammalian Blood Coagulation Factors)
Edited by LASZLO LORAND AND KENNETH G. MANN

VOLUME 224. Molecular Evolution: Producing the Biochemical Data
Edited by ELIZABETH ANNE ZIMMER, THOMAS J. WHITE, REBECCA L. CANN, AND ALLAN C. WILSON

VOLUME 225. Guide to Techniques in Mouse Development
Edited by PAUL M. WASSARMAN AND MELVIN L. DEPAMPHILIS

VOLUME 226. Metallobiochemistry (Part C: Spectroscopic and Physical Methods for Probing Metal Ion Environments in Metalloenzymes and Metalloproteins)
Edited by JAMES F. RIORDAN AND BERT L. VALLEE

VOLUME 227. Metallobiochemistry (Part D: Physical and Spectroscopic Methods for Probing Metal Ion Environments in Metalloproteins)
Edited by JAMES F. RIORDAN AND BERT L. VALLEE

VOLUME 228. Aqueous Two-Phase Systems
Edited by HARRY WALTER AND GÖTE JOHANSSON

VOLUME 229. Cumulative Subject Index Volumes 195–198, 200–227

VOLUME 230. Guide to Techniques in Glycobiology
Edited by WILLIAM J. LENNARZ AND GERALD W. HART

VOLUME 231. Hemoglobins (Part B: Biochemical and Analytical Methods)
Edited by JOHANNES EVERSE, KIM D. VANDEGRIFF, AND ROBERT M. WINSLOW

VOLUME 232. Hemoglobins (Part C: Biophysical Methods)
Edited by JOHANNES EVERSE, KIM D. VANDEGRIFF, AND ROBERT M. WINSLOW

VOLUME 233. Oxygen Radicals in Biological Systems (Part C)
Edited by LESTER PACKER

VOLUME 234. Oxygen Radicals in Biological Systems (Part D)
Edited by LESTER PACKER

VOLUME 235. Bacterial Pathogenesis (Part A: Identification and Regulation of Virulence Factors)
Edited by VIRGINIA L. CLARK AND PATRIK M. BAVOIL

VOLUME 236. Bacterial Pathogenesis (Part B: Integration of Pathogenic Bacteria with Host Cells)
Edited by VIRGINIA L. CLARK AND PATRIK M. BAVOIL

VOLUME 237. Heterotrimeric G Proteins
Edited by RAVI IYENGAR

VOLUME 238. Heterotrimeric G-Protein Effectors
Edited by RAVI IYENGAR

VOLUME 239. Nuclear Magnetic Resonance (Part C)
Edited by THOMAS L. JAMES AND NORMAN J. OPPENHEIMER

VOLUME 240. Numerical Computer Methods (Part B)
Edited by MICHAEL L. JOHNSON AND LUDWIG BRAND

VOLUME 241. Retroviral Proteases
Edited by LAWRENCE C. KUO AND JULES A. SHAFER

VOLUME 242. Neoglycoconjugates (Part A)
Edited by Y. C. LEE AND REIKO T. LEE

VOLUME 243. Inorganic Microbial Sulfur Metabolism
Edited by HARRY D. PECK, JR., AND JEAN LEGALL

VOLUME 244. Proteolytic Enzymes: Serine and Cysteine Peptidases
Edited by ALAN J. BARRETT

VOLUME 245. Extracellular Matrix Components
Edited by E. RUOSLAHTI AND E. ENGVALL

VOLUME 246. Biochemical Spectroscopy
Edited by KENNETH SAUER

VOLUME 247. Neoglycoconjugates (Part B: Biomedical Applications)
Edited by Y. C. LEE AND REIKO T. LEE

VOLUME 248. Proteolytic Enzymes: Aspartic and Metallo Peptidases
Edited by ALAN J. BARRETT

VOLUME 249. Enzyme Kinetics and Mechanism (Part D: Developments in Enzyme Dynamics)
Edited by DANIEL L. PURICH

VOLUME 250. Lipid Modifications of Proteins
Edited by PATRICK J. CASEY AND JANICE E. BUSS

VOLUME 251. Biothiols (Part A: Monothiols and Dithiols, Protein Thiols, and Thiyl Radicals)
Edited by LESTER PACKER

VOLUME 252. Biothiols (Part B: Glutathione and Thioredoxin; Thiols in Signal Transduction and Gene Regulation)
Edited by LESTER PACKER

VOLUME 253. Adhesion of Microbial Pathogens
Edited by RON J. DOYLE AND ITZHAK OFEK

VOLUME 254. Oncogene Techniques
Edited by PETER K. VOGT AND INDER M. VERMA

VOLUME 255. Small GTPases and Their Regulators (Part A: Ras Family)
Edited by W. E. BALCH, CHANNING J. DER, AND ALAN HALL

Volume 256. Small GTPases and Their Regulators (Part B: Rho Family)
Edited by W. E. Balch, Channing J. Der, and Alan Hall

Volume 257. Small GTPases and Their Regulators (Part C: Proteins Involved in Transport)
Edited by W. E. Balch, Channing J. Der, and Alan Hall

Volume 258. Redox-Active Amino Acids in Biology
Edited by Judith P. Klinman

Volume 259. Energetics of Biological Macromolecules
Edited by Michael L. Johnson and Gary K. Ackers

Volume 260. Mitochondrial Biogenesis and Genetics (Part A)
Edited by Giuseppe M. Attardi and Anne Chomyn

Volume 261. Nuclear Magnetic Resonance and Nucleic Acids
Edited by Thomas L. James

Volume 262. DNA Replication
Edited by Judith L. Campbell

Volume 263. Plasma Lipoproteins (Part C: Quantitation)
Edited by William A. Bradley, Sandra H. Gianturco, and Jere P. Segrest

Volume 264. Mitochondrial Biogenesis and Genetics (Part B)
Edited by Giuseppe M. Attardi and Anne Chomyn

Volume 265. Cumulative Subject Index Volumes 228, 230–262

Volume 266. Computer Methods for Macromolecular Sequence Analysis
Edited by Russell F. Doolittle

Volume 267. Combinatorial Chemistry
Edited by John N. Abelson

Volume 268. Nitric Oxide (Part A: Sources and Detection of NO; NO Synthase)
Edited by Lester Packer

Volume 269. Nitric Oxide (Part B: Physiological and Pathological Processes)
Edited by Lester Packer

Volume 270. High Resolution Separation and Analysis of Biological Macromolecules (Part A: Fundamentals)
Edited by Barry L. Karger and William S. Hancock

Volume 271. High Resolution Separation and Analysis of Biological Macromolecules (Part B: Applications)
Edited by Barry L. Karger and William S. Hancock

Volume 272. Cytochrome P450 (Part B)
Edited by Eric F. Johnson and Michael R. Waterman

Volume 273. RNA Polymerase and Associated Factors (Part A)
Edited by Sankar Adhya

VOLUME 274. RNA Polymerase and Associated Factors (Part B)
Edited by SANKAR ADHYA

VOLUME 275. Viral Polymerases and Related Proteins
Edited by LAWRENCE C. KUO, DAVID B. OLSEN, AND STEVEN S. CARROLL

VOLUME 276. Macromolecular Crystallography (Part A)
Edited by CHARLES W. CARTER, JR., AND ROBERT M. SWEET

VOLUME 277. Macromolecular Crystallography (Part B)
Edited by CHARLES W. CARTER, JR., AND ROBERT M. SWEET

VOLUME 278. Fluorescence Spectroscopy
Edited by LUDWIG BRAND AND MICHAEL L. JOHNSON

VOLUME 279. Vitamins and Coenzymes (Part I)
Edited by DONALD B. MCCORMICK, JOHN W. SUTTIE, AND CONRAD WAGNER

VOLUME 280. Vitamins and Coenzymes (Part J)
Edited by DONALD B. MCCORMICK, JOHN W. SUTTIE, AND CONRAD WAGNER

VOLUME 281. Vitamins and Coenzymes (Part K)
Edited by DONALD B. MCCORMICK, JOHN W. SUTTIE, AND CONRAD WAGNER

VOLUME 282. Vitamins and Coenzymes (Part L)
Edited by DONALD B. MCCORMICK, JOHN W. SUTTIE, AND CONRAD WAGNER

VOLUME 283. Cell Cycle Control
Edited by WILLIAM G. DUNPHY

VOLUME 284. Lipases (Part A: Biotechnology)
Edited by BYRON RUBIN AND EDWARD A. DENNIS

VOLUME 285. Cumulative Subject Index Volumes 263, 264, 266–284, 286–289

VOLUME 286. Lipases (Part B: Enzyme Characterization and Utilization)
Edited by BYRON RUBIN AND EDWARD A. DENNIS

VOLUME 287. Chemokines
Edited by RICHARD HORUK

VOLUME 288. Chemokine Receptors
Edited by RICHARD HORUK

VOLUME 289. Solid Phase Peptide Synthesis
Edited by GREGG B. FIELDS

VOLUME 290. Molecular Chaperones
Edited by GEORGE H. LORIMER AND THOMAS BALDWIN

VOLUME 291. Caged Compounds
Edited by GERARD MARRIOTT

VOLUME 292. ABC Transporters: Biochemical, Cellular, and Molecular Aspects
Edited by SURESH V. AMBUDKAR AND MICHAEL M. GOTTESMAN

Volume 293. Ion Channels (Part B)
Edited by P. Michael Conn

Volume 294. Ion Channels (Part C)
Edited by P. Michael Conn

Volume 295. Energetics of Biological Macromolecules (Part B)
Edited by Gary K. Ackers and Michael L. Johnson

Volume 296. Neurotransmitter Transporters
Edited by Susan G. Amara

Volume 297. Photosynthesis: Molecular Biology of Energy Capture
Edited by Lee McIntosh

Volume 298. Molecular Motors and the Cytoskeleton (Part B)
Edited by Richard B. Vallee

Volume 299. Oxidants and Antioxidants (Part A)
Edited by Lester Packer

Volume 300. Oxidants and Antioxidants (Part B)
Edited by Lester Packer

Volume 301. Nitric Oxide: Biological and Antioxidant Activities (Part C)
Edited by Lester Packer

Volume 302. Green Fluorescent Protein
Edited by P. Michael Conn

Volume 303. cDNA Preparation and Display
Edited by Sherman M. Weissman

Volume 304. Chromatin
Edited by Paul M. Wassarman and Alan P. Wolffe

Volume 305. Bioluminescence and Chemiluminescence (Part C)
Edited by Thomas O. Baldwin and Miriam M. Ziegler

Volume 306. Expression of Recombinant Genes in Eukaryotic Systems
Edited by Joseph C. Glorioso and Martin C. Schmidt

Volume 307. Confocal Microscopy
Edited by P. Michael Conn

Volume 308. Enzyme Kinetics and Mechanism (Part E: Energetics of Enzyme Catalysis)
Edited by Daniel L. Purich and Vern L. Schramm

Volume 309. Amyloid, Prions, and Other Protein Aggregates
Edited by Ronald Wetzel

Volume 310. Biofilms
Edited by Ron J. Doyle

VOLUME 311. Sphingolipid Metabolism and Cell Signaling (Part A)
Edited by ALFRED H. MERRILL, JR., AND YUSUF A. HANNUN

VOLUME 312. Sphingolipid Metabolism and Cell Signaling (Part B)
Edited by ALFRED H. MERRILL, JR., AND YUSUF A. HANNUN

VOLUME 313. Antisense Technology
(Part A: General Methods, Methods of Delivery, and RNA Studies)
Edited by M. IAN PHILLIPS

VOLUME 314. Antisense Technology (Part B: Applications)
Edited by M. IAN PHILLIPS

VOLUME 315. Vertebrate Phototransduction and the Visual Cycle (Part A)
Edited by KRZYSZTOF PALCZEWSKI

VOLUME 316. Vertebrate Phototransduction and the Visual Cycle (Part B)
Edited by KRZYSZTOF PALCZEWSKI

VOLUME 317. RNA–Ligand Interactions (Part A: Structural Biology Methods)
Edited by DANIEL W. CELANDER AND JOHN N. ABELSON

VOLUME 318. RNA–Ligand Interactions (Part B: Molecular Biology Methods)
Edited by DANIEL W. CELANDER AND JOHN N. ABELSON

VOLUME 319. Singlet Oxygen, UV-A, and Ozone
Edited by LESTER PACKER AND HELMUT SIES

VOLUME 320. Cumulative Subject Index Volumes 290–319

VOLUME 321. Numerical Computer Methods (Part C)
Edited by MICHAEL L. JOHNSON AND LUDWIG BRAND

VOLUME 322. Apoptosis
Edited by JOHN C. REED

VOLUME 323. Energetics of Biological Macromolecules (Part C)
Edited by MICHAEL L. JOHNSON AND GARY K. ACKERS

VOLUME 324. Branched-Chain Amino Acids (Part B)
Edited by ROBERT A. HARRIS AND JOHN R. SOKATCH

VOLUME 325. Regulators and Effectors of Small GTPases
(Part D: Rho Family)
Edited by W. E. BALCH, CHANNING J. DER, AND ALAN HALL

VOLUME 326. Applications of Chimeric Genes and Hybrid Proteins
(Part A: Gene Expression and Protein Purification)
Edited by JEREMY THORNER, SCOTT D. EMR, AND JOHN N. ABELSON

VOLUME 327. Applications of Chimeric Genes and Hybrid Proteins
(Part B: Cell Biology and Physiology)
Edited by JEREMY THORNER, SCOTT D. EMR, AND JOHN N. ABELSON

Volume 328. Applications of Chimeric Genes and Hybrid Proteins (Part C: Protein–Protein Interactions and Genomics)
Edited by Jeremy Thorner, Scott D. Emr, and John N. Abelson

Volume 329. Regulators and Effectors of Small GTPases (Part E: GTPases Involved in Vesicular Traffic)
Edited by W. E. Balch, Channing J. Der, and Alan Hall

Volume 330. Hyperthermophilic Enzymes (Part A)
Edited by Michael W. W. Adams and Robert M. Kelly

Volume 331. Hyperthermophilic Enzymes (Part B)
Edited by Michael W. W. Adams and Robert M. Kelly

Volume 332. Regulators and Effectors of Small GTPases (Part F: Ras Family I)
Edited by W. E. Balch, Channing J. Der, and Alan Hall

Volume 333. Regulators and Effectors of Small GTPases (Part G: Ras Family II)
Edited by W. E. Balch, Channing J. Der, and Alan Hall

Volume 334. Hyperthermophilic Enzymes (Part C)
Edited by Michael W. W. Adams and Robert M. Kelly

Volume 335. Flavonoids and Other Polyphenols
Edited by Lester Packer

Volume 336. Microbial Growth in Biofilms (Part A: Developmental and Molecular Biological Aspects)
Edited by Ron J. Doyle

Volume 337. Microbial Growth in Biofilms (Part B: Special Environments and Physicochemical Aspects)
Edited by Ron J. Doyle

Volume 338. Nuclear Magnetic Resonance of Biological Macromolecules (Part A)
Edited by Thomas L. James, Volker Dötsch, and Uli Schmitz

Volume 339. Nuclear Magnetic Resonance of Biological Macromolecules (Part B)
Edited by Thomas L. James, Volker Dötsch, and Uli Schmitz

Volume 340. Drug–Nucleic Acid Interactions
Edited by Jonathan B. Chaires and Michael J. Waring

Volume 341. Ribonucleases (Part A)
Edited by Allen W. Nicholson

Volume 342. Ribonucleases (Part B)
Edited by Allen W. Nicholson

Volume 343. G Protein Pathways (Part A: Receptors)
Edited by Ravi Iyengar and John D. Hildebrandt

Volume 344. G Protein Pathways (Part B: G Proteins and Their Regulators)
Edited by Ravi Iyengar and John D. Hildebrandt

Volume 345. G Protein Pathways (Part C: Effector Mechanisms)
Edited by Ravi Iyengar and John D. Hildebrandt

Volume 346. Gene Therapy Methods
Edited by M. Ian Phillips

Volume 347. Protein Sensors and Reactive Oxygen Species (Part A: Selenoproteins and Thioredoxin)
Edited by Helmut Sies and Lester Packer

Volume 348. Protein Sensors and Reactive Oxygen Species (Part B: Thiol Enzymes and Proteins)
Edited by Helmut Sies and Lester Packer

Volume 349. Superoxide Dismutase
Edited by Lester Packer

Volume 350. Guide to Yeast Genetics and Molecular and Cell Biology (Part B)
Edited by Christine Guthrie and Gerald R. Fink

Volume 351. Guide to Yeast Genetics and Molecular and Cell Biology (Part C)
Edited by Christine Guthrie and Gerald R. Fink

Volume 352. Redox Cell Biology and Genetics (Part A)
Edited by Chandan K. Sen and Lester Packer

Volume 353. Redox Cell Biology and Genetics (Part B)
Edited by Chandan K. Sen and Lester Packer

Volume 354. Enzyme Kinetics and Mechanisms (Part F: Detection and Characterization of Enzyme Reaction Intermediates)
Edited by Daniel L. Purich

Volume 355. Cumulative Subject Index Volumes 321–354

Volume 356. Laser Capture Microscopy and Microdissection
Edited by P. Michael Conn

Volume 357. Cytochrome P450, Part C
Edited by Eric F. Johnson and Michael R. Waterman

Volume 358. Bacterial Pathogenesis (Part C: Identification, Regulation, and Function of Virulence Factors)
Edited by Virginia L. Clark and Patrik M. Bavoil

Volume 359. Nitric Oxide (Part D)
Edited by Enrique Cadenas and Lester Packer

Volume 360. Biophotonics (Part A)
Edited by Gerard Marriott and Ian Parker

Volume 361. Biophotonics (Part B)
Edited by Gerard Marriott and Ian Parker

VOLUME 362. Recognition of Carbohydrates in Biological Systems (Part A)
Edited by YUAN C. LEE AND REIKO T. LEE

VOLUME 363. Recognition of Carbohydrates in Biological Systems (Part B)
Edited by YUAN C. LEE AND REIKO T. LEE

VOLUME 364. Nuclear Receptors
Edited by DAVID W. RUSSELL AND DAVID J. MANGELSDORF

VOLUME 365. Differentiation of Embryonic Stem Cells
Edited by PAUL M. WASSAUMAN AND GORDON M. KELLER

VOLUME 366. Protein Phosphatases
Edited by SUSANNE KLUMPP AND JOSEF KRIEGLSTEIN

VOLUME 367. Liposomes (Part A)
Edited by NEJAT DÜZGÜNEŞ

VOLUME 368. Macromolecular Crystallography (Part C)
Edited by CHARLES W. CARTER, JR., AND ROBERT M. SWEET

VOLUME 369. Combinational Chemistry (Part B)
Edited by GUILLERMO A. MORALES AND BARRY A. BUNIN

VOLUME 370. RNA Polymerases and Associated Factors (Part C)
Edited by SANKAR L. ADHYA AND SUSAN GARGES

VOLUME 371. RNA Polymerases and Associated Factors (Part D)
Edited by SANKAR L. ADHYA AND SUSAN GARGES

VOLUME 372. Liposomes (Part B)
Edited by NEJAT DÜZGÜNEŞ

VOLUME 373. Liposomes (Part C)
Edited by NEJAT DÜZGÜNEŞ

VOLUME 374. Macromolecular Crystallography (Part D)
Edited by CHARLES W. CARTER, JR., AND ROBERT W. SWEET

VOLUME 375. Chromatin and Chromatin Remodeling Enzymes (Part A)
Edited by C. DAVID ALLIS AND CARL WU

VOLUME 376. Chromatin and Chromatin Remodeling Enzymes (Part B)
Edited by C. DAVID ALLIS AND CARL WU

VOLUME 377. Chromatin and Chromatin Remodeling Enzymes (Part C)
Edited by C. DAVID ALLIS AND CARL WU

VOLUME 378. Quinones and Quinone Enzymes (Part A)
Edited by HELMUT SIES AND LESTER PACKER

VOLUME 379. Energetics of Biological Macromolecules (Part D)
Edited by JO M. HOLT, MICHAEL L. JOHNSON, AND GARY K. ACKERS

VOLUME 380. Energetics of Biological Macromolecules (Part E)
Edited by JO M. HOLT, MICHAEL L. JOHNSON, AND GARY K. ACKERS

VOLUME 381. Oxygen Sensing
Edited by CHANDAN K. SEN AND GREGG L. SEMENZA

VOLUME 382. Quinones and Quinone Enzymes (Part B)
Edited by HELMUT SIES AND LESTER PACKER

VOLUME 383. Numerical Computer Methods (Part D)
Edited by LUDWIG BRAND AND MICHAEL L. JOHNSON

VOLUME 384. Numerical Computer Methods (Part E)
Edited by LUDWIG BRAND AND MICHAEL L. JOHNSON

VOLUME 385. Imaging in Biological Research (Part A)
Edited by P. MICHAEL CONN

VOLUME 386. Imaging in Biological Research (Part B)
Edited by P. MICHAEL CONN

VOLUME 387. Liposomes (Part D)
Edited by NEJAT DÜZGÜNEŞ

VOLUME 388. Protein Engineering
Edited by DAN E. ROBERTSON AND JOSEPH P. NOEL

VOLUME 389. Regulators of G-Protein Signaling (Part A)
Edited by DAVID P. SIDEROVSKI

VOLUME 390. Regulators of G-Protein Signaling (Part B)
Edited by DAVID P. SIDEROVSKI

VOLUME 391. Liposomes (Part E)
Edited by NEJAT DÜZGÜNEŞ

VOLUME 392. RNA Interference
Edited by ENGELKE ROSSI

VOLUME 393. Circadian Rhythms
Edited by MICHAEL W. YOUNG

VOLUME 394. Nuclear Magnetic Resonance of Biological Macromolecules (Part C)
Edited by THOMAS L. JAMES

VOLUME 395. Producing the Biochemical Data (Part B)
Edited by ELIZABETH A. ZIMMER AND ERIC H. ROALSON

VOLUME 396. Nitric Oxide (Part E)
Edited by LESTER PACKER AND ENRIQUE CADENAS

VOLUME 397. Environmental Microbiology
Edited by JARED R. LEADBETTER

VOLUME 398. Ubiquitin and Protein Degradation (Part A)
Edited by RAYMOND J. DESHAIES

Volume 399. Ubiquitin and Protein Degradation (Part B)
Edited by Raymond J. Deshaies

Volume 400. Phase II Conjugation Enzymes and Transport Systems
Edited by Helmut Sies and Lester Packer

Volume 401. Glutathione Transferases and Gamma Glutamyl Transpeptidases
Edited by Helmut Sies and Lester Packer

Volume 402. Biological Mass Spectrometry
Edited by A. L. Burlingame

Volume 403. GTPases Regulating Membrane Targeting and Fusion
Edited by William E. Balch, Channing J. Der, and Alan Hall

Volume 404. GTPases Regulating Membrane Dynamics
Edited by William E. Balch, Channing J. Der, and Alan Hall

Volume 405. Mass Spectrometry: Modified Proteins and Glycoconjugates
Edited by A. L. Burlingame

Volume 406. Regulators and Effectors of Small GTPases: Rho Family
Edited by William E. Balch, Channing J. Der, and Alan Hall

Volume 407. Regulators and Effectors of Small GTPases: Ras Family
Edited by William E. Balch, Channing J. Der, and Alan Hall

Volume 408. DNA Repair (Part A)
Edited by Judith L. Campbell and Paul Modrich

Volume 409. DNA Repair (Part B)
Edited by Judith L. Campbell and Paul Modrich

Volume 410. DNA Microarrays (Part A: Array Platforms and Web-Bench Protocols)
Edited by Alan Kimmel and Brian Oliver

Volume 411. DNA Microarrays (Part B: Databases and Statistics)
Edited by Alan Kimmel and Brian Oliver

Volume 412. Amyloid, Prions, and Other Protein Aggregates (Part B)
Edited by Indu Kheterpal and Ronald Wetzel

Volume 413. Amyloid, Prions, and Other Protein Aggregates (Part C)
Edited by Indu Kheterpal and Ronald Wetzel

Volume 414. Measuring Biological Responses with Automated Microscopy
Edited by James Inglese

Volume 415. Glycobiology
Edited by Minoru Fukuda

Volume 416. Glycomics
Edited by Minoru Fukuda

VOLUME 417. Functional Glycomics
Edited by MINORU FUKUDA

VOLUME 418. Embryonic Stem Cells
Edited by IRINA KLIMANSKAYA AND ROBERT LANZA

VOLUME 419. Adult Stem Cells
Edited by IRINA KLIMANSKAYA AND ROBERT LANZA

VOLUME 420. Stem Cell Tools and Other Experimental Protocols
Edited by IRINA KLIMANSKAYA AND ROBERT LANZA

VOLUME 421. Advanced Bacterial Genetics: Use of Transposons and Phage for Genomic Engineering
Edited by KELLY T. HUGHES

VOLUME 422. Two-Component Signaling Systems, Part A
Edited by MELVIN I. SIMON, BRIAN R. CRANE, AND ALEXANDRINE CRANE

VOLUME 423. Two-Component Signaling Systems, Part B
Edited by MELVIN I. SIMON, BRIAN R. CRANE, AND ALEXANDRINE CRANE

VOLUME 424. RNA Editing
Edited by JONATHA M. GOTT

VOLUME 425. RNA Modification
Edited by JONATHA M. GOTT

VOLUME 426. Integrins
Edited by DAVID CHERESH

VOLUME 427. MicroRNA Methods
Edited by JOHN J. ROSSI

VOLUME 428. Osmosensing and Osmosignaling
Edited by HELMUT SIES AND DIETER HAUSSINGER

VOLUME 429. Translation Initiation: Extract Systems and Molecular Genetics
Edited by JON LORSCH

VOLUME 430. Translation Initiation: Reconstituted Systems and Biophysical Methods
Edited by JON LORSCH

VOLUME 431. Translation Initiation: Cell Biology, High-Throughput and Chemical-Based Approaches
Edited by JON LORSCH

VOLUME 432. Lipidomics and Bioactive Lipids: Mass-Spectrometry–Based Lipid Analysis
Edited by H. ALEX BROWN

Volume 433. Lipidomics and Bioactive Lipids: Specialized Analytical Methods and Lipids in Disease
Edited by H. Alex Brown

Volume 434. Lipidomics and Bioactive Lipids: Lipids and Cell Signaling
Edited by H. Alex Brown

Volume 435. Oxygen Biology and Hypoxia
Edited by Helmut Sies and Bernhard Brüne

Volume 436. Globins and Other Nitric Oxide-Reactive Protiens (Part A)
Edited by Robert K. Poole

Volume 437. Globins and Other Nitric Oxide-Reactive Protiens (Part B)
Edited by Robert K. Poole

Volume 438. Small GTPases in Disease (Part A)
Edited by William E. Balch, Channing J. Der, and Alan Hall

Volume 439. Small GTPases in Disease (Part B)
Edited by William E. Balch, Channing J. Der, and Alan Hall

Volume 440. Nitric Oxide, Part F Oxidative and Nitrosative Stress in Redox Regulation of Cell Signaling
Edited by Enrique Cadenas and Lester Packer

Volume 441. Nitric Oxide, Part G Oxidative and Nitrosative Stress in Redox Regulation of Cell Signaling
Edited by Enrique Cadenas and Lester Packer

Volume 442. Programmed Cell Death, General Principles for Studying Cell Death (Part A)
Edited by Roya Khosravi-Far, Zahra Zakeri, Richard A. Lockshin, and Mauro Piacentini

Volume 443. Angiogenesis: *In Vitro* Systems
Edited by David A. Cheresh

Volume 444. Angiogenesis: *In Vivo* Systems (Part A)
Edited by David A. Cheresh

Volume 445. Angiogenesis: *In Vivo* Systems (Part B)
Edited by David A. Cheresh

Volume 446. Programmed Cell Death, The Biology and Therapeutic Implications of Cell Death (Part B)
Edited by Roya Khosravi-Far, Zahra Zakeri, Richard A. Lockshin, and Mauro Piacentini

Volume 447. RNA Turnover in Bacteria, Archaea and Organelles
Edited by Lynne E. Maquat and Cecilia M. Arraiano

VOLUME 448. RNA Turnover in Eukaryotes: Nucleases, Pathways and Analysis of mRNA Decay
Edited by LYNNE E. MAQUAT AND MEGERDITCH KILEDJIAN

VOLUME 449. RNA Turnover in Eukaryotes: Analysis of Specialized and Quality Control RNA Decay Pathways
Edited by LYNNE E. MAQUAT AND MEGERDITCH KILEDJIAN

VOLUME 450. Fluorescence Spectroscopy
Edited by LUDWIG BRAND AND MICHAEL L. JOHNSON

VOLUME 451. Autophagy: Lower Eukaryotes and Non-Mammalian Systems (Part A)
Edited by DANIEL J. KLIONSKY

VOLUME 452. Autophagy in Mammalian Systems (Part B)
Edited by DANIEL J. KLIONSKY

VOLUME 453. Autophagy in Disease and Clinical Applications (Part C)
Edited by DANIEL J. KLIONSKY

VOLUME 454. Computer Methods (Part A)
Edited by MICHAEL L. JOHNSON AND LUDWIG BRAND

VOLUME 455. Biothermodynamics (Part A)
Edited by MICHAEL L. JOHNSON, JO M. HOLT, AND GARY K. ACKERS (RETIRED)

VOLUME 456. Mitochondrial Function, Part A: Mitochondrial Electron Transport Complexes and Reactive Oxygen Species
Edited by WILLIAM S. ALLISON AND IMMO E. SCHEFFLER

VOLUME 457. Mitochondrial Function, Part B: Mitochondrial Protein Kinases, Protein Phosphatases and Mitochondrial Diseases
Edited by WILLIAM S. ALLISON AND ANNE N. MURPHY

VOLUME 458. Complex Enzymes in Microbial Natural Product Biosynthesis, Part A: Overview Articles and Peptides
Edited by DAVID A. HOPWOOD

VOLUME 459. Complex Enzymes in Microbial Natural Product Biosynthesis, Part B: Polyketides, Aminocoumarins and Carbohydrates
Edited by DAVID A. HOPWOOD

VOLUME 460. Chemokines, Part A
Edited by TRACY M. HANDEL AND DAMON J. HAMEL

VOLUME 461. Chemokines, Part B
Edited by TRACY M. HANDEL AND DAMON J. HAMEL

VOLUME 462. Non-Natural Amino Acids
Edited by TOM W. MUIR AND JOHN N. ABELSON

VOLUME 463. Guide to Protein Purification, 2nd Edition
Edited by RICHARD R. BURGESS AND MURRAY P. DEUTSCHER

Volume 464. Liposomes, Part F
Edited by Nejat Düzgüneş

Volume 465. Liposomes, Part G
Edited by Nejat Düzgüneş

Volume 466. Biothermodynamics, Part B
Edited by Michael L. Johnson, Gary K. Ackers, and Jo M. Holt

Volume 467. Computer Methods Part B
Edited by Michael L. Johnson and Ludwig Brand

Volume 468. Biophysical, Chemical, and Functional Probes of RNA Structure, Interactions and Folding: Part A
Edited by Daniel Herschlag

Volume 469. Biophysical, Chemical, and Functional Probes of RNA Structure, Interactions and Folding: Part B
Edited by Daniel Herschlag

Volume 470. Guide to Yeast Genetics: Functional Genomics, Proteomics, and Other Systems Analysis, 2nd Edition
Edited by Gerald Fink, Jonathan Weissman, and Christine Guthrie

Volume 471. Two-Component Signaling Systems, Part C
Edited by Melvin I. Simon, Brian R. Crane, and Alexandrine Crane

Volume 472. Single Molecule Tools, Part A: Fluorescence Based Approaches
Edited by Nils G. Walter

Volume 473. Thiol Redox Transitions in Cell Signaling, Part A Chemistry and Biochemistry of Low Molecular Weight and Protein Thiols
Edited by Enrique Cadenas and Lester Packer

Volume 474. Thiol Redox Transitions in Cell Signaling, Part B Cellular Localization and Signaling
Edited by Enrique Cadenas and Lester Packer

Volume 475. Single Molecule Tools, Part B: Super-Resolution, Particle Tracking, Multiparameter, and Force Based Methods
Edited by Nils G. Walter

Volume 476. Guide to Techniques in Mouse Development, Part A Mice, Embryos, and Cells, 2nd Edition
Edited by Paul M. Wassarman and Philippe M. Soriano

Volume 477. Guide to Techniques in Mouse Development, Part B Mouse Molecular Genetics, 2nd Edition
Edited by Paul M. Wassarman and Philippe M. Soriano

Volume 478. Glycomics
Edited by Minoru Fukuda

VOLUME 479. Functional Glycomics
Edited by MINORU FUKUDA

VOLUME 480. Glycobiology
Edited by MINORU FUKUDA

VOLUME 481. Cryo-EM, Part A: Sample Preparation and Data Collection
Edited by GRANT J. JENSEN

VOLUME 482. Cryo-EM, Part B: 3-D Reconstruction
Edited by GRANT J. JENSEN

VOLUME 483. Cryo-EM, Part C: Analyses, Interpretation, and Case Studies
Edited by GRANT J. JENSEN

VOLUME 484. Constitutive Activity in Receptors and Other Proteins, Part A
Edited by P. MICHAEL CONN

VOLUME 485. Constitutive Activity in Receptors and Other Proteins, Part B
Edited by P. MICHAEL CONN

VOLUME 486. Research on Nitrification and Related Processes, Part A
Edited by MARTIN G. KLOTZ

VOLUME 487. Computer Methods, Part C
Edited by MICHAEL L. JOHNSON AND LUDWIG BRAND

VOLUME 488. Biothermodynamics, Part C
Edited by MICHAEL L. JOHNSON, JO M. HOLT, AND GARY K. ACKERS

VOLUME 489. The Unfolded Protein Response and Cellular Stress, Part A
Edited by P. MICHAEL CONN

VOLUME 490. The Unfolded Protein Response and Cellular Stress, Part B
Edited by P. MICHAEL CONN

VOLUME 491. The Unfolded Protein Response and Cellular Stress, Part C
Edited by P. MICHAEL CONN

VOLUME 492. Biothermodynamics, Part D
Edited by MICHAEL L. JOHNSON, JO M. HOLT, AND GARY K. ACKERS

VOLUME 493. Fragment-Based Drug Design
Tools, Practical Approaches, and Examples
Edited by LAWRENCE C. KUO

VOLUME 494. Methods in Methane Metabolism, Part A
Methanogenesis
Edited by AMY C. ROSENZWEIG AND STEPHEN W. RAGSDALE

VOLUME 495. Methods in Methane Metabolism, Part B
Methanotrophy
Edited by AMY C. ROSENZWEIG AND STEPHEN W. RAGSDALE

VOLUME 496. Research on Nitrification and Related Processes, Part B
Edited by MARTIN G. KLOTZ AND LISA Y. STEIN

VOLUME 497. Synthetic Biology, Part A
Methods for Part/Device Characterization and Chassis Engineering
Edited by CHRISTOPHER VOIGT

VOLUME 498. Synthetic Biology, Part B
Computer Aided Design and DNA Assembly
Edited by CHRISTOPHER VOIGT

VOLUME 499. Biology of Serpins
Edited by JAMES C. WHISSTOCK AND PHILLIP I. BIRD

VOLUME 500. Methods in Systems Biology
Edited by DANIEL JAMESON, MALKHEY VERMA, AND HANS V. WESTERHOFF

VOLUME 501. Serpin Structure and Evolution
Edited by JAMES C. WHISSTOCK AND PHILLIP I. BIRD

VOLUME 502. Protein Engineering for Therapeutics, Part A
Edited by K. DANE WITTRUP AND GREGORY L. VERDINE

VOLUME 503. Protein Engineering for Therapeutics, Part B
Edited by K. DANE WITTRUP AND GREGORY L. VERDINE

SECTION FOUR

PEPTIDES

CHAPTER ONE

Stapled Peptides for Intracellular Drug Targets

Gregory L. Verdine* *and* Gerard J. Hilinski†

Contents

1. Introduction 4
2. All-Hydrocarbon Stapled α-Helical Peptides 6
3. The Design of Stapled Peptides 9
4. Stapled Peptide Synthesis 12
5. Olefin Metathesis 16
6. N-terminal and Internal Modifications 16
7. Cleavage of Stapled Peptides from the Solid Support 19
8. Stapled Peptide Purification 20
9. Biophysical Characterization 20
10. Cell Permeability 23
11. *In Vitro* Target Interaction and Activity Assays 26
12. *In Vivo* Efficacy 27
13. Strategies for Stapled Peptide Optimization 28
14. Summary 30
Acknowledgments 30
References 30

Abstract

Proteins that engage in intracellular interactions with other proteins are widely considered among the most biologically appealing yet chemically intractable targets for drug discovery. The critical interaction surfaces of these proteins typically lack the deep hydrophobic involutions that enable potent, selective targeting by small organic molecules, and their localization within the cell puts them beyond the reach of protein therapeutics. Considerable interest has therefore arisen in next-generation targeting molecules that combine the broad target recognition capabilities of protein therapeutics with the robust cell-penetrating ability of small molecules. One type that has shown promise in early-stage studies is hydrocarbon-stapled α-helical peptides, a novel class of synthetic miniproteins locked into their bioactive α-helical fold through the

* Departments of Chemistry and Chemical Biology, Stem Cell and Regenerative Biology, and Molecular and Cellular Biology, Harvard University, and Program in Cancer Chemical Biology, Dana - Farber Cancer Institute, Boston, Massachusetts, USA
† Department of Stem Cell and Regenerative Biology, Harvard University, Cambridge, Massachusetts, USA

Methods in Enzymology, Volume 503
ISSN 0076-6879, DOI: 10.1016/B978-0-12-396962-0.00001-X

site-specific introduction of a chemical brace, an all-hydrocarbon staple. Stapling can greatly improve the pharmacologic performance of peptides, increasing their target affinity, proteolytic resistance, and serum half-life while conferring on them high levels of cell penetration through endocytic vesicle trafficking. Here, we discuss considerations crucial to the successful design and evaluation of potent stapled peptide interactions, our intention being to facilitate the broad application of this technology to intractable targets of both basic biologic interest and potential therapeutic value.

1. Introduction

The two broad classes of well-established therapeutic agents—small molecules and protein therapeutics (also known as biologics)—are each severely limited in their target range, albeit in different ways (Hopkins and Groom, 2002, 2003; Russ and Lampel, 2005; Verdine and Walensky, 2007). Small molecules are able to target with high affinity and specificity only the approximately 10% of human proteins that possess a hydrophobic pocket on their surface. Protein therapeutics are able to access only the approximately 10% of all human targets that are accessible to the cell exterior. Many targets of significant potential therapeutic value contribute to human disease pathology through engagement in protein–protein interactions characterized by an extensive association between the complementary surfaces of two proteins (Betzi *et al.*, 2009; Chene, 2006; Fry and Vassilev, 2005; Jones and Thornton, 1996; Verdine and Walensky, 2007). When such interactions take place outside the cell, they are usually targetable by protein therapeutics. However, when such interactions take place inside the cell, they are beyond the reach of biologics. Intracellular protein–protein interactions are also by and large untargetable by small molecules, because the interaction surfaces are relatively flat and distributed over a large contact surface area. The limitations of the two current therapeutic classes have fueled efforts along two major fronts to discover next-generation therapeutic modalities. One front involves the discovery of address labels that promote the active, endosomal uptake and release of molecules that are not otherwise cell-penetrant, including peptides, proteins, nucleic acids, and others. The other front entails the discovery of new types of targeting molecules that enter cells through either passive diffusion or active transport and can access targets that neither small molecules nor proteins can.

It is not our intention here to comprehensively review cargo delivery technologies being developed to ferry protein therapeutics across the cell membrane, but we will mention a few examples we find particularly noteworthy. It has been discovered that protein therapeutics can adsorb onto cell-penetrating polybutylcyanoacrylate nanoparticles and undergo cointernalization (Hasadsri *et al.*, 2009). Additionally, moieties known to undergo receptor-mediated internalization can be covalently coupled to biological macromolecules such as synthetic antibody fragments, conferring upon them access to the cytosol through a hijacked receptor (Rizk *et al.*, 2009). A recent example demonstrated high levels of brain exposure gained by co-opting import across the blood–brain barrier via the transferring/transferrin receptor pathway (Atwal *et al.*, 2011; Yu *et al.*, 2011). Cell-penetrating peptides (CPPs) have also been extensively studied as cargo carriers. CPPs are relatively short stretches of contiguous amino acids, typically rich in lysine or arginine, that feature an intrinsic capacity for cell penetration (Heitz *et al.*, 2009). Some of the most well-known CPPs are HIV-1 $TAT_{48–60}$ (Vives *et al.*, 1997), $Antennapedia_{43–58}$/Penetratin (Derossi *et al.*, 1994), and polyarginine (Wender *et al.*, 2000). The fusion of a CPP sequence to a biological macromolecule grants it entry into the intracellular milieu (Heitz *et al.*, 2009; van den Berg and Dowdy, 2011). It was recently discovered that a positively "supercharged" version of green fluorescent protein, which carries a +36 net charge and is inherently cell penetrating, can quickly and efficiently deliver fusion protein cargoes into the cell via an endosomal mechanism (Cronican *et al.*, 2010; Lawrence *et al.*, 2007; McNaughton *et al.*, 2009).

With respect to fundamentally new types of targeting molecules, the majority of the effort to date has focused on nucleic acids that target sequence-complementary cellular RNAs through either antisense or RNA interference mechanisms. One antisense molecule, Vitravene®, has been approved by the FDA, and a second, Mipomersen, is in Phase III clinical testing. Dozens of additional nucleic acid therapeutics are at various earlier stages of clinical development (Alvarez-Salas, 2008; Di Cresce and Koropatnick, 2010; Mescalchin and Restle, 2011). Systemic bioavailability continues to be a challenge with nucleic acid therapeutics.

A rapidly emerging class of next-generation drugs is cell-penetrating miniproteins, molecules designed to combine the synthetic manipulability and cell-penetrating ability of small molecules with the three-dimensionality and versatile target recognition ability of biologics. As a class, cell-penetrating miniproteins possess some kind of scaffold that both stabilizes the folded structure of the molecule and facilitates cellular uptake. Though many different strategies have been reported for producing conformationally stabilized miniproteins of diverse structures, few of these have been shown to afford high levels of cell penetration, and of those that are cell-penetrant; to our knowledge, only one has been demonstrated safe and efficacious *in vivo*. That class, all-hydrocarbon-stapled α-helical peptides, is the subject of this chapter.

The α-helix is the single most common element of protein secondary structure in protein–protein interfaces (Guharoy and Chakrabarti, 2007; Henchey *et al.*, 2008; Jochim and Arora, 2009); indeed, this element has received the greatest attention in attempts to develop miniproteins. Peptides as a class have historically been relegated to a niche role in chemical genetics and drug discovery, owing to their poor binding affinity, susceptibility to proteolytic degradation, rapid clearance by renal filtration, and ineffective cell penetration (Jenssen and Aspmo, 2008). Most of these problems trace their origins in one way or another to the high conformational instability of peptides, which introduces an energetic (entropic) penalty to binding of the target, renders the amide bonds vulnerable to proteolytic attack, decreases cell penetration by increasing backbone desolvation energy, and facilitates passage through glomerular pores that filter out polar molecules. The remarkable tendency of α-helical peptides to undergo conformational nucleation, a property first recognized by Isabella Karle (Karle, 2001; Karle *et al.*, 1990), indicated that conformational stabilization introduced in one small stretch of a peptide could migrate some distance along the polypeptide chain, perhaps stabilizing the entire structure. All efforts to design cell-penetrating α-helical miniproteins have employed helix stabilization and nucleation.

2. All-Hydrocarbon Stapled α-Helical Peptides

The all-hydrocarbon α-helix staple (Fig. 1.1) (Schafmeister *et al.*, 2000) combines two distinct conformational stabilization strategies previously found individually to induce α-helical structure, namely, α,α-disubstitution and macrocyclic bridge formation (these strategies are reviewed in Henchey *et al.*, 2008). The introduction of a hydrocarbon staple has been found in numerous examples to confer high levels of α-helical content, and this is associated with a 5- to 5000-fold increase in target affinity, strong protection from proteolytic degradation, robust cell-penetration by endocytic vesicle trafficking, extension of *in vivo* half-life, and specific antagonism of protein–protein interactions in cultured cells (see references accompanying Table 1.1). Importantly, stapled peptides have been shown safe, efficacious, and biological pathway specific in animal models of human cancer driven by Bcl-2 overexpression (Walensky *et al.*, 2004) and constitutive activation of the NOTCH transcriptional activation pathway (Moellering *et al.*, 2009). The latter example has provided a long-sought experimental precedent for therapeutic intervention via direct antagonism of a transcription factor, important regulatory proteins previously considered to be "undruggable."

In design terms, the all-hydrocarbon stapling system is unique not only in its combination of two distinct helix-stabilizing strategies but also in its avoidance of heteroatoms in the macrocyclic bridge, a feature intended to facilitate cell penetration. Fortunately, Grubbs and coworkers had already shown that a powerful new C—C bond-forming reaction, ruthenium-mediated olefin metathesis, was fully compatible with solid-phase peptide synthesis (SPPS) (Blackwell and Grubbs, 1998), hence this reaction was

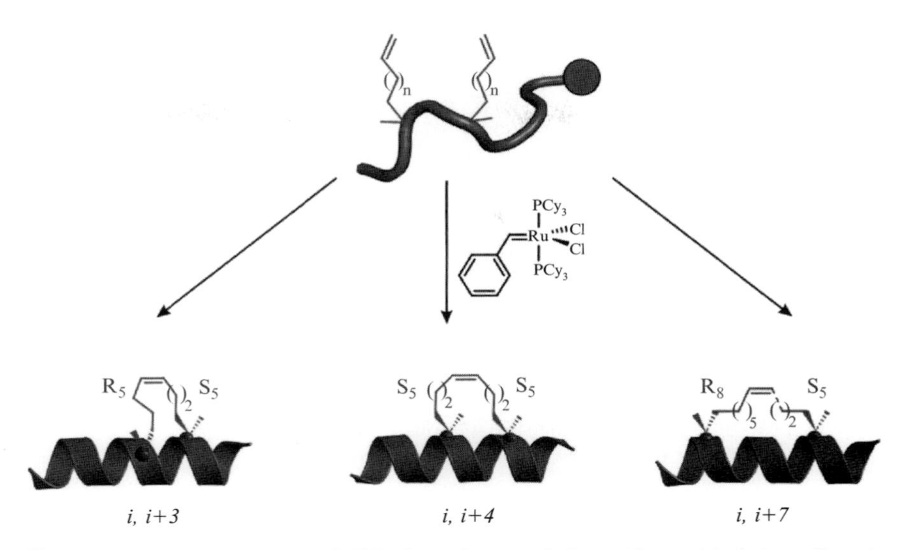

Figure 1.1 The three types of all-hydrocarbon stapled peptides. α-Methyl, α-alkenylglycine cross-linking amino acids are incorporated during solid-phase peptide synthesis. An *i*, *i*+3 stapled peptide requires one unit of R_5 at the *i* position and one unit of S_5 at the *i*+3 position. An *i*, *i*+4 stapled peptide requires two units of S_5 incorporated at the relative positions *i* and *i*+4. An *i*, *i*+7 stapled peptide requires one unit of R_8 at the *i* position and one unit of S_5 at the *i*+7 position. Resin-bound peptide is treated with Grubbs I olefin metathesis catalyst to produce a cross-link between the two nonnatural amino acids, resulting in a stapled peptide that is braced in an α-helical conformation. (For color version of this figure, the reader is referred to the Web version of this chapter.)

employed in our system. Incorporation of a staple into a peptide thus entails the incorporation of two appropriately spaced α-methyl,α-alkenylglycine residues, having defined stereochemical configuration and alkene chain length (details below), followed by ruthenium-mediated olefin metathesis on the synthesis resin and then release from the resin and deprotection to yield the stapled peptide.

Multiple types of all-hydrocarbon staples have been optimized through systematic variation of the stereochemistry, length, and relative disposition of the olefin-bearing nonnatural amino acids (Kim and Verdine, 2009; Kim *et al.*, 2010; Schafmeister *et al.*, 2000). The optimized versions have proven portable from peptide to peptide, though screening of the location of the staple within a given peptide is still necessary to optimize its overall performance. An *i*, *i*+7 staple (where the residue directly C-terminal to residue *i* is designated *i*+1, etc.) spans two successive α-helical turns, while an *i*, *i*+3 or *i*, *i*+4 staple spans approximately one α-helical turn (Kim *et al.*, 2010; Schafmeister *et al.*, 2000). Using the naming convention X_Y, where X is the stereochemistry at the α-carbon (Cahn-Ingold-Prelog S and R designations) and Y is the length of the alkenyl side chain, the placement of an

Table 1.1 A list of bioactive all-hydrocarbon stapled peptides

Stapled peptide	Sequence	Source protein	Target protein(s)	Reference(s)
BID $SAHB_A$	EDIIRNIARHLA$\mathbf{S}_5$VGD$\mathbf{S}_5$$N_L$DRSIW	BID	BCL-2, $BCL\text{-}X_L$, BAX	Walensky *et al.* (2004), Walensky *et al.* (2006)
BAD $SAHB_A$	NLWAAQRYGRELR$\mathbf{S}_5$$N_LSD\mathbf{S}_5$FVDSFKK	BAD	$BCL\text{-}X_L$	Walensky *et al.* (2006)
BIM $SAHB_A$	IWIAQELR$\mathbf{S}_5$IGD$\mathbf{S}_5$FNAYYARR	BIM	$BCL\text{-}X_L$, BAX	Walensky *et al.* (2006), Gavathiotis *et al.* (2008)
SAH-p53-8	QSQQTF$\mathbf{R}_8$NLWRLL$\mathbf{S}_5$QN	p53	hDM2, hDMX	Bernal *et al.* (2007), Bernal *et al.* (2010)
NYAD-1	ITF$\mathbf{S}_5$DLL$\mathbf{S}_5$YYGP	(Phage display)	HIV-1 CA	Zhang *et al.* (2008)
SAHM1	ERLRRRI$\mathbf{S}_5$LCR$\mathbf{S}_5$HHST	MAML	NOTCH/CSL	Moellering *et al.* (2009)
SAH-$gp41_{(626–662)}$	N_LTW$\mathbf{S}_5$EWD$\mathbf{S}_5$EINNYTSLIHSLIEESQNQ$\mathbf{S}_5$EKN$\mathbf{S}_5$QELLE	HIV-1 gp41	HIV-1 gp41	Bird *et al.* (2010)
MCL-1 $SAHB_A$	KALETLR$\mathbf{S}_5$VGD$\mathbf{S}_5$VQRNHETAF	MCL-1	MCL-1	Stewart *et al.* (2010)
SAH-apoA-I	VLESFKVS$\mathbf{R}_8$LSALEES_5TKKLNTQ	apoA-I	ABCA1	Sviridov *et al.* (2011)

S_5 and R_8 are abbreviations for α-methyl, α-alkenylglycine cross-linking amino acids. N_L is an abbreviation for norleucine.

S_5 cross-linking amino acid at both the *i* and *i*+4 positions results in an optimally stabilized *i*, *i*+4 stapled peptide. To obtain the optimally stabilized *i*, *i*+7 stapled peptide, R_8 is placed at the *i* position, whereas S_5 is placed at the *i*+7 position. Finally, the optimally stabilized *i*, *i*+3 stapled peptide is obtained by placement of R_5 at the *i* position and S_5 at the *i*+3 position. (Fig. 1.1)

The incorporation of multiple staples into a single peptide is also possible. Placement of two independent *i*, *i*+4 staples separated by several helix turns into a relative of the HIV-1 fusion inhibitor enfuvirtide resulted in greater resistance to proteolysis and conferred oral bioavailability on the reinforced peptide (Bird *et al.*, 2010). In addition, Kim and colleagues demonstrated that two staples separated by as few as two intervening amino acid residues can be simultaneously incorporated by virtue of highly stereospecific olefin metathesis reactions directed by the helical peptide scaffold, providing additional modes of helical stabilization (Kim *et al.*, 2010).

3. The Design of Stapled Peptides

Protein–protein interactions that have been successfully disrupted by stapled peptides *in vivo* are characterized by a ligand-target pair. The ligand possesses an α-helical motif, borne on a protein or (better yet a) peptide, that docks into a shallow cleft on the surface of the target. Stapled peptide inhibitors represent "dominant-negative" versions of this docking helix optimized through synthetic modification (i.e., stapling) and sequence alteration to penetrate cells and compete effectively with the intracellular version of the ligand protein. When designing a stapled peptide, it is imperative to position the cross-linking amino acids such that the intended contact interface with the target protein(s) remains intact. For this reason, the design process is simplified by both atomic-resolution structures of the protein–protein interaction and mutagenesis data for residues at or near the binding interface. In a recent article, Walensky and coworkers gave a detailed exposition on how the design of $SAHB_A$, a stapled peptide based on the BH3 domain of the proapoptotic BID protein (Walensky *et al.*, 2004), was guided by consideration of the high-resolution crystal structure of an α-helical peptide from a homologous protein bound to $BCL\text{-}X_L$ (Bird *et al.*, 2008). Other successful applications of stapled peptide inhibitors have to date also benefited from the use of high-resolution structures to direct the placement of the staple and, ultimately, the identification of the most potent inhibitor (Bernal *et al.*, 2007; Bird *et al.*, 2010; Moellering *et al.*, 2009; Stewart *et al.*, 2010; Zhang *et al.*, 2008). If a structure is not available for a system of interest, but there is reason to believe that the ligand is α-helical when bound to the target, then other information such as Ala-scanning or residue conservation can be used as a basis for positionally biasing a stapling approach. In the absence of such information, it may be necessary to synthesize and screen a panel of stapled peptides encompassing most or all candidate stapling positions.

The following illustrates the design process used in a recent study targeting NOTCH. A high-resolution structure of an approximately 60-amino acid α-helical portion of the MAML transcriptional coactivator protein bound in a cleft at the interface of CSL and the NOTCH oncoprotein (Nam *et al.*, 2006) guided the structure-based rational design of a potent stapled peptide inhibitor of NOTCH function (Moellering *et al.*, 2009). As it was already known that the recombinant α-helical MAML domain could act as a dominant-negative inhibitor of NOTCH-dependent transcriptional activation (Maillard *et al.*, 2004; Weng *et al.*, 2003), Moellering *et al.* set out to determine if a much smaller portion of this MAML peptide, constrained into an α-helical structure by all-hydrocarbon stapling, could produce a cell-penetrating antagonist of the NOTCH pathway.

Due to the fact that the α-helical MAML peptide spans approximately 60 amino acids, there were many potential positions into which cross-linking amino acids could have been incorporated without disrupting critical residues at the protein–protein interface. Examining the NOTCH–CSL–MAML ternary complex, it was possible to hypothesize which amino acids in MAML were likely to be dispensable for binding to the NOTCH–CSL binary complex (Fig. 1.2A). The MAML peptide residues that contact neither NOTCH nor CSL were mapped onto the linear sequence of the MAML peptide in order to identify pairs of residues with the correct relative spacing such that they could be replaced to form one of the three types of all-hydrocarbon staples.

It is also worth noting that a high-resolution structure can illuminate positions at which introduction of a rigid α-helical structure might be deleterious toward binding. The structure of the ternary NOTCH–CSL–MAML complex revealed the presence of a significant kink in the center of the MAML peptide at serine 46 and proline 47, two helix-destabilizing residues. A staple that spans this region might be expected to induce a more canonical α-helical structure than what promotes optimal binding, and the placement of a staple at this position would be expected to have a negative effect on binding. Upon analysis of the atomic-resolution structure of the NOTCH–CSL–MAML complex, stapling positions anticipated to be compatible with maintenance of essential binding contacts were identified. Subsequently, the 60-mer MAML peptide was divided into smaller segments of approximately 12–20 amino acids, and a panel of analogs bearing all-hydrocarbon staples was synthesized and tested for inhibition of complex assembly. Ultimately, the i, i+4 stapled peptide SAHM1, which spanned 16 residues in the N-terminal portion of the 60-mer MAML peptide and

A

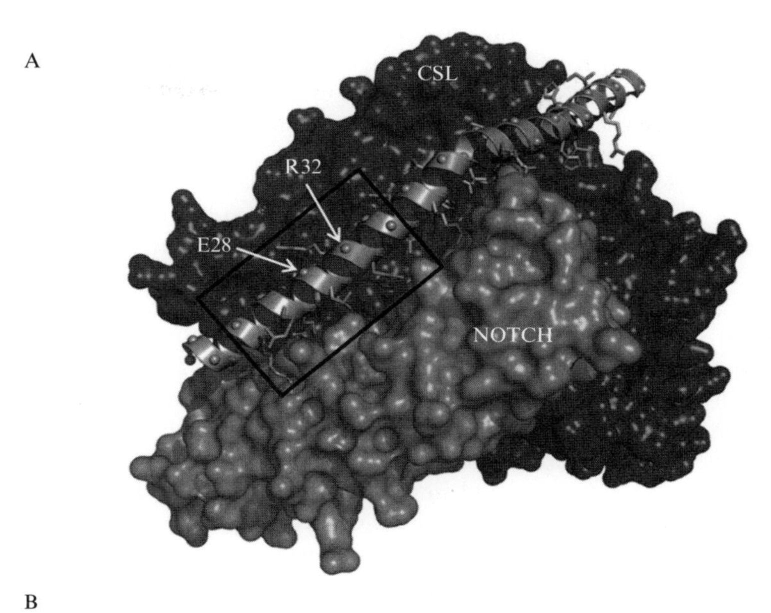

B

	16				20										30										40			
MAML	H	S	A	V	M	E	R	L	R	R	R	I	E	L	C	R	R	H	H	S	T	C	E	A	R	Y	E	A
SAHM1						E	R	L	R	R	R	I	S_5	L	C	R	S_5	H	H	S	T							

							50										60										70
MAML	V	S	P	E	R	L	E	L	E	R	Q	H	T	F	A	L	H	Q	R	C	I	Q	A	K	A	K	R
SAHM1																											

Figure 1.2 The design of stapled peptides targeting the NOTCH/CSL binary complex. (A) A kinked α-helix from the MAML transcriptional coactivator protein binds to a cleft created by the NOTCH/CSL complex. The α-carbons of the MAML residues not expected to interact with the NOTCH/CSL complex are shown as spheres. The region of MAML from which the bioactive SAHM1 stapled peptide was designed is indicated with a box. MAML residues E28 and R32 were replaced with S_5 cross-linking amino acids to form SAHM1. PDB: 2F8X. (B) Residues 16–70 of the MAML coactivator protein are shown, along with the sequence of the SAHM1 stapled peptide. Residues in red appear to contact the NOTCH/CSL complex, whereas residues in black do not appear to be involved in the interface. Residues 45 and 46 of the MAML protein (shaded gray) correspond to the kink in the MAML α-helix. (For interpretation of the references to color in this figure legend, the reader is referred to the Web version of this chapter.)

featured the replacement of glutamate 28 and arginine 32 with S_5 cross-linking amino acids, was identified as a potent antagonist of complex assembly (Moellering *et al.*, 2009) (Fig. 1.2B). Though SAHM1 is less than one-third the length of the dominant-negative MAML peptide, it inhibited the interaction of the dominant-negative MAML peptide with the NOTCH–CSL binary complex with an IC_{50} of approximately 4 μ*M*.

After identifying a stapled peptide that interacts with the target, an important parallel aspect is to design a negative control stapled peptide that can be used to confirm target-specific effects *in vitro* and *in vivo*. The design of a negative control stapled peptide is highly context dependent, and thus it is difficult here to provide a general design proscription. In some cases, residues previously validated as important for target interaction can be mutated nonconservatively to produce a negative control stapled peptide (Bernal *et al.*, 2007; Moellering *et al.*, 2009; Walensky *et al.*, 2004). If residues in the parent sequence of the stapled peptide have not been explicitly validated as determinants of binding specificity, structure-based design can be utilized to introduce mutations that would be expected to disrupt key interactions. During the design of a negative control, it is important to consider the effects that any engineered mutations might have on other properties of the stapled peptide. For example, a negative control stapled peptide with a significantly decreased capacity for cell penetration relative to the corresponding active stapled peptide would not be an ideal specificity control for cell-based functional assays, as any perceived difference in activity could be due to a significantly lower peak intracellular concentration. Often, the alteration of properties such as intracellular uptake that could render a negative control stapled peptide a poor tool for specificity determination must be minimized through an iterative process of design and characterization similar to the process used in identification of a stapled peptide with affinity for the desired target.

4. Stapled Peptide Synthesis

Stapled peptides are typically synthesized by SPPS using amino acids with acid-labile side chain protecting groups (when necessary) and a base-labile fluorenylmethoxycarbonyl (Fmoc) protecting group on the backbone amine (for additional details, see Kim *et al.*, 2011). *N*-α-Fmoc-protected amino acids are often offered with a choice of side chain protecting groups; for standard SPPS of stapled peptides, the side chain protecting groups indicated in Table 1.2 are employed. Reagents are typically purchased from Sigma–Aldrich or EMD Chemicals and used as received. The *N*-α-Fmoc-protected cross-linking amino acids, R_8 and S_5, can be purchased from Okeanos Technology (Beijing) or synthesized as previously described (Bird *et al.*, 2008; Schafmeister *et al.*, 2000; Williams and Im, 1991; Williams *et al.*, 2003). Alternatively, due to the use of liquid ammonia and sodium or lithium metal in a dangerous dissolving metal reduction step, it is recommended that the Fmoc-protected α-methyl, α-alkenyl cross-linking amino acids are synthesized using the Ala-Ni(II)–BPB complex method (Qiu *et al.*, 2000; Zhang *et al.*, 2008).

Rink Amide MBHA resin (100–200 mesh, loading 0.4–0.8 mmol/g) is used as a solid support for the synthesis of stapled peptides with an amidated

Table 1.2 Acid-labile side chain protecting groups used in Fmoc-based solid-phase peptide synthesis of stapled peptides

Amino acid	3-Letter code	1-Letter code	Side chain protecting group
Alanine	Ala	A	N/A
Cysteine	Cys	C	Trityl (Trt)
Aspartic acid	Asp	D	*tert*-Butyl (OtBu)
Glutamic acid	Glu	E	*tert*-Butyl (OtBu)
Phenylalanine	Phe	F	N/A
Glycine	Gly	G	N/A
Histidine	His	H	Trityl (Trt)
Isoleucine	Ile	I	N/A
Lysine	Lys	K	*tert*-Butoxy (Boc)
Leucine	Leu	L	N/A
Methionine	Met	M	N/A
Asparagine	Asn	N	Trityl (Trt)
Proline	Pro	P	N/A
Glutamine	Gln	Q	Trityl (Trt)
Arginine	Arg	R	Pentamethyldihydrobenzofurane (Pbf)
Serine	Ser	S	*tert*-Butyl (OtBu)
Threonine	Thr	T	*tert*-Butyl (OtBu)
Valine	Val	V	N/A
Tryptophan	Trp	W	*tert*-Butoxy (Boc)
Tyrosine	Tyr	Y	*tert*-Butyl (OtBu)

The indicated side chain protecting groups are used during standard solid-phase peptide synthesis. The protecting group abbreviation is given in parentheses. N/A (nonapplicable) indicates that no side chain protecting group is necessary due to the presence of a side chain that is inert to the conditions of standard Fmoc-based solid-phase peptide synthesis.

C-terminus.[1] A Vac-Man® Laboratory Vacuum Manifold (Promega) or similar multiport apparatus can be used for the manual parallel synthesis of a panel of stapled peptides. The waste reservoir is connected to the house vacuum, and each port that will be used for peptide synthesis is fitted with a solvent resistant 3-way stopcock (Bio-Rad) connected to a nitrogen stream and a disposable polypropylene chromatography column (Bio-Rad). When preparing a panel of stapled peptides for initial screening, a synthetic scale in the range of 10–25 μmol typically yields a sufficient amount of stapled peptide for biophysical characterization, *in vitro* target interaction assays,

[1] Peptides with C-terminal acids can be synthesized using a solid support such as Wang resin and a modified version of the procedures given for the synthesis of peptides with C-terminal amides. Peptides with amidated C-termini are often desired because of the protection offered against some endogenous proteases.

and evaluation of cell penetration and efficacy in cell-based activity assays. This synthetic scale can be accommodated using a chromatography column with a 2 mL bed volume; chromatography columns with different bed volumes can be employed if necessitated by the synthetic scale. The protocol described below is for a 25 μmol synthesis.

For a 25 μmol scale synthesis, 0.038 g of Rink Amide MBHA resin is added to each chromatography column. Approximately 1 mL of *N*-methyl-2-pyrrolidone (NMP) is added and the resin is bubbled under nitrogen for at least 30 min to achieve thorough swelling of the resin.[2] The peptide chain is then elongated from C- to N-terminus by a repeating cycle of *N*-α-Fmoc deprotection and subsequent coupling of an *N*-α-Fmoc-protected amino acid to the nascent peptide chain (Fig. 1.3). (*Note*: The resin is derivatized with an Fmoc protecting group, so it must be deprotected using the procedure below prior to coupling of the C-terminal amino acid.) Extensive washing of the resin is performed after each instance of deprotection and coupling, and all steps are performed with nitrogen bubbling to ensure equal contact of all beads with the reaction solution. A detailed description of one cycle of deprotection and coupling is described below.

(1) Deprotect *N*-α-Fmoc by 3×10 min treatments with 1 mL of 25% (v/v) piperidine in NMP.
(2) Wash resin with 5×1 min treatments with 1 mL of NMP.
(3) Couple activated amino acid to the deprotected amine by one treatment with the following solution (activated amino acid solution is premixed prior to addition to the reaction vessel, approximate coupling time is based on the guidelines below):

Activated amino acid solution (natural amino acids)

0.375 mL of 0.4 *M* *N*-α-Fmoc-protected amino acid in NMP (6 equiv)
0.375 mL of 0.38 *M* HCTU coupling reagent[3] in NMP (5.7 equiv)
0.052 mL of DIPEA (*N*,*N*-diisopropylethylamine) (12 equiv)

Activated amino acid solution (cross-linking amino acids)

0.250 mL of 0.4 *M* *N*-α-Fmoc-protected amino acid in NMP (4 equiv)
0.250 mL of 0.38 *M* HCTU coupling reagent in NMP (3.8 equiv)
0.035 mL of DIPEA (8 equiv)

Coupling time

Natural amino acids (not including sterically hindered amino acids): 45 min

[2] In all cases where NMP is indicated as the solvent, *N*,*N*-dimethylformamide (DMF) can also be used if desired.

[3] HCTU=(2-(6-chloro-1H-benzotriazole-1-yl)-1,1,3,3-tetramethylaminium hexafluorophosphate).

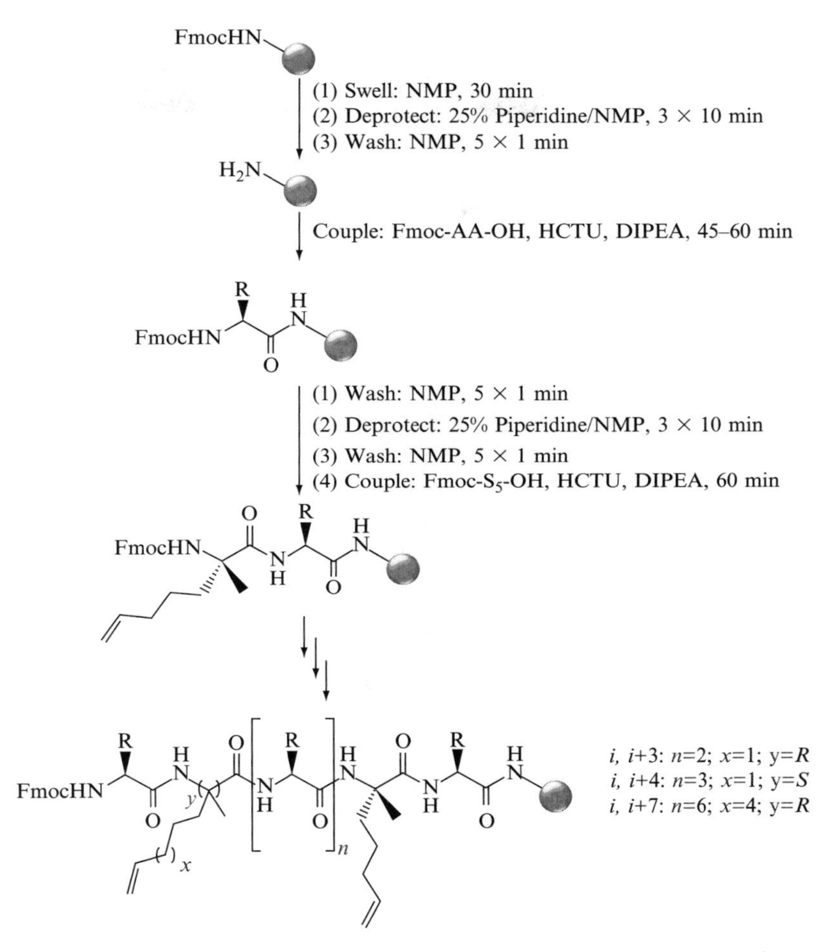

Figure 1.3 Fmoc-based solid-phase peptide synthesis (SPPS) of all-hydrocarbon stapled peptides. Rink Amide MBHA resin is swollen in NMP and the Fmoc protecting group is removed by treatment with 25% piperidine in NMP. An Fmoc amino acid is activated by HCTU and coupled to the resin in the presence of DIPEA. The cycle of Fmoc deprotection followed by coupling of an activated amino acid is repeated to elongate the peptide. α-Methyl,α-alkenylglycine cross-linking amino acids are incorporated as shown to produce one of the three types of stapled peptides. (For color version of this figure, the reader is referred to the Web version of this chapter.)

Sterically hindered natural amino acids (His, Ile, Pro, Thr, Trp, Val): 60 min

Cross-linking amino acids: 60 min

(4) Wash resin with 5×1 min treatments with 1 mL of NMP.
(5) Repeat steps 1–4 for each amino acid.

Peptide synthesis can be paused overnight after completion of a coupling step. The resin is washed with NMP as previously described, and then the following procedure is used to dry and shrink the resin:

(1) Wash resin with 3×1 min treatments with 1 mL of dichloromethane (DCM).
(2) Add 1 mL methanol to the resin and bubble to dryness with $N_2(g)$.

5. Olefin Metathesis

Metathesis of the cross-linking amino acids is performed on the solid phase at any point after the incorporation of the two cross-linking amino acids, but it must be completed prior to labeling with fluorescein isothiocyanate. Grubbs' first-generation metathesis catalyst (benzylidene-bis(tricyclohexylphosphine)dichlororuthenium, Grubbs I) is dissolved to 10 m*M* in dry 1,2-dichloroethane (DCE) immediately prior to use. After the resin is swollen by bubbling in 1 mL of dry DCE for 30 min, approximately 1 mL of 10 m*M* catalyst solution is added (assuming a 25 μmol scale synthesis) and the solution is bubbled for 2 h. Intermittent addition of additional DCE is necessary due to evaporation throughout the course of the treatment. After completion of the treatment, the resin is washed once by bubbling for 1 min in 1 mL of DCE. Fresh catalyst solution is then added and an additional 2-h metathesis reaction is performed. Following this second treatment, the resin is washed three times for 1 min with 1 mL of DCE followed by two washes for 1 min with DCM. One milliliter of methanol is then added and subsequently bubbled to dryness to shrink the resin.

6. N-terminal and Internal Modifications

The N-terminus of stapled peptides can be deprotected prior to cleavage from the resin, resulting in a free amine. If the presence of a charged moiety is undesired, the deprotected N-terminus can alternatively be acetylated prior to cleavage by treatment with 30 equiv of acetic anhydride and 60 equiv of DIPEA in NMP for 45 min (Kim *et al.*, 2010). Other modifications can be incorporated either internally or on the N-terminus depending on the particular assays that will be utilized for their evaluation. When incorporating modifications, it is important to consider if the placement of the modification would be expected to interfere with the ability of the stapled peptide to interact with the target as intended.

Stapled peptide modifications typically fall into two categories: a fluorescent label or an affinity tag. Two of the most common moieties appended to the N-terminus of stapled peptides are fluorescein, which can be used for studies of intracellular uptake and biophysical characterization, and biotin, which can be used for biophysical characterization and assessment of *in vitro* target interaction. It is generally desired to include a flexible molecular spacer to isolate the modification from the core of the stapled peptide, decreasing the possibility that the modification will have a deleterious effect on the interaction of the stapled peptide with the target. A flexible molecular spacer and modification can be serially installed, or the modification can be purchased in a form that includes a preincorporated spacer. The structures of selected N-terminal modifications are displayed in Table 1.3.

As demonstrated in previous instances of stapled peptide development, fluorescein and rhodamine have been successfully incorporated at the

Table 1.3 The structures of selected modifications upon installation at the N-terminus of a stapled peptide

N-terminal modification(s)	Structure of coupled modification(s)
Acetate cap	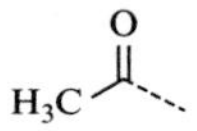
Biotin	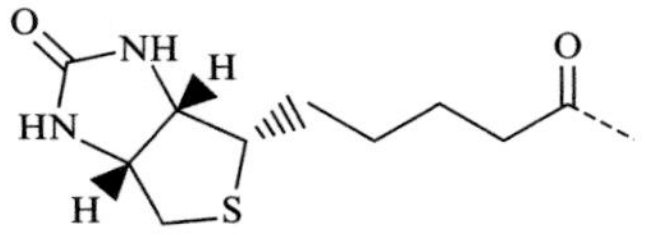
(1) β-Alanine (2) fluorescein-5-isothiocyanate	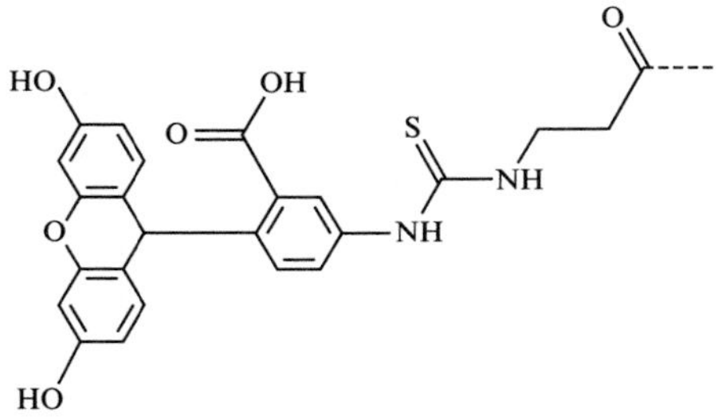
(1) β-Alanine (2) rhodamine B isothiocyanate	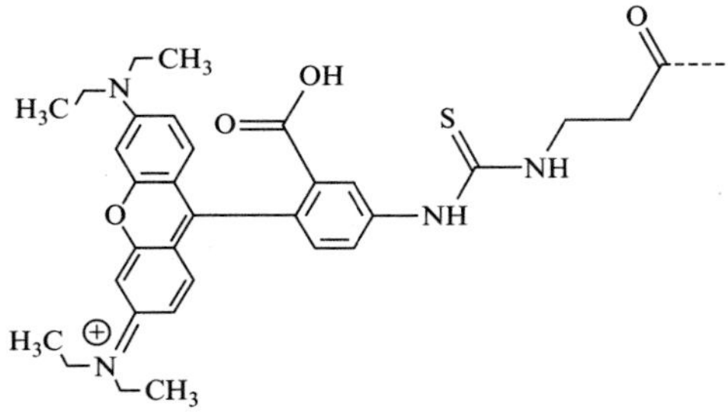

The dashed line represents the bond to the N-terminal backbone amine.

N-terminus of a stapled peptide (Bernal *et al.*, 2007; Bird *et al.*, 2008; Moellering *et al.*, 2009; Pitter *et al.*, 2008; Stewart *et al.*, 2010; Walensky *et al.*, 2004, 2006; Zhang *et al.*, 2008, 2011). A flexible molecular spacer such as *N*-β-Fmoc-β-alanine is first coupled to the deprotected N-terminus of the stapled peptide using the standard amino acid coupling protocol followed by deprotection. Subsequently, any fluorescent molecule with a single amine-reactive functional group can be coupled to the free N-terminus. Fluorescent molecules that are functionalized with an isothiocyanate moiety (e.g., fluorescein-5-isothiocyanate (FITC) and rhodamine B isothiocyanate) can be purchased from a variety of manufacturers. It is important to note that acid-mediated cleavage of a peptide labeled on the N-terminus via an isothiocyanate-based derivatization will result in removal of the label and the first amino acid through the Edman degradation pathway; this undesired side-reaction can be avoided by the inclusion of a non-α-amino acid spacer such as β-alanine (Jullian *et al.*, 2009). Labeling of a deprotected N-terminus with an isothiocyanate-functionalized fluorophore can be achieved on the solid phase by treating the resin with an excess molar amount of the desired label, dissolved in *N*,*N*-dimethylformamide (DMF) in the presence of excess DIPEA. Generally, the samples are protected from light and the reaction is carried out overnight while gently shaking in a sealed vessel in order to ensure proper mixing and to prevent evaporation of volatile material such as DIPEA. After completion of the treatment, the resin is subjected to multiple washes with DMF followed by multiple washes with diethyl ether to effect removal of unreacted components. The resin can then be dried by bubbling in methanol as previously described.[4]

Stapled peptides can be biotinylated on the N-terminus using either a single- or multistep process. Biotinylation reagents that include a flexible molecular spacer, such as *N*-biotinyl-NH-$(PEG)_2$-COOH (Novabiochem) can be installed directly, using the standard HCTU/DIPEA coupling conditions, onto the free N-terminal amine of a stapled peptide with or without an additional β-alanine spacer. Conversely, a flexible molecular spacer of choice, bearing an Fmoc-protected amine, can be incorporated onto the N-terminus of the stapled peptide followed by installation of biotin. Biotin can then be coupled to a free N-terminal amine using the standard HCTU/DIPEA coupling conditions.

If incorporation of a modification such as FITC or biotin on the N-terminus might be expected to have a deleterious effect on the ability of the stapled peptide to interact with the target, an appealing alternative is to install the modification at an internal position. The decision to include an internal modification rather than an N-terminal modification must be

[4] Possibly due to poisoning of the catalyst by the thiourea produced as a result of incorporation of an isothiocyanate-functionalized fluorescent label, olefin metathesis using the Grubbs I metathesis catalyst must be performed prior to labeling with the activated fluorescent molecules described here.

made prior to synthesis of the primary sequence of the stapled peptide, as internal modification requires the incorporation of *N*-α-Fmoc-protected amino acids, usually either lysine or cysteine, with side chain protecting groups that can be selectively cleaved to allow functionalization. As an internal modification must be installed at a position on the stapled peptide that is not involved in the intended interface with the target, it is important to consult high-resolution structural information or the results of amino acid mutagenesis experiments to choose an optimal site for internal modification.

Incorporation of *N*-α-Fmoc-Lysine(Mmt)-OH followed by selective deprotection of the Mmt (monomethoxytrityl) protecting group allows for subsequent installation of the modification via the unmasked ε-amine of the lysine. The Mmt protecting group is orthogonal not only to the Fmoc protecting group that masks the N-terminal amine but also to the standard acid-labile side chain protecting groups (such as *tert*-butyl or trityl) that are normally used in Fmoc-based SPPS. The Mmt group can be selectively deprotected by treatment with 1% trifluoroacetic acid (TFA) in DCM with minimal deprotection of other side chain protecting groups or cleavage from the resin (Bourel *et al.*, 2000; Matysiak *et al.*, 1998). Having achieved selective deprotection of the ε-amine of the internal lysine residue, a modification such as FITC can then be coupled to the free amine as previously described.

A stapled peptide can be subjected to internal biotinylation in a manner analogous to that described for internal labeling with FITC. However, the commercial availability of *N*-α-Fmoc-protected amino acids with a biotinylated side chain, such as Fmoc-Lys(biotinyl-ε-aminocaproyl)-OH (Novabiochem), provides a more efficient means by which to achieve internal biotinylation of a stapled peptide. This type of amino acid can simply be coupled according to normal protocols at the desired position during synthesis of the stapled peptide, allowing incorporation of an affinity tag in one step.

7. Cleavage of Stapled Peptides from the Solid Support

To effect cleavage of the completed stapled peptide from the resin and simultaneous deprotection of the acid-labile side chain protecting groups, dried resin is treated for 3 h, with mixing by nutation, with approximately 1 mL (per 20 mg peptide-bound resin) of a freshly made TFA-based cleavage cocktail. The standard cleavage cocktail (recipe shown below) is used for stapled peptides that do not contain cysteine or methionine. A version of this cleavage cocktail (shown below) that contains 1,2-ethanedithiol (EDT) is preferred for peptides that contain cysteine, methionine, or a biotin moiety.

Standard cleavage cocktail

95% TFA/2.5% triisopropylsilane (TIS)/2.5% H_2O

Cleavage cocktail (cysteine or methionine-containing peptides)

94% TFA/1% TIS/2.5% H_2O/2.5% EDT

After completion of the 3-h cleavage reaction, the cocktail is evaporated under a gentle stream of $N_{2(g)}$ until approximately 100–200 μL of solution remains. The crude peptide is precipitated by addition of 5–6 mL of diethyl ether, and the solution is then flash cooled in $N_{2(l)}$ for approximately 10 s followed by brief vortexing. After incubation at −20 °C for approximately 15 min, the crude peptide is pelleted by centrifugation at 3000×*g* for 15 min at 4 °C. The ether supernatant is decanted, and the pelleted crude stapled peptide is allowed to air dry.

8. Stapled Peptide Purification

The crude peptide is typically dissolved in approximately 1 mL of 1:1 CH_3CN:H_2O, and the resin is subsequently removed by filtration through a 0.2-μm syringe filter.[5] The peptide is then purified by reverse-phase high-performance liquid chromatography (HPLC) using a C18 column and a mobile phase gradient of H_2O and CH_3CN, each supplemented with 0.1% (v/v) TFA. The purity of the HPLC fractions is monitored by LC/MS and fractions containing the product are pooled and concentrated. Purified stapled peptides are dissolved in a minimal amount of dimethylsulfoxide (DMSO) and quantified by UV/VIS absorbance if the compound contains a chromophore (usually tryptophan, tyrosine, or FITC). Stapled peptides that do not contain a chromophore can be quantified by amino acid analysis. Stock solutions of 1–10 m*M* in DMSO are prepared and stored at 4 or −20 °C.

9. Biophysical Characterization

A diverse portfolio of experiments has previously been utilized to perform an initial evaluation of not only the extent of α-helical stabilization but also the degree to which introduction of an all-hydrocarbon staple

[5] The composition of the 1:1 CH_3CN:H_2O solution can be altered to achieve complete dissolution of the stapled peptide. As reverse-phase HPLC will be used to purify the stapled peptide, it is desirable to keep the percentage of CH_3CN in the sample as low as possible in order to ensure that the compound is retained on the column upon injection.

enhances interaction with the target. Circular dichroism (CD) spectroscopy (Greenfield and Fasman, 1969), an established method for analyzing the secondary structure of peptides and proteins in aqueous solution, has been adopted as the preferred method of determining the extent of α-helical stabilization imparted by all-hydrocarbon staples (Kim and Verdine, 2009; Kim *et al.*, 2010; Schafmeister *et al.*, 2000). Previously reported CD spectroscopy data for stapled peptides targeting the hDM2 E3 ubiquitin ligase are displayed in Fig. 1.4A (Bernal *et al.*, 2007).

Peptides that exist in an unstructured conformation exhibit a CD spectrum that is characterized by a strong minimum at 195 nm, whereas peptides that exist in an α-helical conformation exhibit a CD spectrum with dual minima at 208 and 222 nm (Greenfield and Fasman, 1969). If the concentration of the solution used to acquire the CD spectrum is known, the α-helical content of a given peptide can be quantitatively determined from the CD signal measured at 222 nm by first calculating the mean residue ellipticity and comparing that to the mean residue ellipticity calculated for a 100% α-helical peptide of the same length, as described in detail in Bird *et al.* (2008).

To evaluate the α-helicity of stapled peptides using CD spectroscopy, unlabeled compounds are typically dissolved to a concentration of approximately 50 μM in deionized water or a weakly buffered potassium phosphate solution. Modulation of the pH using 0.1% TFA or the addition of an organic cosolvent such as acetonitrile can be performed if necessary to increase stapled peptide solubility, but the manipulation of the solution to nonphysiological conditions should be avoided if possible in order to allow meaningful correlation to data from biological assays.

Initial evaluation of the ability of a panel of stapled peptides to interact with a target is most efficiently performed using purified recombinant proteins. Two types of analysis in particular, fluorescence polarization (FP) and surface plasmon resonance (SPR), have been used successfully to quantitatively measure the binding constants describing the interaction of stapled peptides with a recombinant protein target (Bernal *et al.*, 2007; Moellering *et al.*, 2009; Walensky *et al.*, 2004; Zhang *et al.*, 2008). In an FP binding assay, a stapled peptide labeled with a fluorophore such as FITC and the recombinant target protein are combined in solution in a 96- or 384-well plate and then exposed to an excitation source compatible with the particular fluorophore that is used. The polarization of the emitted fluorescence is then measured, with the extent of polarization correlated to the fraction of the stapled peptide that is bound to the target. Modified versions of the FP assay can be utilized in order to determine the ability of an unlabeled stapled peptide to disrupt the interaction between a native peptide and a protein target. In this type of competition assay, the protein of interest is loaded with a fluorescently labeled native peptide and subsequently exposed to unlabeled stapled peptide. The decrease in the polarization of

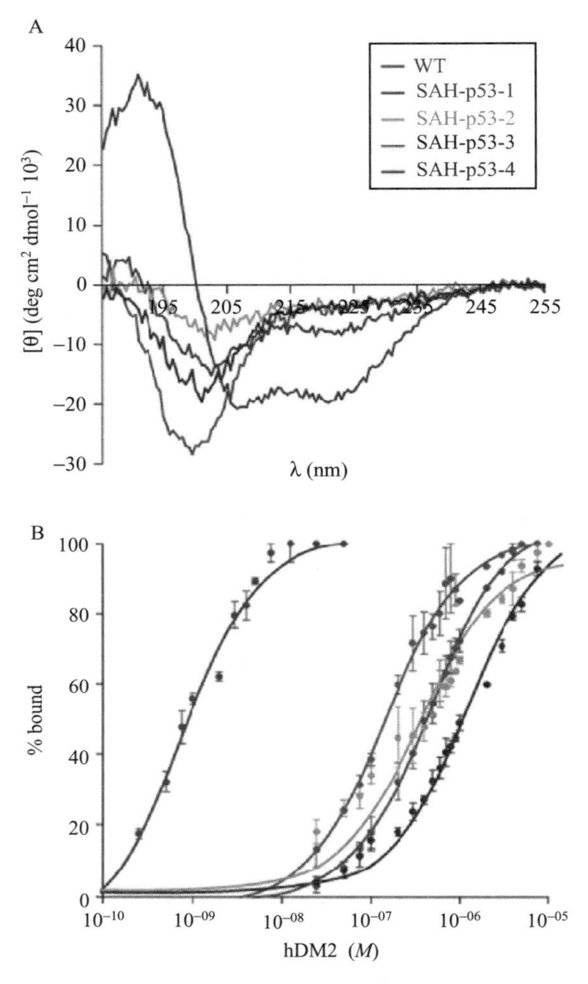

Figure 1.4 The biophysical characterization of stapled peptides. (A) Stapled peptides based on the sequence of a p53 α-helix that interacts with a hydrophobic cleft on the surface of the hDM2 E3 ubiquitin ligase were analyzed by circular dichroism spectroscopy. The unmodified wild-type (WT) peptide adopts a disordered structure, with a minimum at approximately 200 nm. The four stapled peptides (SAH-p53-x) exhibit varying degrees of α-helicity. SAH-p53-4 exhibits the most significant α-helical character, indicated by dual minima at 208 and 222 nm. The stapled peptides were analyzed in H_2O at concentrations between 10 and 50 μ*M* using a JASCO J-715 spectropolarimeter. (B) FITC-labeled SAH-p53 stapled peptides were assessed for binding to recombinant hDM2 by fluorescence polarization. SAH-p53-4 binds to hDM2 with a K_d of approximately 1 n*M*, a 400-fold improvement compared to the K_d of the unmodified wild-type peptide. Reprinted with permission from Bernal, F., Tyler, A. F., *et al.* (2007). Reactivation of the p53 tumor suppressor pathway by a stapled p53 peptide. *J. Am. Chem. Soc.* **129**(9), 2456–2457. Copyright 2007 American Chemical Society. See reference for experimental details. (For color version of this figure, the reader is referred to the Web version of this chapter.)

the fluorescence emission from the labeled native peptide is measured as a function of the increase in concentration of the unlabeled stapled peptide, and an IC_{50} for disruption of the interaction between a native peptide and its protein target can be determined. Previously reported FP assay binding data for stapled peptides targeting a hydrophobic cleft on the hDM2 E3 ubiquitin ligase are displayed in Fig. 1.4B (Bernal *et al.*, 2007).

In an SPR binding assay, a biotinylated stapled peptide is immobilized on a streptavidin-coated chip and subsequently exposed to a recombinant protein of interest. The interaction of the stapled peptide with the target protein is then measured to give a detailed illustration of the binding kinetics that governs the interaction. Additionally, SPR can be used to examine the extent to which a stapled peptide is able to inhibit the formation of a complex between two or more proteins (Moellering *et al.*, 2009). In this type of assay, a multiprotein complex is immobilized on a sample chip via an affinity tag. The immobilized complex is then exposed to the stapled peptide of interest, and its effect on the association and dissociation kinetics of the multiprotein complex is evaluated.

The FP binding assay is preferred for high-throughput screening, while SPR, due to the expense of the sample chips and the low-throughput nature of the experiment, is perhaps better suited for in-depth analysis of the binding kinetics of a limited number of stapled peptides.

The decrease in proteolytic susceptibility enjoyed by stapled peptides relative to their native peptide counterparts provides a level of *in vivo* stability that is characteristic of small molecule therapeutics. The proteolytic susceptibility of stapled peptides has been measured using three types of assays that depend on HPLC- or MS-based detection of intact stapled peptide as a function of length of sample exposure: incubation with a purified protease such as trypsin, *ex vivo* serum incubation, and *in vivo* serum incubation (Bird *et al.*, 2008; Schafmeister *et al.*, 2000; Walensky *et al.*, 2004). Bird *et al.*, (2008) have described in detail the application of these three types of proteolytic susceptibility assays to the SAHB (stabilized α-helix of BCL-2 domains) series of stapled peptides; their protocols can be extended to stapled peptides of all types. Each of these three types of protease stability assays features a level of operational complexity that directly correlates with the level with which it approximates the environment a therapeutic would experience upon administration.

10. Cell Permeability

The cell permeability of stapled peptides has previously been evaluated using flow cytometry and confocal fluorescence microscopy, both of which require the utilization of fluorescently labeled stapled peptides (Bernal *et al.*, 2007; Bird *et al.*, 2008; Moellering *et al.*, 2009; Walensky *et al.*, 2004,

2006). Figure 1.5 contains cell permeability data for SAH-p53 stapled peptides obtained from both flow cytometry and confocal fluorescence microscopy (Bernal *et al.*, 2007). Bird *et al.*, (2008) have provided a detailed description of specific procedures that can be used for the evaluation of cell permeability by each of these methods. Flow cytometry-based cell

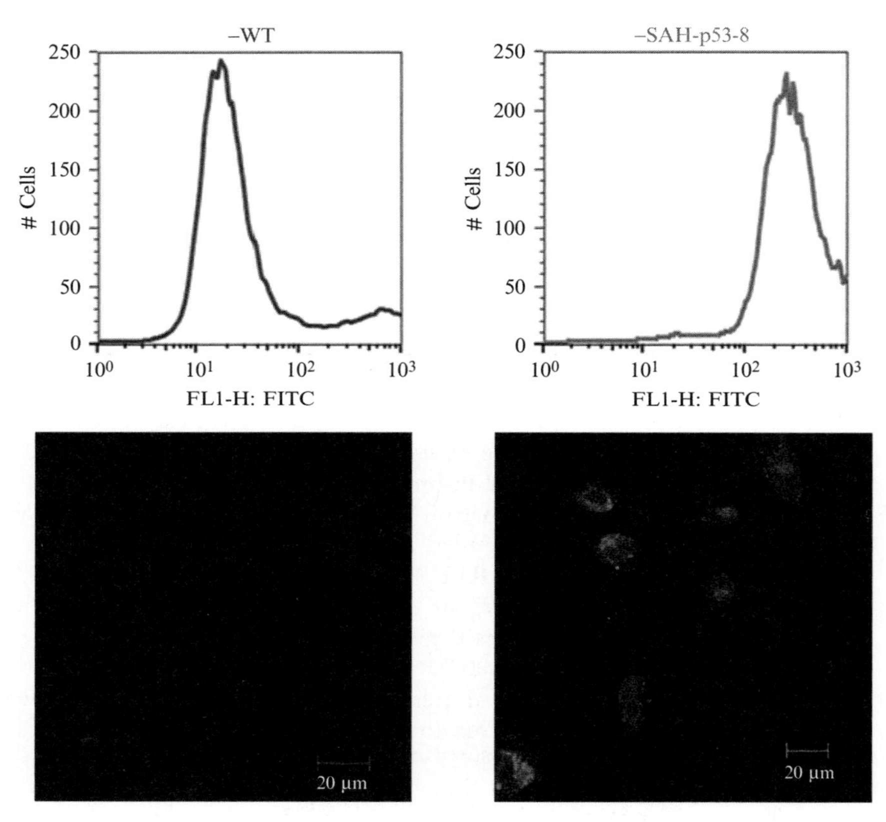

Figure 1.5 Flow cytometry and confocal fluorescence microscopy are complementary methods for assessment of the cell permeability of stapled peptides. In both cases, SJSA-1 cells were treated with the indicated FITC-labeled peptide for 4h at 37°C. Top: Flow cytometry indicates that an unmodified wild-type (WT) peptide targeting the p53 binding cleft on hDM2 does not accumulate inside cells, whereas SAH-p53-8, an *i*, *i*+7 stapled peptide targeting the same cleft on hDM2, exhibits significant intracellular accumulation. Bottom: Confocal fluorescence microscopy images indicating the degree of cell permeability of the WT peptide and the SAH-p53-8 stapled peptide (green). The nuclei are visible as a result of treating the cells with DAPI nuclear stain (blue). Reprinted with permission from Bernal, F., Tyler, A. F., *et al.* (2007). Reactivation of the p53 tumor suppressor pathway by a stapled p53 peptide. *J. Am. Chem. Soc.* 129(9), 2456–2457. Copyright 2007 American Chemical Society. See reference for experimental details. (For interpretation of the references to color in this figure legend, the reader is referred to the Web version of this chapter.)

permeability assays are advantageous for their high-throughput nature; cells can be grown and treated in a 96-well plate format, enabling the high-throughput analysis of the cell permeability of an extensive panel of stapled peptides in parallel. Another advantage of flow cytometry is that a large number of cells, typically more than 1×10^4, are analyzed for intracellular fluorescence, allowing the evaluation of cell permeability over a very large sample size. One disadvantage of a flow cytometry-based assay is that lack of a visual output renders it incapable of producing any information about subcellular localization. Further, although cell surface proteins can be removed via trypsinization prior to analysis in an effort to release stapled peptides that are merely stuck to the cell surface, flow cytometry-based analysis can occasionally result in misleading information regarding whether a stapled peptide has permeated the cell membrane and entered the cytosol.

In the absence of highly specialized instrumentation that allows confocal imaging of cells in a multi-well plate format, confocal fluorescence microscopy is a low-throughput method for analyzing the cell permeability of stapled peptides. Further, the number of cells in each treatment condition that can be analyzed using confocal fluorescence microscopy is orders of magnitude smaller than the number analyzed by flow cytometry. Because it is possible to detect emission from multiple fluorophores simultaneously and map the results onto a phase contrast image of a cell at high magnification and resolution, confocal fluorescence microscopy can provide a wealth of detailed visual information regarding the mechanism of entry and subcellular localization of stapled peptides. Multicolor confocal fluorescence microscopy utilizing TRITC (tetramethylrhodamine isothiocyanate)-dextran endosomal marker, for example, has demonstrated that the uptake of stapled peptides occurs through fluid-phase pinocytosis (Bernal *et al.*, 2007; Walensky *et al.*, 2004). Additionally, confocal fluorescence microscopy performed with dual treatment of cells with FITC-labeled stapled peptides and a mitochondrion-specific fluorescent marker confirmed that the $SAHB_A$ stapled peptide, as expected due to the mitochondrial-localization of its protein targets, is targeted to the correct organelle (Walensky *et al.*, 2004).

One notable drawback of confocal fluorescence microscopy is that it provides only a qualitative assessment of cell permeability. Recently, quantitative epifluorescence microscopy has been adopted as a means of quantitatively evaluating the cell permeability of stapled peptides without sacrificing the ability to visualize stapled peptides inside the cell (Q. Chu, R. E. Moellering, and G. L. Verdine, unpublished results). Quantitative epifluorescence microscopy enables a direct and quantitative comparison of the extent of cellular uptake of stapled peptides in a 384-well plate format and provides the opportunity to quantify distribution such as nuclear versus cytoplasmic localization when combined with a nuclear stain such as DAPI (4′,6-diamidino-2-phenylindole).

11. *In Vitro* Target Interaction and Activity Assays

To confirm that a stapled peptide interacts with the target in the context of the cellular environment, it can be evaluated for the ability to interact with a target protein by *in vitro* immunoprecipitation (IP) and pull-down assays (Bernal *et al.*, 2007, 2010; Gavathiotis *et al.*, 2008; Moellering *et al.*, 2009; Pitter *et al.*, 2008; Stewart *et al.*, 2010; Walensky *et al.*, 2006).

IP/pull-down assays can be performed in a variety of ways depending on the reagents that are available and the specific biological pathway that is being studied. For example, after treating cells with FITC-labeled stapled peptides, the cells can be lysed and the proteins that are interacting with the stapled peptides can be immunoprecipitated by adding α-FITC antibody and protein A/G-sepharose to the lysate (Bernal *et al.*, 2007, 2010; Pitter *et al.*, 2008; Walensky *et al.*, 2006). Similarly, biotinylated stapled peptides can be immobilized on streptavidin-agarose beads and subsequently incubated with whole-cell lysates (Moellering *et al.*, 2009). In both cases, the captured proteins can then be resolved by denaturing polyacrylamide gel electrophoresis and Western blotting can be used to confirm that the stapled peptide interacts with the intended target protein.

Another variant of the IP assay uses an antibody that binds to the protein target in order to show that addition of the stapled peptide is able to block IP of the intact protein–protein complex. In one such example, cells treated with a stapled peptide designed to inhibit the interaction between MCL-1 and BAK were lysed and the lysate was subsequently incubated with α-MCL-1 antibody (Stewart *et al.*, 2010). Protein A/G-sepharose beads were then added to the lysate to capture the α-MCL-1 antibodies and Western blotting was used to demonstrate that the treatment of cells with stapled peptide inhibitor prevented co-IP of BAK by MCL-1. A similar type of experiment was performed to show that a MAML-derived stapled peptide was able to disrupt the protein–protein interaction between MAML1 and the intracellular domain of NOTCH (Moellering *et al.*, 2009).

The type of *in vitro* activity assays that are used to demonstrate that a stapled peptide is capable of producing a biological effect vary widely depending on the particular interaction that has been targeted for inhibition. In the case of stapled peptides intended to disrupt a protein–protein interaction that is important for the survival and proliferation of cancer cells, it is useful to examine the effect of the stapled peptide on the proliferation of a relevant cancer cell line. There are many commercially available kits, such as the CellTiter-Glo Luminescent Cell Viability Assay (Promega), that allow for efficient measurement of cell viability in a 96-well plate format. Though this type of cell viability assay does not couple the inhibition of cell proliferation to the inhibition of a specific cellular pathway, it is often an ideal assay for high-throughput screening of a panel of stapled peptides to identify those with a biological effect.

Pathway-specific activity assays can determine if a stapled peptide exhibits any observed biological function through a specific interaction with the intended target. For example, the BID $SAHB_A$ peptide was intended to induce apoptosis by inhibiting the sequestration of proapoptotic BCL-2 family proteins (Walensky *et al.*, 2004). Walensky and colleagues treated Jurkat T cells with BID $SAHB_A$ and were able to induce apoptosis in a dose-dependent manner; the dependence of this effect on the intended pathway was confirmed by showing that overexpression of the antiapoptotic BCL-2 protein was able to confer resistance to the BID $SAHB_A$ stapled peptide.

In cases where a stapled peptide is designed to inhibit a protein–protein interaction essential for transcriptional activation, reporter gene assays and RT-PCR are ideal methods for assessing *in vitro* efficacy. The inhibitory effect of SAHM1 treatment on NOTCH-dependent transcriptional activation was examined using both types of assay (Moellering *et al.*, 2009). In one experiment, cancer cells were cotransfected with a plasmid directing constitutive expression of Renilla luciferase and a plasmid bearing a firefly luciferase gene engineered to be transcriptionally regulated by NOTCH. A standard dual-luciferase assay demonstrated inhibition of NOTCH-dependent transcriptional activation. In another experiment, T-ALL (T-cell acute lymphoblastic leukemia) cells featuring deregulated NOTCH were treated with the SAHM1 peptide and RT-PCR was performed to demonstrate a dose-dependent inhibition of expression of NOTCH-dependent target genes. In both the reporter gene assay and the RT-PCR assay, the effect of a designed negative control stapled peptide which had been shown to lack the capacity to inhibit the assembly of the active NOTCH–CSL–MAML ternary complex was examined, and it confirmed that the effect observed by SAHM1 could be attributed to disruption of the NOTCH–CSL–MAML ternary complex formation.

12. *In Vivo* Efficacy

A final step toward validating the potential therapeutic utility of the designed compound includes evaluating its efficacy in an animal model of disease. In three cases, stapled peptides designed to disrupt protein–protein interactions important for the progression of cancer were demonstrated to inhibit tumor progression in relevant murine models of cancer (Bernal *et al.*, 2010; Moellering *et al.*, 2009; Walensky *et al.*, 2004). To ensure maximal bioavailability of the stapled peptides in these proof-of-principle *in vivo* experiments, the stapled peptides are administered by intraperitoneal or intravenous injection. In addition to observing the overall effect on tumor burden, animals treated with compound can be dissected postmortem and their tissues can be further subjected to histological and gene expression analysis to obtain a more complete picture of the effect of the stapled peptides on the progression of disease *in vivo*.

13. Strategies for Stapled Peptide Optimization

As stapled peptide-mediated disruption of intracellular protein–protein interactions is a recent advancement, the successful implementation of stapled peptide technology to target diverse biological pathways is a difficult undertaking. The development of a stapled peptide inhibitor of a protein–protein interaction is typically a multistep process involving iterative optimization of high-affinity binding and other properties critical for the production of a molecule with potential as an *in vivo* therapeutic. Our experiences with stapled peptides have identified two properties in particular, cell permeability and aqueous solubility, which often require optimization.

The SAH-p53 stapled peptides, which prevent the interaction of the hDM2 and hDMX E3 ubiquitin ligases with the p53 tumor suppressor protein, present a case in which a lead stapled peptide that exhibited poor cell permeability was subjected to sequence optimization by iterative mutagenesis to produce a cell permeable variant that retained the ability to bind to the target with high affinity (Bernal *et al.*, 2007). In an initial round of stapled peptide development, Bernal and colleagues identified an i, $i+7$ stapled peptide, SAH-p53-4, that bound to hDM2 with a subnanomolar K_d. This stapled peptide, which featured a net charge of -2 at physiological pH, displayed poor cell permeability. Noting that CPPs such as HIV-1 TAT feature significant net positive charge and reasoning that a net negative charge may prevent association of the stapled peptide with the negatively charged phospholipid bilayer of the cell membrane in the initial stages of cell penetration, they introduced a series of charge-altering mutations into SAH-p53-4. (Table 1.4) Using high-resolution structural information (Kussie *et al.*, 1996) to guide optimization, Bernal and colleagues replaced a glutamic acid and aspartic acid that were putatively uninvolved in binding to the target with glutamine and asparagine, respectively. The resulting stapled peptide, SAH-p53-5, experienced no loss in affinity for hDM2 but, presumably due to the loss of a net negative charge, demonstrated significant cell permeability. This structure-based manipulation of charged residues demonstrates a successful strategy for optimization of the cell permeability of stapled peptides without sacrificing binding affinity.

The introduction of an all-hydrocarbon staple can significantly decrease the aqueous solubility of a peptide for multiple reasons. First and foremost, the installation of a hydrocarbon staple results in the addition of a significantly

Table 1.4 Optimization of stapled peptides for cell permeability

Compound	Sequence ★=R_8; ★=S_5	Charge at pH 7.4	Cell permeable	K_d (n*M*)
WT	Ac-LSQETFSDLWKLLPEN-NH2	−2	No	410±19
SAH-p53-4	Ac-LSQETF*DLWKLL*EN-NH2	−2	No	0.92±0.11
SAH-p53-5	Ac-LSQETF*NLWKLL*QN-NH2	0	Yes	0.80±0.05

Incorporation of an *i*, *i*+7 staple into a peptide derived from the sequence of the hDM2-binding portion of the p53 tumor suppressor protein produces SAH-p53-4, which binds to hDM2 with an affinity enhancement of more than 400-fold. SAH-p53-4, however, did not exhibit significant cell permeability, possibly due to a charge of −2 at physiologic pH. An aspartic acid and glutamic acid in SAH-p53-4 were replaced with asparagine and glutamine, respectively, to produce SAH-p53-5, a stapled peptide with no formal charge at physiologic pH. These two mutations did not significantly affect binding to hDM2 but markedly enhanced the cell permeability. Adapted with permission from Bernal, F., Tyler, A. F., *et al.* (2007). Reactivation of the p53 tumor suppressor pathway by a stapled p53 peptide. *J. Am. Chem. Soc.* **129**(9), 2456–2457. Copyright 2007 American Chemical Society. (For color version of this table, the reader is referred to the Web version of this chapter.)

hydrophobic patch to one face of the α-helical peptide. Owing directly to the incorporation of this intrinsically hydrophobic moiety, it would be expected that a stapled peptide would exhibit decreased aqueous solubility compared to its parent unmodified peptide. Additionally, the induction of α-helical structure results in the burial of the hydrophilic amide backbone in the core of the folded conformation. While the increased hydrophobicity achieved by masking the hydrophilic peptide backbone provides an explanation for the increased capability for penetration through the hydrophobic interior of the cell membrane, it necessarily results in a decrease in the hydrophilicity and, therefore, the aqueous solubility of the stapled peptide.

The introduction of an all-hydrocarbon staple into a peptide that is designed to interact with a target through hydrophobic interactions can result in a stapled peptide that exhibits insufficient aqueous solubility to allow full characterization in biophysical assays as well as *in vitro* target interaction and activity assays (G. J. Hilinski and G. L. Verdine, unpublished results). A similar strategy used for the optimization of cell permeability can be used for the optimization of aqueous solubility. Specifically, structure-based design can be used to replace hydrophobic amino acids that are not expected to participate in interaction with the target with polar or positively charged (but not negatively charged) residues that would be expected to increase the hydrophilicity of the stapled peptide. The stapled peptide variants can then be evaluated for interaction with the target, and those variants that retain binding affinity can then be evaluated for solubility in a physiologically relevant solution such as phosphate-buffered saline. In the case that such mutational engineering does not achieve the desired effect, N-terminal or internal modifications can also be utilized as a means of increasing solubility. For example, a variety of monodisperse pegylation reagents designed for increasing the aqueous solubility of peptides are commercially available from a number of suppliers (e.g., Novabiochem, Anaspec). These reagents feature a defined number of ethylene glycol units and either a carboxylic acid or other reactive moiety such as maleimide that facilitates efficient incorporation during SPPS.

Indeed, the introduction of mutations engineered to increase cell permeability or aqueous solubility without decreasing binding affinity may be unsuccessful. Additionally, as high-resolution structures are merely a snapshot of an interaction in an environment that does not necessarily replicate the environment in which the interaction takes place *in vivo*, even structure-based design of mutations may result in unexpected decreases in binding affinity. Likewise, it is possible that amino acids that do not appear to be significantly involved in a particular binding interaction may indeed impart a degree of target specificity *in vivo*. If this is the case, optimization of properties such as solubility and cell permeability may be detrimental to the *in vivo* target specificity exhibited by the original stapled peptide.

14. Summary

The ability of stapled peptides to inhibit intracellular protein–protein interactions that have previously been intractable to inhibition with traditional small molecule or protein-based therapeutics suggests that they may define a novel therapeutic modality. There is still much to learn regarding the development of stapled peptides as potent and specific inhibitors of protein–protein interactions, but each new report of their successful application provides information vital to the continued development of these unique molecules. It is hoped that the information contained here will increase the accessibility of stapled peptides and provides researchers with the tools to confidently employ this technology to tackle the most daunting therapeutic targets.

ACKNOWLEDGMENTS

We thank T.N. Grossman for assistance with Fig. 1.1 and Y.W. Kim for editorial assistance. Work on stapled peptides in the Verdine laboratory has been generously supported by grants from the High-tech Fund of the Dana-Farber Cancer Institute, GlaxoSmithKline and Johnson & Johnson.

REFERENCES

Alvarez-Salas, L. M. (2008). Nucleic acids as therapeutic agents. *Curr. Top. Med. Chem.* **8**(15), 1379–1404.

Atwal, J. K., Chen, Y., *et al.* (2011). A therapeutic antibody targeting BACE1 inhibits amyloid-{beta} production in vivo. *Sci. Transl. Med.* **3**(84), 84ra43.

Bernal, F., Tyler, A. F., *et al.* (2007). Reactivation of the p53 tumor suppressor pathway by a stapled p53 peptide. *J. Am. Chem. Soc.* **129**(9), 2456–2457.

Bernal, F., Wade, M., *et al.* (2010). A stapled p53 helix overcomes HDMX-mediated suppression of p53. *Cancer Cell* **18**(5), 411–422.

Betzi, S., Guerlesquin, F., *et al.* (2009). Protein-protein interaction inhibition (2P2I): Fewer and fewer undruggable targets. *Comb. Chem. High Throughput Screen.* **12**(10), 968–983.

Bird, G. H., Bernal, F., *et al.* (2008). Synthesis and biophysical characterization of stabilized alpha-helices of BCL-2 domains. *Methods Enzymol.* **446,** 369–386.

Bird, G. H., Madani, N., *et al.* (2010). Hydrocarbon double-stapling remedies the proteolytic instability of a lengthy peptide therapeutic. *Proc. Natl. Acad. Sci. USA* **107**(32), 14093–14098.

Blackwell, H. E., and Grubbs, R. H. (1998). Highly efficient synthesis of covalently cross-linked peptide helices by ring-closing metathesis. *Ange. Chem. Int. Ed.* **37**(23), 3281–3284.

Bourel, L., Carion, O., *et al.* (2000). The deprotection of Lys(Mtt) revisited. *J. Pept. Sci.* **6**(6), 264–270.

Chene, P. (2006). Drugs targeting protein-protein interactions. *ChemMedChem* **1**(4), 400–411.

Cronican, J. J., Thompson, D. B., *et al.* (2010). Potent delivery of functional proteins into Mammalian cells in vitro and in vivo using a supercharged protein. *ACS Chem. Biol.* **5**(8), 747–752.

Derossi, D., Joliot, A. H., *et al.* (1994). The third helix of the Antennapedia homeodomain translocates through biological membranes. *J. Biol. Chem.* **269**(14), 10444–10450.

Di Cresce, C., and Koropatnick, J. (2010). Antisense treatment in human prostate cancer and melanoma. *Curr. Cancer Drug Targets* **10**(6), 555–565.

Fry, D. C., and Vassilev, L. T. (2005). Targeting protein-protein interactions for cancer therapy. *J. Mol. Med.* **83**(12), 955–963.

Gavathiotis, E., Suzuki, M., *et al.* (2008). BAX activation is initiated at a novel interaction site. *Nature* **455**(7216), 1076–1081.

Greenfield, N., and Fasman, G. D. (1969). Computed circular dichroism spectra for the evaluation of protein conformation. *Biochemistry* **8**(10), 4108–4116.

Guharoy, M., and Chakrabarti, P. (2007). Secondary structure based analysis and classification of biological interfaces: Identification of binding motifs in protein-protein interactions. *Bioinformatics* **23**(15), 1909–1918.

Hasadsri, L., Kreuter, J., *et al.* (2009). Functional protein delivery into neurons using polymeric nanoparticles. *J. Biol. Chem.* **284**(11), 6972–6981.

Heitz, F., Morris, M. C., *et al.* (2009). Twenty years of cell-penetrating peptides: From molecular mechanisms to therapeutics. *Br. J. Pharmacol.* **157**(2), 195–206.

Henchey, L. K., Jochim, A. L., *et al.* (2008). Contemporary strategies for the stabilization of peptides in the alpha-helical conformation. *Curr. Opin. Chem. Biol.* **12**(6), 692–697.

Hopkins, A. L., and Groom, C. R. (2002). The druggable genome. *Nat. Rev. Drug Discov.* **1**(9), 727–730.

Hopkins, A. L., and Groom, C. R. (2003). Target analysis: A priori assessment of druggability. *Ernst Schering Res. Found. Workshop* (42), 11–17.

Jenssen, H., and Aspmo, S. I. (2008). Serum stability of peptides. *Methods Mol. Biol.* **494,** 177–186.

Jochim, A. L., and Arora, P. S. (2009). Assessment of helical interfaces in protein-protein interactions. *Mol. Biosyst.* **5**(9), 924–926.

Jones, S., and Thornton, J. M. (1996). Principles of protein-protein interactions. *Proc. Natl. Acad. Sci. USA* **93**(1), 13–20.

Jullian, M., Hernandez, A., *et al.* (2009). N-terminus FITC labeling of peptides on solid support: The truth behind the spacer. *Tetrahedron Lett.* **50**(3), 4p.

Karle, I. L. (2001). Controls exerted by the Aib residue: Helix formation and helix reversal. *Biopolymers* **60**(5), 351–365.

Karle, I. L., Flippen-Anderson, J. L., *et al.* (1990). Modular design of synthetic protein mimics. Crystal structures, assembly, and hydration of two 15- and 16-residue apolar, leucyl-rich helical peptides. *J. Am. Chem. Soc.* **112,** 9350–9356.

Kim, Y. W., and Verdine, G. L. (2009). Stereochemical effects of all-hydrocarbon tethers in i, i+4 stapled peptides. *Bioorg. Med. Chem. Lett.* **19**(9), 2533–2536.

Kim, Y. W., Kutchukian, P. S., *et al.* (2010). Introduction of all-hydrocarbon i, i+3 staples into alpha-helices via ring-closing olefin metathesis. *Org. Lett.* **12**(13), 3046–3049.

Kim, Y. W., Grossmann, T. N., *et al.* (2011). Synthesis of all-hydrocarbon stapled alpha-helical peptides by ring-closing olefin metathesis. *Nat. Protoc.* **6**(6), 761–771.

Kussie, P. H., Gorina, S., *et al.* (1996). Structure of the MDM2 oncoprotein bound to the p53 tumor suppressor transactivation domain. *Science* **274**(5289), 948–953.

Lawrence, M. S., Phillips, K. J., *et al.* (2007). Supercharging proteins can impart unusual resilience. *J. Am. Chem. Soc.* **129**(33), 10110–10112.

Maillard, I., Weng, A. P., *et al.* (2004). Mastermind critically regulates Notch-mediated lymphoid cell fate decisions. *Blood* **104**(6), 1696–1702.

Matysiak, S., Boldicke, T., *et al.* (1998). Evaluation of monomethoxytrityl and dimethoxytrityl as orthogonal amino protecting groups in Fmoc solid phase peptide synthesis. *Tetrahedron Lett.* **39**(13), 1733–1734.

McNaughton, B. R., Cronican, J. J., *et al.* (2009). Mammalian cell penetration, siRNA transfection, and DNA transfection by supercharged proteins. *Proc. Natl. Acad. Sci. USA* **106**(15), 6111–6116.

Mescalchin, A., and Restle, T. (2011). Oligomeric nucleic acids as antivirals. *Molecules* **16**(2), 1271–1296.

Moellering, R. E., Cornejo, M., *et al.* (2009). Direct inhibition of the NOTCH transcription factor complex. *Nature* **462**(7270), 182–188.

Nam, Y., Sliz, P., *et al.* (2006). Structural basis for cooperativity in recruitment of MAML coactivators to Notch transcription complexes. *Cell* **124**(5), 973–983.

Pitter, K., Bernal, F., *et al.* (2008). Dissection of the BCL-2 family signaling network with stabilized alpha-helices of BCL-2 domains. *Methods Enzymol.* **446,** 387–408.

Qiu, W., Soloshonok, V. A., *et al.* (2000). Convenient, large-scale asymmetric synthesis of enantiomerically pure trans-cinnamylglycine and -alpha-alanine. *Tetrahedron* **56**(17), 2577–2582.

Rizk, S. S., Luchniak, A., *et al.* (2009). An engineered substance P variant for receptor-mediated delivery of synthetic antibodies into tumor cells. *Proc. Natl. Acad. Sci. USA* **106**(27), 11011–11015.

Russ, A. P., and Lampel, S. (2005). The druggable genome: An update. *Drug Discov. Today* **10**(23–24), 1607–1610.

Schafmeister, C. E., Po, J., *et al.* (2000). An all-hydrocarbon cross-linking system for enhancing the helicity and metabolic stability of peptides. *J. Am. Chem. Soc.* **122**(24), 5891–5892.

Stewart, M. L., Fire, E., *et al.* (2010). The MCL-1 BH3 helix is an exclusive MCL-1 inhibitor and apoptosis sensitizer. *Nat. Chem. Biol.* **6**(8), 595–601.

Sviridov, D. O., Ikpot, I. Z., *et al.* (2011). Helix stabilization of amphipathic peptides by hydrocarbon stapling increases cholesterol efflux by the ABCA1 transporter. *Biochem. Biophys. Res. Commun.* **410**(3), 446–451.

van den Berg, A., and Dowdy, S. F. (2011). Protein transduction domain delivery of therapeutic macromolecules. *Curr. Opin. Biotechnol.* **22**(6), 888–893.

Verdine, G. L., and Walensky, L. D. (2007). The challenge of drugging undruggable targets in cancer: Lessons learned from targeting BCL-2 family members. *Clin. Cancer Res.* **13**(24), 7264–7270.

Vives, E., Brodin, P., *et al.* (1997). A truncated HIV-1 Tat protein basic domain rapidly translocates through the plasma membrane and accumulates in the cell nucleus. *J. Biol. Chem.* **272**(25), 16010–16017.
Walensky, L. D., Kung, A. L., *et al.* (2004). Activation of apoptosis in vivo by a hydrocarbon-stapled BH3 helix. *Science* **305**(5689), 1466–1470.
Walensky, L. D., Pitter, K., *et al.* (2006). A stapled BID BH3 helix directly binds and activates BAX. *Mol. Cell* **24**(2), 199–210.
Wender, P. A., Mitchell, D. J., *et al.* (2000). The design, synthesis, and evaluation of molecules that enable or enhance cellular uptake: Peptoid molecular transporters. *Proc. Natl. Acad. Sci. USA* **97**(24), 13003–13008.
Weng, A. P., Nam, Y., *et al.* (2003). Growth suppression of pre-T acute lymphoblastic leukemia cells by inhibition of notch signaling. *Mol. Cell. Biol.* **23**(2), 655–664.
Williams, R. M., and Im, M. N. (1991). Asymmetric-synthesis of monosubstituted and alpha, alpha-disubstituted alpha-amino-acids via diastereoselective glycine enolate alkylations. *J. Am. Chem. Soc.* **113**(24), 9276–9286.
Williams, R. S., DeMong, P., Chen, D., and Zhai, D. (2003). Asymmetric synthesis of N-tert-butoxycarbonyl α-amino acids. Synthesis of (5S,6R)-4-tert-butoxycarbonyl-5,6-diphenylmorpholin-2-one. *Org. Synth.* **80,** 18–30.
Yu, Y. J., Zhang, Y., *et al.* (2011). Boosting brain uptake of a therapeutic antibody by reducing its affinity for a transcytosis target. *Sci. Transl. Med.* **3**(84), 84ra44.
Zhang, H., Zhao, Q., *et al.* (2008). A cell-penetrating helical peptide as a potential HIV-1 inhibitor. *J. Mol. Biol.* **378**(3), 565–580.
Zhang, H., Curreli, F., *et al.* (2011). Antiviral activity of alpha-helical stapled peptides designed from the HIV-1 capsid dimerization domain. *Retrovirology* **8**(1), 28.

CHAPTER TWO

Mapping of Vascular ZIP Codes by Phage Display

Tambet Teesalu,[*,†,1] Kazuki N. Sugahara,[†,1] *and* Erkki Ruoslahti[*,†]

Contents

1. Introduction 36
   1.1. Vascular ZIP codes: Molecular basis and exploration by phage display 36
   1.2. CendR peptides for tissue penetrative cocomposition targeting 40
2. Methods 41
   2.1. T7 phage display: General notes and protocols 41
   2.2. General considerations in phage biopanning 46
   2.3. *Ex vivo* phage display 46
   2.4. *In vivo* phage display 48
   2.5. Postscreening phage auditions 49
   2.6. *In vivo* homing of synthetic peptides 50
   2.7. Identification of the peptide receptors 51
3. Concluding Remarks and Perspectives 52
Acknowledgments 53
References 53

Abstract

Each organ and pathology has a unique vascular ZIP code that can be targeted with affinity ligands. *In vivo* peptide phage display can be used for unbiased mapping of the vascular diversity. Remarkably, some of the peptides identified by such screens not only bind to target vessels but also elicit biological responses. Recently identified tissue-penetrating CendR peptides trigger vascular exit and parenchymal spread of a wide range of conjugated and coadministered payloads. This review is designed to serve as a practical guide for

[*] Center for Nanomedicine, Sanford-Burnham Medical Research Institute at UCSB, Santa Barbara, California, USA
[†] Cancer Center, Sanford-Burnham Medical Research Institute, La Jolla, California, USA
[1] Equal contribution.

Methods in Enzymology, Volume 503
ISSN 0076-6879, DOI: 10.1016/B978-0-12-396962-0.00002-1

researchers interested in setting up *ex vivo* and *in vivo* phage display technology. We focus on T7 coliphage platform that our lab prefers to use due to its versatility, physical resemblance of phage particles to clinical nanoparticles, and ease of manipulation.

1. Introduction

1.1. Vascular ZIP codes: Molecular basis and exploration by phage display

Vascular endothelial cells lining the lumen of blood vessels are critically involved in many physiological processes (e.g., blood-parenchymal exchange, immune responses, hemostasis) and diseases (e.g., atherosclerosis, carcinogenesis, thrombosis, sepsis, and vascular leak syndromes; Cines *et al.*, 1998; De Caterina and Libby, 2007). Tumor vessels grow as the tumor grows, and this process—angiogenesis—makes tumor blood vessels distinct from normal resting blood vessels (Hanahan and Folkman, 1996; Ruoslahti, 2002). Abnormal features of tumor vasculature include large and variable intervascular distances, complex and tortuous branching patterns, lack of lymphatic vessel drainage, and irregular vascular phenotypes related to immaturity, angiogenic capacity, size, perfusion, and permeability (Carmeliet and Jain, 2000). Differences in anatomy, structure, and functional status of blood vessels are mirrored in the molecular diversity of vascular endothelial cells, including in the luminal side that is readily accessible to circulating probes (Allen and Cullis, 2004; Jain and Gerlowski, 1986; Ruoslahti, 2002). These systemically accessible receptors, vascular ZIP codes, can be used for synaphic (affinity-based) targeting using ligands such as peptides, antibodies, and aptamers for selective delivery of diagnostic and therapeutic cargo (Ruoslahti, 2002, 2004; Ruoslahti *et al.*, 2010).

Peptides are powerful molecular ligands for targeting tumor microvessels. Rational design and structure–activity studies have led to development of peptide ligands that associate with tumor vessels such as octreotide (analogue of somatostatin used clinically for imaging of somatostatin receptor-positive tumors; Olsen *et al.*, 1995), bombesin/gastrin-releasing peptide (for detection and treatment of several types of carcinoma; Zhang *et al.*, 2004), and neurotensin analogs (Achilefu *et al.*, 2003). However, this approach relies on preexisting knowledge on the expression and biology of the peptide receptors and on assumptions on receptors' subcellular distribution and density, on the physical properties and fluid dynamics of target vascular beds, and on optimal affinity (avidity) of peptides necessary for efficient targeting. In contrast, *in vivo* phage display, first reported by our group in 1996 (Pasqualini and Ruoslahti, 1996), allows unbiased exploration of vascular diversity (Ruoslahti, 2002). Phage display is a powerful method for peptide library screening that provides a physical linkage between peptides (i.e., the phenotype), which are displayed on the surface of a bacteriophage particle, and the encoding DNA (genotype; Smith and Petrenko, 1997). For *in vivo* biopanning, genetically encoded phage peptide libraries are injected into the systemic circulation, followed by removal of free phage by perfusion, quantification of retained phage in target and control organs, and its amplification for subsequent round of selection. *In vivo* biopanning combines positive selection at the target tissue with negative selection (elimination of nonspecific, pan-cell binding phage) at other vascular beds. *In vivo* screening of phage libraries has yielded a variety of homing peptides that are specific for endothelia of normal organs, tumor vasculature, and tumor cells (Arap *et al.*, 1998; Laakkonen *et al.*, 2002; Lee *et al.*, 2007; Ruoslahti, 2004; Simberg *et al.*, 2007; Zhang *et al.*, 2006). Examples of angiogenesis-associated vascular molecules include certain growth factor receptors, cell adhesion molecules, and intracellular proteins aberrantly expressed at the cell surface in tumors (Christian *et al.*, 2003; Oh *et al.*, 2004; Ruoslahti, 2002). Another set of vascular markers in tumors appears to be unrelated to angiogenesis; these markers can be specific for a tumor type, stage in tumor development, or location (Hoffman *et al.*, 2003; Joyce *et al.*, 2003). Tumor lymphatics are also specialized, as they express markers that are not present in the lymphatics of normal tissues, or in tumor blood vessels (Fogal *et al.*, 2008; Laakkonen *et al.*, 2002; Zhang *et al.*, 2006). Generally, the homing peptides appear to recognize conserved regions in proteins, such as the antigen-binding site in antibodies, the substrate-binding site in enzymes, and the ligand-binding site in receptors. For example, a phage library screen with an integrin that uses the RGD tripeptide motif as its binding site in extracellular matrix proteins yielded essentially only (31/32) RGD peptides and peptides that were clearly RGD mimics (Koivunen *et al.*, 1993). Thus, the ligand-binding site in this receptor was overwhelmingly preferred as a binding site for peptides. As a consequence of their preferential binding conserved sites in proteins, peptides rarely show species specificity. As vascular receptor expression pattern may be different between mice and men, *in vivo* mapping of human vasculature could reveal targets that cannot be identified by screening in mice (Arap *et al.*, 2002b; Krag *et al.*, 2006). A summary of selected peptides derived from biopanning screens and their targeting specificities is listed in Table 2.1.

The primary applications of vascular homing peptides are related to targeted delivery of conjugated therapeutic and imaging payloads to tumors. Cytotoxic anticancer drugs cause side effects in normal tissues that prevent application of the drugs at doses sufficient to eliminate the entire tumor (Perry, 2001; Souhami and Tobias, 2005). Moreover, under the selective pressure of a toxic therapy, genetic diversity within tumors leads to outgrowth of drug-resistant cells and consequent cancer recurrence (Chabner and Roberts, 2005), especially when some tumor cells encounter drug

Table 2.1 Selected homing peptides and their *in vivo* targeting specificity

	Peptide sequence	*In vivo* homing specificity (designation, optional)	Reference
1.	gSMSIARL	Normal prostate	Arap *et al.* (2002a)
2.	gVSFLEYR	Normal prostate	Arap *et al.* (2002a)
3.	CNGRC	Angiogenic vessels, various locations	Arap *et al.* (1998)
4.	CREAGRKAC	Tramp lymphatics	Zhang *et al.* (2006)
5.	CAGRRSAYC	Tramp premalignant lymphatics	Zhang *et al.* (2006)
6.	CRAKSKVAC	Pan-endothelial homer (dysplastic skin)	Zhang *et al.* (2006)
7.	CRRETAWAC	$\alpha5\beta1$ integrin	Koivunen *et al.* (1994)
8.	CRPPR	Heart	Zhang *et al.* (2005)
9.	CPKTRRVPC	Heart	Zhang *et al.* (2005)
10.	CGLIIQKNEC	Blood clot (CLT1)	Pilch *et al.* (2006)
11.	CNAGESSKNC	Blood clot (CLT2)	Pilch *et al.* (2006)
12.	CARSKNKDC	Wound (CAR)	Järvinen and Ruoslahti (2007)
13.	CRKDKC	Wound (CRK)	Järvinen and Ruoslahti (2007)
14.	GLNGLSSADPSSD WNAPAEEWGNW VDEDRASLLKSQE PISNDQKVSDDDK EKGEGALPTGKSK	Lung homing domain of metadherin	Brown and Ruoslahti (2004)
15.	CGNKRTRGC	Tumor lymphatics, tumor macrophages, and tumor cells in hypoxic areas (Lyp-1)	Laakkonen *et al.* (2002)
16.	CREKA	Angiogenic vessels (CREKA)	Simberg *et al.* (2007)
17.	CTTHWGFTLC	Gelatinase in angiogenic vessels	Koivunen *et al.* (1999)
18.	CSRPRRSEC	Dysplastic skin	Hoffman *et al.* (2003)

Table 2.1 (*Continued*)

	Peptide sequence	*In vivo* homing specificity (designation, optional)	Reference
19.	CGKRK	Squamous cell carcinoma	Hoffman *et al.* (2003)
20.	CLSDGKRKC	Lymphatics in C8161 melanoma	Zhang *et al.* (2006)
21.	CNRRTKAGC	Lyp2/K14HPV16 dysplastic skin lesions	Zhang *et al.* (2006)
22.	RPARPAR	Prototypic CendR peptide	Teesalu *et al.* (2009)
23.	CRGDKGPDC	iRGD (a prototypic tissue-penetrating peptide), different tumors	Sugahara *et al.* (2009)
24.	CAGALCY	Brain	Fan *et al.* (2007)
25.	GGGGGGG	Control 1	Teesalu *et al.* (2009)
26.	CGGGGGGGC	Control 2	Sugahara *et al.* (2009)
27.	SSVDKLAALE	Insertless phage	Hoffman *et al.* (2004)

The T7Select415-1b phage displaying the listed peptides are available upon request.

concentrations insufficient for cell killing (Hambley and Hait, 2009). Coupled to a tumor-homing peptide, a variety of anticancer drugs and nanoparticles can be delivered to tumors (Arap *et al.*, 1998; Chen *et al.*, 2001; Curnis *et al.*, 2000; Ellerby *et al.*, 1999; Hamzah *et al.*, 2008; Karmali *et al.*, 2009). This means better efficacy and reduced side effects.

Over the years, the *in vivo* biopanning technology has evolved to become useful for applications beyond synaphic vascular delivery. Identification of receptors for homing peptides has provided new tumor markers and revealed signaling pathways that can be applied to target tumor growth/malignancy (Brown and Ruoslahti, 2004; Fogal *et al.*, 2008). Some of the vascular homing peptides have antitumor activity on their own (Laakkonen *et al.*, 2002) or are capable of modulating important biological processes such as clot formation (Simberg *et al.*, 2007). Our recent work has uncovered a class of tissue-penetrating peptides with exciting applications for delivery of both coupled and coadministered payloads deep into target tissue parenchyma (Sugahara *et al.*, 2009; Teesalu *et al.*, 2009). Some tissue-penetrating peptides are capable of reaching specific cell populations in the target parenchyma, as we have demonstrated recently for LyP-1 peptide and macrophages in atherosclerotic plaques (Hamzah *et al.*, 2011).

1.2. CendR peptides for tissue penetrative cocomposition targeting

Phage display libraries are generally oriented: the foreign peptides are expressed at the N- or C-terminal distal end of phage surface proteins. The positioning was generally not thought to be important for synaphic targeting, although in other systems, examples of position dependent-binding peptides are known (Songyang *et al.*, 1997). We have recently used C-terminal display of foreign peptides on the surface of T7 phage to identify and characterize a new class of position-dependent synaphic targeting peptides that must have a free C-terminal carboxyl group for activity (Sugahara *et al.*, 2009; Teesalu *et al.*, 2009). The defining features of these C-end rule or CendR peptides are (i) R/KXXR/K recognition motif, (ii) requirement for C-terminal exposure of the motif for activity, (iii) conversion of internal cryptic CendR motifs into active, C-terminal ones through proteolytic cleavage, and (iv) neuropilin-1 (NRP-1) dependence of tissue penetration (Sugahara *et al.*, 2009; Teesalu *et al.*, 2009). The CendR receptor, NRP-1, is a pleiotropic cell surface receptor with essential roles in angiogenesis, regulation of vascular permeability, and development of the nervous system (Geretti and Klagsbrun, 2007; Miao and Klagsbrun, 2000). VEGF-A165 and some other ligands of NRP-1 possess a C-terminal CendR sequence that interacts with the b1 domain of NRP-1 and causes cellular internalization and vascular leakage (Becker *et al.*, 2005). CendR peptides have similar effects, particularly when made multivalent through coupling to a molecular scaffold or a particle (Teesalu *et al.*, 2009). CendR peptides are able to take cargo up to a nanoparticle size deep into extravascular tissue. In studying a prototypic tumor-specific CendR peptide, iRGD (CRGDKGPDC, cryptic CendR element underlined), we showed that CendR peptides can be target specific and that it is not necessary to couple a cargo to the CendR peptide for targeted delivery; the free peptide induces tissue permeability in the tumor and allows a coinjected drug or nanoparticle to extravasate and penetrate into tumor parenchyma (Sugahara *et al.*, 2009, 2010). This latter concept is illustrated by robust increase in accumulation and efficacy of a clinically approved breast cancer drug, trastuzumab (Herceptin, an antibody against the HER2 receptor expressed on breast tumor cells), upon coinjection with iRGD peptide (Sugahara *et al.*, 2010).

The CendR mechanism has important implications for peptide-mediated systemic delivery. Most importantly, CendR peptides can be used to overcome a major limitation of synaphic targeting—that the maximum capacity of the delivery is limited by the availability of receptors for the targeting probe. Whereas targeting may greatly enhance the concentration and activity of a drug at the target tissue (Karmali *et al.*, 2009; Liu *et al.*, 2007;

Weissleder *et al.*, 2005), the effect is strongest at low concentrations of the drug and tends to disappear when higher doses are used (Karmali *et al.*, 2009). The apparent reason is that the receptors in tumor endothelial cells and tumor cells can only handle a limited quantity of a targeted compound, and the rest is handled the same way as the free drug (Ruoslahti *et al.*, 2010). The CendR system appears to circumvent this limitation (Sugahara *et al.*, 2010). In addition, the fact that the drug does not have to be conjugated to the peptide means that once a peptide such as iRGD has been clinically validated, its use in combination with other drugs will be greatly simplified.

2. Methods

Below is a list of protocols that we use for *in vivo* and *ex vivo* biopanning experiments. Additional technical details can be found in Hoffman *et al.* (2004) and Trepel *et al.* (2008).

2.1. T7 phage display: General notes and protocols

2.1.1. Media for phage amplification

Carbenicillin: To prepare 1000× Stock, dissolve carbenicillin in water at 50mg/ml. Store at −20°C.

LB Broth: Dissolve 10g Bacto-tryptone, 5g yeast extract, 10g NaCl in 1L of deionized water; adjust pH to 7.5 with NaOH, autoclave. Store at 4°C.

LB agar plates: Combine LB media with Bacto-Agar (BD Bioscience, final concentration 1.5%), autoclave to sterilize and dissolve the agar. Cool the LB-Agar down to 55°C, add antibiotics, and pour the solution on bacterial plates (15ml/100mm plate). Store at 4°C.

Top agar: Combine LB media with agarose (Sigma A5093, final concentration of agar 0.7%), autoclave to sterilize and dissolve the agarose. Store short term on waterbath at 60°C.

2.1.2. General recommendations

Cross-contamination of phage preparation is the most common cause of failure in phage display. Every effort should be made to avoid such issues. The use of bleached and autoclaved glassware, autoclaved or sterile-filtered solutions, and filter pipette tips is recommended. The gloves should be frequently changed during phage manipulations, and the bench and automatic pipettes should be regularly bleached and wiped with 70% ethanol. The solutions must be routinely tested for the presence of phage contamination, and the libraries and individual phage stocks should be regularly subjected to sequence confirmation.

2.1.3. Phage vectors and *Escherichia coli* strains

For biopanning, we prefer to use T7 bacteriophage (T7select, Novagen) display system that allows expression of foreign peptides fused to the C-terminus of the major coat protein 10A. T7 is a lytic phage with fewer constraints on displayed peptide sequences than filamentous phage. T7 has similar shape, size (55nm diameter), and targeting peptide density (up to 415 peptides/particle) as commonly used preclinical nanoparticles. Peptide density onT7 coat can be readily modulated (from 1–415 peptides/particle) by combining commercially available T7 vectors of various strength of coat protein promoter with different *E. coli* strains (BL21 for high-copy display and BLT5403 and BLT5615 for low copy display). BLT5403 and BLT5616 are derived from BL21 cells by transformation with ampicillin-resistant expression plasmids that drive production of extra wild-type coat protein (under T7 and lacUV5 promoter, respectively). Empirically, we have found that medium-copy peptide display (ca. 200 foreign peptides per phage particle) achieved by amplifying T7Select415-1b phage in BLT5403 bacterial strain provides an optimal balance for most biopanning experiments. In this configuration, the presence of the wild-type coat protein helps to neutralize the possible deleterious effects of exogenous peptides on the phage fitness, whereas the displayed peptide is present at sufficient density for high-avidity binding.

2.1.4. Construction of phage libraries and individual peptide phage

Phage libraries and individual phage can be readily constructed by ligating the chemically synthesized and annealed oligonucleotides into T7Select vector arms (Novagen), followed by *in vitro* capsid packaging according to manufacturer's recommendations. To encode for amino acids of random libraries, we use NNK codons (K=T or G) that cover all potential amino acids and avoid two stop codons (TAA and TGA). For individual phage cloning, oligonucleotide sequences are obtained by back-translation (*E. coli* codon usage optimization). To be compatible with commercial *Eco*RI- and *Hin*dIII-cleaved T7 vector arms, the oligonucleotides must be synthesized with complementary overhanging ends and be 5′-phosphorylated. An example of oligonucleotides' design for a disulfide constrained cyclic peptide library is presented in Table 2.2.

For cloning, complementary oligonucleotides are first diluted in ultrapure water at 100n*M*, heated at 95°C for 5min, followed by slow (1–2h) cooling to room temperature for annealing. For ligation, 1μl of annealed oligonucleotides is combined with 1μg of phage vector arms and 80 units of T4 DNA ligase in 1× ligation buffer and incubated overnight at 15°C. The following day, 5μl of ligation mix is combined with 25μl of T7 packaging mixture (Novagen) and incubated for 2h at room temperature to package phage genomic DNA to phage particles.

Table 2.2 Design of oligonucleotides for cyclic CX7C library design in T7Select vectors (Novagen)

A. Aligned oligonucleotides and corresponding CX_7C peptide sequence													
Sense 5′–3′	AAT	TCT	TGC	NNK	NNK	NNK	NNK	NNK	NNK	NNK	TGC	TA	*Hin*dIII
Antisense 3′–5′	*Eco*RI	GA	ACG	NNM	NNM	NNM	NNM	NNM	NNM	NNM	ACG	ATT	CGA
Aminoacids	N	S	C	X1	X2	X3	X4	X5	X6	X7	C	Stop	–

B. Oligonucleotides to be ordered for CX_7C library construction
cx7c sense oligonucleotide 5′–3′: 5′Phos-AATTCTTGCNNKNNKNNKNNKNNKNNKNNKTGCTA
cx7c antisensesense oligonucleotide 5′–3′: 5′Phos-AGCTTAGCAMNNMNNMNNMNNMNNMNNMNNGCAAG

2.1.5. T7 phage amplification and purification

To prepare phage stocks low in bacterial contaminants, the phage is purified by combination of PEG-8000 precipitation and cesium chloride gradient centrifugation. A day before phage amplification, a single colony of BLT5403 cells from a freshly streaked selective plate is inoculated in 20 of LB medium containing 50 μg/ml carbenicillin (LB/carbenicillin) and grown in 100 ml Erlenmeyer flask in 37 °C bacterial shaker at 240 rpm overnight. Next morning, the stationary culture is diluted 1/100 in LB/carbenicillin prewarmed at 37 °C and incubated in shaker at 240 rpm and 37 °C until the OD_{600} of the culture reaches 0.5 ($\sim 1\times10^8$ cells/ml). The bacteria are inoculated with the phage at multiplicity of infection 0.001–0.01 (i.e., 100–1000 bacterial cells per each pfu), and incubation in shaker is continued until the decrease of the cloudiness of the culture, indicative of bacterial lysis, occurs (ca. 90 min postinoculation). 26 ml of the lysate is combined with 3 ml of 5 *M* NaCl and centrifuged for 10 min at 10,000 rpm and 4 °C in ss34 rotor in Sorvall superspeed centrifuge to clear the lysate. For PEG8000 precipitation, 8.4 ml of 50%PEG8000 in phosphate-buffered saline (PBS) is thoroughly mixed with 29 ml of cleared lysate, incubated on ice for 30 min, and centrifuged for 10 min at 8000 rpm at 4 °C in ss34 rotor in Sorvall superspeed centrifuge to collect the phage. Supernatant is decanted, and inverted tubes are drained on a paper towel for 15 min. The phage pellet resuspended in 1.5 ml of PBS for gradient centrifugation.

For density gradient purification, the $CsCl_2$ gradient is prepared by layering the mix of 62.5% $CsCl_2$ (percent weight/weight) with PBS using pipet tips with ends cutoff in Beckman 344057 ultracentrifugation tubes (in the loading order): (1) $CsCl_2$; 0.25 ml; (2) $CsCl_2$/PBS—2/1; 1.5 ml; (3) $CsCl_2$/PBS—1/1; 1.5 ml; (4) $CsCl_2$/PBS—1/2; 0.25 ml. 1.5 ml of phage in PBS is loaded on the top of the $CsCl_2$ gradient and ultracentrifuged using sw50.1 swing-out rotor (Sorvall) at 40,000 rpm and 22 °C for 35 min. The phage band (visible best under top illumination on the dark background) will appear between layers (2) and (3). The phage is collected using a 3-ml disposable syringe and 22G1 needle and dialyzed using Slide-A-Lyzer 10,000 kDa cut-off dialysis cassettes (Pierce) against two changes of PBS (1 h and 500× volume excess each change). $CsCl_2$-purified phage can be stored at 4 °C for at least a week without a loss of titer. However, we recommend using freshly prepared phage for naïve library screening experiments to avoid bias against unstable peptides.

2.1.6. T7 titration and sequencing

For phage titration, we use BLT5615 bacteria (Novagen), as this strain gives more clear and uniform plaques than BL21 or BLT5403. First, duplicate serial dilutions of phage are prepared in LB. Subsequently, 100 μl of diluted phage is combined in 15-ml falcon tubes with 500 μl of BLT5615 cells

(OD600) and 6ml of top agar containing 2 m*M* isopropyl thiogalactoside and plated on 100-mm LB/carbenicillin agar plates. Plaques will appear as transparent "holes" in bacterial lawn at about 3h at 37 °C or overnight at room temperature. Phage plaques will be counted to estimate starting phage concentration.

For sequencing, individual plaques are first resuspended in 30 μl of PBS (96-well PCR plates are convenient for phage resuspension and medium-term storage), and 1 μl of this mix is combined with 14 μl of PCR Supermix (Invitrogen) containing 0.5 μ*M* primers that flank the peptide-encoding region of the phage genome (oligonucleotide sequences: upstream 5′-AGCGGACCAGATTATCGCTA-3′; downstream: 5′-AACCCCT-CAAGACCCGTTTA-3′) and subjected to PCR (1 cycle of 72 °C for 10 min+95 °C for 5 min; 35 cycles of 94 °C for 50 s+50 °C for 1 min+72 °C for 1 min; hold at 72 °C for 10 min; hold at 4 °C). After amplification, the PCR products are sequenced using the upstream primer and translated to amino acid sequence.

2.1.7. Detection and amplification of noninfectious phage

T7 phage is inactivated at low pH and by exposure to certain proteases (Sugahara *et al.*, 2009; Teesalu *et al.*, 2009) (and Tambet Teesalu and Kazuki N. Sugahara, unpublished data). During selections on live cells and tissues, some bound T7 phage may become inactivated and remain undetectable in infectivity-based assays. We use Taqman qPCR-based assays for quantification of the phage copy number independent of its infectious state. To design Taqman primers and probes, we use Beacon Express software (Premier Biosoft International). To quantify the total bound phage, we use a set of probe and primers complementary to the constant region of the phage genome (probe: 5′-FAM-AGAAGTTCCGCACCTCACCGCTGG-BHQ1-3′; upstream primer: 5′-CCGCAACGTTATGGGCTTTG-3′; downstream primer: 5′-CTCACCTTTATTGGCAGGGAAG-3′). Quantitative PCR reaction mix is prepared by combining 2 μl of DNA (prepared from cells or tissues by extraction with DNeasy Blood and Tissue Mini kit; Qiagen), 12.5 μl of 2× Taqman PCR master mix (Applied Biosystems), 900 nmol of each primer, and a 200-nmol probe in a final volume of 25 μl. The thermal cycling conditions in IQ-5 real-time PCR system (BioRad) are as follows: 2 min at 50 °C, 10 min at 95 °C followed by 40 repeats of 15 s at 92 °C, and 1 min at 60 °C. Data collection is performed during each annealing phase. During each run, we include a standard dilution of the phage DNA with known quantity to permit phage copy number quantification, a negative control (water), and a DNA sample from cells or tissue not exposed to phage. Measurements of copy number are taken three times, and the mean of these values is used for further analysis. In addition to universal probe-primers set above, we have designed primer/probe sets specific for individual peptides for multiplex qPCR assays for simultaneous quantification of multiple phage in the same sample.

A DNA-based back-cloning approach is useful for avoiding bias in phage recovery during library screens. We first use PCR on total DNA prepared from target cells or organs to amplify 77bp (for CX7C library) peptide-encoding region of phage genomic DNA (up, 5′-GTGATGCTCGGG-GATCCGAATT-3′; down, 5′-GTTACTCGAGTGCGGCCGCAA-3′). PCR products are cleaved with *Eco*RI and *Hin*dIII, purified with Qiaquick nucleotide removal kit (cutoff 17bp, Qiagen), ligated into T7Select vector arms, and packaged to phage particles as described in Section 2.1.4.

2.2. General considerations in phage biopanning

Design of a screen for homing peptides requires careful consideration of target organ: its localization, structure, vascularization, and size. *In vivo* selection will subtract the phage displaying peptides with affinity to nontarget vascular beds and enrich for phage specifically retained in target vascular beds. For subtractive aspect to be effective, the target vascular bed should not be the first (or among the first) that injected phage will encounter. For example, to select for lung-homing peptides, the phage should be administered in the left ventricle of the heart, rather than tail vein. For all but most highly vascularized target tissues, we recommend *in vivo* selections to be preceded by *ex vivo* selection (phage binding to dissociated cells prepared from target tissue). *Ex vivo* selection will decrease the risk of nonspecific loss of homing phage due to uptake by the reticuloendothelial system and to increase chances that good binders can interact with target tissue. *Ex vivo* selections can be readily tailored to target particular tissue and subcellular compartment. The stringency of *ex vivo* selections can be modulated by changing the washing conditions. In addition, cell sorting can be used to render *ex vivo* selection step more selective for the target cell population, and acid wash can be included to obtain peptides suitable for intracellular delivery (T7 phage is inactivated at $pH<4$; Teesalu *et al.*, 2009). Microfluidics technology can be applied to minimize the number of cells required for biopanning, achieve readily tunable stringency, and increase reproducibility of the selections (Wang *et al.*, 2011). During the *ex vivo* selection, the diversity of peptide libraries collapses (typically by about four orders of magnitude, from 10^8–10^9 to 10^4–10^5). This means that during subsequent *in vivo* selection (phage input range from 10^9 to 10^{10}), each phage variant will be present in more than 10^4 copies, thereby significantly decreasing the possibility of accidental loss of the specific homing phage.

2.3. *Ex vivo* phage display

2.3.1. Preparation of cell suspension

We use mechanical tissue dissociation for preparation of cell suspensions of solid tissues for *ex vivo* binding assays. The tissue is minced with a razor blade into small fragments and treated with Medimachine (BD Bioscience) for further dissociation. This procedure yields a mixture of small cell aggregates and single cells that can be directly used for phage binding. If only single cells are required (e.g., for cell sorting), the cell suspension may be further treated with collagenase I (in DMEM at final concentration 1 mg/ml, 1 h at 37 °C) and passed through cell strainer. Dissociated cells are washed with DMEM containing 1% bovine serum albumin (DMEM–BSA) in 50 ml Falcon tube, collected by centrifugation at 400 g for 5 min, and resuspended in DMEM–BSA at 10–50 mg starting tissue wet weight per milliliter.

2.3.2. Phage binding and washes

For phage binding, we combine 10^{10} pfu of naïve library (i.e., 100 copies of each individual phage in case of typical library diversity of 10^8), cell suspension corresponding to 10–50 mg wet weight of starting tissue, and in DMEM–BSA to a final volume of 10 ml. In subsequent rounds of selection (after the diversity of library has collapsed), 10^9 pfu are used per binding reaction. The quantity of starting tissue will depend on its characteristics such as cellularity, vascularization, abundance of target cell population, etc. Libraries and cells are incubated on a gently rocking platform for 1 h at 4 °C (for binding) or 37 °C (for internalization). Subsequently, tubes are placed on ice and washed four times by centrifugation at 400 × *g* for 5 min with 15 ml DMEM–BSA (fresh tubes are used for each wash). For removal and inactivation of surface-bound phage, the second wash can be replaced with optional acid wash with 500 m*M* NaCl, 0.1 *M* glycine, 1% BSA, pH 2.5. After last wash, the cell pellet is lysed in 1 ml LB containing 1% Nonidet P-40 (LB-NP40) and titrated. To express phage binding data, we normalize the bound phage number to the tissue wet weight and express the bound phage number against background binding of control phage clones (such as insertless phage, phage expressing heptaglycine, or a scrambled peptide) or naïve library as "Binding fold control" on *y*-axis.

2.3.3. Background blocking in *ex vivo* phage binding assays

T7 phage can be readily rendered noninfective (while preserving its binding properties) by limited exposure to the ultraviolet light. If a background binding of phage is a concern, an optional 1 h blocking with 10-fold excess of inactivated control phage can be performed prior to the library binding. Briefly, a control phage (e.g., insertless phage or phage expressing heptaglycine) is diluted in PBS containing 1% BSA at 10^{11} pfu/ml and 5 ml of phage is treated in uncovered 5 cm Petri dish with UV light to inactivate the phage. For phage inactivation, we use 15 min incubation at 15-cm distance from the UV lamp of the PCR hood (Airclean 600 PCR Workstation, ISC-BioExpress). Conditions for alternative UV light sources should be determined empirically to achieve complete elimination of live phage.

2.3.4. Expected outcome

Recovered phage concentration from an *ex vivo* selection ranges from 10^3 to 10^6 pfu/ml. We perform typically 1–3 rounds of *ex vivo* selection. For each round, we will determine the enrichment of phage binding to cells derived from target and control organs. After each round, a small set (i.e., 24 clones) of recovered phage is sequenced to assess library quality and predominant motifs. As *ex vivo* selections do not comprise a built-in negative selection, in most cases, the selected pool shows increased binding to cells derived from control organs. The "sticky" phage displaying pan-binding peptides will be eliminated during subsequent *in vivo* selection.

2.4. *In vivo* phage display

Prior to initiating screens using naïve phage libraries, we recommend performing an *in vivo* biopanning procedure using a set of control phage and a phage of established *in vivo* homing specificity. Intravenous administration of T7 phage expressing a control heptaglycine (G7) peptide and an active CendR phage (RPARPAR) in normal mice of various strains will yield preferential binding of RPARPAR phage to lung and heart vasculature (30- to 100-fold excess over control (Teesalu *et al.*, 2009). These recombinant phages (along the other peptide phage clones listed in Table 2.1) are upon request available from our lab.

2.4.1. Phage dosing

For *in vivo* display, we inject 10^9–10^{10} pfu of T7 phage into the mouse tail vein. The mouse is placed in a restrainer, and phage (diluted in less than 200 μl of PBS) is slowly injected using an insulin syringe. Prior to injection, the tail veins, located on the sides of the tail, are dilated by immersing the tail in the warm water. After injection, the phage is allowed to circulate for 10–30 min.

2.4.2. Anesthesia, perfusion, and dissection

Five minutes prior perfusion, the mouse is anesthetized (e.g., by intraperitoneal injection of the 1.25%, w/v, avertin at 15 μl/g body weight). Depth of anesthesia is verified by the absence of toe-pinch reflex and mouse is fixed on a single-use styrofoam platform (e.g., styrofoam racks of falcon tubes) using bent needles. To gain access to the heart, a horizontal incision is made in the abdominal skin right below the diaphragm followed by a longitudinal incision in the skin of the thoracic area to expose the rib cage. The skin in the abdominal area is detached from the body wall, and a horizontal incision is made into the peritoneum to access the abdominal cavity. The diaphragm is cut along the rib cage taking care not to damage internal organs. Rib cage is cut on both sides along the sternum (at ca. 3 mm laterally, not to damage the two internal thoracic arteries) to expose the heart. For perfusion, the right atrium is clipped, and a butterfly vacutainer needle is inserted into the left ventricle. Perfusion with PBS (or DMEM) is performed by gravity flow tubing (at rate of ca. 2ml/min) until the organs and major veins (such as vena cava) are desanguinated (typically over 5–10min). To remove blood from the pulmonary circulation, a brief (1min) perfusion through the right ventricle is performed. The organs are collected, rinsed with PBS from a spray bottle, weighted, and placed in 1–4ml LB-NP40 in 14ml snap-cap falcon tubes on ice for phage extraction. Dissected tissues are homogenized using handheld homogenizer with disposable plastic hard-tissue probes (Omni International Inc.).

2.4.3. Expected outcome

Typical recovered phage concentration from an *in vivo* selection ranges from 10^2 to 10^7 pfu/ml. For each round, we will determine the enrichment of phage binding to cells derived from target and control organs. After each round, a small set (i.e., 24 clones) of recovered phage is sequenced to assess library quality and to detect any predominant motifs. As the peptide library is generally short, the simple use of a spreadsheet is often sufficient to compare and align the isolated peptides. Alternatively, multiple sequence alignment tools, such as Clustal-W, may be required to align the peptide sequences (Higgins *et al.*, 1996). To represent the consensus peptide sequences, Weblogo can be used to compile and visualize consensus profiles of the selected peptides such that the height of an amino acid in the sequence logo is indicative of its relative frequency at that position (Crooks *et al.*, 2004). Clones with common motifs are evaluated in single clone or phage pool play-off assays as described in Section 2.5.

2.4.4. Alternative strategies of *in vivo* phage display

In vivo display with washes. If the target tissue cannot be adequately perfused due to dysfunctional vasculature (as often in the case of tumors), the remaining free phage in blood causes high background. In this case, we recommend homogenizing the tissues first in DMEM–BSA followed by 1–5 washes as during *ex vivo* phage display, lysis in LB-NP40 and titration. For selection of tissue-specific internalizing phage, a low-pH wash step can be included as during *ex vivo* display selections.

2.5. Postscreening phage auditions

2.5.1. Titration-based phage audition

To evaluate the target specificity and strength of peptides derived from biopanning experiments, individual candidate phage is tested for *ex vivo* binding and *in vivo* homing. Technically, the assays are performed as during *ex vivo* and *in vivo* library screens above, and the recovered phage titer is normalized to control phage. To exclude the effect of possible secondary mutations on phage recovery, the phage must be recloned (as outlined in Section 2.1.4) prior to final evaluation.

For medium-throughput evaluation of candidate phage, it may be advantageous to use an assay that we have termed "play-off phage display." A pool of about 10 equally represented phage displaying candidate and control peptides is subjected to biopanning experiment, and the representation of each phage in input and output pools is determined by sequencing. The data are graphed with phage sequences on the x-axis and output/input ratio on the y-axis. Play-off screen with its built-in controls is best suited for *in vivo* situation, in which the interanimal variability is an important concern. Play-off display is also useful for finding homing peptides among already characterized peptides prior to starting screens of naive libraries.

2.5.2. Titration-independent phage audition

Due to phage sensitivity to low pH and enzymatic degradation, titration-based phage audition should be complemented with techniques that do not depend on the presence of live phage.

For quantitation of the phage in tissue extracts, we use quantitative PCR as outlined in Section 2.1.7. This assay can be multiplexed to include negative and positive control phage along with the tester phage in the same biopanning experiment.

To qualitatively evaluate phage homing and determine its tissue distribution, we use immunostaining of tissue sections with anti-T7 antibodies. Briefly, the tissues are fixed in 4% paraformaldehyde in PBS (pH 7.4) at 4° C overnight, followed by 3×1 h washes at 4 °C, cryprotection in 30% sucrose in PBS at 4 °C (until tissues sink, 24–72 h), freezing the tissues in cryomolds in optimal cutting temperature compound, and preparation 5–10 μm cryosections on Superfrost Plus slides. Immunostaining steps (all performed in humid chamber at room temperature) are (1) blocking the nonspecific binding by incubating sections in DAKO antibody diluent, (2) incubation of the sections with in-house prepared rabbit-anti T7 antibody in antibody diluent (1:100), (3) washes with PBS containing 0.05% Tween-20 (PBST) to remove excess primary antibody, (4) incubation with a commercial fluorophore-labeled anti-rabbit antibody, (5) washes with PBST and nuclear staining with 5 μg/ml DAPI, and (6) mounting in the fluorescence-compatible aqueous mounting medium.

2.6. *In vivo* homing of synthetic peptides

After peptide phage validation, synthetic peptides can be tested for homing as fluorescent peptides and as targeting ligands on the surface of synthetic nanoparticles.

We typically synthesize the peptides in the following configuration: (1) N-terminal fluorophore or moiety used for nanoparticle conjugation; (2) a linker, such as 6-aminohexanoic acid, to serve as inert spacer between the peptide and reporter; and (3) peptide itself with appropriate C-terminal modification (carboxylic acid or amide). As in phage the C-terminus of the peptide is free, it is advisable to first evaluate peptides with free C-termini. Peptides are purified by HPLC to 90–95% purity and validated by mass spectral analysis.

For *in vivo* peptide homing studies, it is advisable to use fluorescent labels that have high extinction coefficient and high fluorescence quantum yield and are stable to photobleaching and to low pH, such as the Alexa Fluor series. An alternative is to rely on fluorescence-independent secondary detection, such as use of anti-FITC antibodies (Sugahara *et al.*, 2009). Typically, 100 μl of 1 m*M* fluorophore-labeled peptide in PBS is intravenously injected into a 25-g mouse and allowed to circulate for 1 h. The circulation time may be adjusted according to the peptide characteristics and whether or not the mouse is anesthetized at the time of peptide injection. In general, the circulation time should be prolonged in anesthetized mice because renal clearance is decreased. To visualize patent blood vessels, 15 min prior to perfusion mice can be injected with 100 μl of 2 mg/ml lectin (e.g., wheat germ agglutinin labeled with a fluorophore that is of different color than the label on the peptide). The mouse is anesthetized and perfused as described in Section 2.4.2. Tissues are collected, macroscopically observed under UV light (Illuminatool Bright Light System LT-9900), and processed for immunofluorescence as described in Section 2.5.2. For quantitation of the extent of peptide accumulation and spreading within tissues, we stain tissue sections immunohistochemically with antifluorophore antibodies followed by slide scanning with the Scanscope CM-1 scanner and analysis with the ImageScope software (Aperio Technologies, Vista, CA; Fogal *et al.*, 2008).

2.7. Identification of the peptide receptors

It is important to identify the receptors of homing peptides, as (i) the receptor expression in target tissue or pathology may vary and affect clinical responses to peptide-mediated drug delivery; (ii) the receptor may become an important vascular marker or target for drug intervention on its own (as shown for P32, a receptor of LyP1 peptide (Fogal *et al.*, 2008); (iii) a better understanding of the receptor biology may give mechanistic clues for improvement of the targeting system; and (iv) receptors can be targeted using alternative affinity ligands.

To identify the homing peptide receptors, we use affinity chromatography with peptides immobilized on solid matrix, followed by extensive washes, elution of bound proteins using an excess of free peptide, and

MALDI-TOF-based protein identification (Teesalu *et al.*, 2009). Various covalent and noncovalent systems of coupling the peptides to the resins can be used (e.g., Pierce Sulfolink coupling gel for cysteine-tagged peptides and streptavidin Sepharose for biotinylated peptides). Tissue (or cells) are homogenized on ice using hand-held homogenizer in equal weight of 2× tissue lysis buffer (PBS containing 400 m*M* *n*-octyl-beta-D-glucopyranoside, 2 m*M* $MgSO_4$, 2 m*M* $MnCl_2$, 2 m*M* $CaCl_2$, and 2× EDTA-free protease inhibitors cocktail tablets; Roche Biochemicals). The homogenate is stored on a shaker at 4 °C for 6 h followed by centrifugation at 14,000 rpm for 30 min at 4 °C in Eppendorf centrifuge. The supernatant is combined with peptide-coupled affinity matrix (3 ml of lysate per 0.5 ml of matrix in the case of Sulfolink gel) that has been equilibrated in 1× tissue lysis buffer and incubated on rotary shaker at 4 °C overnight. Next morning, the tissue lysate/affinity matrix mixture is transferred to the chromatography column and the column is washed 10 times with 10 column volumes of 1× tissue lysis buffer, followed by elution by applying 1 column volume fractions of 1× tissue lysis buffer containing 2 m*M* free peptide (total 10 elution fractions). Aliquots from each wash are used for protein quantitation and SDS-PAGE analysis. Protein gels are stained by SilverSnap SDS-PAGE silver staining kit (Pierce), and differential bands are subjected to protein identification using mass-spectroscopy.

Candidate receptors are validated using biochemical, immunological, and cell-based assays. To confirm the peptide-receptor binding, the putative receptor proteins will be captured onto plastic wells as a purified protein (if commercially available) or through antibodies, and binding of phage displaying the relevant peptides (and scrambled control peptides) will be measured. In addition, colocalization studies of the labeled peptide (or a phage) and a candidate receptor can be performed. The functional role of receptor can be probed by expression modulation studies and changes in the binding and uptake of peptide or in its subcellular distribution if the modulated protein is the primary receptor (e.g., Fogal *et al.*, 2008; Teesalu *et al.*, 2009). Rescue of peptide and phage retention upon by reintroducing the receptor to silenced cells can be used to confirm that the effect of the siRNA knockdown is specific to the receptor (e.g., Fogal *et al.*, 2008).

3. Concluding Remarks and Perspectives

In vivo phage display has become established as a useful technology for unbiased exploration of the vascular diversity. Homing peptides can be used for synaphic delivery of drugs, imaging agents, and theranostic compounds to target vascular beds. Bioactive homing peptides include CendR peptides that trigger vascular exit and tissue penetration at the target tissues.

In the future, *in vivo* biopanning in combination with comprehensive determination of entire phage-displayed peptide selectome by next generation DNA sequencing technology will further enhance the throughput and power of the technology.

ACKNOWLEDGMENTS

We thank all the past and present members of the Ruoslahti laboratory for contributions to the development of *in vivo* phage display technology. This work was supported by NIH Grant R01 CA152327 and Department of Defense Grant W81XWH-08-1-0727.

REFERENCES

Achilefu, S., Srinivasan, A., Schmidt, M. A., Jimenez, H. N., Bugaj, J. E., and Erion, J. L. (2003). Novel bioactive and stable neurotensin peptide analogues capable of delivering radiopharmaceuticals and molecular beacons to tumors. *J. Med. Chem.* **46,** 3403–3411.

Allen, T. M., and Cullis, P. R. (2004). Drug delivery systems: Entering the mainstream. *Science* **303,** 1818–1822.

Arap, W., Pasqualini, R., and Ruoslahti, E. (1998). Cancer treatment by targeted drug delivery to tumor vasculature in a mouse model. *Science* **279,** 377–380.

Arap, W., Haedicke, W., Bernasconi, M., Kain, R., Rajotte, D., Krajewski, S., Ellerby, H. M., Bredesen, D. E., Pasqualini, R., and Ruoslahti, E. (2002a). Targeting the prostate for destruction through a vascular address. *Proc. Natl. Acad. Sci. USA* **99,** 1527–1531.

Arap, W., Kolonin, M. G., Trepel, M., Lahdenranta, J., Cardo-Vila, M., Giordano, R. J., Mintz, P. J., Ardelt, P. U., Yao, V. J., Vidal, C. I., Chen, L., Flamm, A., *et al.* (2002b). Steps toward mapping the human vasculature by phage display. *Nat. Med.* **8,** 121–127.

Becker, P. M., Waltenberger, J., Yachechko, R., Mirzapoiazova, T., Sham, J. S., Lee, C. G., Elias, J. A., and Verin, A. D. (2005). Neuropilin-1 regulates vascular endothelial growth factor-mediated endothelial permeability. *Circ. Res.* **96,** 1257–1265.

Brown, D. M., and Ruoslahti, E. (2004). Metadherin, a cell surface protein in breast tumors that mediates lung metastasis. *Cancer Cell* **5,** 365–374.

Carmeliet, P., and Jain, R. K. (2000). Angiogenesis in cancer and other diseases. *Nature* **407,** 249–257.

Chabner, B. A., and Roberts, T. G., Jr. (2005). Timeline: Chemotherapy and the war on cancer. *Nat. Rev. Cancer* **5,** 65–72.

Chen, Y., Xu, X., Hong, S., Chen, J., Liu, N., Underhill, C. B., Creswell, K., and Zhang, L. (2001). RGD-Tachyplesin inhibits tumor growth. *Cancer Res.* **61,** 2434–2438.

Christian, S., Pilch, J., Akerman, M. E., Porkka, K., Laakkonen, P., and Ruoslahti, E. (2003). Nucleolin expressed at the cell surface is a marker of endothelial cells in angiogenic blood vessels. *J. Cell Biol.* **163,** 871–878.

Cines, D. B., Pollak, E. S., Buck, C. A., Loscalzo, J., Zimmerman, G. A., McEver, R. P., Pober, J. S., Wick, T. M., Konkle, B. A., Schwartz, B. S., Barnathan, E. S., McCrae, K. R., *et al.* (1998). Endothelial cells in physiology and in the pathophysiology of vascular disorders. *Blood* **91,** 3527–3561.

Crooks, G. E., Hon, G., Chandonia, J. M., and Brenner, S. E. (2004). WebLogo: A sequence logo generator. *Genome Res.* **14,** 1188–1190.

Curnis, F., Sacchi, A., Borgna, L., Magni, F., Gasparri, A., and Corti, A. (2000). Enhancement of tumor necrosis factor alpha antitumor immunotherapeutic properties by targeted delivery to aminopeptidase N (CD13). *Nat. Biotechnol.* **18,** 1185–1190.

De Caterina, R., and Libby, P. (2007). Endothelial Dysfunctions and Vascular Disease. Blackwell Futura, Malden, MA (xiv, 416p).

Ellerby, H. M., Arap, W., Ellerby, L. M., Kain, R., Andrusiak, R., Rio, G. D., Krajewski, S., Lombardo, C. R., Rao, R., Ruoslahti, E., Bredesen, D. E., and Pasqualini, R. (1999). Anti-cancer activity of targeted pro-apoptotic peptides. *Nat. Med.* **5,** 1032–1038.

Fan, X., Venegas, R., Fey, R., van der Heyde, H., Bernard, M. A., Lazarides, E., and Woods, C. M. (2007). An in vivo approach to structure activity relationship analysis of peptide ligands. *Pharm. Res.* **24,** 868–879.

Fogal, V., Zhang, L., Krajewski, S., and Ruoslahti, E. (2008). Mitochondrial/cell-surface protein p32/gC1qR as a molecular target in tumor cells and tumor stroma. *Cancer Res.* **68,** 7210–7218.

Geretti, E., and Klagsbrun, M. (2007). Neuropilins: Novel targets for anti-angiogenesis therapies. *Cell Adh. Migr.* **1,** 56–61.

Hambley, T. W., and Hait, W. N. (2009). Is anticancer drug development heading in the right direction? *Cancer Res.* **69,** 1259–1262.

Hamzah, J., Nelson, D., Moldenhauer, G., Arnold, B., Hammerling, G. J., and Ganss, R. (2008). Vascular targeting of anti-CD40 antibodies and IL-2 into autochthonous tumors enhances immunotherapy in mice. *J. Clin. Invest.* **118,** 1691–1699.

Hamzah, J., Kotamraju, V. R., Seo, J. W., Agemy, L., Fogal, V., Mahakian, L. M., Peters, D., Roth, L., Gagnon, M. K., Ferrara, K. W., and Ruoslahti, E. (2011). Specific penetration and accumulation of a homing peptide within atherosclerotic plaques of apolipoprotein E-deficient mice. *Proc. Natl. Acad. Sci. USA.* **108,** 7154–7159.

Hanahan, D., and Folkman, J. (1996). Patterns and emerging mechanisms of the angiogenic switch during tumorigenesis. *Cell* **86,** 353–364.

Higgins, D. G., Thompson, J. D., and Gibson, T. J. (1996). Using CLUSTAL for multiple sequence alignments. *Methods Enzymol.* **266,** 383–402.

Hoffman, J. A., Giraudo, E., Singh, M., Zhang, L., Inoue, M., Porkka, K., Hanahan, D., and Ruoslahti, E. (2003). Progressive vascular changes in a transgenic mouse model of squamous cell carcinoma. *Cancer Cell* **4,** 383–391 PMID.

Hoffman, J. A., Laakkonen, P., Porkka, K., Bernasconi, M., and Ruoslahti, E. (2004). In vivo and ex vivo selections using phage-displayed libraries. *In* "Phage Display: A Practical Approach," (H. B. Clackson, ed.), p. 171. Oxford University Press, Oxford, UK.

Jain, R. K., and Gerlowski, L. E. (1986). Extravascular transport in normal and tumor tissues. *Crit. Rev. Oncol. Hematol.* **5,** 115–170.

Järvinen, T. A., and Ruoslahti, E. (2007). Molecular changes in the vasculature of injured tissues. *Am. J. Pathol.* **171,** 702–711.

Joyce, J. A., Laakkonen, P., Bernasconi, M., Bergers, G., Ruoslahti, E., and Hanahan, D. (2003). Stage-specific vascular markers revealed by phage display in a mouse model of pancreatic islet tumorigenesis. *Cancer Cell* **4,** 393–403.

Karmali, P. P., Kotamraju, V. R., Kastantin, M., Black, M., Missirlis, D., Tirrell, M., and Ruoslahti, E. (2009). Targeting of albumin-embedded paclitaxel nanoparticles to tumors. *Nanomedicine* **5,** 73–82.

Koivunen, E., Gay, D. A., and Ruoslahti, E. (1993). Selection of peptides binding to the $\alpha_5\beta_1$ integrin from phage display library. *J. Biol. Chem.* **268,** 20205–20210.

Koivunen, E., Wang, B., and Ruoslahti, E. (1994). Isolation of a highly specific ligand for the alpha 5 beta 1 integrin from a phage display library. *J. Cell Biol.* **124,** 373–380.

Koivunen, E., Arap, W., Valtanen, H., Rainisalo, A., Medina, O. P., Heikkila, P., Kantor, C., Gahmberg, C. G., Salo, T., Konttinen, Y. T., Sorsa, T., Ruoslahti, E., *et al.* (Check for Year). Tumor targeting with a selective gelatinase inhibitor. *Nat. Biotechnol.* **17,** 768–774.

Krag, D. N., Shukla, G. S., Shen, G. P., Pero, S., Ashikaga, T., Fuller, S., Weaver, D. L., Burdette-Radoux, S., and Thomas, C. (2006). Selection of tumor-binding ligands in cancer patients with phage display libraries. *Cancer Res.* **66,** 7724–7733.

Laakkonen, P., Porkka, K., Hoffman, J. A., and Ruoslahti, E. (2002). A tumor-homing peptide with a targeting specificity related to lymphatic vessels. *Nat. Med.* **8,** 751–755.

Lee, S. M., Lee, E. J., Hong, H. Y., Kwon, M. K., Kwon, T. H., Choi, J. Y., Park, R. W., Kwon, T. G., Yoo, E. S., Yoon, G. S., Kim, I. S., Ruoslahti, E., *et al.* (2007). Targeting bladder tumor cells in vivo and in the urine with a peptide identified by phage display. *Mol. Cancer Res.* **5,** 11–19.

Liu, Z., Cai, W., He, L., Nakayama, N., Chen, K., Sun, X., Chen, X., and Dai, H. (2007). In vivo biodistribution and highly efficient tumour targeting of carbon nanotubes in mice. *Nat. Nanotechnol.* **2,** 47–52.

Miao, H. Q., and Klagsbrun, M. (2000). Neuropilin is a mediator of angiogenesis. *Cancer Metastasis Rev.* **19,** 29–37.

Oh, P., Li, Y., Yu, J., Durr, E., Krasinska, K. M., Carver, L. A., Testa, J. E., and Schnitzer, J. E. (2004). Subtractive proteomic mapping of the endothelial surface in lung and solid tumours for tissue-specific therapy. *Nature* **429,** 629–635.

Olsen, J. O., Pozderac, R. V., Hinkle, G., Hill, T., O'Dorisio, T. M., Schirmer, W. J., Ellison, E. C., and O'Dorisio, M. S. (1995). Somatostatin receptor imaging of neuroendocrine tumors with indium-111 pentetreotide (Octreoscan). *Semin. Nucl. Med.* **25,** 251–261.

Pasqualini, R., and Ruoslahti, E. (1996). Organ targeting in vivo using phage display peptide libraries. *Nature* **380,** 364–366.

Perry, M. C. (2001). The Chemotherapy Source Book. Lippincott Williams & Wilkins, Philadelphia (xiii, 1003p).

Pilch, J., Brown, D. M., Komatsu, M., Jarvinen, T. A., Yang, M., Peters, D., Hoffman, R. M., and Ruoslahti, E. (2006). Peptides selected for binding to clotted plasma accumulate in tumor stroma and wounds. *Proc. Natl. Acad. Sci. USA* **103,** 2800–2804.

Ruoslahti, E. (2002). Specialization of tumour vasculature. *Nat. Rev. Cancer* **2,** 83–90.

Ruoslahti, E. (2004). Vascular zip codes in angiogenesis and metastasis. *Biochem. Soc. Trans.* **32,** 397–402.

Ruoslahti, E., Bhatia, S. N., and Sailor, M. J. (2010). Targeting of drugs and nanoparticles to tumors. *J. Cell Biol.* **188,** 759–768.

Simberg, D., Duza, T., Park, J. H., Essler, M., Pilch, J., Zhang, L., Derfus, A. M., Yang, M., Hoffman, R. M., Bhatia, S., Sailor, M. J., and Ruoslahti, E. (2007). Biomimetic amplification of nanoparticle homing to tumors. *Proc. Natl. Acad. Sci. USA* **104,** 932–936.

Smith, G. P., and Petrenko, V. A. (1997). Phage display. *Chem. Rev.* **97,** 391–410.

Songyang, Z., Fanning, A. S., Fu, C., Xu, J., Marfatia, S. M., Chishti, A. H., Crompton, A., Chan, A. C., Anderson, J. M., and Cantley, L. C. (1997). Recognition of unique carboxyl-terminal motifs by distinct PDZ domains. *Science* **275,** 73–77.

Souhami, R. L., and Tobias, J. S. (2005). Cancer and Its Management. Blackwell Publishers, Malden, MA (x, 533p).

Sugahara, K. N., Teesalu, T., Karmali, P. P., Kotamraju, V. R., Agemy, L., Girard, O. M., Hanahan, D., Mattrey, R. F., and Ruoslahti, E. (2009). Tissue-penetrating delivery of compounds and nanoparticles into tumors. *Cancer Cell* **16,** 510–520.

Sugahara, K. N., Teesalu, T., Karmali, P. P., Kotamraju, V. R., Agemy, L., Greenwald, D. R., and Ruoslahti, E. (2010). Coadministration of a tumor-penetrating peptide enhances the efficacy of cancer drugs. *Science* **328,** 1031–1035.

Teesalu, T., Sugahara, K. N., Kotamraju, V. R., and Ruoslahti, E. (2009). C-end rule peptides mediate neuropilin-1-dependent cell, vascular, and tissue penetration. *Proc. Natl. Acad. Sci. USA* **106,** 16157–16162.

Trepel, M., Pasqualini, R., and Arap, W. (2008). Screening phage-display peptide libraries for vascular targeted peptides. *Methods Enzymol.* **445,** 83–106 (Chapter 4).

Wang, J., Liu, Y., Teesalu, T., Sugahara, K. N., Kotamrajua, V. R., Adams, J. D., Ferguson, B. S., Gong, Q., Oh, S. S., Csordas, A. T., Cho, M., Ruoslahti, E., Xiao, Y., and Soh, H. T. (2011). Selection of phage-displayed peptides on live adherent cells in microfluidic channels. *Proc. Natl. Acad. Sci. USA.* **108,** 6909–6914.

Weissleder, R., Kelly, K., Sun, E. Y., Shtatland, T., and Josephson, L. (2005). Cell-specific targeting of nanoparticles by multivalent attachment of small molecules. *Nat. Biotechnol.* **23,** 1418–1423.

Zhang, H., Chen, J., Waldherr, C., Hinni, K., Waser, B., Reubi, J. C., and Maecke, H. R. (2004). Synthesis and evaluation of bombesin derivatives on the basis of pan-bombesin peptides labeled with indium-111, lutetium-177, and yttrium-90 for targeting bombesin receptor-expressing tumors. *Cancer Res.* **64,** 6707–6715.

Zhang, L., Hoffman, J. A., and Ruoslahti, E. (2005). Molecular profiling of heart endothelial cells. *Circulation* **112,** 1601–1611.

Zhang, L., Giraudo, E., Hoffman, J. A., Hanahan, D., and Ruoslahti, E. (2006). Lymphatic zip codes in premalignant lesions and tumors. *Cancer Res.* **66,** 5696–5706.

CHAPTER THREE

Engineering Cyclic Peptide Toxins

Richard J. Clark* *and* David J. Craik†

Contents

1. Introduction 58
2. Peptide Design 59
3. Peptide Synthesis 60
4. Structural Analysis 66
5. Stability Assays 68
   5.1. Proteolytic stability assay 68
   5.2. Simulated gastric fluid assay 68
   5.3. Simulated intestinal fluid assay 69
   5.4. Serum stability assay 69
6. Special Considerations for Cyclic Peptides 69
Acknowledgments 71
References 71

Abstract

Peptide-based toxins have attracted much attention in recent years for their exciting potential applications in drug design and development. This interest has arisen because toxins are highly potent and selectively target a range of physiologically important receptors. However, peptides suffer from a number of disadvantages, including poor *in vivo* stability and poor bioavailability. A number of naturally occurring cyclic peptides have been discovered in plants, animals, and bacteria that have exceptional stability and potentially ameliorate these disadvantages. The lessons learned from studies of the structures, stabilities, and biological activities of these cyclic peptides can be applied to the reengineering of toxins that are not naturally cyclic but are amenable to cyclization. In this chapter, we describe solid-phase chemical synthetic methods for the reengineering of peptide toxins to improve their suitability as therapeutic, diagnostic, or imaging agents. The focus is on small disulfide-rich peptides from the venoms of cone snails and scorpions, but the technology is potentially widely applicable to a number of other peptide-based toxins.

* School of Biomedical Sciences, The University of Queensland, Brisbane, Queensland, Australia
† Institute for Molecular Bioscience, The University of Queensland, Brisbane, Queensland, Australia

Methods in Enzymology, Volume 503
ISSN 0076-6879, DOI: 10.1016/B978-0-12-396962-0.00003-3

1. Introduction

Venomous creatures produce a myriad of peptide toxins as a means of defense or prey capture (Terlau and Olivera, 2004). These peptide toxins exhibit a diverse range of biological activities, many of which have potential applications in the pharmaceutical industry (Lewis and Garcia, 2003). However, despite high potency and exquisite selectivity for their biological targets, the potential of these toxins in therapeutic applications is limited by their peptidic nature as they are susceptible to degradation by proteases and generally have low oral bioavailability. Cyclization has been used in the pharmaceutical industry as a strategy for stabilizing and locking the conformation of small peptides in drug design applications (Hruby, 2002). Similarly, some microorganisms produce backbone-cyclized peptides that have been adopted for therapeutic purposes. A well-known example is cyclosporin A, a potent immunosuppressant widely used in organ transplant procedures to suppress organ rejection (Starzl *et al.*, 1981). Until recently, most work on cyclic peptides, either synthetic or natural, has focused on small (<12 amino acids) peptides, and typically these did not contain disulfide bonds. However, with the recent discoveries of a number of larger ribosomally synthesized cyclic peptides (Craik *et al.*, 2006; Daly *et al.*, 2009; Kawai *et al.*, 2004; Selsted, 2004; Trabi and Craik, 2002), it is clear that a cyclization approach can potentially be applied to a range of peptide toxins to improve biopharmaceutical properties.

The naturally occurring cyclic peptides discovered over the past 15 years come from a broad range of organisms, including bacteria, fungi, plants, and mammals (Craik, 2006). Specific examples include the bacteriocins from bacteria (Maqueda *et al.*, 2004), cyclotides from members of the Rubiaceae, Violaceae, Fabaceae, and Cucurbitaceae plant families (Craik *et al.*, 1999; Poth *et al.*, 2011), the sunflower trypsin inhibitor peptide SFTI-1 (Luckett *et al.*, 1999; Mylne *et al.*, 2011), and the mammalian θ-defensins (Cole *et al.*, 2002; Selsted, 2004; Tang *et al.*, 1999). Features common to these cyclic peptides, aside from their head-to-tail cyclic backbones, are their remarkable stability and compact structures. With the stability of natural cyclic peptides in mind, we have shown that backbone cyclization of acyclic peptide toxins can dramatically increase the therapeutic potential of these molecules, which might otherwise be limited by poor *in vivo* stability (Akcan *et al.*, 2011; Clark *et al.*, 2005, 2010; Lovelace *et al.*, 2006, 2011).

In this chapter, we describe the methodologies used in the design, synthesis, and characterization of cyclic peptide toxins and illustrate these methodologies with specific examples from our work. Our research is focused on the introduction of cyclic backbones into disulfide-rich peptide toxins to reengineer them with favorable biopharmaceutical properties. The majority of our work has focused on conotoxins, which are disulfide-rich peptide toxins from the venom of the marine snails of the *Conus* genus (Adams *et al.*, 1999; Terlau and Olivera, 2004). Conotoxins have shown exciting potential for the treatment of neuropathic pain, with one molecule, MVIIA (Prialt^{TM}), currently marketed in the United States and Europe (Miljanich, 2004). We recently expanded the scope of our work to other peptide toxins, including chlorotoxin from the scorpion *Leiurus quinquestriatus*. A labeled analogue of chlorotoxin is currently being developed as a visualizing agent for brain tumors during surgery (Akcan *et al.*, 2011; Veiseh *et al.*, 2007). A variety of plant-based peptides have been used as scaffolds for diagnostic imaging applications (Kimura *et al.*, 2010; Kolmar, 2010; Nielsen *et al.*, 2010). Thus, peptide toxins have applications not only as therapeutics but also as imaging or diagnostic agents.

2. Peptide Design

The concept of backbone cyclization of a peptide toxin is illustrated in Fig. 3.1. Ideally, the three-dimensional structure, determined by either X-ray or nuclear magnetic resonance (NMR) methods, of the peptide is known and can be used as a basis for designing a cyclic analogue. In the absence of structural data for the peptide of interest, a three-dimensional structure from a homologous peptide can be used in the design process. A number of factors should be considered when designing the cyclic peptide, including distance between the termini and their relative orientation, flexibility of the termini, and the position of disulfide bonds and key residues for biological activity. In our work, we have found that the key factor to consider is the distance between the N- and C-termini. If the termini are in close proximity, they can often be joined directly together without affecting the structure and biological activity of the peptide. However, if the termini are further apart, then a linking sequence needs to be incorporated to join the ends of the peptide. The length of this linker sequence is critical; if it is too long or too short, the structure becomes distorted, which can result in reduced activity. In disulfide-rich toxins, an incorrect linker length can also change the folding pathway of the peptide, resulting in the formation of an inactive disulfide isomer.

Figure 3.2 illustrates the relationship between termini distance and linker length for successful cyclization based on studies of a range of peptides (Camarero and Muir, 1999; Camarero *et al.*, 2001; Clark *et al.*, 2005, 2010; Deechongkit and Kelly, 2002; Iwai *et al.*, 2001; Lovelace *et al.*, 2006, 2011; Takahashi *et al.*, 2007; Williams *et al.*, 2005). If the distance between the termini of a peptide toxin (or a homologue) to be cyclized is known, then a suitable linker length can be determined from this relationship. We usually synthesize several analogues where the length of this linker

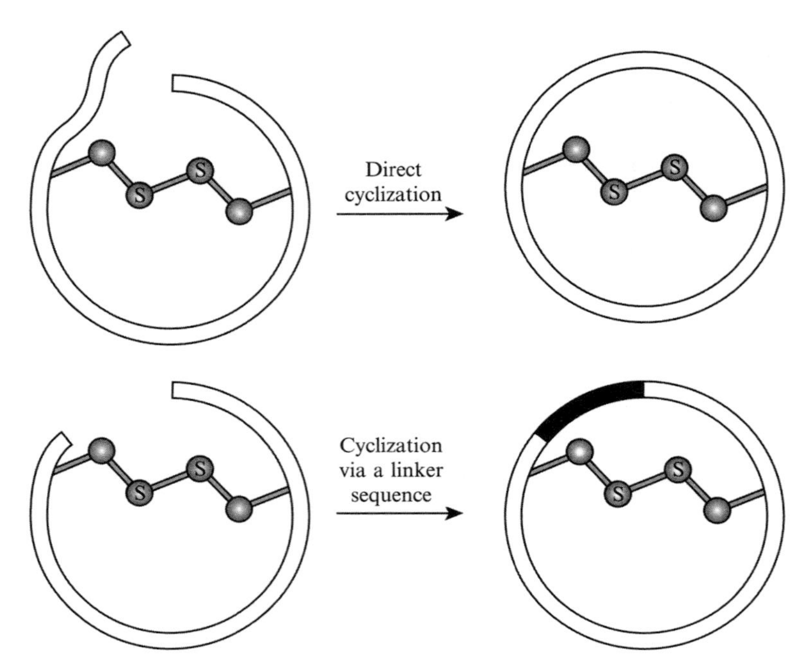

Figure 3.1 The concept of backbone cyclization of peptide toxins. If the termini of a peptide are in close proximity and possess some flexibility, it is possible that backbone cyclization can be achieved by joining the N- and C-termini directly (top). If the termini are far apart or structurally constrained, then a linker sequence can be introduced to span the distance between the termini (bottom).

varies by plus or minus one residue from the predicted value. To date, we have used linker sequences comprising alanine and glycine as these are chemically "inert" residues (Clark *et al.*, 2005, 2010), but in principle, the composition of the linker can be varied across all amino acids, to exert control over conformational or biophysical properties.

3. Peptide Synthesis

To date, the most successful and predominant method for the solid-phase peptide synthesis of cyclized peptide toxins utilizes an intramolecular native chemical ligation strategy (Dawson *et al.*, 1994) and *t*-butoxycarbonyl (Boc) *in situ* neutralization chemistry (Schnölzer *et al.*, 1992), which is therefore the focus of this chapter. A summary of the synthetic scheme is illustrated in Fig. 3.3.

During synthesis, the peptide chain is assembled from the C-terminus to the N-terminus. The native chemical ligation reaction requires an

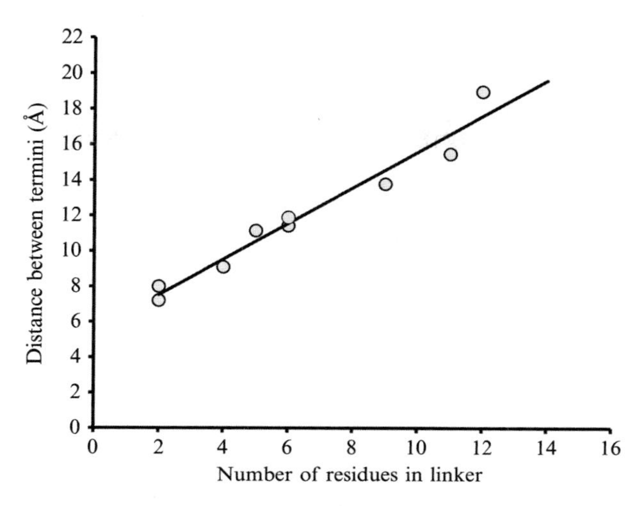

Figure 3.2 Empirical relationship between the distance between N- and C-termini and the number of residues required for successful cyclization of a peptide. The data presented are from the literature and work in our laboratory and include values for dihydrofolate reductase (Takahashi *et al.*, 2007), SH3 domain (Camarero and Muir, 1999; Camarero *et al.*, 2001), GFP (Iwai *et al.*, 2001), α-conotoxin MII (Clark *et al.*, 2005), α-conotoxin Vc1.1 (Clark *et al.*, 2010), α-conotoxin AuIB (Lovelace *et al.*, 2011), χ-conotoxin MrIA (Lovelace *et al.*, 2006), DnaB (Williams *et al.*, 2005), and a PinWW domain (Deechongkit and Kelly, 2002).

N-terminal cysteine, and therefore, the final residue added to the chain must be a cysteine. As cyclic peptides have no N- or C-termini, any point in the sequence that is on the N-terminal side of a cysteine can be selected as the synthetic starting point of the linear precursor. If possible, it is best to select a precursor where the C-terminal residue is sterically small, as this has been shown to be kinetically more favorable (Hackeng *et al.*, 1999).

The peptide is assembled on phenylacetamidomethyl (PAM) resin preloaded with Boc-protected glycine (100–200 mesh size, Peptides International, Louisville, USA), although any preloaded amino acid can be used. Amino acids with *N*-Boc protecting groups and the following side chain protection are used as follows: Arg(Tos), Asn(Xan), Asp(OcHx), Cys (4-MeBzl), Glu(OcHx), Gln(Xan), His(Tos), Lys(2-Cl-Z), Ser(Bzl), Thr (Bzl), Trp(For), and Tyr(2-Br-Z). The resin is soaked in dimethylformamide (DMF) overnight to swell it before beginning the synthesis. The Boc protecting group is removed by treatment of the resin twice with trifluoroacetic acid (TFA) for 1 min each time. It is important to then wash the resin extensively with DMF to remove all traces of TFA before proceeding to the next step.

Four molar equivalents (relative to reactive groups on the resin) of *S*-trityl-β-mercaptopropionic acid are dissolved in 4 mL of a solution of 0.5 *M* *O*-benzotriazole-*N*,*N*,*N*′,*N*′-tetramethyl-uronium-hexafluoro-phosphate

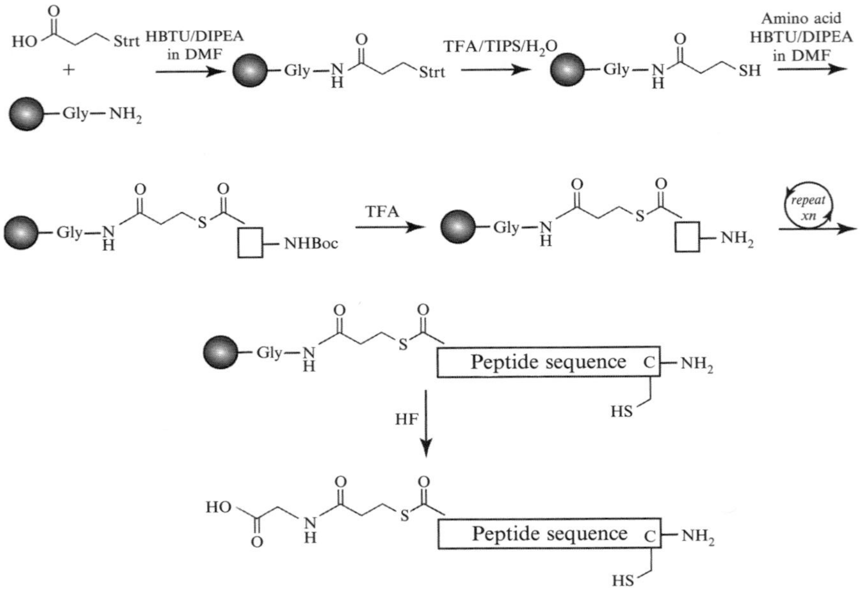

Figure 3.3 A summary of the synthetic strategy used to synthesize the linear precursors of cyclic peptide toxins by solid-phase peptide synthesis with Boc chemistry. Peptides are assembled on preloaded phenylacetamidomethyl (PAM) (e.g., PAM-Gly) resin. *S*-Trityl-β-mercaptopropionic acid is coupled to the resin, using standard amide-coupling conditions, for subsequent generation of the thioester after HF cleavage. After removal of the trityl protecting group, the C-terminal residue of the peptide chain is added using standard coupling conditions. The peptide sequence is then completed, ending in a cysteine residue, and the peptide is cleaved from the resin with anhydrous hydrofluoric acid.

(HBTU) in DMF, and five equivalents of diisopropylethylamine (DIPEA) are added. When the *S*-trityl-β-mercaptopropionic acid has dissolved, the solution is added to the resin and the mixture agitated for 30 min. The efficiency of the reaction is then determined using a ninhydrin test. A small amount of resin (2–4 mg) is removed, washed with 50% dichloromethane (DCM) in methanol, and dried. To the dried resin sample, two drops of 76% (w/w) phenol in ethanol, four drops of 0.2 m*M* KCN in pyridine, and two drops of 0.28 *M* ninhydrin in ethanol are added. A blank, containing no resin, is also prepared. The samples are incubated at 98 °C for 5 min and then 2.8 mL of 60% ethanol in water is added. The samples are briefly spun to settle the resin and the absorbance of the solution is read against the reagent blank at 570 nm. The % coupling is then calculated using the following formula: % coupling $= 100 \times (1 - (A_{570} \times 200/\mathrm{SV} \times \text{mass of resin}))$, where A_{570} is the absorbance value and SV is the number of micromoles of peptide per gram of resin.

If the % coupling is less than 99%, then the resin is washed with DMF and a second coupling of *S*-trityl-β-mercaptopropionic acid is performed.

When the % coupling of *S*-trityl-β-mercaptopropionic acid is >99%, the trityl protecting group is removed by repeated treatment of the resin with 1 min washes of TFA/triisopropylsilane/H_2O (90:5:5) until the deprotection solution becomes colorless. Care should be taken to minimize the amount of air that is drawn through the resin during draining as this can result in oxidation of the free thiol groups generated by the deprotection step.

After thorough washing of the resin with DMF to remove all traces of the deprotection mixture, the first (C-terminal) amino acid is coupled. Four molar equivalents of the appropriate Boc-protected amino acid are dissolved in 4 mL of a solution of 0.5 *M* HBTU in DMF and five molar equivalents of DIPEA are added. The amino acid mixture is then added to the resin and the coupling reaction is allowed to proceed for 10 min. The reaction mixture is then drained, the resin washed with DMF, and a second batch of amino acid/DMF/DIPEA added and the reaction is left for 30 min. This double coupling is done because it is not possible to check the reaction yield using the ninhydrin reaction as the amino acid is being added to a thiol group. The reagents are then drained away and the resin is thoroughly washed with DMF ready for the next amino-acid coupling. The Boc protecting group is then removed by treatment of the resin with TFA for 2×1 min, the resin is washed with DMF to remove all TFA, and then next amino acid is added as described above. The reaction is left for 10 min and the coupling yield is determined using the ninhydrin assay. If the yield is <99%, then the resin is washed with DMF and a second coupling reaction is undertaken. This is repeated until the yield is >99%.

The next amino acid in the sequence is then added following the same protocol of deprotection of the Boc group with TFA and addition of the amino acid using the HBTU/DMF/DIPEA mixture. This process is continued until the peptide chain is complete. If incomplete coupling of an amino acid occurs, despite multiple coupling reactions, then the unreacted amine groups can be blocked by acetylation. To acetylate any free amines, the resin is treated (2×5 min) with a mixture containing 870 μL of acetic anhydride, 470 μL of DIPEA, and 15 mL of DMF.

On completion of the peptide chain assembly, the resin is treated with 1×1 and 1×5 min of TFA to remove the final *N*-Boc group. The resin is then washed thoroughly with DMF followed by treatment with 2×10 mL of 10% DIPEA in DMF for 5 min each time with a DMF wash between each treatment. The resin is then thoroughly washed with DMF followed by DCM and the resin is then dried with vacuum under a blanket of nitrogen.

The peptide is then cleaved from the resin using anhydrous hydrofluoric acid (HF) with *p*-cresol and *p*-thiocresol as scavengers (9:0.8:0.2 (v/v) HF:*p*-cresol:thiocresol) at −5–0 °C for 1.5 h. The HF is then evaporated off under vacuum and cold diethyl ether is added to the reaction to precipitate the peptide. The mixture is then filtered to collect the peptide, which is washed with cold diethyl ether and then dissolved in 50% acetonitrile containing 0.05% TFA and lyophilized.

The crude peptide mixture is purified by reverse phase high-performance liquid chromatography (RP-HPLC). Typically, the peptide is dissolved in water containing 0.05–0.1% TFA, filtered, and then loaded onto a C_{18} RP-HPLC column. The peptide mixture is then eluted with an acetonitrile/water gradient with the percentage of acetonitrile increasing over time. Typically, TFA (0.05–0.1%, v/v) is used in the HPLC buffer as an ion-pairing agent. The eluent is monitored by UV detection at, for example, 214, 230, and/or 280nm, and samples are collected as the peptide elutes. These samples are analyzed by electrospray mass spectrometry (ES-MS) and analytical scale RP-HPLC to determine the presence of the desired product in the sample and its purity. Samples containing the desired peptide at sufficient purity are then combined and lyophilized.

Peptide toxins often contain disulfide bonds, and there are several approaches that can be taken for the formation of the disulfide bonds and the cyclization of the backbone, as illustrated in Fig. 3.4. Both reactions are normally performed under basic conditions (pH$\sim$8) in aqueous buffer. Therefore, the first approach is to undertake a "one-pot" reaction in which the intramolecular native chemical ligation reaction and the disulfide formation occur at the same time. If multiple disulfide bonds are to be formed, the buffer used is one that favors formation of the native disulfide isomer. These buffer conditions can be based on those used for the oxidative folding of the linear peptide or might need to be determined for the cyclic analogue by varying conditions such as salt concentration, reaction time, temperature, the presence of thiol reagents (e.g., glutathione), or the addition of cosolvents and detergents. The second approach is to first perform the native chemical ligation reaction under reducing conditions (e.g., in the presence of tris(2-carboxyethyl)phosphine) to obtain the cyclic peptide, and the disulfide bonds are formed in a second, oxidative, step. The third option is used when undertaking a regioselective approach to form multiple disulfide bonds. Selected cysteine residues with acetamidomethyl (Acm)-protected thiols can be incorporated into the sequence to allow selective formation of the disulfide bonds. Typically, the native chemical ligation reaction and formation of the first disulfide bond are carried out in the same reaction followed by formation of a second disulfide bond under selective conditions.

Using approach 1, the peptide is dissolved in the chosen buffer at a concentration of 0.1–0.5mg/mL and stirred under the appropriate conditions for disulfide formation. Using approach 2, the peptide is dissolved in aqueous buffer (e.g., 0.1 *M* NH_4HCO_3, pH 8) containing 20 m*M* tris (2-carboxyethyl)phosphine at 0.1–0.5mg/mL, stirred at room temperature

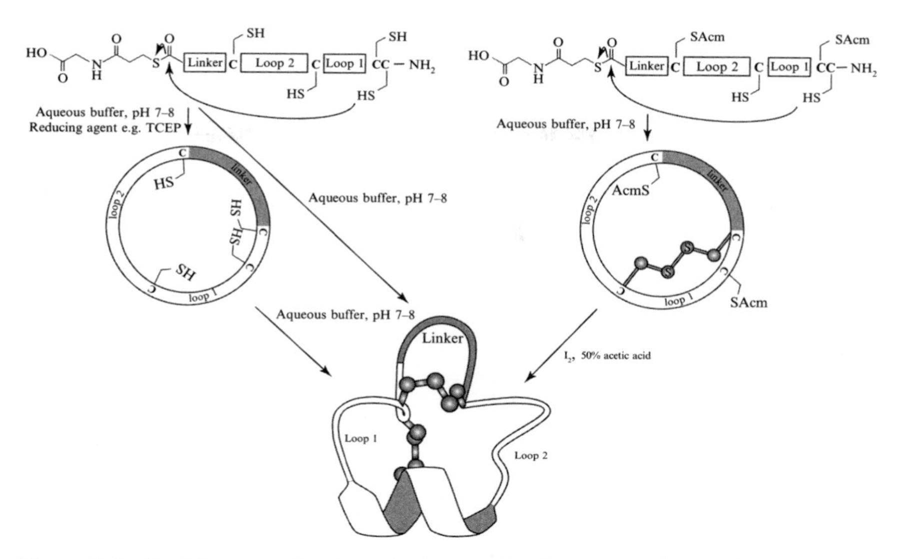

Figure 3.4 Backbone cyclization via intramolecular native chemical ligation and disulfide formation. Backbone cyclization and disulfide formation of a peptide toxin can be achieved via several strategies. Cyclization and oxidation can be achieved in one pot by incubating the peptide in a suitable folding buffer at basic pH. Alternatively, a two-step strategy can be employed where backbone cyclization without disulfide formation is achieved by incubating the peptide in aqueous buffer at basic pH in the presence of a reducing agent (e.g., TCEP). The cyclic reduced peptide can then be isolated and the disulfide bonds can be formed in a second step. Finally, regioselective disulfide bond formation can be used to direct the correct folding of the peptide toxin. In this last approach, cysteines with Acm protection are incorporated into the peptide chain at selected positions. Backbone cyclization and oxidation of the unprotected cysteines are then performed in one reaction followed by Acm deprotection and oxidation of the remaining disulfide (e.g., in an acetic acid solution with iodine). The example shown is that of a generic α-conotoxin framework where there are four cysteine residues that form two disulfide bonds (Dutton and Craik, 2001). The spacing between the cysteine residues defines two backbone loops that vary in their sequence in different α-conotoxins (Millard *et al.*, 2009). The linker region is incorporated into the linear peptide precursor in a position that eventually results in it spanning the N- and C-termini.

for 8–16h, and then purified by RP-HPLC. The cyclic reduced peptide obtained is then redissolved in oxidation buffer to form the disulfide bonds. In the third approach, the peptide is dissolved in aqueous buffer (e.g., 0.1 *M* NH_4HCO_3, pH 8), stirred at room temperature for 8–16h, and then purified by RP-HPLC. The final disulfide bond is then formed by incubation of the peptide (0.1–0.3mg/mL) with iodine in acidic conditions (e.g., 50% aqueous acetic acid) and monitored by RP-HPLC and ES-MS until the desired completely oxidized peptide is formed.

4. Structural Analysis

NMR is the technique of choice for the structural analysis of peptide toxins due to their small size, high solubility, and usually well-defined structures (Daly and Craik, 2009; Marx *et al.*, 2006). Samples are prepared for NMR analysis by dissolving the peptide in either water/deuterated water (9:1, v/v) or water/deuterated acetonitrile/deuterated water (7:2:1, v/v/v) depending on solubility. A 1m*M* solution is sufficient for high-quality spectra, but lower concentrations can be used with NMR spectrometers equipped with high-sensitivity cryoprobes. A range of pH values can be used (~3–7), and pH-induced conformational changes (if present) can be assessed by measuring changes in chemical shifts with pH.

A ^{1}H one-dimensional NMR spectrum is recorded to check on sample quality. Wide dispersion of the NMR signals (Fig. 3.5) is a good indication that the peptide adopts a defined structure in solution. Two-dimensional TOCSY and NOESY spectra are then recorded to allow sequence-specific assignment of individual residues. Mixing times of approximately 80 and 200ms for the TOCSY and NOESY spectra, respectively, are typically used. Spectra are usually recorded for a minimum of two temperatures (e.g., 290 and 298K) because different temperatures can be useful for resolving areas where resonances are overlapped in the amide region at any one temperature.

After sequence specific assignment, a chemical shift analysis can be performed as a preliminary assessment of the structural features of the cyclic peptide and for comparison to the native linear peptide. Secondary αH chemical shifts are the difference between an observed αH chemical shift and that of the corresponding residue in a random coil peptide (Wishart *et al.*, 1992). A comparison of secondary αH shifts between different peptides provides information on the similarity of their structures. Further, an analysis of secondary shift data yields information on the secondary structures present within peptides. A series of negative secondary αH shift values are indicative of helical character, whereas a series of positive values suggest the presence of a β-strand.

To determine the three-dimensional structure of a peptide, distance restraints can be derived from NOESY spectra. Analysis of the NOESY peak intensities can be complicated by spin diffusion, and thus a mixing time of 100ms is generally used for this analysis to minimize this complication. Angle restraints are derived for ϕ angles by measuring αN couplings from either the one-dimensional spectrum or a DQFCOSY spectrum. $\chi 1$ angle restraints are derived from analysis of NOESY peak intensities and measuring αβ couplings from an ECOSY spectrum. Slowly exchanging amide protons are determined by dissolving the freeze-dried peptide in deuterated water and monitoring the exchange of the amide protons with the solvent

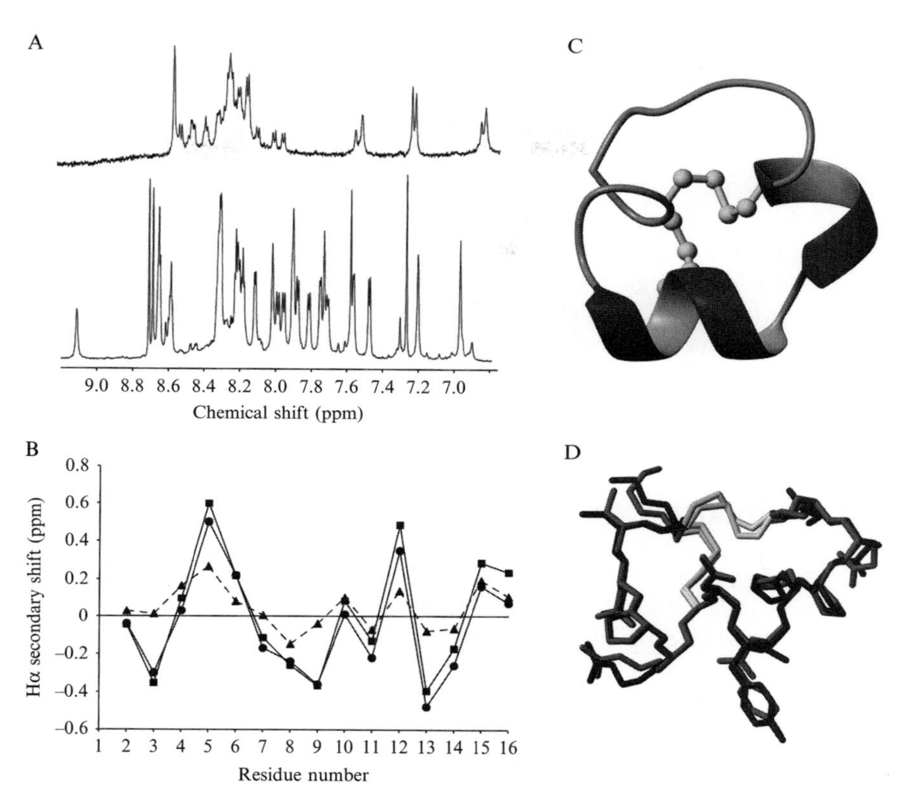

Figure 3.5 Structural characterization of cyclic peptide toxins. NMR is a powerful method for the structural characterization of cyclic peptide toxins. (A) The amide region of ^{1}H NMR spectra for a misfolded (top) and folded cyclic peptide (bottom). (B) Secondary shift analysis is an effective method for comparing structurally related peptides. For example, similar αH secondary shift values for the native α-conotoxin MII (squares) and a cyclic analogue (cMII-7, circles) revealed that they are structurally similar, whereas the analysis for a second cyclic analogue (cMII-5, triangles) showed that it was misfolded (Clark *et al.*, 2005). (C) The three-dimensional structure of cyclized α-conotoxin cVc1.1 illustrating that NMR spectroscopy can be used to determine the three-dimensional structure of cyclic peptide toxins. (D) A backbone overlay of the three-dimensional structures of linear Vc1.1 (dark gray) and cyclic Vc1.1 (black) showing the similarity in structure of the two peptides. The disulfide bonds are highlighted in light gray (Clark *et al.*, 2010).

by recording one dimensional and TOCSY spectra over time. These data are used to generate hydrogen bond restraints (Scanlon *et al.*, 1997).

The distance and angle restraints determined from the NMR data are used to calculate three-dimensional structures consistent with these experimentally derived data. One program for calculating these structures is CNS, and the methods used for this have been described previously (Rosengren *et al.*, 2003). The initial structures are analyzed for the presence of hydrogen bonds. If the hydrogen bond donor is slowly exchanging in the deuterated water exchange experiments, hydrogen-bond restraints can be added to the structure calculations. Calculating structures is an iterative process that improves as additional restraints, such as these hydrogen-bonding restraints, are added and as any errors in the initial assignments or restraints sets are discovered and removed. The structures are then analyzed for violations of the experimental data. If there are violations greater than 3 Å for distance restraints or 3° for angle restraints, the restraints should be checked for errors and the structures recalculated. The final structures are then analyzed in terms of their Ramachandran and energetic statistics. These statistics provide information on the precision and quality of the structures and examples are given in references (Daly *et al.*, 2006; Heitz *et al.*, 2001; Jennings *et al.*, 2005).

5. Stability Assays

A series of straightforward assays can be used to compare the stability of a linear peptide toxin to a cyclic analogue and provide an indication of improvements in drug-like characteristics for the cyclic toxin.

5.1. Proteolytic stability assay

A 1-mg/mL stock solution of each test peptide is prepared in the appropriate buffer (e.g., 100 m*M* ammonium acetate, pH 7.4 for trypsin, chymotrypsin, and endoGluC; 100 m*M* acetic/formic acid for pepsin). A solution of the enzyme (25 μg/mL) is also prepared in the same buffer. An equal volume of peptide and enzyme solutions is then combined and the solution incubated at 37 °C, either in a water bath or in an incubator, for the duration of experiment. The experiment is performed in triplicate for each peptide. At each time point (e.g., 1, 5, 10, 30, 45, 60, 90, 120, 150, 180, 240, 300, 360 min and 24 h), a 5-μL aliquot is taken and quenched with 95 μL of 5% formic acid. For a pepsin digestion, 95 μL of 100 m*M* ammonium acetate (pH 7.4) is used to quench the reaction. Prepare a blank by placing 95 μL of quenching solution in a vial, add 2.5 μL of enzyme, and then add 2.5 μL of peptide. Samples should be stored at 4 °C until they are analyzed by either RP-HPLC or LC/MS.

5.2. Simulated gastric fluid assay

Simulated gastric fluid (SGF) can be prepared as described in the U.S. Pharmacopeia by dissolving 20 mg of NaCl and 32 mg of pepsin (with an activity of 800–2500 units/mg of protein) in 70 μL of concentrated HCl and then diluting the solution to 10 mL with water (final pH~1.2). Peptides are dissolved in SGF at a concentration of 100 μg/mL (in triplicate) and incubated at 37 °C. Fifty-microliter aliquots are taken at desired time points, quenched with 50 μL of 0.2 N sodium carbonate, and analyzed by RP-HPLC. The amount of peptide remaining is determined by measuring the peak area and expressing it as a percentage of the peak area at the 0 h time point.

5.3. Simulated intestinal fluid assay

Simulated intestinal fluid (SIF) is prepared as described in the U.S. Pharmacopeia by dissolving 68 mg of KH_2PO_4 in 250 μL of water followed by the addition of 770 μL of 0.2 N NaOH, 5 mL of water, and 100 μg of porcine pancreatin (Sigma-Aldrich). The pH of the solution is adjusted to 6.8 with NaOH and the solution made up to 10 mL with water. Peptides are dissolved in SIF at a concentration of 100 μg/mL (in triplicate) and incubated at 37 °C. Fifty-microliter aliquots are taken at the desired time points, quenched with 50 μL of 4% aqueous TFA, and analyzed by RP-HPLC. The amount of peptide remaining is determined by measuring the peak area and expressing it as a percentage of the peak area at the 0 h time point.

5.4. Serum stability assay

Pooled human serum (Sigma-Aldrich, H4522, serum from male AB human plasma) is centrifuged at 13,000 rpm for 15 min to separate lipids and the serum subsequently removed from the lipid layer. Each peptide is dissolved in serum at a concentration of 100 μg/mL (in triplicate) and incubated at 37 °C. Forty-microliter aliquots are taken at the desired time points, for example, 0, 0.5, 1, 2, 4, 6, 8, 12, and 24 h. To each serum aliquot, 40 μL of 6 *M* urea is added and the samples are incubated for 10 min at 4 °C. Then, each serum aliquot is quenched with 40 μL of 20% trichloroacetic acid and incubated for another 10 min at 4 °C to precipitate serum proteins. The samples are centrifuged at 14,000 × *g* for 10 min, and 100 μL of the supernatant is analyzed on RP-HPLC or by LC-MS/MS. The amount of peptide remaining can be determined by measuring the peak area and expressing it as a percentage of the peak area at the 0 h time point.

6. Special Considerations for Cyclic Peptides

As a consequence of cyclic peptides having no N- or C-termini, there are special considerations that need to be taken into account in studying them. Standard peptide-sequencing methods such as Edman degradation or

MS-based methods cannot be used directly on cyclic peptides as these techniques require termini. Therefore, the cyclic peptide backbone needs to be cleaved, preferably using an enzymatic approach. The choice of protease for cleaving the sequence is important as the fragments generated from the proteolysis need to be of a suitable length and preferably contain functional groups that are ionizable by protonation (e.g., Arg, His, Lys) to be amenable to sequencing by MS. In addition, if disulfide bonds are present, these need to be reduced and alkylated before enzyme cleavage. A typical approach for preparing a disulfide-rich cyclic peptide for sequencing using trypsin as the protease is as follows: dissolve 10 μg of peptide in 10 μL of 0.1 *M* ammonium bicarbonate buffer (pH 8.1) and add 0.5 μL of freshly prepared 100 μ*M* TCEP; incubate the solution under nitrogen for 30 min at 55 °C and then add 5 μL of trypsin (40 μg/mL); incubate the solution at 37 °C for 3 h and then desalt the sample using solid-phase extraction (e.g., using a ZipTip®, Millipore); and elute the sample with 10 μL of 80% acetonitrile/0.1% formic acid in water and the sample is then ready for MS sequencing.

It is often desirable to label a peptide of interest with, for example, a fluorescent dye or a biotin moiety for use in biological studies. For linear peptides, the N-terminus provides a convenient functional group for labeling by amine-reactive probes such as succinimidyl ester or isothiocyanate derivatives. However, in cyclic peptides, this N-terminal group is missing, and therefore, alternative labeling strategies need to be explored. If there is a lysine present in the sequence of the cyclic peptide, then the side chain amine can be used as a labeling site; however, derivatization of amino acids within the sequence might affect biological activity or the pharmacological profile of the peptide. If no lysine exists within the sequence then one can be introduced at a position that will have minimal impact on the structure, activity, and pharmacology of the peptide. This might require the synthesis of a number of peptides with different point mutations before one with suitable properties is found.

One advantage of working with backbone cyclic peptides is that a resin-splitting strategy can be used to produce libraries of cyclic peptide analogues. This strategy, highlighted in Fig. 3.6, takes advantage of the NCL reaction and the cysteine-rich nature of many peptide toxins, which allows selection of several potential start sites for synthesis of the linear precursor. Specifically, the requirement for an N-terminal Cys residue in the linear precursor sequence can be accommodated by any of the Cys residues because all of the corresponding precursors will lead to the same cyclic structure. Therefore, in studies where multiple peptides with variations in a given loop are to be made, a start point can be selected so that the majority of the sequence can be synthesized in one pot and then the resin split for incorporation of the individual unique sequences toward the end of the chain assembly, maximizing the efficiency of the synthesis.

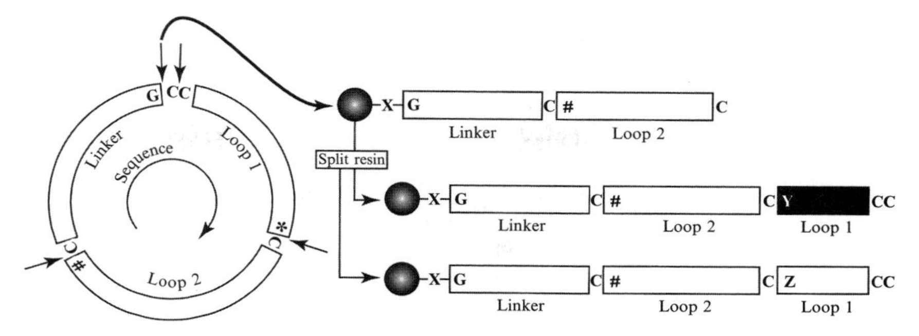

Figure 3.6 Schematic illustration of a combinatorial synthetic strategy used to synthesize cyclic peptide toxin analogues, using the α-conotoxin framework as an example. When multiple cysteines are present, there are a number of possible starting points for the synthesis, at the N-terminal side of each cysteine, which are indicated by arrows in the example shown. The amino acid that would become the C-terminal residue in a corresponding linear precursor is also indicated (i.e., C, G, #, or *). The example shown is for a series of loop 1 analogues that are synthesized by starting at the Gly at the C-terminal end of the linker sequence as shown on the right. The peptide chain, connected to the resin via a thioester moiety (X), is built up on-resin (starting at the C-terminus) until the loop 1 region is reached. The resin is then split and the synthesis of the loop 1 variants is completed separately. The example shown is for two variants (starting with nominal amino acids Y or Z, respectively), but the approach is extendible to more variants and takes advantage of splitting the resin late in the synthesis after assembly of the invariant part of the sequence.

ACKNOWLEDGMENTS

Work in our laboratory on peptide toxins is supported by grants from the National Health & Medical Research Council (Grant ID Nos. APP631457 and APP1010552) and the Australian Research Council. D. J. C. is an NHMRC Professorial Research Fellow (ID No. 569603). R. J. C. is an ARC Future Fellow (ID No. FT100100476).

REFERENCES

Adams, D. J., Alewood, P. F., Craik, D. J., Drinkwater, R. D., and Lewis, R. J. (1999). Conotoxins and their potential pharmaceutical applications. *Drug Dev. Res.* **46,** 219–234.

Akcan, M., Stroud, M. R., Hansen, S. J., Clark, R. J., Daly, N. L., Craik, D. J., and Olson, J. M. (2011). Chemical re-engineering of chlorotoxin improves bioconjugation properties for tumor imaging and targeted therapy. *J. Med. Chem.* **54,** 782–787.

Camarero, J. A., and Muir, T. W. (1999). Biosynthesis of a head-to-tail cyclized protein with improved biological activity. *J. Am. Chem. Soc.* **121,** 5597–5598.

Camarero, J. A., Fushman, D., Sato, S., Giriat, I., Cowburn, D., Raleigh, D. P., and Muir, T. W. (2001). Rescuing a destabilized protein fold through backbone cyclization. *J. Mol. Biol.* **308,** 1045–1062.

Clark, R. J., Fischer, H., Dempster, L., Daly, N. L., Rosengren, K. J., Nevin, S. T., Meunier, F. A., Adams, D. J., and Craik, D. J. (2005). Engineering stable peptide toxins by means of backbone cyclization: Stabilization of the alpha-conotoxin MII. *Proc. Natl. Acad. Sci. USA* **102,** 13767–13772.

Clark, J., Jensen, J., Nevin, S., Brid, C., Adams, D., and Craik, D. (2010). The engineering of an orally active conotoxin for the treatment of neuropathic pain. *Angew. Chem. Int. Ed. Engl.* **49,** 6545–6548.

Cole, A. M., Hong, T., Boo, L. M., Nguyen, T., Zhao, C., Bristol, G., Zack, J. A., Waring, A. J., Yang, O. O., and Lehrer, R. I. (2002). Retrocyclin: A primate peptide that protects cells from infection by T- and M-tropic strains of HIV-1. *Proc. Natl. Acad. Sci. USA* **99,** 1813–1818.

Craik, D. J. (2006). Chemistry. Seamless proteins tie up their loose ends. *Science* **311,** 1563–1564.

Craik, D. J., Daly, N. L., Bond, T., and Waine, C. (1999). Plant cyclotides: A unique family of cyclic and knotted proteins that defines the cyclic cystine knot structural motif. *J. Mol. Biol.* **294,** 1327–1336.

Craik, D. J., Cemazar, M., and Daly, N. L. (2006). The cyclotides and related macrocyclic peptides as scaffolds in drug design. *Curr. Opin. Drug Discov. Dev.* **9,** 251–260.

Daly, N. L., and Craik, D. J. (2009). Structural studies of conotoxins. *IUBMB Life* **61,** 144–150.

Daly, N. L., Clark, R. J., Plan, M. R., and Craik, D. J. (2006). Kalata B8, a novel antiviral circular protein, exhibits conformational flexibility in the cystine knot motif. *Biochem. J.* **393,** 619–626.

Daly, N. L., Rosengren, K. J., and Craik, D. J. (2009). Discovery, structure and biological activities of cyclotides. *Adv. Drug Deliv. Rev.* **61,** 918–930.

Dawson, P. E., Muir, T. W., Clark-Lewis, I., and Kent, S. B. (1994). Synthesis of proteins by native chemical ligation. *Science* **266,** 776–779.

Deechongkit, S., and Kelly, J. W. (2002). The effect of backbone cyclization on the thermodynamics of beta-sheet unfolding: Stability optimization of the PIN WW domain. *J. Am. Chem. Soc.* **124,** 4980–4986.

Dutton, J. L., and Craik, D. J. (2001). α-Conotoxins: Nicotinic acetylcholine receptor antagonists as pharmacological tools and drug leads. *Curr. Med. Chem.* **8,** 327–344.

Hackeng, T. M., Griffin, J. H., and Dawson, P. E. (1999). Protein synthesis by native chemical ligation: Expanded scope by using straightforward methodology. *Proc. Natl. Acad. Sci. USA* **96,** 10068–10073.

Heitz, A., Hernandez, J. F., Gagnon, J., Hong, T. T., Pham, T. T., Nguyen, T. M., Le-Nguyen, D., and Chiche, L. (2001). Solution structure of the squash trypsin inhibitor MCoTI-II. A new family for cyclic knottins. *Biochemistry* **40,** 7973–7983.

Hruby, V. J. (2002). Designing peptide receptor agonists and antagonists. *Nat. Rev. Drug Discov.* **1,** 847–858.

Iwai, H., Lingel, A., and Pluckthun, A. (2001). Cyclic green fluorescent protein produced *in vivo* using an artificially split PI-PfuI intein from Pyrococcus furiosus. *J. Biol. Chem.* **276,** 16548–16554.

Jennings, C. V., Rosengren, K. J., Daly, N. L., Plan, M., Stevens, J., Scanlon, M. J., Waine, C., Norman, D. G., Anderson, M. A., and Craik, D. J. (2005). Isolation, solution structure, and insecticidal activity of kalata B2, a circular protein with a twist: Do Mobius strips exist in nature? *Biochemistry* **44,** 851–860.

Kawai, Y., Kemperman, R., Kok, J., and Saito, T. (2004). The circular bacteriocins gassericin A and circularin A. *Curr. Protein Pept. Sci.* **5,** 393–398.

Kimura, R. H., Miao, Z., Cheng, Z., Gambhir, S. S., and Cochran, J. R. (2010). A dual-labeled knottin peptide for PET and near-Infrared fluorescence imaging of integrin expression in living subjects. *Bioconjug. Chem.* **21,** 436–444.

Kolmar, H. (2010). Engineered cystine-knot miniproteins for diagnostic applications. *Expert Rev. Mol. Diagn.* **10,** 361–368.

Lewis, R. J., and Garcia, M. L. (2003). Therapeutic potential of venom peptides. *Nat. Rev. Drug Discov.* **2,** 790–802.

Lovelace, E. S., Armishaw, C. J., Colgrave, M. L., Wahlstrom, M. E., Alewood, P. F., Daly, N. L., and Craik, D. J. (2006). Cyclic MrIA: A stable and potent cyclic conotoxin with a novel topological fold that targets the norepinephrine transporter. *J. Med. Chem.* **49,** 6561–6568.

Lovelace, E. S., Gunasekera, S., Alvarmo, C., Clark, R. J., Nevin, S. T., Grishin, A. A., Adams, D. J., Craik, D. J., and Daly, N. L. (2011). Stabilization of alpha-conotoxin AuIB: Influences of disulfide connectivity and backbone cyclization. *Antioxid. Redox Signal.* **14,** 87–95.

Luckett, S., Garcia, R. S., Barker, J. J., Konarev, A. V., Shewry, P. R., Clarke, A. R., and Brady, R. L. (1999). High-resolution structure of a potent, cyclic proteinase inhibitor from sunflower seeds. *J. Mol. Biol.* **290,** 525–533.

Maqueda, M., Galvez, A., Bueno, M. M., Sanchez-Barrena, M. J., Gonzalez, C., Albert, A., Rico, M., and Valdivia, E. (2004). Peptide AS-48: Prototype of a new class of cyclic bacteriocins. *Curr. Protein Pept. Sci.* **5,** 399–416.

Marx, U. C., Daly, N. L., and Craik, D. J. (2006). NMR of conotoxins: Structural features and an analysis of chemical shifts of post-translationally modified amino acids. *Magn. Reson. Chem.* **44,** S41–S50.

Miljanich, G. P. (2004). Ziconotide: Neuronal calcium channel blocker for treating severe chronic pain. *Curr. Med. Chem.* **11,** 3029–3040.

Millard, E. L., Nevin, S. T., Loughnan, M. L., Nicke, A., Clark, R. J., Alewood, P. F., Lewis, R. J., Adams, D. J., Craik, D. J., and Daly, N. D. (2009). Inhibition of neuronal nicotinic acetylcholine receptor subtypes by α-conotoxin GID and analogues. *J. Biol. Chem.* **284,** 4944–4951.

Mylne, J. S., Colgrave, M. L., Daly, N. L., Chanson, A. H., Elliott, A. G., McCallum, E. J., Jones, A., and Craik, D. J. (2011). Albumins and their processing machinery are hijacked for cyclic peptides in sunflower. *Nat. Chem. Biol.* **7,** 257–259.

Nielsen, C. H., Kimura, R. H., Withofs, N., Tran, P. T., Miao, Z., Cochran, J. R., Cheng, Z., Felsher, D., Kjaer, A., Willmann, J. K., and Gambhir, S. S. (2010). PET imaging of tumor neovascularization in a transgenic mouse model with a novel Cu-64-DOTA-knottin peptide. *Cancer Res.* **70,** 9022–9030.

Poth, A. G., Colgrave, M. L., Philip, R., Kerenga, B., Daly, N. L., Anderson, M. A., and Craik, D. J. (2011). Discovery of cyclotides in the Fabaceae plant family provides new insights into the cyclization, evolution, and distribution of circular proteins. *ACS Chem. Biol.* **6,** 345–355.

Rosengren, K. J., Daly, N. L., Plan, M. R., Waine, C., and Craik, D. J. (2003). Twists, knots, and rings in proteins. Structural definition of the cyclotide framework. *J. Biol. Chem.* **278,** 8606–8616.

Scanlon, M. J., Naranjo, D., Thomas, L., Alewood, P. F., Lewis, R. J., and Craik, D. J. (1997). The structure of a novel potassium channel toxin, κ-conotoxin PVIIA, from the venom of *Conus purpurascens*: Implications for the mechanism of channel block. *Structure* **5,** 1585–1597.

Schnölzer, M., Alewood, P., Jones, A., Alewood, D., and Kent, S. B. H. (1992). *In situ* neutralization in Boc-chemistry solid phase peptide synthesis. *Int. J. Pept. Protein Res.* **40,** 180–193.

Selsted, M. E. (2004). Theta-defensins: Cyclic antimicrobial peptides produced by binary ligation of truncated alpha-defensins. *Curr. Protein Pept. Sci.* **5,** 365–371.

Starzl, T. E., Klintmalm, G. B., Porter, K. A., Iwatsuki, S., and Schroter, G. P. (1981). Liver transplantation with use of cyclosporin a and prednisone. *N. Engl. J. Med.* **305,** 266–269.

Takahashi, H., Arai, M., Takenawa, T., Sota, H., Xie, Q. H., and Iwakura, M. (2007). Stabilization of hyperactive dihydrofolate reductase by cyanocysteine-mediated backbone cyclization. *J. Biol. Chem.* **282,** 9420–9429.

Tang, Y.-Q., Yuan, J., Ösapay, G., Ösapay, K., Tran, D., Miller, C. J., Ouellette, A. J., and Selsted, M. E. (1999). A cyclic antimicrobial peptide produced in primate leukocytes by the ligation of two truncated a-defensins. *Science* **286,** 498–502.

Terlau, H., and Olivera, B. M. (2004). Conus venoms: A rich source of novel ion channel-targeted peptides. *Physiol. Rev.* **84,** 41–68.

Trabi, M., and Craik, D. J. (2002). Circular proteins—No end in sight. *Trends Biochem. Sci.* **27,** 132–138.

Veiseh, M., Gabikian, P., Bahrami, S. B., Veiseh, O., Zhang, M., Hackman, R. C., Ravanpay, A. C., Stroud, M. R., Kusuma, Y., Hansen, S. J., Kwok, D., Munoz, N. M., *et al.* (2007). Tumor paint: A chlorotoxin:Cy5.5 bioconjugate for intraoperative visualization of cancer foci. *Cancer Res.* **67,** 6882–6888.

Williams, N. K., Liepinsh, E., Watt, S. J., Prosselkov, P., Matthews, J. M., Attard, P., Beck, J. L., Dixon, N. E., and Otting, G. (2005). Stabilization of native protein fold by intein-mediated covalent cyclization. *J. Mol. Biol.* **346,** 1095–1108.

Wishart, D. S., Sykes, B. D., and Richards, F. M. (1992). The chemical shift index: A fast and simple method for the assignment of protein secondary structure through NMR spectroscopy. *Biochemistry* **31,** 1647–1651.

CHAPTER FOUR

Peptide Discovery Using Bacterial Display and Flow Cytometry

Jennifer A. Getz, Tobias D. Schoep, *and* Patrick S. Daugherty

Contents

1. Introduction 76
   1.1. Principles of peptide isolation using bacterial display libraries 76
   1.2. Bacterial peptide display scaffolds 76
   1.3. Bacterial peptide display systems 79
2. Protocols 79
   2.1. Construction of large bacterial display peptide libraries 79
   2.2. Sorting bacterial display libraries to identify peptides that bind a target protein 86
   2.3. Troubleshooting the library sorting 88
   2.4. Measuring the binding affinity of displayed peptides 89
   2.5. Peptide affinity maturation 90
   2.6. Identification of cell-binding peptides 90
   2.7. Protease substrate identification using cellular libraries of peptide substrates (CLiPS) 93
3. Conclusions 94
Acknowledgments 95
References 95

Abstract

Peptides are increasingly used as therapeutic and diagnostic agents. The combination of bacterial cell-surface display peptide libraries with magnetic- and fluorescence-activated cell sorting technologies provides an efficient and highly effective methodology to identify and engineer peptides for a growing number of molecular recognition applications. Here, detailed protocols for both the generation and screening of bacterial display peptide libraries are presented. The methods described enable the discovery and evolutionary optimization of protein-binding peptides, cell-specific peptides, and enzyme substrates for diverse biotechnology applications.

Department of Chemical Engineering, Institute for Collaborative Biotechnologies, University of California, Santa Barbara, California, USA

Methods in Enzymology, Volume 503
ISSN 0076-6879, DOI: 10.1016/B978-0-12-396962-0.00004-5

1. Introduction

1.1. Principles of peptide isolation using bacterial display libraries

Biological display technologies have been applied to identify peptides that bind to various targets including proteins, whole cells, and synthetic materials. Bacterial surface display is a well-established methodology for the discovery and optimization of peptides with desired functions for an expanding group of applications. Typically, bacteria that display target-binding peptides are enriched from large libraries exceeding one billion members using sequential magnetic-activated cell sorting (MACS) and fluorescence-activated cell sorting (FACS) (Fig. 4.1). Thus, bacterial display peptides libraries are distinct from other peptide library technologies, including bacteriophage display, in that FACS instrumentation allows real-time analysis of the properties of peptides displayed on the cell surface during screening; measurable properties include target-binding affinity and specificity, and peptide stability to proteases. The ability to measure these properties and tune the stringency of the screen enables efficient identification of peptides with desired properties for a wide range of purposes (Daugherty, 2007). In particular, bacterial display libraries have been used to identify peptides with potential utility in therapeutic or diagnostic applications (Table 4.1). Here, we describe detailed methods that enable the selection of peptides from bacterial display libraries that bind or otherwise interact with a target.

1.2. Bacterial peptide display scaffolds

A variety of display scaffolds have been exploited to present peptides on the surface of bacteria. Scaffolds have been developed for both Gram-negative and Gram-positive bacteria, although Gram-negative based systems have been more extensively utilized (Jostock and Dubel, 2005). General considerations involved in bacterial display and library screening have been reviewed (Daugherty, 2007; Strauch and Georgiou, 2009), and improvements to display scaffolds and screening methods are rapidly evolving (Binder *et al.*, 2010; Chen *et al.*, 2008; Kronqvist *et al.*, 2008). Peptides can be displayed on the surface of *Escherichia coli* by: insertion into outer membrane proteins (e.g., OmpA, OmpX) (Bessette *et al.*, 2004) or flagellar proteins (FliC) (Lu *et al.*, 1995); fusion to surface exposed N-termini (e.g., IgA protease) (Wentzel *et al.*, 1999); or fusion to both surface exposed N- and C-termini (e.g., circularly permuted OmpX (CPX)) (Rice *et al.*, 2006). The use of terminally fused peptide libraries is especially useful for identifying peptides that are more likely to retain function when removed from

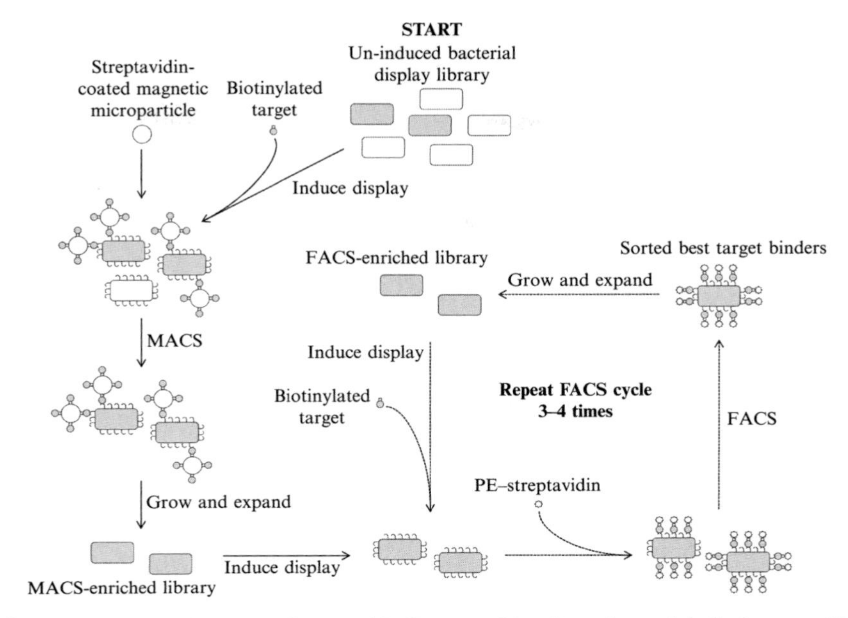

Figure 4.1 Identification of target-binding peptides from bacterial display peptide libraries. Peptide-displaying bacteria are incubated with biotinylated target and streptavidin-functionalized magnetic microparticles, and selected using MACS. The recovered cells are amplified by growth and then screened for binding to labeled targets using FACS. Additional cycles of FACS and regrowth are performed (dotted lines) to enrich bacteria that display peptides with the desired affinity and specificity. PE, phycoerythrin.

their scaffold. Each display scaffold, however, imposes differing constraints upon passenger peptides. Consequently, the methods presented are specific to the host bacterial strain and display scaffold.

Peptides can be displayed in a non-constrained fashion, or in a constrained conformation with one or more stabilizing disulfide bonds. Typically, constrained peptides exhibit increased binding affinity and specificity (Getz *et al.*, 2011; Smith and Petrenko, 1997) since they can adopt fewer conformations when compared to linear peptides. Additionally, the use of structural constraints may better preserve peptide-binding properties when peptides are characterized independent of the display scaffold. Structurally constrained peptide libraries have proven superior for generating high-affinity and high-stability peptides in several studies (Getz *et al.*, 2011; Orning *et al.*, 2006; Uchiyama *et al.*, 2005). Nevertheless, binders exhibiting disulfide constraints can often be directly identified from fully random, non-constrained libraries (Kenrick and Daugherty, 2010; Rojas *et al.*, 2008).

Table 4.1 Applications of bacterial display peptide libraries

Achievement	Potential use	Organism	Scaffold	Example reference
Identification of tumor cell-specific peptides	Tumor targeted delivery	*E. coli*	FliTrx	Zitzmann *et al.* (2005) and Li *et al.* (2008)
Identification of breast tumor cell-specific peptides		*E. coli*	OmpX (CPX)	Dane *et al.* (2009)
Identification of high-affinity VEGF-binding peptides	Peptide therapeutic agonists/antagonists	*E. coli*	OmpX (eCPX)	Kenrick and Daugherty (2010)
Identification of high-affinity TNFα-binding affibodies		*Staphylococcus. carnosus*	Protein A (fragment)	Kronqvist *et al.* (2008)
Identification of high-affinity CTLA-4-binding anticalin		*E. coli*	EspP autotransporter/ lipocalin	Binder *et al.* (2010)
Epitope mapping of antigens	Validation of monoclonal therapeutic antibody specificity	*S. carnosus*	Protein A (fragment)	Rockberg *et al.* (2008)
Identification of protease-activated proligands	Pathology mediated activation of peptide based therapeutics	*E. coli*	OmpX (eCPX)	Thomas and Daugherty (2009)

1.3. Bacterial peptide display systems

Various display systems have been used to generate peptides that bind a target molecule (Jostock and Dubel, 2005; van Bloois *et al.*, 2011). A majority of previously reported bacterial display scaffolds and libraries are not available commercially and must be obtained by request or reconstructed. The protocols presented below are specific to peptide libraries displayed using the eCPX scaffold (Rice and Daugherty, 2008). The eCPX scaffold is a circularly permuted variant of the outer membrane protein OmpX (Rice *et al.*, 2006) with additional mutations that aid passenger presentation on the cell surface (Rice and Daugherty, 2008). Peptide libraries can be displayed at either the N- or C-terminus of the protein and can be utilized for diverse applications (Fig. 4.2). We and others have demonstrated the use of eCPX to identify tumor cell-specific peptides (Dane *et al.*, 2009), generate and affinity mature protein-binding peptides (Kenrick and Daugherty, 2010; Sun *et al.*, 2009), construct protease-activated VEGF-binding ligands (Thomas and Daugherty, 2009), and identify determinants of antibody-epitope binding specificity (Hall and Daugherty, 2009).

2. Protocols

2.1. Construction of large bacterial display peptide libraries

The following protocol provides detailed instructions for constructing bacterial display libraries with the eCPX display scaffold (Rice and Daugherty, 2008). The pBAD33-eCPX plasmid (available from Addgene) includes an arabinose operon (araBAD) promoter and the chloramphenicol acetyltransferase gene to confer antibiotic resistance to chloramphenicol (CM) (Fig. 4.3). The following sections describe the preparation of the vector and insert DNA, performing test ligations, library construction, and production of electrocompetent *E. coli* cells.

2.1.1. Preparation of vector DNA

1. Isolate at least 10 μg of pBAD33-eCPX plasmid at a concentration of approximately 200 ng/μL.
2. Digest the vector DNA with 10 units of *Sfi*I restriction enzyme per microgram of DNA in a 200-μL reaction volume using the appropriate buffer and BSA concentration. Incubate at 50 °C for 4 h.
3. Prepare a 0.7% (w/v) agarose gel; rinse and fill the gel box with TAE buffer. Load DNA ladder into lane 1 and the digestion reaction in the remaining lanes (less than 5 μg each) and run for 45 min at 100 V. SYBER Safe DNA gel stain (Invitrogen) can be used with a blue light

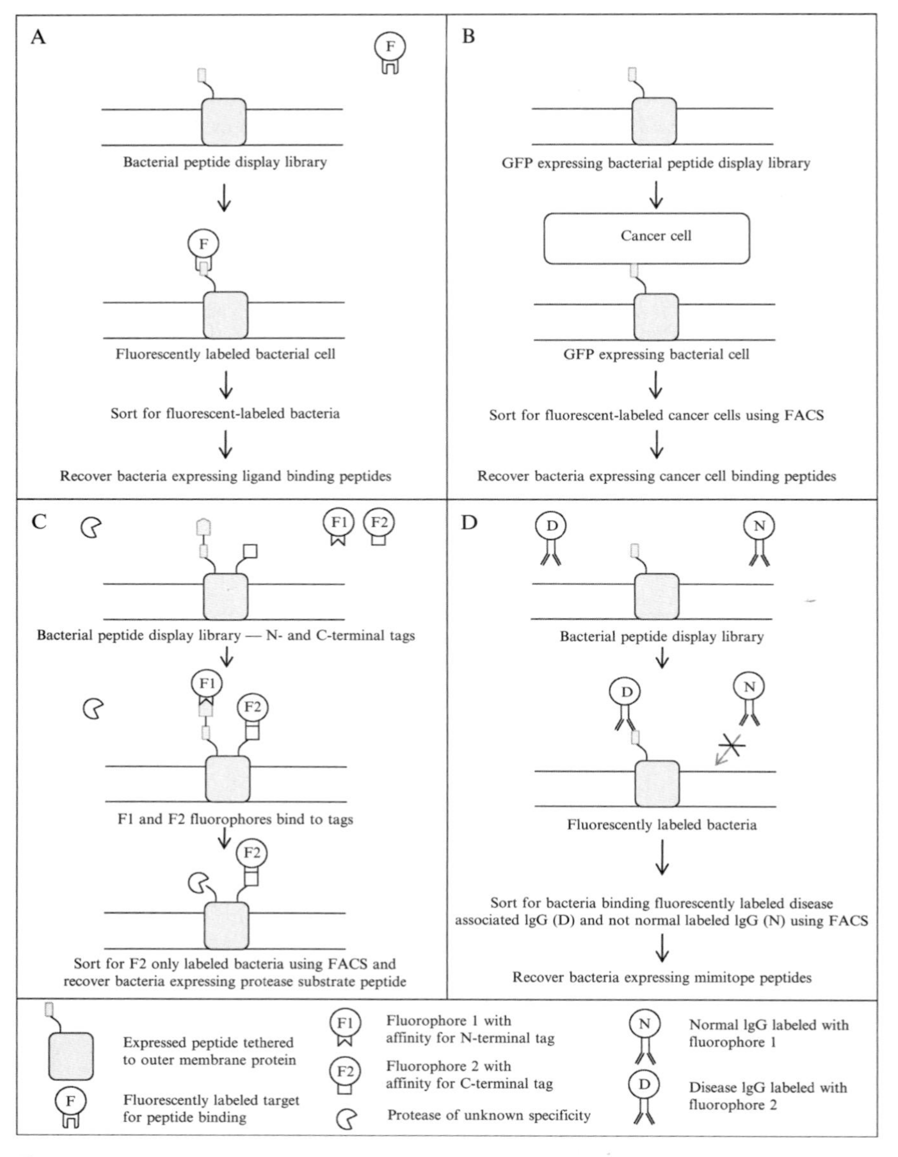

Figure 4.2 Bacterial peptide surface display and analysis strategies. (A) Identification of peptide binders to a target protein. (B) Identification of peptides that bind cultured cells or tissue. (C) Identification of protease substrates using cellular libraries of peptide substrates (CLiPS). (D) Identification of peptides (mimitopes) that bind disease-associated antibodies.

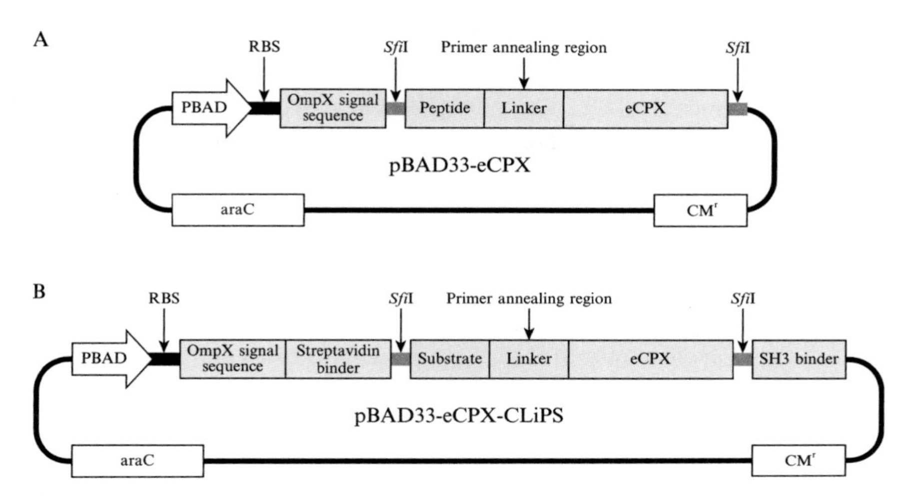

Figure 4.3 Plasmids for bacterial display. (A) Map of plasmid pBAD33-eCPX for expression of peptide libraries fused to the N-terminus of the display scaffold eCPX. The plasmid carries a chloramphenicol resistance marker (CM^r) and an arabinose promoter (PBAD). (B) Map of plasmid pBAD33-eCPX-CLiPS for expression of substrate libraries with tags on both the N- and C-termini of eCPX.

transilluminator to reduce nicking caused by the traditional use of ethidium bromide and UV light.

4. Excise the bands corresponding to the double-cut plasmid (~5400bp) using a razor blade; extract and purify the DNA using the Zymoclean gel DNA recovery kit (each column can bind up to 5 μg of DNA); and elute with a total volume (across all columns) of 50 μL of sterile MilliQ water.
5. For incompletely digested vector DNA, dephosphorylation of the vector can prevent ligation of single-cut plasmid and therefore reduce the amount of template in the library. Add 1/10 volume of 10× Antarctic Phosphatase buffer (NEB) and 1 μL (5 units) of Antarctic Phosphatase per 1–5 μg DNA. Incubate at 37 °C for 60 min and heat inactivate the enzyme for 5 min at 65 °C.

2.1.2. Preparation of insert DNA

To construct a random library on the N-terminus of eCPX, oligonucleotide primers can be designed that anneal to DNA coding the short linker region (GGQSGQ) upstream of eCPX and extend to the *Sfi*I site immediately downstream of the OmpX signal sequence (Fig. 4.3). The reverse primer anneals downstream of the *Sfi*I site on the C-terminal end of eCPX. Degenerate codons of the form NNS (N=A,G,C,T; S=G,C) are utilized to create the randomized region. The forward and reverse primers used to construct a 15-mer peptide library are listed below:

Forward primer. ACTTCCGTAGCTGGCCAGTCTGGCCAGNNS NNSNNSNNSNNSNNSNNSNNSNNSNNSNNSNNS NNSNNSNNSNNSGGAGGGCAGTCTGGGCAG TCTG

Reverse primer. GGCTGAAAATCTTCTCTC

1. Dilute the oligonucleotide primers to 10 μ*M*.
2. For the PCR reaction, mix the following and then aliquot into 10 0.2 mL PCR tubes:
   325 μL sterile MilliQ water
   50 μL 10× DNA polymerase buffer
   5 μL diluted vector DNA (~100 ng total of plasmid)
   15 μL forward (5′) primer (10 μ*M* stock)
   15 μL reverse (3′) primer (10 μ*M* stock)
   50 μL dNTPs (2 m*M* each)
   30 μL $MgSO_4$ (25 m*M*)
   10 μL proofreading, hot start DNA polymerase, 1 unit/μL (KOD Hot-Start Polymerase, EMD Chemicals)
3. Perform a PCR with the following conditions:
   a. 94 °C for 2 min
   b. 94 °C for 30 s
   c. 55 °C for 30 s
   d. 68 °C for 1 min
   e. Repeat steps b through d for 25 cycles
4. Check the insert PCR product (~600–700 bp) by analyzing 5 μL of the reaction on an agarose gel. Purify the remaining reaction using the Zymo DNA Clean and Concentrator kit. The PCR should yield approximately 10 μg of insert DNA.
5. Digest 10 μg (or the amount recovered) of insert DNA with 10 units of *Sfi*I per microgram of DNA for 4 h at 50 °C, purify using the Zymo DNA Clean and Concentrator kit (each column can bind up to 5 μg of DNA), and elute with a total volume (across all columns) of 50 μL of sterile MilliQ water.

2.1.3. Performing test ligations

A protocol for making high efficiency electrocompetent *E. coli* cells is described at the end of Section 2.1.4. The construction of large libraries is dependent upon electrocompetent cells with a transformation efficiency of at least 1×10^{10} CFU/μg pUC18.

1. Determine the concentration of the digested vector and insert DNA. To do this, dilute the DNA 1:50 and measure the absorbance at 260 and 280 nm (the absorbance reading should be in the linear range between 0.01 and 1). Then, calculate the DNA concentration using the following conversion: $A_{260\ \mathrm{nm}}$ of 1.0=50 ng/μL DNA. The DNA purity, as determined by the ratio of absorbance at 260 nm versus 280 nm, should be greater than 1.8. With perfect recovery, the vector and insert concentrations should be approximately 200 ng/μL, but concentrations as low as 50 ng/μL are sufficient.
2. For the test ligations, try multiple molar ratios of insert to vector (*I*: *V*) in the reactions: 0:1 (vector only control), 1:1, 3:1, and 5:1. Use the same mass of vector DNA (~50 ng) for each ligation reaction and adjust the amount of insert to achieve the desired ratio. Also, make sure to add less than 100 ng of total DNA (insert plus vector) per 20 μL ligation reaction. Equation (4.1) can be used to calculate the amount of insert DNA for each reaction:

$$\mathrm{ng}_V \times \frac{\mathrm{bp}_I}{\mathrm{bp}_V} \times I\!:\!V = \mathrm{ng}_I \qquad (4.1)$$

   where ng_V is the mass (ng) of vector in the ligation reaction; bp_I and bp_V are the number of base pairs in the insert and vector DNA, respectively; $I\!:\!V$ is the ratio of insert to vector; and ng_I is the mass (ng) of insert.
3. Each test ligation reaction should have a volume of 20 μL with the following components: 4 μL of 5× T4 DNA ligase buffer, vector DNA, insert DNA, sterile MilliQ water, and 1 μL of T4 DNA ligase (Invitrogen).
4. Incubate ligation reactions at 14 °C for 16 h and heat inactivate the ligase by incubation at 70 °C for 10 min.
5. Desalt 10 μL of the ligation reactions using a membrane filter (Millipore "V" Series, 0.025 μm) on sterile MilliQ water for 2 h. The reaction volume on the surface of the filter will roughly double in this time period.
6. Thaw four 70 μL aliquots of electrocompetent MC1061 cells on ice. Add 10 μL of each chilled, desalted ligation to the cell aliquots and transfer to an ice cold 1 mm electroporation cuvette. Pulse at 1.8 kV, 50 μF, and 100 Ω (Bio-Rad Gene Pulser) and rinse the cuvette twice with 0.75 mL of warm SOC media to recover the bacteria and incubate with shaking at 37 °C for 1 h.
7. Plate 100 μL of 1:10^2, 1:10^3, and 1:10^4 dilutions of each of the transformations (1:10 and 1:10^2 for the vector only control transformation) on LB-agar plates supplemented with 34 μg/mL CM as the selection antibiotic.
8. Incubate overnight at 37 °C and on the following day, count the number of colonies on each of the plates to calculate the total number of transformants. For example, 100 CFU on the 1:10^3 dilution plate would equal

1.2×10^8 CFU/µg (assuming 12.5 ng vector DNA per transformation). The transformation efficiency should be between 10^7 and 10^9 transformants/µg of vector DNA. Determine which insert to vector ratio in the ligation produces the maximum number of transformants. Also, the number of transformants on the vector-only control plate divided by the transformants under optimum ligation conditions can be used to estimate the fraction of background or template DNA that will result in the library. If the library background is greater than 10%, redo the vector digestion with an increased enzyme concentration and use Antarctic Phosphatase to prevent the ligation of single-cut plasmid.

2.1.4. Library construction

1. Using the best insert to vector ratio, scale-up the volume of the ligation reaction (e.g., by 50–100×) depending on the size of the desired library and the amount of digested insert and vector DNA available. Using 5–10 µg of vector DNA in the ligation should be adequate.
2. Divide the larger reaction volume into 100 µL aliquots, incubate for 16 h at 14 °C, and deactivate the ligase by heating to 70 °C for 10 min.
3. To reduce the volume after ligation, precipitate the DNA (divided into 400 µL aliquots) with 1/10 the volume of 3 *M* sodium acetate (pH 5.2), 2 µL oyster glycogen (20 mg/mL), and 2.2 volumes of ethanol (alternatively use Novagen Pellet Paint Co-Precipitant, or use the Zymo DNA Clean and Concentrator kit to skip the precipitation).
4. Mix and incubate for 1–2 h at −80 °C or overnight at −20 °C.
5. Centrifuge at maximum speed for 15 min at 4 °C. Remove the supernatant and allow the pellet to dry until the edges begin to turn translucent.
6. Resuspend the pellet in 50 µL of sterile MilliQ water.
7. Desalt the ligated DNA in ~10–15 µL aliquots using membrane filters floating on sterile MilliQ water for 2 h.
8. Chill aliquots of 0.5–3 µg of desalted DNA on ice (the volume should be <30 µL). Thaw four or more (depending on the number of transformations) 250 µL aliquots of electrocompetent cells on ice and add the desalted DNA as soon as the competent cells are thawed. Transfer the mixture to an ice cold 2 mm electroporation cuvette and pulse at 2.5 kV, 50 µF, and 100 Ω. Rinse the cuvette three times with 1 mL of warm SOC media and transfer the cells to a 125-mL flask containing 7 mL of SOC (10 mL total volume per flask).
9. Repeat procedure for the remaining transformations and incubate the flasks with shaking at 37 °C for 1 h.
10. If all the time constants during electroporation were approximately the same, pool the transformations into a single flask. Plate 100 µL of $1{:}10^3$, $1{:}10^4$, and $1{:}10^5$ dilutions on LB-agar/CM plates. Repeat the series

of dilutions and prepare duplicate plates. Incubate the plates overnight at 37 °C.

11. Add the pooled transformations to a 2-L flask containing 500 mL LB/CM with 0.2% (w/v) sterile-filtered glucose. Incubate the flask at 37 °C until the culture reaches an OD_{600} of 2 (3–8 h depending on library size).
12. Inoculate a sterile 2-L flask containing 400 mL LB/CM and 0.2% glucose with 100 mL of the transformation culture and grow overnight.
13. Centrifuge the remaining 400 mL of culture at 3000 × *g* for 15 min.
14. Resuspend the pellet in 10 mL SOB supplemented with 15% (v/v) glycerol. Aliquot into cryovials and freeze at −80 °C. Each frozen stock should contain at least 10 times the number of independent transformants.
15. The following morning, count the colonies on the plates to determine the library size (number of independent transformants).
16. Isolate plasmid DNA from the overnight culture of the newly constructed library using a maxiprep kit.

2.1.5. Making electrocompetent cells

Day 1

1. Thoroughly rinse centrifuge and Pyrex bottles with MilliQ water and autoclave the following:
   a. 500 mL of media (5 g tryptone, 2.5 g yeast extract, 2 g NaCl) in a 2-L flask.
   b. 2× 500 mL MilliQ water in rinsed bottles.
   c. 300 mL 10% (v/v) glycerol (270 mL water+37.5 g glycerol).
   d. 500 mL centrifuge bottle.
   e. 1.5 mL microcentrifuge tubes (~20).
2. Start a 5 mL overnight culture from a colony of the desired *E. coli* strain. For bacterial display library construction, we use *E. coli* strain MC1061.

Day 2

1. Refrigerate the bottles of water and glycerol and freeze the centrifuge bottle and labeled microcentrifuge tubes at −20 °C.
2. Add 0.2% (w/v) sterile-filtered glucose to 500 mL of media and inoculate with 5 mL of overnight culture.
3. Grow at 37 °C until the OD_{600} reaches 0.6. This occurs after 2–2.5 h for the MC1061 strain.
4. Chill the flask in an ice water bath for 30 min. For the remainder of the steps, it is important to keep the cells on ice.
5. Transfer the cells to the cold 500 mL centrifuge bottle and centrifuge the cells at 3000 × *g* for 20 min at 4 °C.

6. Pour off the supernatant being careful to not lose any of the cell pellet. Add 5–10 mL of ice cold sterile MilliQ water and resuspend the pellet by swirling the centrifuge bottle in an ice bath. Then, add the remaining water (total volume of 500 mL) and centrifuge at 3000 × *g* for 20 min at 4 °C.
7. Repeat step 6.
8. Pour off the supernatant and add 5–10 mL of cold 10% glycerol and resuspend the pellet as before. Fill the bottle halfway (~250 mL) with 10% glycerol and centrifuge at 3000 × *g* for 10 min at 4 °C.
9. Pour off as much liquid as possible without disturbing the pellet and gently resuspend the cell pellet in the remaining glycerol by swirling the centrifuge bottle in an ice bath (~1.5 mL of total liquid).
10. Aliquot 70 μL of cell suspension in 15–20 of the cold 1.5 mL centrifuge tubes and freeze immediately at −80 °C. For library construction, divide into 250 μL aliquots with a cell density of approximately 2×10^{11} cells/mL.

2.1.6. Measurement of cell competency

1. Thaw an aliquot of electrocompetent cells on ice and add 10 pg of pUC18.
2. Transfer to a cold 1 mm electroporation cuvette and pulse at 1.8 kV, 50 μF, and 100 Ω.
3. Rinse the cuvette twice with 0.75 mL of SOC media and grow at 37 °C for 1 h.
4. Plate 100 μL of $1{:}10^2$ and $1{:}10^3$ dilutions of the transformation on LB-agar plates supplemented with 100 μg/mL ampicillin.
5. Grow overnight at 37 °C and count the colonies to determine the transformation efficiency of the electrocompetent cells. A count of 100 colonies on platings of the $1{:}10^2$ dilution corresponds to a transformation efficiency of 1.5×10^{10} CFU/μg pUC18.

2.2. Sorting bacterial display libraries to identify peptides that bind a target protein

Below is a general protocol for labeling peptide-displaying bacteria with a biotinylated target protein. The growth and labeling conditions specified are for the eCPX display scaffold. Since eCPX enables presentation of peptides on either scaffold terminus, an affinity tag can be fused to the C-terminus to enable improved resolution of scaffold expression and binding affinity (Kenrick and Daugherty, 2010). For libraries with a diversity greater than 10^7 members, a magnetic selection (MACS) is used to pre-enrich bacteria which display peptides that bind a target protein and to deplete binders to the secondary label streptavidin. One or two rounds of MACS will reduce the population to a reasonable size for sorting by flow cytometry. Detailed protocols for library screening using MACS and FACS have been described elsewhere (Kenrick *et al.*, 2007).

1. Grow library population or single clone overnight at 37 °C in LB/CM supplemented with 0.2% (w/v) sterile-filtered glucose (glucose is not required for single clones).
2. Subculture at a 1:50 dilution into 5 mL LB/CM and grow for 2 h at 37 °C.
3. Induce for 1 h at 37 °C with 0.04% (w/v) L-arabinose.
4. If working with a library population, determine the amount of cells needed to oversample the calculated library size by 20- to 50-fold. Be sure to use at least 10 μL of cells so that a cell pellet is easily visible following centrifugation.
5. Centrifuge the cells at 3000×*g* for 5 min and remove the supernatant.
6. Resuspend the pellet in 100 μL of the biotinylated target protein solution in PBS supplemented with 0.25% (w/v) bovine serum albumin (BSA). In general, a protein concentration of 100–500 n*M* is a good starting point for the first round of sorting. The concentration of target can be lowered to increase the stringency in later rounds of sorting. Also, prepare unlabeled cells as a negative control.
7. Incubate cells with the target protein at 4 °C for 45 min, centrifuge the cells at 3000×*g* for 5 min, and remove the supernatant. *Note*: To favor specificity for the target, unlabeled protein mixtures representing diverse specificities (e.g., purified human IgG) can be added before or during this step (Hall and Daugherty, 2009).
8. Resuspend the pellet in 100 μL of 15 n*M* streptavidin-*R*-phycoerythrin (SAPE) solution and incubate at 4 °C for 45 min. This step is not required if the protein is directly labeled with a fluorescent dye. Also, a cell sample labeled with SAPE only (skip target protein incubation step) is an important control to test for streptavidin-binding peptides. Neutravidin–phycoerythrin or a fluorescently labeled anti-biotin antibody can be used as an alternative secondary detection agent if streptavidin-binding peptides become enriched during sorting.
9. Centrifuge the cells at 3000×*g* for 5 min, remove the supernatant, and resuspend in PBS (500 μL) for analysis or sorting on the flow cytometer.
10. Load the unlabeled cells (negative control) and adjust the forward scatter (FSC) and side scatter (SSC) detector voltages until the entire cell population is clearly visible on the plot.
11. For the phycoerythrin dye, the flow cytometer should be equipped with a 488-nm laser. Adjust the 576-nm PMT voltage until the mean cell fluorescence of the negative control is at a background level of ~100.
12. On the fluorescence plot, set the sort gate so that it encompasses 0.1–1% of the negative control population. Also, adjust the gate to exclude as many streptavidin-binding peptides as possible.

13. Analyze the remaining samples and proceed to sorting (collect the sorted cells in SOC). Oversample the library population by 5- to-10-fold to ensure representation of all the members.

2.3. Troubleshooting the library sorting

The failure to enrich binders to the target can result from a number of potential causes. If library enrichment of binding clones does not occur, try increasing the concentration of target protein and/or re-sort using a population isolated from an earlier round of selection. Also, determine whether the target is properly biotinylated using a degree of labeling kit (optimal degree of labeling is ~2–4 mol biotin per mol target protein) or with bacteria displaying a known protein-binding peptide, if available. If possible, analyze the positive control bacteria *after* sorting the library population to prevent contamination with the positive control clone; alternatively, run bleach through the system following the control clone to sanitize the tubing and rinse thoroughly with water prior to sorting.

Avidity and rebinding effects can dramatically increase apparent affinities measured on the cell surface when working with homo-multimeric target proteins. To reduce the avidity effects, the induction time can be shortened (minimum of 10 min) to reduce the number of peptides displayed on the bacterial surface. To diminish rebinding effects, a decoy protein that binds to a tag on the C-terminus of eCPX can increase the steric hindrance and prevent the target protein from unbinding and then reattaching to a neighboring peptide (Kenrick and Daugherty, 2010). Finally, when measuring dissociation rate constants, as discussed in the next section, an excess concentration of unlabeled competitor can be used so that once a protein dissociates from the cell surface, it is blocked from rebinding. This strategy is especially useful in later rounds of sorting with a multimeric target for isolating peptides that have slow off rates and therefore higher affinity.

One common problem is the enrichment of peptides that bind to the secondary fluorescent probe. This is especially an issue when using streptavidin, which has four biotin-binding sites that can promote tight binding to multiple peptides on the bacterial surface. The number of streptavidin binders can be reduced through depletion sorting for nonbinding clones to the secondary label, or by using streptavidin-coated magnetic beads to remove streptavidin-binding clones. Also, the secondary label can be alternated between streptavidin–phycoerythrin and either neutravidin–phycoerythrin or a fluorescently tagged anti-biotin antibody in each cycle of sorting. If binding to the secondary label remains a problem, eliminate the secondary labeling step and directly conjugate a fluorescent dye (e.g., AlexaFluor 488) to the target protein. When using protein targets conjugated with a fluorophore, the mean cell fluorescence of the binding clones will be significantly lower compared to that obtained by labeling with streptavidin–phycoerythrin.

2.4. Measuring the binding affinity of displayed peptides

Flow cytometry can be used to determine apparent equilibrium dissociation constants (K_d^{APP}) or dissociation rate constants (k_{off}) for peptides displayed on the cell surface (Fig. 4.4). Dissociation constants are to be considered apparent, since they may differ from K_d values measured in solution due to many factors including avidity and rebinding effects. To measure K_d^{APP}, the cells are labeled with a concentration series of target protein (three- to five-fold titration series). The normalized mean cell fluorescence is plotted as a function of the protein concentration (logarithmic scale), and the data can be fit to Eq. (4.2) to determine the K_d^{APP}:

$$\frac{\mathrm{FL} - \mathrm{FL}_{bg}}{\mathrm{FL}_{max} - \mathrm{FL}_{bg}} = \frac{[\mathrm{T}]}{[\mathrm{T}] + K_d^{APP}} \tag{4.2}$$

FL is the mean cell fluorescence of the bacterial cell population, FL_{bg} is the background cell fluorescence of unlabeled bacteria, FL_{max} is the maximum observed fluorescence assuming that every peptide is bound to a target molecule, $[T]$ is the target protein concentration, and K_d^{APP} is the apparent or cell-surface equilibrium dissociation constant (Boder and Wittrup, 1998).

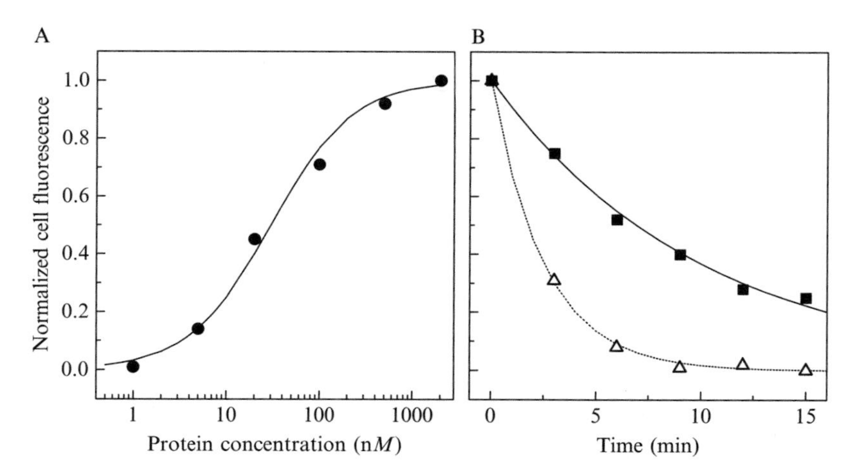

Figure 4.4 Measurement of binding kinetics and affinity. (A) Equilibrium-binding isotherm (hypothetical data) with the normalized cell fluorescence plotted versus the target protein concentration. The data are fit to Eq. (4.2) to calculate the K_d^{APP} for a clonal population of peptide-displaying bacteria. (B) Off-rate analysis (hypothetical data) of two different clones fit to Eq. (4.3) to calculate apparent k_{off} values. A high concentration of unlabeled competitor is added at the initial time point to prevent rebinding of the target protein to the displayed peptide and therefore allows for a more accurate representation of the solution behavior of the peptide.

Another method for ranking individual clones is to determine their relative dissociation rate constants (off rates or k_{off}). Assuming the peptides have similar association rates, their affinity will be strongly influenced by how quickly the target dissociates from the peptide. To quantify the dissociation rate, cells are incubated with saturating concentrations of biotinylated target protein. The cells are then washed once in ice cold buffer and resuspended in an excess of unlabeled competitor to prevent the rebinding of the biotinylated protein to the displayed peptides. At numerous time points, an aliquot is removed and labeled with streptavidin–phycoerythrin for analysis on the flow cytometer. By plotting the normalized cell fluorescence versus time, the dissociation rate constant can be determined using Eq. (4.3):

$$\frac{FL - FL_{bg}}{FL_0 - FL_{bg}} = e^{-k_{off}t} \tag{4.3}$$

FL is the observed mean cell fluorescence, FL_0 is the initial mean cell fluorescence, FL_{bg} is the background cell fluorescence after complete dissociation, t is the time following addition of the unlabeled competitor, and k_{off} is the dissociation rate constant (Boder and Wittrup, 1998).

2.5. Peptide affinity maturation

The affinity of peptides isolated from libraries is typically nonoptimal but can be readily improved using bacterial display in conjunction with FACS. When compared to phage display, the use of cell display and FACS can be advantageous to identify optimal screening conditions and to rapidly characterize the improved clones. For affinity maturation, a second-generation library is typically designed based on peptide sequences identified in the initial screen. The second-generation library is sorted using increased stringency to identify peptide variants with improved affinity for the target. There are several strategies to affinity mature peptides; for example, the focused library can be based on a consensus motif found from the initial library sorting or can be a soft randomization of a single binder. Affinity maturation was recently applied to VEGF binders using bacterial surface display by creating focused libraries based on the naïve 15-mer screen (Kenrick and Daugherty, 2010). Complete affinity maturation protocols have been described elsewhere (Fairbrother *et al.*, 1998; Fleming *et al.*, 2005; Levin and Weiss, 2006; Yu and Smith, 1996).

2.6. Identification of cell-binding peptides

To discover peptide ligands that bind specifically to a given cell type, bacterial display libraries can be rendered intrinsically fluorescent by co-expression of GFP intracellularly (Dane *et al.*, 2006, 2009). Thus, the binding of bacteria to the target cell type can be readily detected by cytometry to enable simple and efficient screening against many different cell types including stem cells (Little *et al.*, 2011). The library can be counter screened against cell lines that have different receptors to improve the specificity of the ligands for a particular cell type. For example, bacterial display was used to identify peptides that specifically recognized human breast cancer cells over normal cells and could also distinguish breast cancer subtypes (Dane *et al.*, 2006, 2009).

To identify cell-binding peptides (Fig. 4.5), a control cell sample containing only the target mammalian cells is needed to properly set the detector voltages to measure background fluorescence levels. In addition, a sample containing only bacteria enables a clear measurement of the green fluorescence signal and enables location of the bacteria on the FSC–SSC plot. A sample of mammalian cells with GFP-expressing bacteria that express only the display scaffold (no displayed peptide) is crucial for both determining the extent to which the bacterial surface interacts with the mammalian cells as well as validating the sort gate. If a high background signal is observed with this negative control, consider adjusting the wash buffer or number of washes to reduce nonspecific binding of the bacteria to the mammalian cells. With an efficient cell-binding peptide, it is possible for 10 or more bacteria to bind to an individual mammalian cell. In addition to cytometry, these samples can be easily imaged using fluorescence microscopy to visualize the interaction of the GFP bacteria with the target cells.

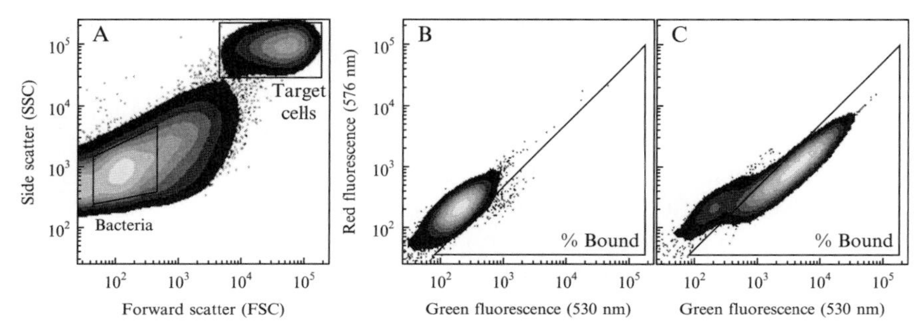

Figure 4.5 Screening for cell-binding peptides. (A) Scatter plot (side vs. forward scatter with both on a logarithmic scale) with the detector voltages adjusted so that the bacteria and target cells are two complete and distinct populations. (B) Bacteria expressing GFP, but without a displayed peptide, show negligible binding to the target cells. This plot only displays events within the target cell gate; the 576-nm fluorescence is due to GFP cross talk with the 576-nm channel (only one fluorophore is present in this system). (C) Bacteria expressing a receptor-specific peptide bind efficiently to the tumor cells as shown by the large increase in green fluorescence. The events in the % bound gate represent target cells that have one or more bacteria bound to their surface, and the majority of tumor cells are labeled with green bacteria due to the peptide ligand.

1. Follow the general growth and induction procedures (steps 1–4) in the Target Protein Sorting protocol (Section 2.2).
2. While the bacteria are growing, harvest the mammalian cells from dishes by scraping, or by using another trypsin-independent dissociation method, to preserve the integrity of cell-surface receptors. Be sure to rinse off any media containing antibiotic using PBS.
3. Count the target cells using a hemocytometer. The cell suspension should have >90% viability as determined by the trypan blue cell-viability test.
4. Take an aliquot of the cell suspension adequate to obtain between 2 and 5×10^5 cells per sample, and spin down the cells in a clinical centrifuge at $150\times g$ for 7 min.
5. Resuspend the cell pellet in PBS (50 μL for each sample).
6. After induction, dilute the bacteria 1:2 and measure the OD_{600}; assume a value of 1 corresponds to 1×10^9 bacterial cells/mL. Determine the volume of bacteria needed for a ratio of 50:1 bacteria to mammalian cells ($1–2.5\times10^7$ bacteria/sample) and aliquot into separate tubes for each sample.
7. Centrifuge the bacteria at $3000\times g$ for 5 min and resuspend the pellet with 50 μL of the mammalian cell suspension.
8. Incubate the mammalian cells and bacteria together for 1 h at 4 °C on an inversion shaker.
9. Wash the sample three times with PBS:
   a. Add 1 mL of PBS and pipet up and down to mix.
   b. Centrifuge the cells at $1500\times g$ for 30 s using a swinging bucket rotor. The mammalian cells should form a pellet, while most unbound bacteria remain in the supernatant.
   c. Remove the supernatant.
   d. Repeat steps (a) through (c) twice.
10. Resuspend the cell pellet in PBS (250 μL) and filter the samples using 35 μm cell-strainer caps (BD Falcon).
11. Load the target cells only and bacteria only control samples, and adjust the FSC and SSC voltages so that both the bacteria and mammalian cell populations are clearly visible on the scatter plot.
12. Plot 576 nm (red) versus 530 nm (green) fluorescence of the target cell population with the graph only displaying events from the target cell gate.
13. Adjust the 530 and 576 nm detector voltages so that the mean fluorescence of the target cell population is set at the background level (~100). Also, draw the sort gate (% bound in Fig. 4.5B) on the 576 versus 530 nm fluorescence plot so that only ~0.1% of the target cells are within the gate. The GFP-expressing bacteria should produce green fluorescent signals that are ~5- to 10-fold above the background autofluorescence of the target cells.

14. Analyze the sample containing target cells with nonbinding fluorescent bacteria (only the display scaffold), and verify that only a small percentage of the cells are within the sort gate. This negative control is important for determining the level of nonspecific binding and for optimizing the wash conditions.
15. Analyze the remaining samples and sort the events in the % bound gate.

2.7. Protease substrate identification using cellular libraries of peptide substrates (CLiPS)

A two-color bacterial display system termed cellular libraries of peptide substrates (CLiPS) was developed to quickly identify novel protease substrates (Boulware *et al.*, 2010, Jabaiah and Daugherty, 2011). For CLiPS, the plasmid pBAD33-eCPX-CLiPS codes a streptavidin-binding peptide (WVCHPMWEVMCLR; Rice *et al.*, 2006) upstream of the substrate library fused to the N-terminus of eCPX (Fig. 4.3B). On the C-terminus of eCPX, a SH3 domain-binding peptide (PAPSIDRSTKPPL; Harkiolaki *et al.*, 2003) was added to verify expression of the full-length construct and enable two color screening for cleaved substrates. The bacteria can be incubated with both SAPE (red probe) and *Alaj*GFP (green probe) fused to the SH3 domain of monocytic adaptor protein (Mona) to separately label both termini of the eCPX construct. For the CLiPS procedure (Fig. 4.6), the selection rounds consist of sorting for bacteria that are efficiently labeled with both the red and green probes at the N- and C-terminus, followed by a round of sorting post-protease treatment where the red fluorescence decreases while the green fluorescence remains high which indicates substrate hydrolysis. Sorting bacteria that are positive for both probes is important to remove substrate peptides that bind to the *Alaj*GFP-Mona fusion protein (green-positive but red-negative). It also removes bacteria that do not properly display the substrate peptide (both green- and red-negative) due to stop codons or frame shift mutations from the library construction.

If a canonical substrate is known, it can make a useful positive control for cleavage and also serve as a comparison for determining the improved cleavage kinetics of the isolated substrates. Also, this substrate can be used to determine the best incubation time and protease concentration for cleavage prior to sorting the library. It is important to pick the best reaction buffer for each protease to ensure maximal activity (usually a Tris/HEPES-based buffer, with cofactors possibly being required).

1. Follow the general growth and induction procedures (steps 1–4) in the Target Protein Sorting protocol (Section 2.2).
2. Incubate the bacteria with protease at room temperature or 37 °C for 1–3 h (skip this step if sorting for labeling only). Multiple protease

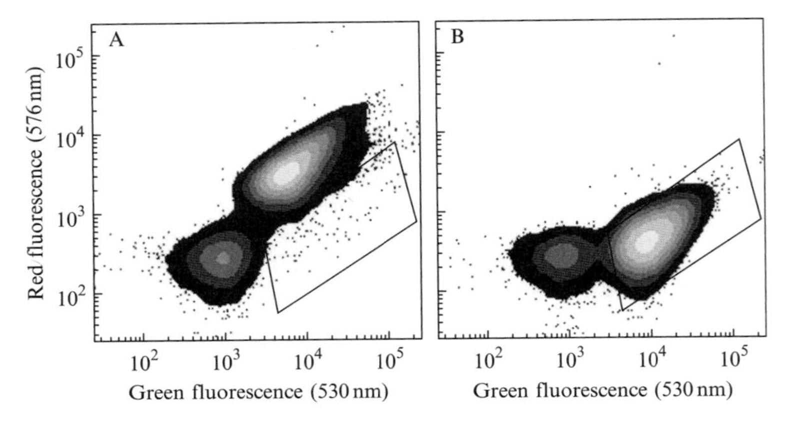

Figure 4.6 Screening for protease substrates. (A) Bacteria that are labeled with a red fluorescent probe (SAPE) which binds to the N-terminus of eCPX and a green fluorescent probe (*Alaj*GFP-Mona) which interacts with the C-terminus. With no protease present, the red- and green-positive cells are sorted to ensure expression of full-length eCPX constructs. (B) The bacteria are incubated with the protease, and the hydrolysis of the substrate causes a large decrease in the red fluorescence (streptavidin-binding peptide is no longer attached to the cell surface) while the green signal remains positive.

concentrations and incubation times can be tested to determine the best conditions for cleavage and sorting.

3. Dilute the sample 100-fold in PBS to stop the reaction, centrifuge the cells at 3000×*g* for 5 min, and remove the supernatant.
4. Label the bacteria with SAPE (50 n*M*) and *Alaj*GFP-Mona fusion (250 n*M*) in PBS (~100 μL).
5. Incubate at 4 °C for 1 h.
6. Centrifuge the cells at 3000×*g* for 5 min and resuspend the pellet in PBS (500 μL) for analysis or sorting on the flow cytometer.
7. Load an unlabeled bacteria control to set the 530 and 576 nm detector PMT voltages, so the mean red and green fluorescence is at the background level (~100).
8. Analyze the bacteria labeled with both the red and green probes (no protease) and draw the sort gate based on this control with the gate extending to lower red fluorescence (Fig. 4.6).
9. Analyze the remaining samples and sort the cell population with reduced red fluorescence.

3. Conclusions

The protocols described above enable the construction and screening of large bacterial display libraries. Given the efficiency of peptide library construction and screening in *E. coli*, distinct libraries can be readily tailored to accomplish diverse experimental objectives; these protocols illustrate methods of peptide ligand engineering for proteins and cell membrane receptors along with protease substrate discovery. In addition, the ability to screen large peptide libraries quantitatively for specific functional properties using FACS provides unique opportunities to adapt the presented techniques for a variety of peptide discovery and design goals. Overall, the flexible nature of bacterial display peptide libraries allows these construction and screening protocols to evolve in order to address a broad range of emerging applications.

ACKNOWLEDGMENTS

We would like to thank Dr. Abeer Jabaiah for the CLiPS data and Dr. Jeffrey Rice for critically reading the protocol.

REFERENCES

Bessette, P. H., Rice, J. J., and Daugherty, P. S. (2004). Rapid isolation of high-affinity protein binding peptides using bacterial display. *Protein Eng. Des. Sel.* **17,** 731–739.

Binder, U., Matschiner, G., Theobald, I., and Skerra, A. (2010). High-throughput sorting of an Anticalin library via EspP-mediated functional display on the Escherichia coli cell surface. *J. Mol. Biol.* **400,** 783–802.

Boder, E. T., and Wittrup, K. D. (1998). Optimal screening of surface-displayed polypeptide libraries. *Biotechnol. Prog.* **14,** 55–62.

Boulware, K. T., Jabaiah, A., and Daugherty, P. S. (2010). Evolutionary optimization of peptide substrates for proteases that exhibit rapid hydrolysis kinetics. *Biotechnol. Bioeng.* **106,** 339–346.

Chen, C. L., Wu, S. C., Tjia, W. M., Wang, C. L., Lohka, M. J., and Wong, S. L. (2008). Development of a LytE-based high-density surface display system in Bacillus subtilis. *Microb. Biotechnol.* **1,** 177–190.

Dane, K. Y., Chan, L. A., Rice, J. J., and Daugherty, P. S. (2006). Isolation of cell specific peptide ligands using fluorescent bacterial display libraries. *J. Immunol. Methods* **309,** 120–129.

Dane, K. Y., Gottstein, C., and Daugherty, P. S. (2009). Cell surface profiling with peptide libraries yields ligand arrays that classify breast tumor subtypes. *Mol. Cancer Ther.* **8,** 1312–1318.

Daugherty, P. S. (2007). Protein engineering with bacterial display. *Curr. Opin. Struct. Biol.* **17,** 474–480.

Fairbrother, W. J., Christinger, H. W., Cochran, A. G., Fuh, C., Keenan, C. J., Quan, C., Shriver, S. K., Tom, J. Y. K., Wells, J. A., and Cunningham, B. C. (1998). Novel peptides selected to bind vascular endothelial growth factor target the receptor-binding site. *Biochemistry* **37,** 17754–17764.

Fleming, T. J., Sachdeva, M., Delic, M., Beltzer, J., Wescott, C. R., Devlin, M., Ladner, R. C., Nixon, A. E., Roschke, V., Hilbert, D. M., and Sexton, D. J. (2005). Discovery of high-affinity peptide binders to BLyS by phage display. *J. Mol. Recognit.* **18,** 94–102.

Getz, J. A., Rice, J. J., and Daugherty, P. S. (2011). Protease-Resistant Peptide Ligands from a Stable Knottin Scaffold Library. *ACS Chem. Biol.* **6,** 837–844.

Hall, S. S., and Daugherty, P. S. (2009). Quantitative specificity-based display library screening identifies determinants of antibody-epitope binding specificity. *Protein Sci.* **18,** 1926–1934.

Harkiolaki, M., Lewitzky, M., Gilbert, R. J. C., Jones, E. Y., Bourette, R. P., Mouchiroud, G., Sondermann, H., Moarefi, I., and Feller, S. M. (2003). Structural basis for SH3 domain-mediated high-affinity binding between Mona/Gads and SLP-76. *EMBO J.* **22,** 2571–2582.

Jabaiah, A., and Daugherty, P. S. (2011). Directed Evolution of Protease Beacons that Enable Sensitive Detection of Endogenous MT1-MMP Activity in Tumor Cell Lines. *Chemistry & Biology* **18,** 392–401.

Jostock, T., and Dubel, S. (2005). Screening of molecular repertoires by microbial surface display. *Comb. Chem. High Throughput Screen.* **8,** 127–133.

Kenrick, S. A., and Daugherty, P. S. (2010). Bacterial display enables efficient and quantitative peptide affinity maturation. *Protein Eng. Des. Sel.* **23,** 9–17.

Kenrick, S., Rice, J., and Daugherty, P. (2007). Flow cytometric sorting of bacterial surface-displayed libraries. *Curr. Protoc. Cytom.* **42,** 4.6.1–4.6.27.

Kronqvist, N., Lofblom, J., Jonsson, A., Wernerus, H., and Stahl, S. (2008). A novel affinity protein selection system based on staphylococcal cell surface display and flow cytometry. *Protein Eng. Des. Sel.* **21,** 247–255.

Levin, A. M., and Weiss, G. A. (2006). Optimizing the affinity and specificity of proteins with molecular display. *Mol. Biosyst.* **2,** 49–57.

Li, W., Lei, P., Yu, B., Wu, S., Peng, J., Zhao, X., Zhu, H., Kirschfink, M., and Shen, G. (2008). Screening and identification of a novel target specific for hepatoma cell line HepG2 from the FliTrx bacterial peptide library. *Acta Biochim. Biophys. Sin. (Shanghai)* **40,** 443–451.

Little, L. E., Dane, K. Y., Daugherty, P. S., Healy, K. E., and Schaffer, D. V. (2011). Exploiting bacterial peptide display technology to engineer biomaterials for neural stem cell culture. *Biomaterials* **32,** 1484–1494.

Lu, Z., Murray, K. S., Van Cleave, V., LaVallie, E. R., Stahl, M. L., and McCoy, J. M. (1995). Expression of thioredoxin random peptide libraries on the Escherichia coli cell surface as functional fusions to flagellin: A system designed for exploring protein-protein interactions. *Nat. Biotechnol.* **13,** 366–372.

Orning, L., Rian, A., Campbell, A., Brady, J., Fedosov, S. N., Bramlage, B., Thompson, K., and Quadros, E. V. (2006). Characterization of a monoclonal antibody with specificity for holo-transcobalamin. *Nutr. Metab. (Lond.)* **3,** 3.

Rice, J. J., and Daugherty, P. S. (2008). Directed evolution of a biterminal bacterial display scaffold enhances the display of diverse peptides. *Protein Eng. Des. Sel.* **21,** 435–442.

Rice, J. J., Schohn, A., Bessette, P. H., Boulware, K. T., and Daugherty, P. S. (2006). Bacterial display using circularly permuted outer membrane protein OmpX yields high affinity peptide ligands. *Protein Sci.* **15,** 825–836.

Rockberg, J., Lofblom, J., Hjelm, B., Uhlen, M., and Stahl, S. (2008). Epitope mapping of antibodies using bacterial surface display. *Nat. Methods* **5,** 1039–1045.

Rojas, G., Pupo, A., Del Rosario Aleman, M., and Vispo, N. S. (2008). Preferential selection of Cys-constrained peptides from a random phage-displayed library by anti-glucitollysine antibodies. *J. Pept. Sci.* **14,** 1216–1221.

Smith, G. P., and Petrenko, V. A. (1997). Phage display. *Chem. Rev.* **97,** 391–410.

Strauch, A., and Georgiou, G. (2009). Mechanistic challenges and engineering applications of protein export in E. Coli. *In* "Systems Biology and Biotechnology of Escherichia coli," (S. Y. Lee, ed.), pp. 327–349. Springer, London.

Sun, J. J., Abdeljabbar, D. M., Clarke, N., Bellows, M. L., Floudas, C. A., and Link, A. J. (2009). Reconstitution and engineering of apoptotic protein interactions on the bacterial cell surface. *J. Mol. Biol.* **394,** 297–305.

Thomas, J. M., and Daugherty, P. S. (2009). Proligands with protease-regulated binding activity identified from cell-displayed prodomain libraries. *Protein Sci.* **18,** 2053–2059.

Uchiyama, F., Tanaka, Y., Minari, Y., and Tokui, N. (2005). Designing scaffolds of peptides for phage display libraries. *J. Biosci. Bioeng.* **99,** 448–456.

van Bloois, E., Winter, R. T., Kolmar, H., and Fraaije, M. W. (2011). Decorating microbes: Surface display of proteins on Escherichia coli. *Trends Biotechnol.* **29,** 79–86.

Wentzel, A., Christmann, A., Kratzner, R., and Kolmar, H. (1999). Sequence requirements of the GPNG beta-turn of the Ecballium elaterium trypsin inhibitor II explored by combinatorial library screening. *J. Biol. Chem.* **274,** 21037–21043.

Yu, J. N., and Smith, G. P. (1996). Affinity maturation of phage-displayed peptide ligands. *Comb. Chem.* **267,** 3–27.

Zitzmann, S., Kramer, S., Mier, W., Mahmut, M., Fleig, J., Altmann, A., Eisenhut, M., and Haberkorn, U. (2005). Identification of a new prostate-specific cyclic peptide with the bacterial FliTrx system. *J. Nucl. Med.* **46,** 782–785.

SECTION FIVE

SCAFFOLDS

CHAPTER FIVE

Designed Ankyrin Repeat Proteins (DARPins): From Research to Therapy

Rastislav Tamaskovic, Manuel Simon, Nikolas Stefan, Martin Schwill, *and* Andreas Plückthun

Contents

1. Introduction 102
   1.1. Background 102
   1.2. Properties of repeat proteins 103
   1.3. Properties of DARPins 103
   1.4. Selection technologies for DARPin libraries 106
   1.5. DARPins as pure proteins 107
2. Applications of DARPins 107
   2.1. DARPins in diagnostics 107
   2.2. DARPins in tumor targeting 108
   2.3. Approaches to targeted tumor therapy 109
   2.4. DARPins for viral retargeting 111
   2.5. DARPins in other approaches 112
   2.6. DARPins in the clinic 112
3. Protocols for DARPins in Biomedical Applications 112
   3.1. Stoichiometric cysteine labeling with maleimide-coupled fluorescent probes 113
   3.2. Stoichiometric N-terminal labeling with succinimidyl-ester coupled fluorescent probes 114
   3.3. Quantitative PEGylation of DARPins 115
   3.4. Introduction to "Click chemistry" 117
   3.5. Expression of "clickable" DARPins 118
   3.6. IMAC purification of "clickable" DARPins 119
   3.7. Analysis of "clickable" DARPins 119
   3.8. Site-specific PEGylation of DARPins using Cu-free "click chemistry" 120
   3.9. Purification of "click" PEGylated DARPins 120
   3.10. Measurement of DARPin binding affinity to whole cells 121
   3.11. Determination of the dissociation constant by equilibrium titration on cells 122

Department of Biochemistry, University of Zurich, Winterthurerstrasse, Zurich, Switzerland

Methods in Enzymology, Volume 503
ISSN 0076-6879, DOI: 10.1016/B978-0-12-396962-0.00005-7

3.12. Determination of kinetic parameters of binding on cells 124
3.13. Expression and purification of DARPin–toxin fusion proteins 127
Acknowledgment 129
References 129

Abstract

Designed ankyrin repeat proteins (DARPins) have been developed into a robust and versatile scaffold for binding proteins. High-affinity binders are routinely selected by ribosome display and phage display. DARPins have entered clinical trials and have found numerous uses in research, due to their high stability and robust folding, allowing many new molecular formats. We summarize the DARPin properties and highlight some biomedical applications. Protocols are given for labeling with dyes and polyethylene glycol, for quantitatively measuring binding to cell surface receptors by kinetics and thermodynamics, and for exploiting new engineering opportunities from using "click chemistry" with nonnatural amino acids.

1. Introduction

1.1. Background

To embark on developing a new protein scaffold class as a general binding module, there has to be a strong motivation, especially when carried out in a laboratory with a long-standing focus on antibody engineering and technology (Glockshuber *et al.*, 1990; Skerra and Plückthun, 1988). As the designed ankyrin repeat protein (DARPin) technology was developed in an academic setting (Binz *et al.*, 2003; Forrer *et al.*, 2003), the driving force was the desire to create a scaffold with superior technological properties.

Through the design of a fully synthetic antibody library (Knappik *et al.*, 2000) and the development of ribosome display (Hanes and Plückthun, 1997; Hanes *et al.*, 1998), which incorporates affinity maturation and directed evolution directly in the workflow, the basic operation of both natural antibody generation and somatic mutation had been replicated in the laboratory. Ironically, the antibody molecule was then no longer needed, and the technology had become independent of its roots. From extensive experiments with the engineering of antibodies and their fragments (Wörn and Plückthun, 2001), it had become evident that the molecules themselves contain some technical limitations.

Today, recombinant antibody scFv or Fab fragments are usually converted back to an IgG for therapeutic applications (Plückthun and Moroney, 2005), and the biophysical properties (or "developability") of a particular antibody are an important consideration. When using the antibody fragments in more ambitious formats (as fusions to other aggregation-prone proteins, in multimers, or in the absence of disulfide bonds), the limitations in their biophysical properties become even more apparent.

1.2. Properties of repeat proteins

Repeat proteins appeared very attractive as a choice for a general binding protein. Repeat proteins (Kobe and Kajava, 2000) contain modules, whose number can be chosen freely, which stack up on each other to create a rigid protein domain. They have different architectures, but within one family, they use modules of almost identical structure but with individual surfaces to specifically bind their target. After engineering work on the repeat proteins had started (Binz *et al.*, 2003; Forrer *et al.*, 2003; Stumpp *et al.*, 2003), the discovery that jawless vertebrates use an adaptive immune system made of leucine-rich repeat proteins (Pancer *et al.*, 2004) came as a surprising validation of the concept.

Repeat proteins seem to follow rules which can be derived from biophysical considerations: a large interaction surface is usually a prerequisite for tight binding. Such a binding interface should be rigid, in order to not lose entropy upon binding, which would otherwise decrease the overall achievable binding free energy (equivalent to a loss in affinity). Also, the surface should be modular and thus "patches" should be individually exchangeable or modifiable by affinity maturation. Furthermore, the protein should also not require disulfides, to have the option of expressing it in the bacterial cytoplasm, and to later introduce unique cysteines for site-specific coupling with drugs, fluorescent labels, or polyethylene glycol (PEG), just to name a few.

1.3. Properties of DARPins

Ankyrin repeat proteins (Bork, 1993; Li *et al.*, 2006) are built from tightly joined repeats of usually 33 amino acid residues. Each repeat forms a structural unit consisting of a β-turn followed by two antiparallel α-helices (Fig. 5.1), and up to 29 consecutive repeats can be found in a single protein (Walker *et al.*, 2000). Yet, ankyrin repeat domains usually consist of four to six repeats, which stack onto each other, leading to a right-handed solenoid structure with a continuous hydrophobic core and a large solvent accessible surface (Kobe and Kajava, 2000; Sedgwick and Smerdon, 1999).

We exploited the huge available sequence information of the repeats within the protein family (different protein variants each with several repeats) by using a "consensus" strategy (Forrer *et al.*, 2004). The underlying assumption is that residues important for maintaining the fold will be more conserved and thus show up prominently in an alignment, while residues involved in interactions of individual members of the protein family with their specific target will not be conserved. The ankyrin repeats seem to belong to one predominant single family, such that consensus design is comparatively straightforward.

The repeat modules are held together by a hydrophobic interface, and thus the first (N-capping repeat or N-cap) and last repeat (C-capping repeat

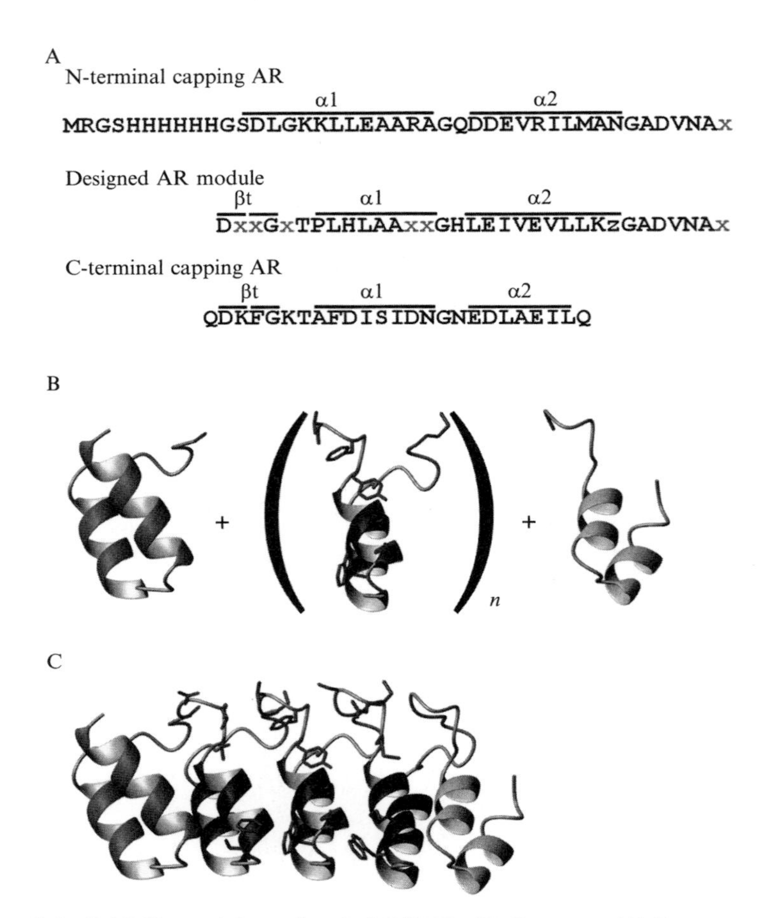

Figure 5.1 DARPin modules and typical DARPin 3D Structure. (A) Sequences of the N-terminal capping ankyrin repeat (AR), the designed AR module, and the C-terminal capping AR. The secondary structure elements are indicated above the sequences. The designed AR module consists of 26 defined framework residues, 6 randomized potential interaction residues (red x, any of the 20 natural amino acids except cysteine, glycine, or proline), and 1 randomized framework residue (z, any of the amino acids asparagine, histidine, or tyrosine). The designed AR module was derived via sequence and structure consensus analyses. (B) Schematic representation of the library generation of designed AR proteins (DARPins). Note that this assembly is represented on the protein level, whereas the real library assembly is on the DNA level. By assembling an N-terminal capping AR (green, left), varying numbers of the designed AR module (blue, middle), and a C-terminal capping AR (cyan, right), combinatorial libraries of DARPins of different repeat numbers were generated (side chains of the randomized potential interaction residues are shown in stick-mode in red). (C) Ribbon representation of the selected MBP binding DARPin off7 (colors as in B). This binder is derived from a library consisting of an N-terminal capping AR, three designed AR modules, and a C-terminal capping AR. Figure reproduced from Binz *et al.* (2004). (See Color Insert.)

or C-cap) have to be special and need to present a hydrophilic outside surface exposed to the solvent. In the original design, both caps were taken from a natural protein (Binz *et al.*, 2003). More recently, this C-cap has been redesigned, based on molecular dynamics calculations, to make it more similar to the consensus, and it has been experimentally verified that the new C-cap is indeed much more resistant against thermal and denaturant-induced unfolding (Interlandi *et al.*, 2008). Crystallography (Kramer *et al.*, 2010) and NMR (Wetzel *et al.*, 2010) show that this is due to better packing of the C-cap against the internal repeats.

The "full-consensus" DARPins, where all the residues are chosen from consensus considerations, show remarkable properties. They express very well in *Escherichia coli* as soluble monomers, their stability increases with length, and those with more than three internal repeats are resistant to denaturation by boiling or guanidine hydrochloride. Full denaturation requires high temperature in 5 *M* guanidine hydrochloride (Wetzel *et al.*, 2008). Hydrogen/deuterium exchange experiments of DARPins with two and three internal repeats indicate that this high stability of the full-consensus ankyrin repeat proteins is due to the strong coupling between repeats. Some amide protons require more than a year to exchange at 37 °C (Wetzel *et al.*, 2010), highlighting the extraordinary stability of the proteins. The location of these very slowly exchanging protons indicates a very stable core structure in the DARPins that combines hydrophobic shielding with favorable electrostatic interactions.

We can consider these full-consensus DARPins as the hypothetical diversification point of a library (even though, historically, the library had been constructed first; Binz *et al.*, 2003). Thus, when diverging from a point of extremely high stability, many changes in the protein, necessary for function but potentially detrimental to stability, can be tolerated, and the outcome is still a very good protein. It appears that the experimental results confirm this hypothesis (see e.g., Binz *et al.*, 2003, 2006; Kohl *et al.*, 2003; Wetzel *et al.*, 2008).

A second reason, besides stability, for basing the DARPin library on a consensus ankyrin, as opposed to a particular naturally occurring ankyrin, was to make the library modules self-compatible (Binz *et al.*, 2003) (Fig. 5.1A and B) such that they can be assembled in any order. Such a designed repeat library module comprises fixed and variable positions. The fixed positions mainly reflect conserved framework positions, while the six variable positions mainly reflect nonconserved surface-exposed residues that can be potentially engaged in interactions with the target, as they are located on the target-binding (concave) face. The theoretical diversities of the DARPin libraries are 5.2×10^{15} or 3.8×10^{23} for two- or tree-module binders, respectively, and the actual sizes of the libraries are equal to the number of different molecules present. We can estimate them as 10^{12} in ribosome display (Plückthun, 2011) and 10^{10} in phage display (Steiner *et al.*, 2008).

1.4. Selection technologies for DARPin libraries

DARPins have been selected from the synthetic libraries by ribosome display and phage display. Ribosome display is a potent *in vitro* method to select and evolve proteins or peptides from a naïve library with very high diversity to bind to any chosen target of interest (Hanes and Plückthun, 1997; Hanes *et al.*, 1998, 2000a; Mattheakis *et al.*, 1994). A key feature of ribosome display, in contradistinction to most other selection technologies, is that it incorporates PCR into the procedure and thus allows a convenient incorporation of a diversification ("randomization") step. Thereby, ribosome display allows refinement and affinity maturation not only of preexisting binders (Dreier *et al.*, 2011; Hanes *et al.*, 2000b; Luginbühl *et al.*, 2006; Zahnd *et al.*, 2007b) but also of the whole pool during selection from a complex library, if desired. It appears that, as the DARPins fold very well, also in cell free translation, binders are enriched somewhat faster than binders from a comparable scFv ribosome display library (Dreier and Plückthun, 2010; Hanes *et al.*, 2000a).

Using ribosome display, DARPins have been evolved to bind various targets with affinities all the way down to the picomolar range (Amstutz *et al.*, 2005; Binz *et al.*, 2004; Dreier *et al.*, 2011; Huber *et al.*, 2007; Schweizer *et al.*, 2007; Veesler *et al.*, 2009; Zahnd *et al.*, 2006, 2007b). The theoretical considerations for designing efficient off-rate selection experiments were recently summarized (Zahnd *et al.*, 2010a).

DARPins have also been selected by phage display. In filamentous phage display using fusions to the minor coat protein g3p, the protein of interest is first produced as a membrane-bound intermediate, with the domains of interest secreted to the *E. coli* periplasm but remaining still attached to the inner membrane. The g3p fusion is then taken up by the extruding phage. As DARPins fold very fast in the cytoplasm (Wetzel *et al.*, 2008), they fold before they can be transported across the membrane via the posttranslational Sec system, the normal way of secreting *E. coli* proteins. Thus, very low display rates would be observed using Sec-dependent signal sequences. However, *E. coli* has another secretion system, the signal recognition particle (SRP) dependent one, which is essentially cotranslational (Bibi, 2011; Fekkes and Driessen, 1999). Using phagemids with SRP-dependent signal sequences, phage display selection of DARPins leads to enrichment just as fast as phage display of slow folding proteins, for example, scFv fragments, using Sec-dependent phage display (Steiner *et al.*, 2006). Binders with subnanomolar K_D could be obtained from the phage display library without affinity maturation for a variety of targets (Steiner *et al.*, 2008). For completeness, we also mention that previous attempts to achieve functional display of g3p fusions via the Tat route have proven unsuccessful (Dröge *et al.*, 2006; Nangola *et al.*, 2010; Paschke and Höhne, 2005), as the full-length p3 protein may itself be incompatible with the Tat system. However, a truncated version of p3 can support Tat-mediated phage display (Speck *et al.*, 2011).

1.5. DARPins as pure proteins

Most DARPins express well in *E. coli* in soluble form constituting up to 30% of total cellular protein (up to 200 mg per 1 l of shake flask culture), and in the fermenter, multigram quantities per liter culture can be achieved (www.molecularpartners.com). Purification is thus straightforward, and for laboratory use, IMAC purification is the standard method used (Binz *et al.*, 2004). Usually, in the initial research, only few milligrams are needed, and thus the extreme overloading of an IMAC column of small capacity leads to very pure protein in a single step, as most *E. coli* contaminants are thereby competed out.

Additional purification steps are of course required when the protein is derivatized (e.g., with PEG, or fluorescent dyes) (see Section 3). For animal experiments, still higher purity is needed and absence of endotoxins needs to be secured, requiring additional washing steps and chromatography for endotoxin removal (Section 3.13).

Not only the full-consensus DARPin molecules but also most library members showed high thermodynamic stability during unfolding induced by heat or denaturants (Binz *et al.*, 2004; Kohl *et al.*, 2003) and can be brought to very high protein concentrations.

2. Applications of DARPins

2.1. DARPins in diagnostics

DARPins have been tested for their suitability in quantitative immunohistochemistry (Theurillat *et al.*, 2010), which requires high specificity in complex tissue. A DARPin specific for epidermal growth factor receptor 2 (HER2) with picomolar affinity was compared to an FDA-approved rabbit monoclonal antibody in paraffin-embedded tissue sections in tissue microarrays. HER2 gene amplification status is an important criterion to determine the optimal therapy in breast cancer. As an external reference, the HER2 amplification status was determined by fluorescence *in situ* hybridization. It was found that the DARPin detects a positive HER2 amplification status with similar sensitivity and significantly higher specificity than the FDA-approved antibody (Theurillat *et al.*, 2010). Thus, DARPins have the desired specificity characteristics for diagnostic pathology.

2.2. DARPins in tumor targeting

High affinity and specificity for a tumor cell marker is necessary but not sufficient for successful tumor targeting. Another very important issue is the quantitative enrichment at the tumor site. This, in turn, depends on the molecular format. Using DARPins, the influence of affinity and size on the efficiency of targeting was systematically investigated. DARPins are very small (molecular weight, MW 15–18kDa) yet can easily be PEGylated (see Sections 3.3 and 3.8). For a DARPin conjugated to a PEG molecule with a nominal MW of 20kDa, the hydrodynamic properties correspond to a MW of about 250–350kDa (Chapman, 2002; Kubetzko *et al.*, 2005), and thus the effect of tumor targeting over a large size range could be studied. Furthermore, different point mutants binding to the same epitope of HER2 were available from the directed evolution by ribosome display, spanning affinities from 280n*M* to 90p*M*.

These studies showed (Zahnd *et al.*, 2010b) that there are *two* parameter regions for efficient tumor accumulation. Perhaps at first somewhat unexpected, unmodified small DARPins were found to accumulate rather well, directly proportional to affinity, with 8% ID/g after 24h for a 90-p*M* binder in an SK-OV-3 subcutaneous mouse xenograft model. No evidence for a barrier effect was observed. The small DARPins were cleared from the blood extremely rapidly such that very high tumor to blood ratios (60:1) were measured. For bivalent DARPins (measured avidity on cells: 10p*M*, see Section 3.12 for the methodology), a *lower* accumulation in the tumor was seen than for the monovalent counterpart (which already had a $K_D \approx 90$ p*M*), suggesting that smaller size is more important than very high avidity. Similarly, fusing a nonbinding DARPin to the anti-HER2 DARPin (K_D remains about 90p*M*) gave the same *lowered* uptake compared to the single-domain DARPin. This lower uptake value is consistent with similar numbers obtained for antibody scFv fragments in the same tumor model (Adams *et al.*, 1993; Willuda *et al.*, 2001), and scFv fragments have the same MW as the DARPin–DARPin fusions. Thus, a very small MW seems to be a virtue, provided affinity is picomolar, thereby defining the first parameter set for efficient tumor accumulation.

The second parameter region of high tumor accumulation was found with PEGylated DARPins. The PEGylated DARPins accumulated more slowly and even better (13% ID/g after 24h), and as might be expected from their larger size, their blood clearance was much slower, leading to smaller tumor to blood ratios. Interestingly, the importance of affinity was diminished, with the DARPin of K_D=90 p*M* not showing a great advantage over the one with K_D=1 n*M*.

How can we rationalize these data? First, it should be emphasized that they are in excellent agreement with elegant modeling studies of Wittrup and colleagues (Schmidt and Wittrup, 2009; Thurber *et al.*, 2008), which have independently lead to very similar conclusions. If a very pronounced dependence of extravasation on MW is assumed, and that its MW cutoff is lower than that of renal filtration, then a molecule of intermediate MW would be filtered through the kidney, but would still not extravasate well. However, a molecule of small MW needs to bind to its receptor on the tumor very tightly, or it will be washed out rapidly. This affinity requirement is not as strong for very large, PEGylated molecules, which reside in the serum for much longer times. In contrast, medium sized molecules (such as scFv fragments) are still being cleared rapidly through the kidney, without reaching the tumor fast enough, because of their slower extravasation.

A series of elegant studies on quantifying tumor accumulation of mono- and multivalent scFv fragments have been reported, which have been summarized to suggest that very high affinity is disadvantageous for tumor targeting (Adams *et al.*, 2001). It should be noted, however, that in these investigations, iodine was used as a label which is removed by dehalogenases upon internalization. Thus, a higher affinity or avidity leading to more internalization will lead to *less* remaining label in the tumor (Rudnick *et al.*, 2011). In contrast, Zahnd *et al.* (2010b) used a residualizing Tc label, which will not be removed and will thus be counted, no matter whether the protein has become internalized or whether it remains on the surface. It is thus important to point out that the affinity optimum may be different for proteins that deliver a cargo to the cell (affinity should be as high as possible) versus those that should remain on the surface (there is an affinity optimum).

2.3. Approaches to targeted tumor therapy

DARPins have been used as targeting proteins in preclinical tumor models. Two examples, both using the epithelial cell adhesion molecule (EpCAM) as the target, shall be mentioned to illustrate the potential of DARPins. EpCAM is a homophilic cell adhesion molecule of 39–42 kDa, consisting of an extracellular domain with an epidermal growth factor- and a human thyroglobulin-like domain, and a short cytoplasmic domain (Trzpis *et al.*, 2007; van der Gun *et al.*, 2010). Its processing by regulated intramembrane proteolysis releases a cytoplasmic tail that activates the Wnt signaling pathway and induces transcription of c-myc and cyclins. EpCAM is an attractive tumor-associated target, as it is expressed at low levels on basolateral cell surfaces of some normal epithelia, while high levels of homogeneously distributed EpCAM are detectable on cells of epithelial tumors. Recently,

EpCAM was also identified as a marker of cancer-initiating cells (Trzpis *et al.*, 2007; van der Gun *et al.*, 2010). The favorable properties of EpCAM for cancer therapy are currently exploited in phase II clinical trials with a scFv-exotoxin A (ETA) immunotoxin (Biggers and Scheinfeld, 2008; Di Paolo *et al.*, 2003; Hussain *et al.*, 2006, 2007), which we developed previously (Di Paolo *et al.*, 2003).

There has been a lot of interest in attempting to employ small-interfering RNA (siRNA) for tumor control, but specific delivery to tumors and efficient cellular uptake of nucleic acids remain major challenges for gene-targeted cancer therapies. An anti-EpCAM DARPin was used as a carrier for siRNA complementary to the bcl-2 mRNA, a pro-apoptotic factor (Winkler *et al.*, 2009). To achieve complexation, the DARPin was genetically fused to protamine, and about 4–5 molecules siRNA could be bound per protamine.

Bivalent binders with a C-terminal leucine zipper (leading to a tail-to-tail fusion) lead to an avidity gain, while head-to-tail fusions (fused via a gly-ser linker) did not, suggesting that the former may have better matched the geometry of EpCAM on the cell (Winkler *et al.*, 2009). For all tested constructs (but best for the leucine-zipper constructs), a decrease in bcl-2 expression was observed at the mRNA and the protein level; this resulted in a significant sensitization of EpCAM-positive MCF-7 cells toward doxorubicin. Indeed, this sensitization was not observed in EpCAM-negative cells, indicating that siRNA uptake is receptor dependent (Winkler *et al.*, 2009).

Another EpCAM-specific DARPin was produced as a fusion toxin with *Pseudomonas aeruginosa* ETA and expressed in the cytoplasm of *E. coli* (Martin-Killias *et al.*, 2011; Stefan *et al.*, 2011) (cf. Section 3.13). While the DARPins have no cysteines, the disulfides in the toxin part formed spontaneously, and the protein was monomeric, yielding up to 90 mg after purification from 1 l of culture from a simple *E. coli* shake flask. The DARPin–ETA fusion was highly cytotoxic against various EpCAM-positive tumor cell lines with IC_{50} values less than 0.005 p*M*. This effect was competed by free DARPin, but not by unspecific DARPins. Upon systemic administration in athymic mice, the DARPin–ETA fusion efficiently localized to EpCAM-positive tumors to achieve maximum accumulation 48–72 h after injection, whereas an irrelevant control fusion toxin did not accumulate. Tumor targeting with the DARPin–ETA resulted in a strong antitumor response in mice bearing two different EpCAM-positive tumor xenografts, including complete regressions in some animals (Martin-Killias *et al.*, 2011). Thus, DARPin–ETA fusions deserve attention for clinical development.

While these examples have illustrated the possibility of using DARPins for the delivery of a payload to a tumor, the facile engineering of DARPins might also be exploited to create a multivalent DARPin with biological activity by itself, using either cross-linking on the same cell or between cells.

2.4. DARPins for viral retargeting

The facile engineering of DARPins, undoubtedly due to their robust folding in various fusion formats, has also been used in viral retargeting. Adenoviruses (Ads) are a family of nonenveloped viruses which contain a double-stranded DNA genome, and they have been developed as prototypes of gene vectors for gene therapy (Amalfitano and Parks, 2002), genetic immunizations (Sullivan *et al.*, 2003), and molecular-genetic imaging (Yeh *et al.*, 2011). Many studies have been reported to alter the native tropism of the virus to make Ad-mediated transgene delivery to desired cell targets both more efficient and target specific. Most studies focused on modifications of one of the major components of the Ad capsid, the fiber protein (Nicklin *et al.*, 2005). Ideally, this change in virus tropism should be achieved with technology that can realistically be scaled up and be applicable to any cell surface receptor. For this purpose, a strategy was developed with a bispecific adapter which can be produced in *E. coli* (Dreier *et al.*, 2011). In its most efficient form, it consists of four DARPins in tandem, three of which bind and wrap around the trimeric knob domain at the distal end of the protruding Ad fibers (while also blocking binding to the natural receptor CAR), while the fourth DARPin binds to the cellular target of interest. This strategy was tested with a HER2-binding DARPin. The adaptors showed a significant increase in cell-specific transduction, measured by luciferase activity (Dreier *et al.*, 2011). This new strategy of altering the natural tropism of Ad5 with rationally designed adapters holds great promise for future developments in gene therapy.

While the Ad genome remains episomal, lentiviral vectors lead to stable integration and transgene expression in nondividing cells. Again, the challenge is to achieve cell-specific retargeting by recognition of a specific surface antigen. Cell entry is dependent on two viral glycoproteins, hemagglutinin (H) and fusion protein (F), for example, in measles virus (MV) (Yanagi *et al.*, 2006). By using lentiviral vectors expressing MV-H and MV-F, and specifically creating a variety of fusions MV-H to HER2-specific DARPins, infection of HER2 expressing cells could be obtained (Münch *et al.*, 2011). All H-DARPin fusion proteins were efficiently expressed on the cell surface and incorporated into lentiviral vectors at a more uniform rate than scFvs, perhaps because of the more robust folding within the fusion protein. The vectors only transduced HER2-positive cells, while HER2-negative cells remained untransduced. Highest titers were observed with one particular anti-HER2 DARPin binding to the membrane distal domain of HER2. When these DARPin-carrying viral vectors were applied *in vivo* systemically in a mouse tumor xenograft model, exclusive gene expression was observed in HER2 positive tumor tissue, while control vectors mainly transduced cells in spleen and liver (Münch *et al.*, 2011). Thus, DARPins are a promising route to engineer the specificity of lentiviral vectors for therapy.

2.5. DARPins in other approaches

We will only mention briefly other approaches of potential therapeutic significance. For example, DARPins have been selected to bind to IgE or its receptor FcεRIα (Baumann *et al.*, 2010; Eggel *et al.*, 2009, 2011). DARPins have also been selected against several kinases (Amstutz *et al.*, 2005, 2006; Bandeiras *et al.*, 2008), and this may provide the basis for novel sensor systems to be able to follow internal signaling events, to monitor the success of therapeutic agents. It can be expected that the broad formatting options of DARPins will be advantageous for future developments.

2.6. DARPins in the clinic

DARPins with very high affinity have been selected against vascular endothelial growth factor VEGF (www.molecularpartners.com) and were developed for intraocular formulation. They have entered two clinical phase I/II trials in two ocular indications, wet age-related macular degeneration (wet AMD) (clinicaltrials.gov/ct2/show/NCT01086761) and diabetic macular edema (clinicaltrials.gov/ct2/show/NCT01042678).

At the time of writing, first data from the wet AMD trial are released (Wolf *et al.*, 2011), a phase I/II, open-label, noncontrolled, multicentre trial. The clinical study with DARPin MP0112 assessed the safety and preliminary efficacy measured by visual acuity, fluorescein angiography, and color fundus photography during 16weeks. MP0112 is an extremely potent VEGF inhibitor and, because of its superior biophysical properties, can be concentrated to extremely high molar concentrations, and shows a long ocular half-life. It is suggested from these studies that dosing frequency in patients may be reduced three to fourfold, compared to current standard therapy with ranibizumab (Lucentis), an anti-VEGF Fab fragment. DARPin MP0112 was found safe and well tolerated. DARPin MP0112 represents a very promising new anti-VEGF treatment option with potential in various retinal diseases. It is a direct demonstration how the biophysical properties of the protein translate to benefit for the patient.

3. Protocols for DARPins in Biomedical Applications

The selection of DARPins from libraries using ribosome display (Dreier and Plückthun, 2011; Zahnd *et al.*, 2007a) or phage display (Steiner *et al.*, 2008) as well as their straightforward biochemical characterization (Binz *et al.*, 2003, 2004) have been described elsewhere. Thus, we will concentrate on methods for coupling DARPins with fluorescent labels and PEG, using both conventional coupling reactions and the introduction of nonnatural amino acids for the use of "click chemistry." We then describe the use of the labeled proteins for quantitating their binding on cells using kinetic methods. Finally, we describe the expression of DARPin–toxin fusion proteins.

3.1. Stoichiometric cysteine labeling with maleimide-coupled fluorescent probes

By design, DARPins do not have any cysteine residues. Using site-directed mutagenesis, a cysteine is typically inserted in front or just behind the unstructured N-terminal $RGSH_6$ tag, or at the C-terminus of the DARPin, without affecting the stability or affinity (Fig. 5.1C).

For quantitative labeling with a maleimide-containing fluorophore, any disulfides that may have spontaneously formed need to be first reduced to free thiols. The reduction and the subsequent coupling reactions are therefore performed in a nitrogen atmosphere in a glove box to avoid reoxidation of the free thiols, which is a prerequisite to approximate quantitative yields. The following protocol describes the batch reaction for preparing 1 mg stoichiometrically labeled DARPin. DARPins, which are expressed and purified by Ni-NTA as described before (Binz *et al.*, 2004), are diluted in 500 μl of degassed Tris-buffered saline (TBS; 25 m*M* Tris(hydroxymethyl)-aminomethane (Tris), 150 m*M* sodium chloride, pH 7.5) to a concentration of 100 μ*M*. A 50 m*M* stock solution of Tris(2-carboxyethyl)phosphine (TCEP-HCl, Thermo Fisher Scientific) is freshly prepared in water (it will have a pH of ca. pH 2.5) and needs to be neutralized by the addition of potassium hydroxide. While thiol-containing reductants (such as dithiothreitol or β-mercaptoethanol) need to be separated before coupling, low concentrations of TCEP hardly interfere with the coupling reaction of small fluorophores (Getz *et al.*, 1999), and a twofold molar excess of fresh TCEP over protein is sufficient to completely reduce all cysteine residues in 30 min at room temperature at the concentrations specified.

Alexa Fluor-488-C5-maleimide (Invitrogen) is dissolved in dimethylformamide (DMF) (10 mg/ml) by sonication and can be stored over months at −80 °C. About 7 μl of this solution is subsequently added in a twofold molar excess to the protein solution (100 μ*M*, 500 μl), and the coupling is performed for 1 h at room temperature.

All subsequent purification steps are carried out under normal atmosphere while the labeled protein is constantly protected from light. The reaction mix is desalted on a NAP-5 bench top column (GE Healthcare) in phosphate-buffered saline (PBS) pH 7.1 to remove more than 98% of the unreacted Alexa Fluor-488 dye.

Alexa Fluor-488 and other small dyes, which introduce an additional negative charge to the labeled conjugate, are suitable for achieving separation of the conjugate from the unlabeled DARPin by ion exchange chromatography (IEX). For separation of labeling products, a Mono-Q 5/50 GL column (GE Healthcare), equilibrated with Buffer A (50 m*M* Tris–HCl, pH 8.3, 10 m*M* NaCl), is used and all labeling species are loaded until the initial salt peak has been eluted. For a DARPin of unknown elution behavior, a linear elution gradient of Buffer B (50 m*M* Tris–HCl, pH 8.3, 1 *M* NaCl) is applied over 1 h. Once the elution behavior of the labeled DARPin is known, the gradient can be replaced by an optimized step gradient. The elution is monitored by the absorbance at 280 nm for the DARPin and 495 nm for the Alexa Fluor-488 dye, and fractions of 500 μl are collected on an ÄKTA Explorer FPLC instrument (GE Healthcare). The unlabeled DARPin, with a high 280 nm and no detectable 495 nm absorbance, elutes first and is followed by the labeled DARPin, which shows high absorbance at both wavelengths.

Peak fractions of the labeled species are pooled and, after desalting, the labeling efficiency is measured in a spectrophotometer by determining the molar ratio of Alexa Fluor-488 to DARPin within the conjugate, $[\text{Alexa}_{488}]/[\text{DARPin}]$, where

$$[\text{Alexa}_{488}] = \frac{A_{495}}{\varepsilon_{\text{Alexa@495}}} \quad \text{and} \quad [\text{DARPin}] = \frac{A_{280} - (A_{495} \times 0.11)}{\varepsilon_{\text{DARPin}}}$$

Here, $\varepsilon_{\text{Alexa@495}}$ is the molar absorbance (=71,000) of the dye at 495 nm (where the protein absorption is negligible) and $\varepsilon_{\text{DARPin}}$ is the molar absorbance of at the DARPin at 280 nm. The term $A_{495}\times 0.11$ is the absorbance of the dye at 280 nm, which is calculated from its absorbance measured at 495 nm.

For DARPins with few aromatic residues, the Alexa Fluor-488 dye may contribute much of the absorbance at 280 nm, and consequently, the labeling efficiency is difficult to determine. Thus, an alternative method is to resolve the labeled fraction by electrophoresis on a 20% sodium dodecyl sulfate polyacrylamide gel (SDS-PAGE). DARPins labeled with Alexa Fluor-488 will shift toward higher mass and can be identified with a fluorescence camera, while the ratio of labeled to unlabeled form of the protein can be estimated by staining with Coomassie Brilliant Blue.

3.2. Stoichiometric N-terminal labeling with succinimidyl-ester coupled fluorescent probes

To circumvent the need for cysteine mutagenesis, DARPins can also be selectively labeled at the N-terminal amino group. Because of the favorable orientation of the N-cap of a DARPin (Fig. 5.1C), an N-terminally coupled fluorophore is unlikely to interfere with the binding properties of the DARPin. The acylation of primary amines occurs essentially through the nonprotonated form and is thus dependent on the respective pK_a value.

While the N-terminal amine has a pK_a of 8, the ε-amino group of a fully exposed lysine side chain has a pK_a of 10.5 (even though buried ones can be significantly lower). There are typically several lysine residues in the protein and their combined reaction with the dye is thus significant even at lower pH. Therefore, reaction conditions (time, pH, and excess of dye) must be set to favor selectivity of the N-terminal amino group labeling and not the yield of the conjugate.

Purified DARPin is diluted to a concentration of 100 μ*M* in 500 μl PBS, pH 7.1. Alexa Fluor-488 carboxylic acid succinimidyl ester (Invitrogen), dissolved in DMF, is added in a threefold molar excess over protein and the coupling reaction is incubated for 80 min at room temperature. The coupling reaction is terminated by rebuffering in TBS on a NAP-5 column.

Monolabeled DARPins (which will be predominantly labeled at the N-terminus) can be purified, similar to cysteine-labeled DARPins, by anion exchange chromatography (Section 3.1). The elution profile is monitored at 280 and 495 nm and will show multiple peaks. The most prominent peak is likely to represent the N-terminally labeled fraction, while the other peaks correspond to different lysine-labeled species. Unlabeled, monolabeled, and multiply labeled protein can be distinguished based on the ratio of A_{280} and A_{495}.

If the assignment of the N-terminally labeled fraction is not unambiguous, a deliberate labeling at higher pH may further help to distinguish between the N-terminal and the lysine-labeled DARPin conjugates. For this purpose, all amines of the DARPin are coupled at pH 8.1 and the products are analyzed analogously by IEX. In the chromatogram overlay of both reactions, the relative amount of N-terminally labeled DARPin is diminished at pH 8.1, while all lysine-labeled DARPin fractions will increase. Exact sequence information of the DARPin modifications can be obtained by analyzing the fractions by means of peptide fragmentation on tandem MS/MS mass spectrometry.

3.3. Quantitative PEGylation of DARPins

Large polymeric moieties such as coupled PEG may interfere with binding of DARPins to their epitopes, but only if their attachment point is very close to the paratope. Thus, the coupling of PEG must be site-specific and the attachment point must be remote. In this case, no effect on the off-rate should be found, and only a small effect on the on-rate, for reasons established elsewhere (Kubetzko *et al.*, 2005).

The covalent coupling of maleimide-functionalized PEG to cysteine residues is the most convenient method for site-specific PEGylation. In contrast to small fluorophores, the reactivity of functionalized PEG decreases with higher concentrations and higher MW of the polymer due to the increasing viscosity of the reaction mix.

Here, we provide a protocol for reacting DARPins in solution with a 40-kDa, 2-branched maleimide-PEG, (SUNBRIGHT GL2-400MA, NOF Corporation). The DARPin is purified according to Binz *et al.* (2004) and diluted to a concentration of 200 μ*M* in 2.5 ml degassed PBS, pH 7.1.

The maleimide coupling is performed under oxygen-free conditions in a glove box with a nitrogen or argon atmosphere, analogously to the cysteine labeling with fluorophores (Section 3.1). For the PEGylation of DARPins, DTT instead of TCEP is used as reducing agent, due to the straightforward removal of DTT by desalting on a bench top column, which may be inefficient for TCEP (Shafer *et al.*, 2000). While low concentrations of TCEP do not significantly hinder the coupling of small dyes (Section 3.1), TCEP measurably affects the rate of the PEGylation reaction, and at long reaction times, side reactions can occur.

A 1 *M* stock solution of DTT is prepared freshly in degassed H_2O right before use, and 12.5 μl of DTT stock solution is added to the protein solution (200 μ*M*, 2.5 ml), vortexed, and incubated for 1 h at room temperature. Meanwhile, a PD-10 gravity-flow column (GE Healthcare) is equilibrated with degassed and N_2-flushed PBS, pH 7.1. DTT is removed from the protein by elution with 3.5 ml PBS pH 7.1. A twofold molar excess of a freshly prepared solution of maleimide-PEG is added directly into the reaction mix, and coupling is performed for 4 h at room temperature.

All following steps are carried out again in normal atmosphere. Free PEG and uncoupled DARPins can be efficiently removed from PEGylated DARPins by anion exchange chromatography. An ÄKTA Explorer FPLC system equipped with a Mono-Q 5/50 GL column is suitable to purify a total of up to 20 mg PEGylated DARPin per run. The reaction mix is diluted threefold to reduce ionic strength and loaded on the Mono-Q column equilibrated in Buffer A (50 m*M* Tris–HCl, pH 8.3, 10 m*M* NaCl), and the reaction mix is loaded. After unbound PEG has eluted in the flow through, a linear salt gradient of Buffer B (50 m*M* Tris–HCl, pH 8.3, 1 *M* NaCl) is applied over 1 h to efficiently separate the reaction products. The absorbance profile is monitored at 280 nm for DARPin, and fractions of 500 μl are collected. Due to charge shielding or the steric interference of PEG with the DARPin-column interaction, the PEGylated DARPins will elute before the non-PEGylated DARPins from the Mono-Q column. For identification of the different reaction products, the collected fractions are resolved on a 12% SDS-PAGE and stained with Coomassie Brilliant Blue and, optionally, with barium iodide solution for visualization of PEG (Kurfürst, 1992). The PEG-coupled DARPin run at a higher MW than the calculated one of the conjugate. On gel filtration, they even show a hydrodynamic radius corresponding to an apparent MW of about 300,000 Da (Chapman, 2002; Kubetzko *et al.*, 2005).

As the Tris buffer used in the IEX purification is toxic, it needs to be exchanged prior to *in vivo* applications. Therefore, PEGylated DARPins are rebuffered in PBS pH 7.4 by size exclusion chromatography on an FPLC system equipped with a Superdex200 10/300 GL column (GE Healthcare).

3.4. Introduction to "Click chemistry"

The azide-alkyne Huisgen cycloaddition ("Click chemistry") offers an orthogonal coupling reaction that does not interfere with functionalities present in proteins (Fig. 5.2). The formation of triazole can be catalyzed by Cu(I) (Deiters *et al.*, 2004; Wang *et al.*, 2003) or by the use of a strained alkyne (Debets *et al.*, 2010). This strategy especially allows the additional use of a cysteine residue to directionally couple two different ligands site-specifically (e.g., PEG, and a radioligand or fluorophore). This approach does require, however, the introduction of either an alkyne or an azido functionality into the protein by the use of nonnatural amino acids. The group of D. Tirrell has shown that the methionine analogs azidohomoalanine

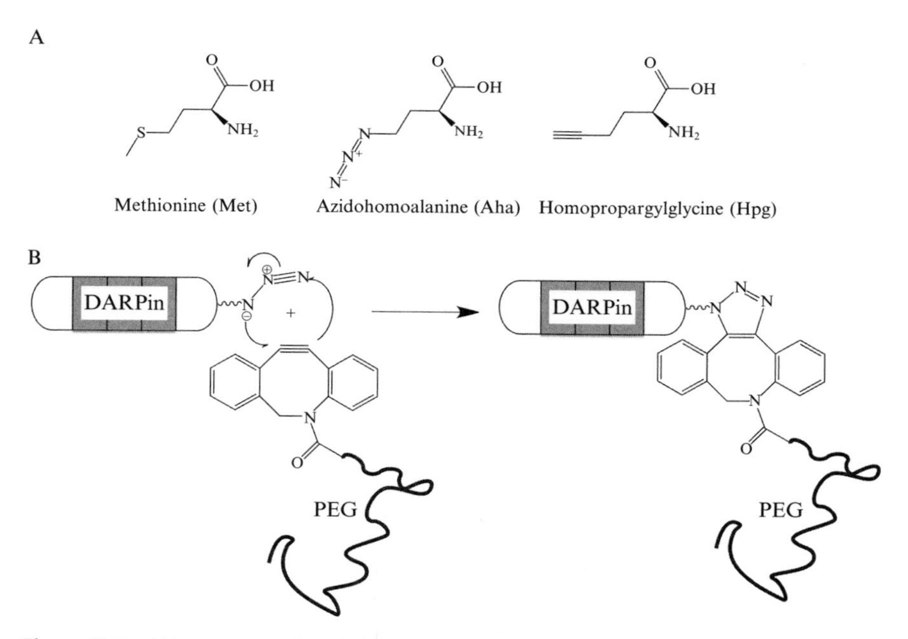

Figure 5.2 (A) Amino acid analogs which can be incorporated as a methionine analog in *E. coli*. (B) Azide-alkyne Huisgen cycloaddition ("click chemistry") of an azidohomoalanine residue, incorporated instead of the initiator methionine (the sole Met in the protein), with aza-dibenzocyclooctyne (DBCO), which is available coupled to polyethylene glycol (PEG).

(Aha) or homopropargylglycine (Hpg) can be incorporated in recombinant proteins in *E. coli* in high yields by the use of a Met auxotroph strain (Kiick *et al.*, 2002; Teeuwen *et al.*, 2009).

DARPins are ideal proteins for the N-terminal modification with methionine analogs *in vivo* as only few (mostly only one) internal methionines are present in the protein backbone. The single methionine present in the N-cap is not required and can be routinely exchanged (Simon *et al.*, 2012). The initiator Met, which is not cleaved in front of the subsequent Arg (of the $RGSH_6$ tag), is thus the only position where a clickable methionine analog will be incorporated. If a unique Met is desired at another position, the choice of a small second amino acid (Ala, Gly, Ser) leads to efficient cleavage of the initiator Met (Wang *et al.*, 2008).

3.5. Expression of "clickable" DARPins

DARPins can be conveniently labeled with the unnatural amino acids Aha or Hpg using a metabolic incorporation strategy. In a first step, the methionine in the conserved N-cap of the DARPins at position 34 can be exchanged for leucine using universal Quikchange^{TM} (Stratagene, La Jolla, USA) primers without altering the biochemical properties of the DARPin molecule. The resulting M34L DARPin mutant is subcloned into a vector containing the *lacI* gene under the stronger $lacI^q$ promoter control and transformed into the methionine auxotroph *E. coli* B-strain B834 DE3 (F^- *ompT* $hsdS_B$ $(r_B^- m_B^-)$ *gal dcm met* (DE3)) (Novagen) for expression. An overnight culture is used to inoculate 2×YT medium supplemented with 1% glucose and 100 μg/ml ampicillin at an OD_{600} of 0.1. Cells are cultivated for 2–3 h at 37 °C with agitation. Once an optical density of 1.0–1.2 is reached, the expression cultures are centrifuged (5000×*g*, 10 min, 4 °C) and the cells washed thoroughly for three times using ice-cold 0.9% NaCl solution by resuspension with a pipette in order to deplete all extracellular methionine. The cells are constantly cooled on ice to stop further cell growth during the wash procedure. Finally, the cells are reinoculated in M9 minimal medium (SelenoMethionine Medium Base plus Nutrient Mix [a nutrient mix with 19 amino acids], Molecular Dimensions Ltd., UK), containing 100 μg/ml ampicillin, and are supplemented further with Hpg (Chiralix, Nijmegen, Netherlands) or Aha (Bapeks, Riga, Latvia) (40 mg/l) from a sterile filtered stock solution. The expression cultures are incubated for 15 min at 30 °C in a shaker in order to additionally deplete all intracellular methionine pools of *E. coli*. The cells are subsequently induced using 1 m*M* isopropyl-β-D-thio-galactopyranoside (IPTG) and incubated for 4 h at 30 °C for expression of "clickable" DARPins. Finally, all cells are pelleted by centrifugation, resuspended in PBS, and pelleted again. The resulting pellets are snap-frozen and stored at −80 °C.

3.6. IMAC purification of "clickable" DARPins

The pellets are thawed on ice with HBS_W buffer (50 m*M* Hepes, 150 m*M* NaCl, 20 m*M* imidazole, pH 8.0) supplemented with 2 mg/l lysozyme and lysed for 3–4 times using a French® pressure cell press (Aminco) at 1200 psi. The lysate is centrifuged (28,000 × *g*, 4 °C, 1 h) and the supernatants are applied on pre-packed bench top Ni-NTA immobilized metal ion affinity chromatography (IMAC) columns (Qiagen). The proteins are washed with 20 column volumes (CV) of HBS_LS (50 m*M* Hepes, 20 m*M* NaCl, 20 m*M* imidazole, pH 8.0) followed by HBS_HS (50 m*M* Hepes, 1 *M* NaCl, 20 m*M* imidazole, pH 8.0) washes for 20 CV. Next, the IMAC columns are again washed with HBS_W (see above, 10 CVs) and the proteins are eluted using PBS_E (PBS, 300 m*M* imidazole, pH 7.4). The fractions are quantified using a Nanodrop spectrophotometer (Thermo Scientific). The proteins are thoroughly dialyzed against PBS overnight, aliquoted, snap-frozen, and stored for "click"-labeling. Usual protein yields are 25–30 mg DARPin per liter expression culture depending on the clone and expression time after induction.

3.7. Analysis of "clickable" DARPins

Incorporation of the nonnatural amino acids Aha or Hpg can be analyzed using N-terminal protein sequencing (Edman degradation). Briefly, 5–15 μl of the protein sample is diluted in 100 μl 0.1% TFA and loaded on a Prosorb Sample Preparation Cartridge (ABI), washed two times with 0.1% TFA, and transferred onto a PVDF membrane. The N-terminal amino acid composition is then analyzed using a PROCISE cLC 492 device. The nonnatural amino acids appear as new peaks that cannot be assigned to methionine: Whereas methionine gives a maximum signal at a retention time of 17.5 min, the maximum of either Hpg or Aha is approximately 16 min. Only very low amounts of methionine, in the noise range, can be detected after successful incorporation of the nonnatural amino acids.

The "clickable" DARPins can in addition be quantified via amino acid hydrolysis using the AccQ Tag Ultra kit (Waters). Ten microliters of the protein sample is dried followed by a vapor phase hydrolysis in 6 *M* HCl for 24 h at 110 °C under argon. The dry sample is taken up in 50 μl borate buffer and 10 μl are derivatized according to the manufacturer's recommendations, and the amino acids are separated on a UPLC system (Waters). Additionally, norvaline is taken as an internal standard and transferrin as a control protein. The analysis normally reveals no methionine present in the "clickable" DARPins after hydrolysis as a result of the substitution by the methionine surrogates.

3.8. Site-specific PEGylation of DARPins using Cu-free "click chemistry"

PEGylation of DARPins at the N-terminus is performed by using aza-dibenzocyclooctyne (DBCO)-PEG-20kDa (Click Chemistry Tools) for Cu-free labeling (Debets *et al.*, 2010). "Clickable" azido-DARPins (containing an N-terminal Aha) (100 μ*M* in PBS) are mixed with two- to threefold molar excess of DBCO-PEG-20kDa from a 5 m*M* DBCO-PEG-20kDa stock in PBS. At these concentrations, after 3–4 h at room temperature or overnight at 4 °C (without the necessity of agitation), the yield reaches about 70–80% (Fig. 5.3). Increasing the temperature leads to increased reaction rates. The maximal yield after extended reaction times (48 h at room temperature) is usually about 85%, where the residual non-PEGylated 15% can be assigned to DARPin molecules whose N-terminal methionine substitute Aha has been posttranslationally cleaved off by the *E. coli* methionine aminopeptidase *in vivo* (Wang *et al.*, 2008), as detected by N-terminal sequencing.

The reaction is not depending on particular buffers, can be performed in a broad pH range, and can even be used in combination with buffers containing imidazole from the elution of IMAC purification straight away. The PEGylated product is detected as a band shift of the DARPins using 15% SDS-PAGE as described above for the cys-maleimide PEGylation.

3.9. Purification of "click" PEGylated DARPins

Anion exchange is used to separate PEGylated from non-PEGylated DARPins and free DBCO-PEG (Seely and Richey, 2001), as described for the cys-maleimide PEGylation (Section 3.3). In order to reduce viscosity and ionic strength, the proteins are diluted in Buffer A (50 m*M* Hepes, 20 m*M*

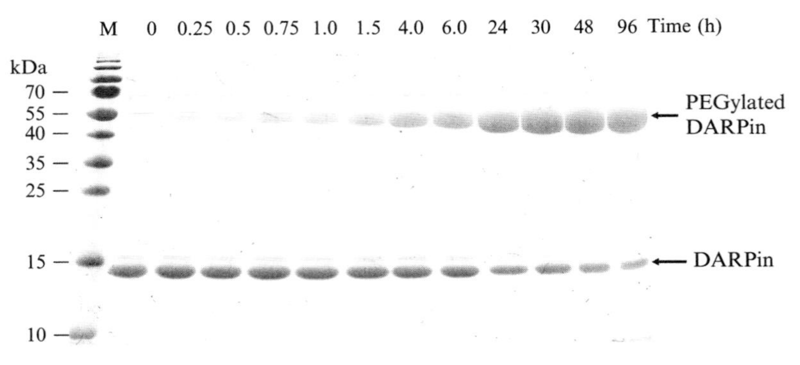

Figure 5.3 Time course of the Cu(I)-independent "click reaction": reaction of a DARPin containing an N-terminal Aha with DBCO-PEG-20kDa (15% SDS-PAGE, stained with Coomassie Brilliant Blue).

NaCl, pH 8.5) and loaded on a MonoQ 5/10 column (GE Healthcare) connected to a pre-equilibrated ÄKTA Explorer FPLC system (GE Healthcare). The proteins are eluted using a linear gradient of Buffer B (50 m*M* Hepes, 1 *M* NaCl, pH 8.5). Whereas the non-PEGylated DARPins remain bound to the anion exchange column, the PEGylated DARPins are eluted early as a result of the covalent attachment of PEG and can be quantitatively eluted at lower salt concentrations as described elsewhere (Seely and Richey, 2001). Using this method, unreacted DBCO-PEG-20kDa is eluted in the flow through and can thus be separated from the proteins. The fractions containing the PEGylated DARPins are pooled and desalted or dialyzed against PBS for further use.

3.10. Measurement of DARPin binding affinity to whole cells

Frequently, DARPins binding to surface receptors have been obtained by selection to the purified receptor, and thus it is important to determine the affinity to the receptor in the natural context on intact viable cells. Binding might be influenced by the local membrane environment, the accessibility of particular epitopes, and posttranslational modifications of the receptors.

Unfortunately, there are technical challenges with many methods when applying them to viable cells. The most common method, the measurement of equilibrium binding of radioactively labeled ligand by radioimmunoassay, has a large experimental error because of the high number of washing steps and the difficulty in discriminating between intact cells and damaged cells with permeable membranes. We therefore elaborated methodology based on fluorescence-activated cell sorting (FACS). By gating on forward and side scatter, we can selectively measure binding to live cells.

To obtain correct values for ligand binding parameters from surface receptors such as HER2, internalization has to be minimized. We thus routinely preincubate the cells with 0.1% sodium azide (blocking ATP-driven endocytosis) for 30 min at 37 °C before incubating with DARPins and have this concentration of sodium azide present during all steps. It was found that by this means, >95% inhibition of HER2 internalization is achieved while keeping reasonable viability of BT474 cells. Under these conditions, DARPins completely dissociate over time, with almost no fluorescence remaining associated with the cells after extended time. However, it has to be noted that the efficient inhibition of internalization could be highly cell-type- and receptor dependent. For other cell lines or receptor–ligand systems, different classes of inhibitors of endocytosis, for example, monodansyl cadaverine, phenylarsine oxide, chlorpromazine, amantadine, monensin, or K^+-depletion, may be the preferred choice (Vercauteren *et al.*, 2010). It is therefore recommended to define the most suitable conditions for suppression of endocytosis in every particular experimental setup.

3.11. Determination of the dissociation constant by equilibrium titration on cells

Equilibrium titration experiments can in principle be used to quantify the number of receptors on the cell surface as well as for determination of ligand affinity. It should be noted, however, that this method is limited to K_D values in the nanomolar range and cannot be used for tighter-binding ligands (see Section 3.12).

Since reaching equilibrium is a prerequisite for the accurate determination of dissociation constants, it is important that incubation with the labeled ligand proceeds for a sufficient period of time, which may range from a few minutes to several hours, depending on ligand affinity, temperature, and analyte concentration. In a typical equilibrium titration experiment, cells (e.g., BT474 or SKOV3 overexpressing the HER2 receptor) are incubated at concentration of 1×10^6 cells/ml with the labeled ligand, for example, DARPin-Alexa Fluor-488 conjugates, at concentration varying from 50 pM to 100 nM at room temperature for 1 h or longer, if required (Fig. 5.4). The incubation is maintained under conditions with minimized

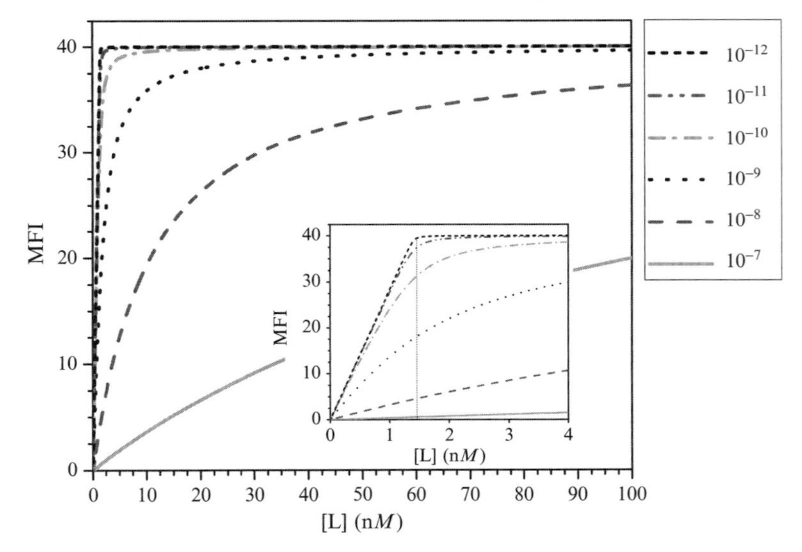

Figure 5.4 Equilibrium titration of cellular receptors with labeled ligands. Simulated data of mean fluorescent intensity are shown according to Eq. (5.2), as they are typically obtained when measuring binding of a fluorescently labeled DARPin by FACS. It can be seen that a reliable determination of K_D by curve fitting is only possible for affinities weaker than 1 nM, otherwise the curves become practically identical (independent of K_D), especially when considering scatter in the data. Conversely, with high-affinity data, the equivalence point allows precise determination of the receptor concentration (insert, vertical line indicating the equivalence point). When the number of cells is known, the receptor number per cell is directly obtained.

receptor internalization in PBS containing 1% (w/v) BSA and 0.1% (w/v) sodium azide (abbreviated PBSBA). For each concentration point, a 1-ml cell-DARPin suspension is subjected to FACS analysis on the FL1 detector (530/20 band pass filter) of an appropriate instrument equipped with a 488-nm argon laser. Before measurements, the cells are briefly washed once with ice-cold PBS to suppress the background arising from the unspecific binding of labeled ligand. If necessary, control samples are preincubated with 100-fold excess of unlabeled DARPins, and the recorded values are subsequently subtracted from total binding to correct for unspecific background.

The data are usually analyzed by nonlinear regression to a simplified binding isotherm:

$$\mathrm{MFI} = \mathrm{MFI}_{\max} \frac{[\mathrm{L}]}{[\mathrm{L}] + K_{\mathrm{D}}} \tag{5.1}$$

where MFI is the measured mean fluorescence intensity (corrected for the background), $\mathrm{MFI}_{\max}$ the background-corrected plateau value corresponding to saturated receptors, [L] the concentration of the labeled ligand, and K_D the dissociation constant to be determined. Equation (5.1) does not take ligand depletion into account, which occurs with tightly binding ligands at stoichiometric ratios with receptor. A more accurate version is the following equation

$$\mathrm{MFI} = \frac{\mathrm{MFI}_{\max}}{[\mathrm{C}]_{\max}} \left(\frac{[\mathrm{R}] + [\mathrm{L}] + K_{\mathrm{D}}}{2} - \sqrt{\left(\frac{[\mathrm{R}] + [\mathrm{L}] + K_{\mathrm{D}}}{2} \right)^2 - [\mathrm{L}]\cdot[\mathrm{R}]} \right) \tag{5.2}$$

where the variables are as in Eq. (5.1), and [R] is the molar concentration of receptor. The value $\mathrm{MFI}_{\max}/[\mathrm{C}]_{\max}$ is a proportionality constant determined from the fit, relating the maximal response to the maximal concentration of receptor–ligand complex C.

It has to be noted that determinations of dissociation constants by equilibrium titration possesses some inherent limitations. For instance, if the analyte concentration is far above its K_D, only inaccurate affinity determinations are possible, as the receptor–ligand complexes are saturated upon titration (Fig. 5.4). For high-affinity binders (K_D below 100 p*M*), this problem cannot be simply solved by dilution, as the sensitivity limits of most detection methods will be reached. However, if the ligand affinity is too low (K_D above 100 n*M*), the rate of dissociation is usually too high for ligand to remain bound to the receptor during washing steps and analysis. Although this problem can be mitigated by an indirect measurement setup employing the unlabeled ligand of interest in combination with a labeled high-affinity competitor, such a high-affinity binder is often not available. Hence, accurate measurements for equilibrium dissociation constant determination can be obtained with ligand concentrations in the range of the K_D, and experimental constraints demand that this has to be around 1–100 n*M*.

However, equilibrium titration with labeled ligand can be readily used to estimate (from the equivalence point in the titration) the total number of receptors expressed on cells (Zahnd *et al.*, 2010b). This is possible because the receptor concentration equals the concentration of the titrating ligand at the equivalence point of a binding isotherm recorded under conditions where the molar concentration of both the receptor and the ligand are sufficiently above the K_D. The situation with $K_D \ll [R]$ and $K_D > [R]$ is shown with simulated data in Fig. 5.4.

3.12. Determination of kinetic parameters of binding on cells

Measuring affinities of picomolar ligands by equilibrium titration is very inaccurate, inasmuch as for sensitivity reasons it would require ligand concentrations in the range far above the K_D value. Thus, it is preferable to determine the dissociation constant indirectly as the ratio of the dissociation rate constant and the association rate constant. Also, the knowledge of these kinetic parameters often contains additional valuable information.

The rate of association, defined as

$$\frac{d[C]}{dt} = k_{on} \cdot [L] \cdot [R] \tag{5.3}$$

where [L] is the DARPin ligand, [R] is the uncomplexed receptor, [C] is the receptor–ligand complex, and k_{on} (in units of $M^{-1}s^{-1}$) is the association rate constant, can be determined by following the binding of labeled DARPin to cells during defined time intervals. The determination of k_{on} does require an estimate of the receptor number per cell and knowledge of the cell number, which is best achieved by titration at high concentrations (Fig. 5.4, inset). Note that [C] is proportional to MFI (more precisely, the receptor-specific part of the signal)

The rate of dissociation is defined as

$$-\frac{d[C]}{dt} = k_{off} \cdot [C] \tag{5.4}$$

where k_{off} is the dissociation rate constant (units of s^{-1}). At equilibrium, the sum of both rates is zero, and thus

$$K_D = \frac{[L] \cdot [R]}{[C]} = \frac{k_{off}}{k_{on}} \tag{5.5}$$

For on-rate determinations, BT474 cells are incubated at a concentration of 1×10^6 cells/ml with 2.5, 7.5, and 22.5 n*M* DARPin-Alexa Fluor-488 conjugates in PBSBA at room temperature for defined time intervals, ranging from 1 to 60 min. For each time point, a 1-ml aliquot of cells is withdrawn and subjected to FACS. Since the applied concentrations of the labeled ligand conjugates are very low, and since the time resolution of the measurement is to be maintained to ensure the accuracy of the on-rate determination, the samples are processed without further washing. For each time point, at least 10^4 intact cells (gated as a uniform population on a FSC/SSC scatter plot) are counted, and the MFI is recorded. Parameters are fitted with a monoexponential equation

$$\mathrm{MFI} = \mathrm{MFI}_{\mathrm{max}}\left(1 - \mathrm{e}^{-k_{\mathrm{obs}}t}\right) \tag{5.6}$$

where k_{obs} is the observed association rate constant, and

$$k_{\mathrm{obs}} = k_{\mathrm{on}} \cdot [\mathrm{L}] + k_{\mathrm{off}} \tag{5.7}$$

In an association-type experiment where k_{off} is not immeasurably small, the increment of MFI combines the ligand binding with the concomitant ligand dissociation and thus, more precisely, describes the rate of reaching equilibrium in a given receptor–ligand system. As a consequence, the actual k_{on} can only be determined in conjunction with the preceding measurement of the dissociation rate constant k_{off}. Typically, the ligands display on-rate constants in the range of 10^5 to $10^6\,M^{-1}\,\mathrm{s}^{-1}$. An example of an on-rate determination experiment is shown in Fig. 5.5A.

To perform off-rate determinations, the receptors are saturated with the labeled ligand, which is subsequently allowed to dissociate, and the amount of remaining cell-bound ligand is repeatedly monitored by FACS at defined time points, usually spanning the interval from 1 min to several hours, depending on dissociation rate. To prevent rebinding of dissociated ligand to the free receptor, which would prevent a correct determination of k_{off}, a large excess (100-fold) of unlabeled ligand should be included in the reaction, which will block each unoccupied receptor after dissociation of the labeled ligand. For instance, BT474 cells at a concentration of 1×10^6 cells/100 μl are saturated with 50 n*M* anti-HER2 DARPin-Alexa Fluor-488 conjugates for 60 min on ice in PBSBA. Thereafter, the cells are washed extensively (at least three times) with ice-cold PBSBA to remove unbound labeled DARPin. The cells are then diluted 10-fold to a final concentration of 1×10^6 cells/ml in PBSBA and gently agitated at room temperature for the duration of the measurement. To prevent reassociation of dissociated DARPin-Alexa Fluor-488 conjugate, an excess of competitor (100 n*M* unlabeled DARPin) must be added. Under these conditions, complete loss of fluorescence over time should be seen, indicating that

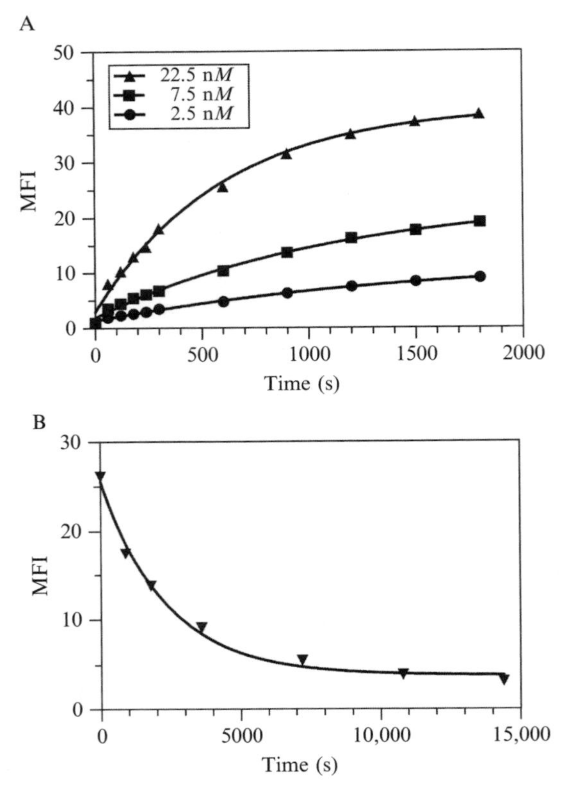

Figure 5.5 Association and dissociation kinetics of DARPin-Alexa Fluor-488 conjugate, binding to HER2 on BT474 cells. The mean fluorescence intensity (MFI), as recorded by FACS, is monitored as a function of time. (A) The DARPin-HER2 association is monitored at three DARPin-Alexa Fluor-488 concentrations as indicated. The association rate constant is evaluated with Eqs. (5.6) and (5.7). (B) DARPin dissociation of BT474 cells preloaded with DARPin-Alexa Fluor-488 after dilution in fresh buffer in the presence of unlabeled competitor DARPin. The dissociation rate constant is evaluated with Eq. (5.8). The K_D is determined with Eq. (5.5).

both rebinding and receptor internalization have been successfully prevented. Before FACS measurements, the cells are briefly washed once with ice-cold PBS. A typical dissociation experiment is shown in Fig. 5.5B.

Collected data are fitted with the following equation

$$\mathrm{MFI} = (\mathrm{MFI}_{\mathrm{max}} - \mathrm{MFI}_{\infty})\mathrm{e}^{-k_{\mathrm{off}}t} \qquad (5.8)$$

where $\mathrm{MFI}_{\mathrm{max}}$ is the initial value, and MFI_{∞} is the plateau value after long dissociation.

In principle, the experiment can be carried out also in the reverse fashion, by preincubating the cells (1×10^6 cells/100 μl) with 50 n*M* unlabeled DARPin. Upon washing cells three times to remove unbound DARPins, the dissociation is started by competing with 100 n*M* of Alexa Fluor-488-conjugated DARPin. In this case, the dissociation of the unlabeled DARPin is measured by the increase of fluorescence, as dissociation is much slower than the very rapid association of the labeled DARPin under these conditions. Irrespective of experimental setup, the dissociation rates for potent receptor-binding DARPins are typically in the range of 10^{-3}–10^{-5} s^{-1}.

3.13. Expression and purification of DARPin–toxin fusion proteins

Potential applications for DARPins include the targeted delivery of protein toxins (Martin-Killias *et al.*, 2011). *P. aeruginosa* ETA (Siegall *et al.*, 1989) has been widely used for this purpose. The DARPin replaces the natural N-terminal domain of the toxin, which mediates binding to the α_2 macroglobulin receptor. DARPins are fused via a 12-amino acid linker $(GSG_4)_2$ to the 40-kDa truncated form of $ETA_{252–608}$KDEL (ETA″), which was cloned as described (Di Paolo *et al.*, 2003; Wels *et al.*, 1992). ETA″ comprises residues Glu252–Pro608 (numbering of the mature protein), fused to a C-terminal His_6 tag followed by KDEL (denoted $ETA_{252–608}$KDEL or ETA″). For purification and detection, the construct in addition contains an MRGS-His_6 tag at the N-terminus.

The DARPin–toxin fusion proteins are expressed in soluble form in the cytoplasm of *E. coli*, using a vector derived from pQE30 (Qiagen), but containing the *lac* repressor gene under the control of the stronger *lacI^q* promoter. The DARPin–ETA″ fusion proteins are expressed in soluble form in the *E. coli* strain BL21 (DE3) (Stratagene). Precultures are grown in 2×YT broth (16 g/l tryptone, 10 g/l yeast extract, 5 g/l NaCl) containing 100 μg/ml ampicillin and 1% glucose, with shaking overnight at 37 °C. Cultures are diluted in Terrific Broth (Tartoff and Hobbs, 1987) (12 g/l tryptone, 24 g/l yeast extract, 4 ml/l glycerol, 17 m*M* KH_2PO_4, and 72 m*M* K_2HPO_4) containing 50 μg/ml ampicillin and 0.1% glucose to OD_{600} 0.1 and grown at 37 °C. Upon reaching OD_{600} of 1.2, the temperature is reduced to 30 °C and recombinant protein production is induced with 1 m*M* IPTG. Cells are harvested 5 h after induction, washed once with TBS_{400} (50 m*M* Tris–HCl, 400 m*M* NaCl, pH 8.0, at 4 °C), snap-frozen in liquid N_2, and stored at −80 °C or used directly for downstream processing.

For purification, the bacteria are resuspended in TBS_{400}_W (50 m*M* Tris–HCl, pH 7.4, 400 m*M* NaCl, 20 m*M* imidazole) and lysed with a TS 1.1-kW cell disruptor (Constant Systems Ltd.) using a pressure of 35 MPa. The cell lysate is centrifuged (48,000×*g*, 30 min at 4 °C) and filtered (pore size 0.22 μm) prior to purifying the fusion toxins from the supernatant by a gravity-flow Ni-NTA superflow column (Qiagen). After loading, the column is washed with 10 CV of low-salt TBS_W (50 m*M* Tris–HCl, pH 7.4, 50 m*M* NaCl, 20 m*M* imidazole) and high-salt TBS_W (50 m*M* Tris–HCl, pH 7.4, 1 *M* NaCl, 20 m*M* imidazole) to remove unspecifically bound material, followed by 10 CV of PBS pH 7.4 containing 20 m*M* imidazole. DARPin–ETA″ is eluted with PBS containing 250 m*M* imidazole and dialyzed twice against PBS at 4 °C. The protein yield after dialysis is up to 90 mg/l of bacterial culture.

For *in vivo* applications, the DARPin–ETA″ fusion toxins are further purified to eliminate endotoxin. To this end, an additional washing step is performed during IMAC purification using 150 column volumes PBS containing 20 m*M* imidazole and 0.1% Triton-X-114, a nonionic detergent which efficiently solubilizes endotoxin at a temperature below its cloud point of 23 °C (Reichelt *et al.*, 2006; Zimmerman *et al.*, 2006). Next, the monomeric fraction of DARPin–ETA″ is separated by size exclusion chromatography using a Superdex-200 10/300 GL column (GE Healthcare) and further depleted of residual endotoxin by passages over an EndoTrap Red affinity column (Hyglos). The final endotoxin content is determined using the Limulus amebocyte lysate endochrome kit (Charles River).

The DARPin contains no cysteine by design, but the ETA″ possesses two disulfide bonds. As the fusion protein is expressed in the reducing cytoplasm of *E. coli*, it is important to verify that proper disulfides have indeed formed. It is noteworthy that their formation normally proceeds spontaneously with excellent yield (Martin-Killias *et al.*, 2011) (cf. Section 2.3), presumably by air oxidation.

SS-bond formation can be quantified according to Hansen *et al.* (2007). Like in the classic Ellman test (Ellman, 1959), the free thiols of L-cysteine are detected by incubation for 30 min at room temperature with 0.36 m*M* 4,4′-dithiodipyridine (4-DPS), a reagent which stoichiometrically reacts with free thiols forming two 4-thiopyridone (4-TP) molecules. The reaction is stopped by the addition of HCl to a final concentration of 0.2 *M*. However, unlike in the classic Ellman test, here 4-TP is assayed by analyzing 20-μl aliquots by reverse phase HPLC on a 250 mm NUCLEOSIL® 120–5 C_{18} column (Macherey-Nagel), using isocratic elution at a flow rate of 1 ml/min in 50 m*M* potassium acetate, pH 4.0. Peaks are detected at 324 nm.

First, a series of standards from 1 to 133 μ*M* cysteine is prepared in 100 m*M* citrate, 0.2 m*M* ethylenediaminetetraacetic acid, and 6 *M* urea at pH 4.5. The standard curve can be calculated from the peak integral of the 4-TP peaks.

The protein of interest is treated with 4-DPS in the presence of 8 *M* urea to detect the free thiols in the protein. In parallel, the same procedure is applied to a sample which had been reduced with sodium borohydride (BH), a strong reducing reagent, to expose all apparent thiols (Hansen *et al.*, 2007). To avoid pressure buildup in the reduced sample tubes, caused by hydrogen development, lids are pierced with a needle. Excess BH is rapidly and quantitatively removed using 1.8 *M* HCl, and the lower pH at the same time preserves the integrity of the thiol groups. It is important to keep the incubation time in presence 4-DPS identical for the reduced and nonreduced samples by adding 4-DPS to both samples at the same time to assure accuracy. Analysis of DARPin–ETA″ typically revealed less than 10% free thiols, suggesting over 90% correctly formed disulfide bridges.

ACKNOWLEDGMENT

Work in the author's laboratory on establishing the methods was supported by the Swiss National Science Foundation and Swiss Anti-Cancer League (Krebsliga Schweiz; KFS 02448-08-2009).

REFERENCES

Adams, G. P., McCartney, J. E., Tai, M. S., Oppermann, H., Huston, J. S., Stafford, W. F., 3rd, Bookman, M. A., Fand, I., Houston, L. L., and Weiner, L. M. (1993). Highly specific in vivo tumor targeting by monovalent and divalent forms of 741F8 anti-c-erbB-2 single-chain Fv. *Cancer Res.* **53,** 4026–4034.

Adams, G. P., Schier, R., McCall, A. M., Simmons, H. H., Horak, E. M., Alpaugh, R. K., Marks, J. D., and Weiner, L. M. (2001). High affinity restricts the localization and tumor penetration of single-chain Fv antibody molecules. *Cancer Res.* **61,** 4750–4755.

Amalfitano, A., and Parks, R. J. (2002). Separating fact from fiction: Assessing the potential of modified adenovirus vectors for use in human gene therapy. *Curr. Gene Ther.* **2,** 111–133.

Amstutz, P., Binz, H. K., Parizek, P., Stumpp, M. T., Kohl, A., Grütter, M. G., Forrer, P., and Plückthun, A. (2005). Intracellular kinase inhibitors selected from combinatorial libraries of designed ankyrin repeat proteins. *J. Biol. Chem.* **280,** 24715–24722.

Amstutz, P., Koch, H., Binz, H. K., Deuber, S. A., and Plückthun, A. (2006). Rapid selection of specific MAP kinase-binders from designed ankyrin repeat protein libraries. *Protein Eng. Des. Sel.* **19,** 219–229.

Bandeiras, T. M., Hillig, R. C., Matias, P. M., Eberspaecher, U., Fanghänel, J., Thomaz, M., Miranda, S., Crusius, K., Pütter, V., Amstutz, P., Gulotti-Georgieva, M., Binz, H. K., *et al.* (2008). Structure of wild-type Plk-1 kinase domain in complex with a selective DARPin. *Acta Crystallogr. D Biol. Crystallogr.* **64,** 339–353.

Baumann, M. J., Eggel, A., Amstutz, P., Stadler, B. M., and Vogel, M. (2010). DARPins against a functional IgE epitope. *Immunol. Lett.* **133,** 78–84.

Bibi, E. (2011). Early targeting events during membrane protein biogenesis in Escherichia coli. *Biochim. Biophys. Acta* **1808,** 841–850.

Biggers, K., and Scheinfeld, N. (2008). VB4-845, a conjugated recombinant antibody and immunotoxin for head and neck cancer and bladder cancer. *Curr. Opin. Mol. Ther.* **10,** 176–186.

Binz, H. K., Stumpp, M. T., Forrer, P., Amstutz, P., and Plückthun, A. (2003). Designing repeat proteins: Well-expressed, soluble and stable proteins from combinatorial libraries of consensus ankyrin repeat proteins. *J. Mol. Biol.* **332,** 489–503.

Binz, H. K., Amstutz, P., Kohl, A., Stumpp, M. T., Briand, C., Forrer, P., Grütter, M. G., and Plückthun, A. (2004). High-affinity binders selected from designed ankyrin repeat protein libraries. *Nat. Biotechnol.* **22,** 575–582.

Binz, H. K., Kohl, A., Plückthun, A., and Grütter, M. G. (2006). Crystal structure of a consensus-designed ankyrin repeat protein: Implications for stability. *Proteins* **65,** 280–284.

Bork, P. (1993). Hundreds of ankyrin-like repeats in functionally diverse proteins: Mobile modules that cross phyla horizontally? *Proteins* **17,** 363–374.

Chapman, A. P. (2002). PEGylated antibodies and antibody fragments for improved therapy: A review. *Adv. Drug Deliv. Rev.* **54,** 531–545.

Debets, M. F., van Berkel, S. S., Schoffelen, S., Rutjes, F. P., van Hest, J. C., and van Delft, F. L. (2010). Aza-dibenzocyclooctynes for fast and efficient enzyme PEGylation via copper-free (3+2) cycloaddition. *Chem. Commun. (Camb.)* **46,** 97–99.

Deiters, A., Cropp, T. A., Summerer, D., Mukherji, M., and Schultz, P. G. (2004). Site-specific PEGylation of proteins containing unnatural amino acids. *Bioorg. Med. Chem. Lett.* **14,** 5743–5745.

Di Paolo, C., Willuda, J., Kubetzko, S., Lauffer, I., Tschudi, D., Waibel, R., Plückthun, A., Stahel, R. A., and Zangemeister-Wittke, U. (2003). A recombinant immunotoxin derived from a humanized epithelial cell adhesion molecule-specific single-chain antibody fragment has potent and selective antitumor activity. *Clin. Cancer Res.* **9,** 2837–2848.

Dreier, B., and Plückthun, A. (2010). Ribosome display, a technology for selecting and evolving proteins from large libraries. *Methods Mol. Biol.* **687,** 283–306.

Dreier, B., and Plückthun, A. (2011). Ribosome display: A technology for selecting and evolving proteins from large libraries. *Methods Mol. Biol.* **687,** 283–306.

Dreier, B., Mikheeva, G., Belousova, N., Parizek, P., Boczek, E., Jelesarov, I., Forrer, P., Plückthun, A., and Krasnykh, V. (2011). Her2-specific multivalent adapters confer designed tropism to adenovirus for gene targeting. *J. Mol. Biol.* **405,** 410–426.

Dröge, M. J., Boersma, Y. L., Braun, P. G., Buining, R. J., Julsing, M. K., Selles, K. G., van Dijl, J. M., and Quax, W. J. (2006). Phage display of an intracellular carboxylesterase of Bacillus subtilis: Comparison of Sec and Tat pathway export capabilities. *Appl. Environ. Microbiol.* **72,** 4589–4595.

Eggel, A., Baumann, M. J., Amstutz, P., Stadler, B. M., and Vogel, M. (2009). DARPins as bispecific receptor antagonists analyzed for immunoglobulin E receptor blockage. *J. Mol. Biol.* **393,** 598–607.

Eggel, A., Buschor, P., Baumann, M. J., Amstutz, P., Stadler, B. M., and Vogel, M. (2011). Inhibition of ongoing allergic reactions using a novel anti-IgE DARPin-Fc fusion protein. *Allergy* **66,** 961–968.

Ellman, G. L. (1959). Tissue sulfhydryl groups. *Arch. Biochem. Biophys.* **82,** 70–77.

Fekkes, P., and Driessen, A. J. (1999). Protein targeting to the bacterial cytoplasmic membrane. *Microbiol. Mol. Biol. Rev.* **63,** 161–173.

Forrer, P., Stumpp, M. T., Binz, H. K., and Plückthun, A. (2003). A novel strategy to design binding molecules harnessing the modular nature of repeat proteins. *FEBS Lett.* **539,** 2–6.

Forrer, P., Binz, H. K., Stumpp, M. T., and Plückthun, A. (2004). Consensus design of repeat proteins. *ChemBioChem* **5,** 183–189.

Getz, E. B., Xiao, M., Chakrabarty, T., Cooke, R., and Selvin, P. R. (1999). A comparison between the sulfhydryl reductants tris(2-carboxyethyl)phosphine and dithiothreitol for use in protein biochemistry. *Anal. Biochem.* **273,** 73–80.

Glockshuber, R., Malia, M., Pfitzinger, I., and Plückthun, A. (1990). A comparison of strategies to stabilize immunoglobulin Fv-fragments. *Biochemistry* **29,** 1362–1367.

Hanes, J., and Plückthun, A. (1997). *In vitro* selection and evolution of functional proteins by using ribosome display. *Proc. Natl. Acad. Sci. USA* **94,** 4937–4942.

Hanes, J., Jermutus, L., Weber-Bornhauser, S., Bosshard, H. R., and Plückthun, A. (1998). Ribosome display efficiently selects and evolves high-affinity antibodies *in vitro* from immune libraries. *Proc. Natl. Acad. Sci. USA* **95,** 14130–14135.

Hanes, J., Jermutus, L., and Plückthun, A. (2000a). Selecting and evolving functional proteins *in vitro* by ribosome display. *Methods Enzymol.* **328,** 404–430.

Hanes, J., Schaffitzel, C., Knappik, A., and Plückthun, A. (2000b). Picomolar affinity antibodies from a fully synthetic naive library selected and evolved by ribosome display. *Nat. Biotechnol.* **18,** 1287–1292.

Hansen, R. E., Ostergaard, H., Norgaard, P., and Winther, J. R. (2007). Quantification of protein thiols and dithiols in the picomolar range using sodium borohydride and 4,4′-dithiodipyridine. *Anal. Biochem.* **363,** 77–82.

Huber, T., Steiner, D., Röthlisberger, D., and Plückthun, A. (2007). In vitro selection and characterization of DARPins and Fab fragments for the co-crystallization of membrane proteins: The Na(+)-citrate symporter CitS as an example. *J. Struct. Biol.* **159,** 206–221.

Hussain, S., Plückthun, A., Allen, T. M., and Zangemeister-Wittke, U. (2006). Chemosensitization of carcinoma cells using epithelial cell adhesion molecule-targeted liposomal antisense against bcl-2/bcl-xL. *Mol. Cancer Ther.* **5,** 3170–3180.

Hussain, S., Plückthun, A., Allen, T. M., and Zangemeister-Wittke, U. (2007). Antitumor activity of an epithelial cell adhesion molecule targeted nanovesicular drug delivery system. *Mol. Cancer Ther.* **6,** 3019–3027.

Interlandi, G., Wetzel, S. K., Settanni, G., Plückthun, A., and Caflisch, A. (2008). Characterization and further stabilization of designed ankyrin repeat proteins by combining molecular dynamics simulations and experiments. *J. Mol. Biol.* **375,** 837–854.

Kiick, K. L., Saxon, E., Tirrell, D. A., and Bertozzi, C. R. (2002). Incorporation of azides into recombinant proteins for chemoselective modification by the Staudinger ligation. *Proc. Natl. Acad. Sci. USA* **99,** 19–24.

Knappik, A., Ge, L., Honegger, A., Pack, P., Fischer, M., Wellnhofer, G., Hoess, A., Wölle, J., Plückthun, A., and Virnekäs, B. (2000). Fully synthetic human combinatorial antibody libraries (HuCAL) based on modular consensus frameworks and CDRs randomized with trinucleotides. *J. Mol. Biol.* **296,** 57–86.

Kobe, B., and Kajava, A. V. (2000). When protein folding is simplified to protein coiling: The continuum of solenoid protein structures. *Trends Biochem. Sci.* **25,** 509–515.

Kohl, A., Binz, H. K., Forrer, P., Stumpp, M. T., Plückthun, A., and Grütter, M. G. (2003). Designed to be stable: Crystal structure of a consensus ankyrin repeat protein. *Proc. Natl. Acad. Sci. USA* **100,** 1700–1705.

Kramer, M. A., Wetzel, S. K., Plückthun, A., Mittl, P. R., and Grütter, M. G. (2010). Structural determinants for improved stability of designed ankyrin repeat proteins with a redesigned C-capping module. *J. Mol. Biol.* **404,** 381–391.

Kubetzko, S., Sarkar, C. A., and Plückthun, A. (2005). Protein PEGylation decreases observed target association rates via a dual blocking mechanism. *Mol. Pharmacol.* **68,** 1439–1454.

Kurfürst, M. M. (1992). Detection and molecular weight determination of polyethylene glycol-modified hirudin by staining after sodium dodecyl sulfate-polyacrylamide gel electrophoresis. *Anal. Biochem.* **200,** 244–248.

Li, J., Mahajan, A., and Tsai, M. D. (2006). Ankyrin repeat: A unique motif mediating protein-protein interactions. *Biochemistry* **45,** 15168–15178.

Luginbühl, B., Kanyo, Z., Jones, R. M., Fletterick, R. J., Prusiner, S. B., Cohen, F. E., Williamson, R. A., Burton, D. R., and Plückthun, A. (2006). Directed evolution of an anti-prion protein scFv fragment to an affinity of 1 pM and its structural interpretation. *J. Mol. Biol.* **363,** 75–97.

Martin-Killias, P., Stefan, N., Rothschild, S., Plückthun, A., and Zangemeister-Wittke, U. (2011). A novel fusion toxin derived from an EpCAM-specific designed ankyrin repeat protein has potent antitumor activity. *Clin. Cancer Res.* **17,** 100–110.

Mattheakis, L. C., Bhatt, R. R., and Dower, W. J. (1994). An in vitro polysome display system for identifying ligands from very large peptide libraries. *Proc. Natl. Acad. Sci. USA* **91,** 9022–9026.

Münch, R. C., Mühlebach, M. D., Schaser, T., Kneissl, S., Jost, C., Plückthun, A., Cichutek, K., and Buchholz, C. J. (2011). DARPins: An efficient targeting domain for lentiviral vectors. *Mol. Ther.* **19,** 686–693.

Nangola, S., Minard, P., and Tayapiwatana, C. (2010). Appraisal of translocation pathways for displaying ankyrin repeat protein on phage particles. *Protein Expr. Purif.* **74,** 156–161.

Nicklin, S. A., Wu, E., Nemerow, G. R., and Baker, A. H. (2005). The influence of adenovirus fiber structure and function on vector development for gene therapy. *Mol. Ther.* **12,** 384–393.

Pancer, Z., Amemiya, C. T., Ehrhardt, G. R., Ceitlin, J., Gartland, G. L., and Cooper, M. D. (2004). Somatic diversification of variable lymphocyte receptors in the agnathan sea lamprey. *Nature* **430,** 174–180.

Paschke, M., and Höhne, W. (2005). A twin-arginine translocation (Tat)-mediated phage display system. *Gene* **350,** 79–88.

Plückthun, A. (2012). Ribosome display: A perspective. *Methods Mol. Biol.* **805,** 3–28.

Plückthun, A., and Moroney, S. E. (2005). Modern antibody technology: The impact on drug development. *In* "Modern Biopharmaceuticals," (J. Knäblein, ed.) Vol. 3, pp. 1147–1186. Wiley-VCH, Weinheim (4 vols).

Reichelt, P., Schwarz, C., and Donzeau, M. (2006). Single step protocol to purify recombinant proteins with low endotoxin contents. *Protein Expr. Purif.* **46,** 483–488.

Rudnick, S. I., Lou, J., Shaller, C. C., Tang, Y., Klein-Szanto, A. J., Weiner, L. M., Marks, J. D., and Adams, G. P. (2011). Influence of affinity and antigen internalization on the uptake and penetration of anti-HER2 antibodies in solid tumors. *Cancer Res.* **71,** 2250–2259.

Schmidt, M. M., and Wittrup, K. D. (2009). A modeling analysis of the effects of molecular size and binding affinity on tumor targeting. *Mol. Cancer Ther.* **8,** 2861–2871.

Schweizer, A., Roschitzki-Voser, H., Amstutz, P., Briand, C., Gulotti-Georgieva, M., Prenosil, E., Binz, H. K., Capitani, G., Baici, A., Plückthun, A., and Grütter, M. G. (2007). Inhibition of caspase-2 by a designed ankyrin repeat protein: Specificity, structure, and inhibition mechanism. *Structure* **15,** 625–636.

Sedgwick, S. G., and Smerdon, S. J. (1999). The ankyrin repeat: A diversity of interactions on a common structural framework. *Trends Biochem. Sci.* **24,** 311–316.

Seely, J. E., and Richey, C. W. (2001). Use of ion-exchange chromatography and hydrophobic interaction chromatography in the preparation and recovery of polyethylene glycol-linked proteins. *J. Chromatogr. A* **908,** 235–241.

Shafer, D. E., Inman, J. K., and Lees, A. (2000). Reaction of tris(2-carboxyethyl)phosphine (TCEP) with maleimide and alpha-haloacyl groups: Anomalous elution of TCEP by gel filtration. *Anal. Biochem.* **282,** 161–164.

Siegall, C. B., Chaudhary, V. K., FitzGerald, D. J., and Pastan, I. (1989). Functional analysis of domains II, Ib, and III of Pseudomonas exotoxin. *J. Biol. Chem.* **264,** 14256–14261.

Simon, M., Zangemeister-Wittke, U. and Plückthun, A. (2012). Facile double-functionalization of designed ankyrin repeat proteins using click and thiol chemistries. *Bioconjug. Chem.* (in press).

Skerra, A., and Plückthun, A. (1988). Assembly of a functional immunoglobulin Fv fragment in *Escherichia coli. Science* **240,** 1038–1041.

Speck, J., Arndt, K. M., and Müller, K. M. (2011). Efficient phage display of intracellularly folded proteins mediated by the TAT pathway. *Protein Eng. Des. Sel.* **24,** 473–484.

Stefan, N., Martin-Killias, P., Wyss-Stoeckle, S., Honegger, A., Zangemeister-Wittke, U., and Plückthun, A. (2011). DARPins recognizing the tumor-associated antigen EpCAM selected by phage and ribosome display and engineered for multivalency. *J. Mol. Biol.* **413,** 826–843.

Steiner, D., Forrer, P., Stumpp, M. T., and Plückthun, A. (2006). Signal sequences directing cotranslational translocation expand the range of proteins amenable to phage display. *Nat. Biotechnol.* **24,** 823–831.

Steiner, D., Forrer, P., and Plückthun, A. (2008). Efficient selection of DARPins with sub-nanomolar affinities using SRP phage display. *J. Mol. Biol.* **382,** 1211–1227.

Stumpp, M. T., Forrer, P., Binz, H. K., and Plückthun, A. (2003). Designing repeat proteins: Modular leucine-rich repeat protein libraries based on the mammalian ribonuclease inhibitor family. *J. Mol. Biol.* **332,** 471–487.

Sullivan, N. J., Geisbert, T. W., Geisbert, J. B., Xu, L., Yang, Z. Y., Roederer, M., Koup, R. A., Jahrling, P. B., and Nabel, G. J. (2003). Accelerated vaccination for Ebola virus haemorrhagic fever in non-human primates. *Nature* **424,** 681–684.

Tartoff, K. D., and Hobbs, C. A. (1987). Improved media for growing plasmid and cosmid clones. *Bethesda Res. Labs Focus* **9,** 12–14.

Teeuwen, R. L., van Berkel, S. S., van Dulmen, T. H., Schoffelen, S., Meeuwissen, S. A., Zuilhof, H., de Wolf, F. A., and van Hest, J. C. (2009). "Clickable" elastins: Elastin-like polypeptides functionalized with azide or alkyne groups. *Chem. Commun. (Camb.)* 4022–4024.

Theurillat, J. P., Dreier, B., Nagy-Davidescu, G., Seifert, B., Behnke, S., Zurrer-Hardi, U., Ingold, F., Plückthun, A., and Moch, H. (2010). Designed ankyrin repeat proteins: A novel tool for testing epidermal growth factor receptor 2 expression in breast cancer. *Mod. Pathol.* **23,** 1289–1297.

Thurber, G. M., Schmidt, M. M., and Wittrup, K. D. (2008). Antibody tumor penetration: Transport opposed by systemic and antigen-mediated clearance. *Adv. Drug Deliv. Rev.* **60,** 1421–1434.

Trzpis, M., McLaughlin, P. M., de Leij, L. M., and Harmsen, M. C. (2007). Epithelial cell adhesion molecule: More than a carcinoma marker and adhesion molecule. *Am. J. Pathol.* **171,** 386–395.

van der Gun, B. T., Melchers, L. J., Ruiters, M. H., de Leij, L. F., McLaughlin, P. M., and Rots, M. G. (2010). EpCAM in carcinogenesis: The good, the bad or the ugly. *Carcinogenesis* **31,** 1913–1921.

Veesler, D., Dreier, B., Blangy, S., Lichière, J., Tremblay, D., Moineau, S., Spinelli, S., Tegoni, M., Plückthun, A., Campanacci, V., and Cambillau, C. (2009). Crystal structure of a DARPin neutralizing inhibitor of lactococcal phage TP901-1: Comparison of DARPin and camelid VHH binding mode. *J. Biol. Chem.* **384,** 30718–30726.

Vercauteren, D., Vandenbroucke, R. E., Jones, A. T., Rejman, J., Demeester, J., De Smedt, S. C., Sanders, N. N., and Braeckmans, K. (2010). The use of inhibitors to study endocytic pathways of gene carriers: Optimization and pitfalls. *Mol. Ther.* **18,** 561–569.

Walker, R. G., Willingham, A. T., and Zuker, C. S. (2000). A Drosophila mechanosensory transduction channel. *Science* **287,** 2229–2234.

Wang, Q., Chan, T. R., Hilgraf, R., Fokin, V. V., Sharpless, K. B., and Finn, M. G. (2003). Bioconjugation by copper(I)-catalyzed azide-alkyne [3+2] cycloaddition. *J. Am. Chem. Soc.* **125,** 3192–3193.

Wang, A., Winblade Nairn, N., Johnson, R. S., Tirrell, D. A., and Grabstein, K. (2008). Processing of N-terminal unnatural amino acids in recombinant human interferon-beta in Escherichia coli. *ChemBioChem* **9,** 324–330.

Wels, W., Harwerth, I. M., Mueller, M., Groner, B., and Hynes, N. E. (1992). Selective inhibition of tumor cell growth by a recombinant single-chain antibody-toxin specific for the erbB-2 receptor. *Cancer Res.* **52,** 6310–6317.

Wetzel, S. K., Settanni, G., Kenig, M., Binz, H. K., and Plückthun, A. (2008). Folding and unfolding mechanism of highly stable full-consensus ankyrin repeat proteins. *J. Mol. Biol.* **376,** 241–257.

Wetzel, S. K., Ewald, C., Settanni, G., Jurt, S., Plückthun, A., and Zerbe, O. (2010). Residue-resolved stability of full-consensus ankyrin repeat proteins probed by NMR. *J. Mol. Biol.* **402,** 241–258.

Willuda, J., Kubetzko, S., Waibel, R., Schubiger, P. A., Zangemeister-Wittke, U., and Plückthun, A. (2001). Tumor targeting of mono-, di- and tetravalent anti-p185^{HER-2} miniantibodies multimerized by self-associating peptides. *J. Biol. Chem.* **276,** 14385–14392.

Winkler, J., Martin-Killias, P., Plückthun, A., and Zangemeister-Wittke, U. (2009). EpCAM-targeted delivery of nanocomplexed siRNA to tumor cells with designed ankyrin repeat proteins. *Mol. Cancer Ther.* **9,** 2674–2683.

Wolf, S., Souied, E. H., Mauget-Faysse, M., Devin, F., Patel, M., Wolf-Schnurrbusch, U. E., and Stumpp, M. T. (2011). Phase I MP0112 wet AMD study: Results of a single escalating dose study with DARPin MP0112 in wet AMD. Abstracts of Papers, 2011 Annual Meeting of the Association for Research in Vision and Ophthalmology (ARVO 2011), Fort Lauderdale, FL, Poster 1655.

Wörn, A., and Plückthun, A. (2001). Stability engineering of antibody single-chain F_v fragments. *J. Mol. Biol.* **305,** 989–1010.

Yanagi, Y., Takeda, M., Ohno, S., and Seki, F. (2006). Measles virus receptors and tropism. *Jpn. J. Infect. Dis.* **59,** 1–5.

Yeh, H. H., Ogawa, K., Balatoni, J., Mukhapadhyay, U., Pal, A., Gonzalez-Lepera, C., Shavrin, A., Soghomonyan, S., Flores, L., 2nd, Young, D., Volgin, A. Y., Najjar, A. M., *et al.* (2011). Molecular imaging of active mutant L858R EGF receptor (EGFR) kinase-expressing nonsmall cell lung carcinomas using PET/CT. *Proc. Natl. Acad. Sci. USA* **108,** 1603–1608.

Zahnd, C., Pécorari, F., Straumann, N., Wyler, E., and Plückthun, A. (2006). Selection and characterization of Her2 binding-designed ankyrin repeat proteins. *J. Biol. Chem.* **281,** 35167–35175.

Zahnd, C., Amstutz, P., and Plückthun, A. (2007a). Ribosome display: Selecting and evolving proteins in vitro that specifically bind to a target. *Nat. Methods* **4,** 269–279.

Zahnd, C., Wyler, E., Schwenk, J. M., Steiner, D., Lawrence, M. C., McKern, N. M., Pecorari, F., Ward, C. W., Joos, T. O., and Plückthun, A. (2007b). A designed ankyrin repeat protein evolved to picomolar affinity to Her2. *J. Mol. Biol.* **369,** 1015–1028.

Zahnd, C., Sarkar, C. A., and Plückthun, A. (2010a). Computational analysis of off-rate selection experiments to optimize affinity maturation by directed evolution. *Protein Eng. Des. Sel.* **23,** 175–184.

Zahnd, C., Kawe, M., Stumpp, M. T., de Pasquale, C., Tamaskovic, R., Nagy-Davidescu, G., Dreier, B., Schibli, R., Binz, H. K., Waibel, R., and Plückthun, A. (2010b). Efficient tumor targeting with high-affinity designed ankyrin repeat proteins: Effects of affinity and molecular size. *Cancer Res.* **70,** 1595–1605.

Zimmerman, T., Petit Frère, C., Satzger, M., Raba, M., Weisbach, M., Döhn, K., Popp, A., and Donzeau, M. (2006). Simultaneous metal chelate affinity purification and endotoxin clearance of recombinant antibody fragments. *J. Immunol. Methods* **314,** 67–73.

CHAPTER SIX

Target-Binding Proteins Based on the 10th Human Fibronectin Type III Domain (^{10}Fn3)

Shohei Koide,* Akiko Koide,* *and* Daša Lipovšek†

Contents

1. Introduction 136
2. Library Design 137
3. Choice of Selection Platform 139
4. Phage Display, mRNA Display, and Yeast-Surface Display of ^{10}Fn3-Based Libraries 141
   4.1. Phage display 142
   4.2. mRNA display 144
   4.3. Yeast-surface display 147
   4.4. Combination of phage or mRNA display with yeast-surface display 151
5. Production 151
   5.1. Production of individual ^{10}Fn3 variants in *E. coli* 151
   5.2. High-throughput production of ^{10}Fn3 variants in *E. coli* 152
   5.3. Fusion proteins with ^{10}Fn3 variants 152
   5.4. Expression inside cells 153
6. Conclusion 153
Acknowledgments 154
References 154

Abstract

We describe concepts and methods for generating a family of engineered target-binding proteins designed on the scaffold of the 10th human fibronectin type III domain (^{10}Fn3), an extremely stable, single-domain protein with an immunoglobulin-like fold but lacking disulfide bonds. Large libraries of possible target-binding proteins can be constructed on the ^{10}Fn3 scaffold by diversifying the sequence and length of its surface loops, which are structurally analogous

* Department of Biochemistry and Molecular Biology, The University of Chicago, Chicago, Illinois, USA
† Department of Protein Design, Adnexus, A Bristol-Myers Squibb R&D, Company Waltham, Massachusetts, USA

Methods in Enzymology, Volume 503
ISSN 0076-6879, DOI: 10.1016/B978-0-12-396962-0.00006-9

to antibody complementarity-determining regions. Target-binding proteins with high affinity and specificity are selected from ^{10}Fn3-based libraries using *in vitro* evolution technologies such as phage display, mRNA display, or yeast-surface display. ^{10}Fn3-based target-binding proteins have binding properties comparable to those of antibodies, but they are smaller, simpler in architecture, and more user-friendly; as a consequence, these proteins are excellent building blocks for the construction of multidomain, multifunctional chains. The ease of engineering and robust properties of ^{10}Fn3-based target-binding proteins have been validated by multiple independent academic and industrial groups. In addition to performing well as specific *in vitro* detection reagents and research tools, ^{10}Fn3-based binding proteins are being developed as therapeutics, with the most advanced candidate currently in Phase II clinical trials.

1. Introduction

The recent intense efforts to develop molecular recognition modules based on protein scaffolds other than the immunoglobulins (Binz *et al.*, 2005; Koide, 2010; Skerra, 2007) aim to overcome inherent limitations of the immunoglobulins as molecular recognition modules, including their large size; complex, heterodimeric architecture; and the requirement of correctly formed disulfide bonds. The fundamental assumption behind these protein-engineering efforts is that it is possible to develop molecules with affinity and specificity comparable to those of antibodies by constructing a binding interface on a suitable protein framework, or molecular scaffold, other than an antibody.

Since first reported in this context in 1998 (Koide *et al.*, 1998), the 10th human fibronectin type III domain (^{10}Fn3) has become among the most widely used nonantibody scaffolds for engineering novel binding proteins. A number of ^{10}Fn3-based molecules are being developed for therapeutic applications (Bloom and Calabro, 2009; Lipovsek, 2011), with the most advanced molecule currently in clinical trials (Tolcher *et al.*, 2011).

^{10}Fn3 has several characteristics superior to the immunoglobulin-based systems. It is a member of the immunoglobulin superfamily due to its global β-sandwich fold (Fig. 6.1). The three surface loops proximal to its N-terminus are structurally equivalent to the three antigen-recognition loops, or the complementarity-determining regions (CDRs), of the immunoglobulin variable domain (Koide *et al.*, 1998). However, ^{10}Fn3 lacks a disulfide bond, unlike the canonical immunoglobulin domain, and it has a higher conformational stability, with the thermal transition above 80 °C, and it exhibits reversible and rapid unfolding and refolding (Plaxco *et al.*, 1996). It is ~95 residues long, smaller than the antigen-binding unit (V_HH) of the heavy-chain antibodies (Muyldermans *et al.*, 2001). These desirable features have made ^{10}Fn3 a particularly robust scaffold system compatible

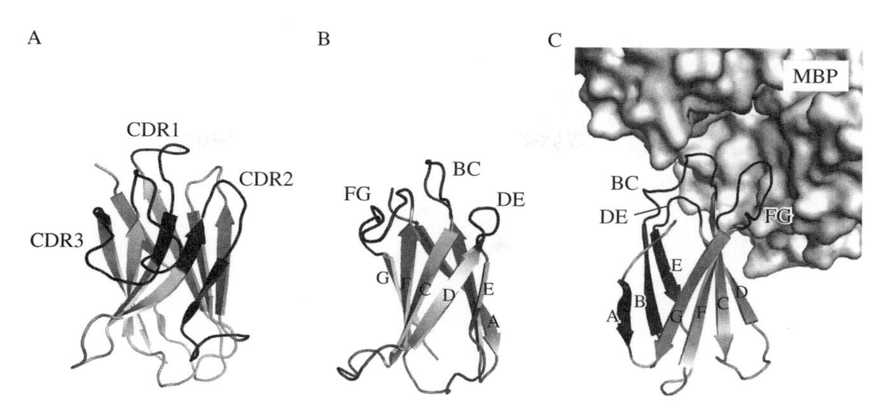

Figure 6.1 Schematic drawing of the crystal structures of the antigen-binding domain of the camelid heavy-chain antibody (V_HH; PDB ID, 1ZVY; A), the ^{10}Fn3 scaffold (PDB ID, 1FNF; B), and a ^{10}Fn3-based binding protein (PDB ID, 3CSB; C). (A) The three complementarity-determining regions (CDRs) are shown in black and also labeled. (B and C) The β-strands of the ^{10}Fn3 scaffold are labeled with A-G and the three loops used for constructing a new binding site are shown in black and also labeled. (C) A portion of the target molecule, maltose-binding protein (MBP), is shown as a surface model.

with diverse molecular display systems and with simple and efficient production methods, as described below.

Depending on their source, ^{10}Fn3-based binding proteins have been identified with several different names. The Koide group, who pioneered the system, has named them "monobodies" (Koide *et al.*, 1998). When first developed in the biotechnology sector, they were called "Trinectins" (Xu *et al.*, 2002). The ^{10}Fn3-based biologics currently in development by Adnexus (Bristol-Myers Squibb) for therapeutic applications are referred to as "Adnectins^{TM}" (Bloom and Calabro, 2009; Lipovsek, 2011). In this chapter, we will refer to these related molecules collectively, regardless of their source, as "^{10}Fn3-based binding proteins."

2. Library Design

The major requirement for a library based on ^{10}Fn3 is that it contains proteins that bind to a wide range of target molecules and that each binding molecule has high affinity and high specificity to its cognate target. Because molecular recognition requires complementarity in both shape and chemistry, achieving such capacity within a limited number of sequences is a formidable challenge.

Clearly, a subset of wild-type ^{10}Fn3 residues needs to be diversified to generate new target-binding surfaces. One needs to consider which positions in ^{10}Fn3 to diversify and what amino-acid residues to allow at those positions.

Typically, the positions diversified in ^{10}Fn3-based libraries comprise between one and three of the ^{10}Fn3 loops that are structurally analogous to antibody CDRs, that is, loops BC, DE, and FG (Fig. 6.1B), the intention being to generate contiguous surfaces for target recognition. Within these broad guidelines, multiple distinct diversification strategies have yielded target-binding proteins with low- to sub-nanomolar affinity (Lipovsek, 2011).

Diversity is typically introduced into a library using synthetic oligonucleotides containing a mixture of nucleotides. The simplest and least expensive random oligonucleotides incorporate a random stretch of nucleotides that can encode any amino acid, such as the so-called NNK codon, where N is an equal mixture of A, T, G, and C, and K is a mixture of T and G. Alternatively, oligonucleotides synthesized from triphosphoramidite or ligated from double-stranded codons can be used to encode a biased amino-acid composition, for example, mimicking side-chain distribution in antibody CDRs (Gilbreth *et al.*, 2008; Wojcik *et al.*, 2010). In addition, amino-acid mixes highly enriched in tyrosine have been shown to be effective (Koide and Sidhu, 2009). In addition to the loop sequence, loop length can be varied by incorporating into the library synthetic oligonucleotides of divergent lengths (Koide *et al.*, 2007).

Some of the general principles to keep in mind when designing ^{10}Fn3-based libraries follow:

(a) It is difficult to predict which library design will work best with a particular target and selection method, so it is helpful to design several different libraries and test them in parallel.
(b) The higher the fraction of ^{10}Fn3 amino-acid residues that are randomized, the larger the number of possible novel target-binding surfaces and the higher the probability of finding a high-affinity binder to a target. On the other hand, more highly diversified libraries are more likely to contain ^{10}Fn3 variants with poor biophysical properties. Also, when the number of possible sequence permutations consistent with a library design is much larger than the physical library size, only a small fraction of all possible sequence permutations (and target-binding surfaces) will be sampled experimentally, and some of the benefit of the wide range of possibilities will be lost.
(c) The starting library tested need not contain the perfect target-binding protein; typically, a target-binding lead molecule selected from a "naive library" serves as the template for the design of a more focused, affinity-maturation library. This two-stage approach is particularly useful when extremely high-affinity binding proteins are required.

3. Choice of Selection Platform

Several molecular display platforms have been established that enable users to generate libraries of large molecular diversity and to select or sort from these libraries to identify the molecules that bind the target of interest (Pluckthun and Mayo, 2007). The three *in vitro* selection methods that have been used extensively to select ^{10}Fn3-based target-binding proteins—phage display, mRNA display, and yeast-surface display—are compared in detail below (Section 4; Table 6.1). Due to the small size, efficient folding, and mostly favorable biophysical properties of ^{10}Fn3-based target-binding proteins, we expect that it would be as straightforward to use other available selection methods such as DNA display, ribosome display, bacterial display, and mammalian-cell display. Thus the choice of the selection method may be at least partially driven by the expertise and equipment available.

Within what is technically feasible, the major parameters to consider when choosing the selection method are accessible library size, target format available, ability to select for more than one desirable property at the same time, and the need for fine discrimination between clones with similar properties.

For example, the completely *in vitro* methods that allow the construction of extremely large libraries, such as mRNA and ribosome display, may be attractive when the expectation is that target-binding variants are going to be extremely rare, or when extremely high affinity is required. On the other hand, these biochemical display methods require significantly more skilled human intervention between selection rounds than do methods that rely on microorganisms to connect genotype to phenotype. Thus a nonspecialist laboratory with a rapid turnover of researchers may find it more efficient to set up phage or yeast display and rely on rediversification of best hits, followed by affinity maturation, to obtain the desired affinity. Naturally, any selection method requires significant expertise and optimization and cannot be successfully implemented as a black-box kit.

All standard selection methods perform well with highly purified, stable, and soluble target that includes a feature enabling specific capture and detection. The presence of aggregated or misfolded species in the target is a major cause for selection of "sticky" ^{10}Fn3 variants with low-affinity, nonspecific binding.

Some of the most popular capture or detection features are biotin, epitope tags, and antibody Fc fragments. Biotinylation, either through chemical conjugation or through biosynthesis, followed by the capture with streptavidin-coupled magnetic beads or by detection with fluorescently conjugated antibodies, is particularly suitable due to its compatibility with multiple selection methods and due to the availability of a wide range

Table 6.1 Characteristics of directed-evolution methods commonly used for the generating of ^{10}Fn3-based binding proteins

	Phage display	mRNA display	Yeast-surface display
Microbial host	*E. coli*	None	*S. cerevisiae*
Particle providing linkage between phenotype and genotype	Phage particle: ^{10}Fn3-variant protein genetically fused to a phage coat protein; DNA that encodes the variant is part of phage genome	Macromolecule: ^{10}Fn3-variant protein covalently linked to mRNA that encodes the variant	Yeast cell: ^{10}Fn3-variant proteins genetically fused to a yeast-surface protein; DNA that encodes the variant on a plasmid inside the yeast cell
Stoichiometry (number of ^{10}Fn3-variant proteins per particle)	1–5 copies per phage particle (1 copy is typical)	1 copy per mRNA molecule	10,000–100,000 copies per yeast cell
Library size	$\leq 10^{11}$	$\leq 10^{13}$	$\leq 10^{8}$
Target capture of binding proteins	On solid support, on beads, or binding in solution followed by bead capture	On beads, or binding in solution followed by bead capture	Binding in solution followed by detection by fluorescent reagents. Bead capture for early-round populations too complex for FACS
Interrogation of phenotype	Selection for tightest binding by population average	Selection for tightest binding by population average	Screening of individual clones according to user-defined criteria
Amplification, manipulation between rounds	Phage growth in *E. coli* host	PCR and chemical modification	Growth of *S. cerevisiae* culture
Selects for	Affinity	Affinity	Affinity, stability
Specialized experimental requirements		Quantification by radioisotope incorporation	Fluorescence-activated cell sorter
Representative references	Wojcik *et al.* (2010)	Xu *et al.* (2002) and Getmanova *et al.* (2006)	Lipovsek *et al.* (2007) and Hackel *et al.* (2010)

of reagents. For example, a new biotin-conjugated reagent with a high affinity to poly-His tag that can be used to biotinylate proteins noncovalently, termed BTtrisNTA, has recently been demonstrated to be compatible with phage-display selection (Koide *et al.*, 2009).

The solution conditions for selection are at least partially dictated by requirements imposed by the target molecule. Some targets such as membrane proteins require a specific additive, for example, detergent, for structural integrity. The ^{10}Fn3 scaffold is compatible with commonly used additives including detergents, reducing reagents, chelating reagents, and osmolytes.

If a recombinant target is not available, it is sometimes possible to use other means of immobilization, such as cells that overexpress the target of interest (Lipes *et al.*, 2008).

Most standard selection methods are designed to enrich clones based solely on their ability to bind to the target, with little discrimination according to their biophysical properties. The exceptions are variations on phage display (Jespers *et al.*, 2004) and standard yeast-surface display. Proteins with higher stability tend to be displayed on yeast surface more efficiently than less stable variants; thus yeast display can be used to select for both target binding and stability. Another advantage of yeast-surface display is that library sorting is performed by screening of individual cells rather than selection of the best clones in a competition with each other, making it easier to discriminate between background binding and signal, and between clones with good and excellent properties.

Given the significant strengths and limitations of each established *in vitro* selection method (Table 6.1), a particularly powerful approach is to combine two different selection methods in series. In the case of ^{10}Fn3-based target-binding proteins, phage display (Gilbreth *et al.*, 2008; Koide *et al.*, 2007) or mRNA display in the first step has been successfully combined with yeast display in the second step (see Section 4.4). In such a combination, the large-library method in the first step is applied for a few rounds, until variants that do not bind the target are removed, and yeast display in the second step is used to remove the variants with poor biophysical properties as well as to retain and amplify the variants with the tightest binding to the target.

4. Phage Display, mRNA Display, and Yeast-Surface Display of ^{10}Fn3-Based Libraries

The three techniques described in this section are well established and documented. Thus we will focus on specific information and considerations pertinent to their applications to ^{10}Fn3-based binding proteins.

4.1. Phage display

4.1.1. Vector design and display optimization

In phage display, a protein of interest is fused to a phage coat protein, typically p3 or p8 (Sidhu *et al.*, 2000). Although ^{10}Fn3 has been successfully displayed using both p3 and p8 systems, most work has been done using p3-based systems that enables monovalent display and consequently more stringent selection. While early studies successfully used standard phage-display systems developed for displaying antibody fragments, recent studies have identified key modifications that significantly improve the efficiency of ^{10}Fn3 display as described below.

In phage display, the protein to be displayed needs to be translocated across the *Escherichia coli* inner membrane into the periplasm through the protein secretion machinery, followed by the incorporation into a phage particle via the fused phage coat protein (Sidhu and Koide, 2007; Sidhu *et al.*, 2000). Because ^{10}Fn3 is highly stable and folds rapidly, the translocation of ^{10}Fn3 using the standard posttranslational signal sequence is inefficient and possibly toxic to the host. Display of ^{10}Fn3 via the cotranslational, signal recognition particle (SRP)-dependent pathway is a few hundredfold more efficient than the posttranslational pathway (Steiner *et al.*, 2006; Wojcik *et al.*, 2010).

An unanticipated consequence of the highly effective display of ^{10}Fn3 was its negative impact on the phage production. A phage displaying a full-length, functional ^{10}Fn3 polypeptide is outgrown by a phage that does not display a fusion protein, for example, because of a termination codon introduced by mutagenesis error or as a part of library design. We have developed a new host in which an additional copy of the *lacI* gene is supplemented via a plasmid, derivative of pCDFDuet-1 (Novagen), to the XL1 Blue strain (Invitrogen), that has made it possible to strictly suppress the expression of the ^{10}Fn3-p3 fusion protein until IPTG is added (Wojcik *et al.*, 2010).

The third parameter that was found important for ^{10}Fn3 display is the conditions for phage propagation. Limiting aeration was important for producing phages with high levels of ^{10}Fn3 display. XL1 Blue with the lacI-encoding pCDFDuet-1 plasmid should be grown in a nonbaffled flask using a slow shaking speed (~75 rpm). Combining these improvements, the display level was improved by ~1000-fold over that with early systems. Because there are only approximately five copies of p3 on the phage surface that define the upper limit for the number of displayed molecules, these results suggest that, in the original vector system using the conventional procedures, fewer than 0.5% (5/1000) of phages displayed a ^{10}Fn3 variant. It is important to determine the effects of these parameters empirically and systematically for each new protein system to be displayed on the phage (Wojcik *et al.*, 2010).

There are a few points for consideration on the ^{10}Fn3 gene used for constructing a library. A library constructed using the Kunkel mutagenesis method (see Section 4.1.2) always contains a substantial fraction (10–50%) of the original, nonmutated template vector. When the three loop regions of ^{10}Fn3 are mutated using three separate oligonucleotides, there are also species in which only one or two loop regions are mutated. Therefore, it is useful to design the template vector in such a way that (i) one can effectively eliminate it from a library and (ii) the nonmutated loops have minimal negative impact on the ability of a mutated loop to bind to the target molecule.

The former point is typically solved by introducing a stop codon in each loop region. However, when a protein-displaying vector reduces phage production as in the case for ^{10}Fn3, such a "stop template" can overwhelm ^{10}Fn3-displaying variants during the phage amplification step. We introduce unique restriction enzyme sites in the BC and FG loop so that the template vector in a library can be selectively digested. A digested vector has much lower transformation efficiency than a circular vector, and consequently, it is effectively eliminated during transformation. The latter point is addressed by eliminating amino acids that are generally detrimental to protein–protein interaction such as Glu and Lys (Koide and Sidhu, 2009; Lo Conte *et al.*, 1999). Combining both considerations, we have replaced the BC, DE, and FG loops with mostly Ser and introduced restriction sites in the BC and FG loops and used such constructs as templates for library construction (Koide *et al.*, 2007).

4.1.2. Library construction

Kunkel mutagenesis is used to construct phage-display libraries (Koide and Koide, 2007; Kunkel *et al.*, 1987; Sidhu *et al.*, 2000). This method is particularly suited for phage display because one can readily produce the required uracil-containing single-stranded DNA from phage particles. Another advantage of this method is that it is straightforward to accurately determine the number of independent sequences in a library from the number of transformants because no PCR amplification is involved.

To introduce amino-acid diversity and loop length diversity in each of the BC, DE, and FG loops, three sets of mutagenic (degenerate) oligonucleotides for the three regions are added in a single mutagenesis reaction mixture. *E. coli* strain SS320 (Sidhu *et al.*, 2000) supplemented with *lacI*-encoding pCDFDuet-1 is used for high efficiency transformation using electroporation, and the transformants are titered to determine the library size. Note that phages are not produced in the first transformation step. The plasmid DNA is recovered from the transformants, and the DNA is digested with restriction enzymes that are unique to the nonmutated BC and FG loops to eliminate nonmutated members in the library. The digested plasmid mixture is used to perform second transformation. This time, the helper phage is added after electroporation, and 0.2 m*M* IPTG is added 30 min after addition of helper phage. After overnight incubation, phages are precipitated using PEG/NaCl solution, suspended in TE, and stored in −20 °C as 50% glycerol solution. Immediately before library sorting, the stored phages are precipitated using PEG/NaCl solution and resuspended in TE in order to remove glycerol. Glycerol interferes with library sorting.

4.1.3. Library sorting

Phage-display libraries are sorted using a biotinylated ligand and streptavidin-coated magnetic beads using procedures detailed previously (Fellouse *et al.*, 2007; Wojcik *et al.*, 2010). The first round of sorting is done using a library solution containing 10^{12} cfu phages in 1 ml with 100 n*M* target that has been already bound to streptavidin-coated magnetic beads. The preformed target-bead complex increases the effective concentration of the target due to the multivalent display of the target on tetrameric streptavidin on the beads. The subsequent rounds of library sorting are performed using a Kingfisher magnetic-bead handler (Thermo Scientific) with reaction volume of 100 μl. Here, a library solution is first mixed with a biotinylated target and then quickly captured with SAV beads, enabling increased stringency of selection. Typically, the target concentration is gradually decreased in sorting rounds, for example, 100, 50, and 20 n*M* for rounds 2, 3, and 4, respectively.

4.1.4. Clone isolation and characterization

After three or four rounds of library sorting, single clones are separated on agar plates and grown individually in 1 ml of media in 96-well deep-well plates with the well capacity of 2 ml. The culture supernatant is used for phage ELISA assays to identify binding-positive clones. One can determine the approximate affinity using competition phage ELISA with a soluble competitor typically included at 100 n*M* (Sidhu *et al.*, 2000).

4.1.5. Affinity maturation

Phage clones obtained from a naive library can be improved by iterative cycles of the introduction of additional amino-acid diversity and selection. While phage display is compatible with standard methods for affinity maturation, including error-prone PCR and gene shuffling, these methods require vector construction that involves restriction digestion and ligation, limiting the size of library. We no longer perform affinity maturation in the phage-display format. Rather, we transfer enriched clones from a phage-display library into the yeast-display format as described under Section 4.4.

4.2. mRNA display

Detailed experimental protocols describing mRNA display in general and its use to select ^{10}Fn3-based target-binding proteins have been published (Baggio *et al.*, 2002; Getmanova *et al.*, 2006; Roberts and Szostak, 1997;

Xu *et al.*, 2002). Here, we provide an overview and highlight the critical parameters and decision points.

4.2.1. Overview

The distinguishing feature of mRNA display is that its unit of selection is a macromolecular entity, the covalent fusion between mRNA and the protein it encodes. To create a library of mRNA–protein fusions, the corresponding DNA library for an mRNA library is assembled from synthetic oligonucleotides, transcribed into mRNA using an *in vitro* transcription system, chemically linked to a DNA oligonucleotide that contains the adaptor molecule puromycin, and subjected to an *in vitro* translation system based on rabbit reticulocyte lysate. After the mRNA is translated into a polypeptide, the puromycin conjugated to the mRNA inserts into the ribosome and is added to the nascent polypeptide chain, as if it were the last amino-acid residue in the chain. The resulting covalent mRNA–protein fusion is reverse-transcribed to add a stabilizing, complementary DNA strand.

4.2.2. Library construction

Because an mRNA-display library is constructed exclusively *in vitro*, it can contain as many as 10^{13} different variants, two to five orders of magnitude higher than what is accessible to methods that require the transformation of a host microorganism. The main limitations on library size in mRNA display are the expense of the reagents used during *in vitro* translation and selection.

Since ^{10}Fn3 contains fewer than 100 amino-acid residues, the DNA segment encoding the library of ^{10}Fn3 variants can be easily assembled by PCR from 8 to 12 overlapped oligonucleotides, some of which contain randomized or partially diversified sequences as described in "Library design." It is advantageous to design the library-construction strategy so that the length of diversified oligonucleotides is minimized, to use the highest quality oligonucleotides available, and to assemble them using a high-fidelity polymerase. The amount of PCR primers and the number of cycles during the amplification of diversified library fragments and full-length library should be limited to obtain the desired number of copies of each clone, while limiting the number of unplanned mutations. As library construction progresses from DNA toward the final form as DNA:mRNA–protein, the yield and amount of amplification should be monitored at each step to ensure that intended library diversity is maintained throughout the process.

4.2.3. Selection

To enrich an mRNA-displayed library of ^{10}Fn3-based variants for clones that bind a target of interest, the library is incubated with the target and the variants that bind to the target are recovered. Typically, a target that is biotinylated or genetically fused to an Fc or a purification tag is captured on magnetic beads coated with streptavidin, protein A, or tag-specific antibody, respectively. The DNA:mRNA–protein molecules whose protein component binds the target remain associated with the bead-captured target, and the nonbinding molecules are removed as flow-through and wash. Alternatively, the target can be conjugated directly to a solid support such as epoxy beads, and the mRNA-fusion library allowed to bind to the beads, and then washed, leaving the fusions with target-biding variants bound to the resin. Next, concentrated base is added in order to degrade the mRNA and thus release from the beads the DNA encoding the ^{10}Fn3 variants bound to the target. High-fidelity PCR is used to amplify the captured DNA, now enriched for sequences encoding target-binding ^{10}Fn3 variants. The enriched population is ready to be transcribed, conjugated to puromycin, and reverse-translated, and to enter the next round of selection. Typically, four to six rounds of mRNA display yield a population highly enriched for target-binding variants of ^{10}Fn3. It is common to increase the stringency of selection in later rounds, either by reducing target concentration (equilibrium selection) or by a prolonged incubation of the DNA:mRNA–protein:target complex with excess, untagged target before capturing on beads (off-rate selection).

Whereas the basic selection protocol is easy to grasp, many of the details require special care and, often, experimental optimization. For example, measures need to be taken to avoid selecting ^{10}Fn3 variants that bind to components of the system other than the target, for example, to the capture beads or streptavidin. This is achieved by subjecting each population of DNA: mRNA–protein fusions to negative selection that is identical to a selection round but lacks the target. The flow-through from the negative selection is then incubated with the target and enriched as described above. In addition, it is also advantageous to alternate between different formats of target and target capture between rounds—for example, each odd-numbered round may employ biotinylated target and capture with streptavidin beads, and each even-numbered round may employ epoxy-immobilized target.

Another sensitive parameter is the initial stringency of the selection (e.g., target concentration in the first selection round) and the rate at which the stringency is increased as the selection progresses (e.g., at what round is the target concentration is lowered, and by how much). Too low a stringency will fail to identify the rare highest-affinity target-binding variants in the library and will instead yield a large number of ^{10}Fn3 variants that bind the target with lower affinity. On the other hand, too high a stringency is likely to lead to a loss of even the highest-affinity binders and will instead yield a larger number of false-positive variants that bind nonspecifically to the more abundant components of the selection system. Early selection rounds, where each library member is present in a small number of copies, are particularly vulnerable to selection stringency that is too high.

4.2.4. Clone isolation and characterization

Once an enriched population of mRNA-displayed ^{10}Fn3 variants shows detectable binding to target under conditions of desired stringency, the eluted DNA mix is PCR-amplified using vector-specific primers and cloned into a pET-based expression vector. Sequencing of DNA from individual colonies identifies unique selected clones and their level of enrichment. The clones of interest are expressed, purified as protein, and characterized; depending on the complexity of the enriched population, this step may involve the purification of a handful of proteins (Section 5.1) or of hundreds of proteins in a high-throughput format (Section 5.2).

4.2.5. Affinity maturation

Typically affinity maturation of mRNA-display-selected ^{10}Fn3 variants is performed on the handful of clones enriched from the naive library that have the best overall balance of relevant properties, such as affinity for the target, specificity, biological activity, thermostability, and solubility. Since the properties of a starting clone are not a fail-safe predictor of the properties of the best affinity-matured descendants, we recommend affinity-maturing, in parallel, several ^{10}Fn3 variants that differ as much as possible from each other in sequence and are thus likely to yield different progeny. Typically an affinity-maturation library is designed by introducing some diversity into the starting clone. For example, two of the three CDR-like loops may be fixed as selected from the naive library, with the third loop rediversified completely, using the same library fragment that had been used to make the naive library. Alternatively, all three loops are rediversified simultaneously but "softly," exploring sequence space adjacent to the parent clone. Such diversification can be achieved by using either error-prone PCR on the regions encoding the loops or oligonucleotide synthesis with a mixture of codons or nucleotides biased toward the parent sequence. Since mRNA display is well suited to high-stringency selections for high affinity, we typically stay with this method for affinity maturation, albeit using much higher stringency of selection. When improvements in both affinity for the target and thermostability are required, a combination of high-stringency mRNA display followed by yeast-surface display may be helpful (Section 4.4).

4.3. Yeast-surface display

Yeast-display methodology is well documented, as applied both to antibody fragments (Chao *et al.*, 2006) and to ^{10}Fn3-based target-binding proteins (Gilbreth *et al.*, 2008; Hackel *et al.*, 2008, 2010; Koide *et al.*, 2007; Lipovsek *et al.*, 2007). Here, we provide an overview and highlight the critical parameters and decision points.

4.3.1. Overview

In yeast display, the unit of selection is a yeast cell that is decorated with tens of thousands of copies of the protein of interest and that carries the plasmid encoding that protein. The plasmid can be shuttled between *Saccharomyces cerevisiae*, for display and sorting, and *E. coli*, for DNA preparation and molecular biology. In the form of yeast display pioneered by the Wittrup group (Chao *et al.*, 2006), each ^{10}Fn3 variant is expressed as a genetic fusion with a native yeast protein found in the cell wall, Aga2p. Aga2p is a domain of the native yeast, an agglutinin mating factor; typically, it is cloned upstream of the sequence encoding the ^{10}Fn3 variant. In addition, an epitope tag, such as c-myc and V5, is engineered immediately downstream from the sequence encoding the ^{10}Fn3 variant. Upon induction, the mating-factor secretory signal peptide directs the fusion protein to be exported from the cell; it is captured on the surface of the yeast cell wall by its binding partner, Aga1p, to which it forms two disulfide bonds. The result is a culture where each yeast cell displays between 10,000 and 100,000 copies of a single ^{10}Fn3 variant. On average, the more thermostable the variant, the larger the number of its molecules on the yeast surface (Hackel *et al.*, 2010).

Alternative versions of yeast-surface display, which use different methods of linking the protein of interest to yeast cell wall (Rakestraw *et al.*, 2011), are also likely to be compatible with the ^{10}Fn3 scaffold and to yield similar correlation between thermostability and display level as the Aga2p-based method.

4.3.2. Library construction

As described above for mRNA display, libraries of ^{10}Fn3 variants are typically assembled from synthetic oligonucleotides encoding both wild-type and diversified sequence, and then PCR-amplified. The resulting linear piece of DNA is simultaneously incorporated into a plasmid and transformed into yeast using homologous recombination: competent *S. cerevisiae* is electroporated with a mixture of linearized vector and insert that encodes the ^{10}Fn3-based library and contains 5′ and 3′ stretches of sequence identical to those of the vector. The yeast DNA-repair system performs the homologous recombination that efficiently joins the two linear DNA fragments into a circular plasmid. Efficiency of yeast transformation typically limits the complexity of yeast-display libraries to the mid-10^8 range, beyond which library construction becomes extremely laborious.

Since yeast plasmids are maintained by auxotrophic markers, the transformed yeast library is typically grown to saturation in selective media, diluted, and regrown several times to decrease the fraction of untransformed, static yeast cells. The culture is then induced in galactose-containing media, labeled, and sorted.

4.3.3. Library sorting

In order to identify and capture the yeast cells that display ^{10}Fn3 variants with desired properties, the culture is incubated with the target of interest and then simultaneously labeled by two sets of reagents: the first set detects the target, and the second set detects the epitope tag expressed at the C-terminus of the ^{10}Fn3 variant. The target is detected using its engineered capture feature; for example, biotinylated target can be detected with streptavidin conjugated with a fluorescent dye, and target formatted as an Fc fusion can be detected with a combination of a primary monoclonal antihuman-IgG antibody and a secondary, anti-mouse-IgG antibody conjugated to a fluorescent dye. Typically the epitope tag fused to the ^{10}Fn3 variant is also detected with a combination of a primary (anti-c-myc) and secondary (fluorescent) antibody. The detection reagents directed at the target and at the C-terminal epitope tag are chosen in such a way that they fluoresce at different wavelengths. As the labeled yeast cells are passed through a fluorescence-activated cell sorter, the intensity of signal at each of the two wavelengths is measured for each cell. The operator then defines the signal range for each wavelength (the gate) for which yeast cells are captured; the rest of the culture is discarded. The captured cells are regrown, relabeled, and resorted; typically, between four and eight sorts are required for the population to converge on a small number of most successful variants.

Throughput of a typical cell sorter allows the above protocol of labeling and cell sorting to be applied to samples of approximately 10^9 cells; if a 10-fold overrepresentation of the library is desired, this corresponds to the library complexity of about 10^8. For sorting significantly larger sample sizes or to minimize expenses associated with the use of a sorter, the first round of sorting can be performed using magnetic beads instead of a flow cytometer. A magnetic bead-based first round is performed in a single dimension, typically using a biotinylated target and streptavidin-coated magnetic beads (Yeung and Wittrup, 2002).

The course and outcome of a yeast-display selection are determined by the absolute and relative stringency of selection for target binding and display level. The parameters that define the binding stringency are target concentration, length of incubation with the target, length of the optional incubation with an excess of target in a format that is not detected by the label (e.g., unbiotinylated target when labeled streptavidin is used in detection), the number and length of the washing steps, and the requirement for target-based signal level imposed by the sorting gate. In combination, these parameters allow for a broad range in stringency of selection based on the equilibrium binding constant or on the kinetics of binding. The parameter that defines the display-level stringency is the requirement for epitope-based signal level imposed by the sorting gate. At one extreme, the gate can include all cells with an epitope-based signal above background, effectively excluding only the yeast cells that show no display of ^{10}Fn3 variants (typically, because they have lost the plasmid); an alternative approach captures the cells with relatively high levels of display, that is, with better than average biophysical properties.

Compared with phage display and mRNA display, yeast-surface display allows more stringency in selection and better discrimination between clones with different binding properties; the interrogation of each cell individually reduces the risk of artifacts. The major remaining risk of overly stringent labeling and sorting conditions is the selection of ^{10}Fn3 variants that bind to a detection reagent such as streptavidin, which tends to be present in higher concentrations than the target of interest. This risk can be mitigated by closely monitoring the amount of binding signal in a control sample that is labeled in the presence of all detection reagents except for the target, and by rotating target format and detection reagents between rounds of sorting.

4.3.4. Clone isolation and characterization

Once an enriched population of yeast-displayed ^{10}Fn3 variants shows detectable binding to target under conditions of desired stringency, single clones can be isolated and characterized in several, complementary methods. When the primary goal is to identify clones with the desired binding affinity and specificity, clones can be isolated by growing yeast cells on agar plate followed by small-scale growth in deep-well microplates. After induction of ^{10}Fn3-Aga2p fusion, labeling with a series of different concentrations of target followed by analytical flow cytometry can yield accurate estimates of the dissociation constant. Similarly, the off rate of a particular variant can be measured by labeling the clonal culture with biotinylated target, incubating with excess unbiotinylated target, and using analytical flow cytometry to determine the fraction of target still bound to yeast at different time points. The DNA sequences of hits can readily be determined using PCR amplification from yeast colonies.

When it is necessary to perform more complex assays such as cell-based functional assays and those for biophysical properties, purified proteins are produced in a small scale. The shuttle-plasmid DNA is extracted from the yeast (e.g., using the Zymoprep™ yeast Plasmid Miniprep II kit by Zymo Research, Irvine, CA) and transferred to *E. coli*. Individual clones are sequenced to identify unique selected clones and their level of enrichment. The selected clones of interest can be characterized as displayed on yeast or expressed in and purified from a bacterial display system. For characterization by yeast display, plasmids encoding a single ^{10}Fn3 variant as the display fusion is used to transform *S. cerevisiae* and analyzed as described above. Characterization of bacterially expressed protein is described in Sections 5.1 and 5.2.

4.3.5. Affinity maturation

When affinity-maturation of ^{10}Fn3 variants is performed by yeast-surface display, the same considerations apply as affinity maturation by mRNA display described above. The major differences are that the library size of yeast-surface display is smaller, which will affect the optimal amount of complexity designed into the affinity-maturation library, and that yeast-surface display allows a parallel selection for tighter binding and favorable biophysical properties.

4.4. Combination of phage or mRNA display with yeast-surface display

In addition to being used on its own, yeast-surface display can be combined with other display methods in order to harness the strengths and minimize the impact of limitations of each individual method. In particular, the combination of yeast-surface display with phage or mRNA display takes advantage of the larger libraries accessible to these methods (see Table 6.1) and of the unique ability of yeast-surface display to finely discriminate clones based on affinity and/or kinetics and to select for ^{10}Fn3-based binding proteins with favorable biophysical properties.

Typically, phage or mRNA display is used for the first two to five rounds of selection and focuses on removing from the library the variants that do not bind to the target of interest. Once population complexity has been reduced sufficiently to be compatible with yeast-surface display, the DNA encoding the enriched population is PCR-amplified with primers that introduce sequence overlaps with the yeast-display plasmid and then homologously recombined into the yeast-display plasmid as described above under library construction. If desired, gene shuffling and/or error-prone PCR may be applied in order to introduce additional diversity. In the second phase of selection, yeast-surface display is applied to select for high display level (and thus for favorable biophysical properties), as well as for high affinity for the target.

5. Production

5.1. Production of individual ^{10}Fn3 variants in *E. coli*

Due to its simple fold, high thermostability, and lack of disulfide bonds, human ^{10}Fn3 and many of its derivatives can be expressed at a high level in *E. coli* and are easy to purify using simple column chromatography. In our experience, strong inducible expression systems such as the T7-promoter-based pET family of vectors (Novagen) provide high yields of stable and soluble ^{10}Fn3-based proteins when used according to the manufacturer's instructions. Less stable variants of ^{10}Fn3 can be expressed in a soluble form by inducing expression at a lower temperature, or else can be expressed as inclusion bodies, solubilized in 6 *M* guanidine hydrochloride, and refolded. Whether dealing with soluble or insoluble expression, an oligohistidine purification tag fused to either terminus of the domain allows metal-chelate chromatography to be used as the first purification step. Further purification can be achieved by size-exclusion or ion-exchange chromatography.

5.2. High-throughput production of ^{10}Fn3 variants in *E. coli*

At times it is desirable to identify ^{10}Fn3-based target-binding proteins with specific attributes that are not explicitly selected for during the selection step and that therefore require the screening of a large number of candidate target-binding proteins. For example, selection methods such as phage display and mRNA display enrich protein variants that bind to the target of interest but typically do not distinguish between variants with different thermostability or different biological effect. If such additional attributes are of critical importance, it is advantageous to stop selection while the enriched population still contains hundreds to thousands of different clones, to express and purify a large number of the selected clones in parallel, and to screen them for the property of interest.

A high-throughput production system for ^{10}Fn3 variants starts with overnight culture growth at 37 °C in 5 ml of selective media in a 24-well plate, followed by 1/25 dilution into 5 ml of fresh selective media, growth to the A600 of 0.6–0.9 at 37 °C, and a 6-h IPTG induction at 30 °C. The cells are lysed in 50 m*M* NaH_2PO_4, 0.5 *M* NaCl, 1× Complete™ Protease Inhibitor Cocktail-EDTA free (Roche), 1 m*M* PMSF, 10 m*M* CHAPS, 40 m*M* Imidazole, 1 mg/ml lysozyme, 30 μg/ml DNAse, 2 μg/ml aprotonin, pH 8. The lysate is filtered on a 96-well filter (GF/D Unifilter, Whatman) and purified in one-step purification on a 96-well plate containing metal-chelate resin such as the Nickel Chelated Swell Gel Discs (Pierce) following manufacturer's instructions. Elution in a 100 μl of PBS buffer containing 20 m*M* EDTA typically yields 10–100 μg of partially purified, screen-ready protein per well.

5.3. Fusion proteins with ^{10}Fn3 variants

Because the ^{10}Fn3 scaffold is a single domain encoded by a single polypeptide and its folding does not depend on disulfide formation or cofactor binding, it has proven to be a robust building block for constructing various fusion proteins. Successful examples of fusion partners include green fluorescent protein, alkaline phosphatase, maltose-binding protein, and VP16 transcription activator in addition to short peptide tags for purification and detection (Gilbreth *et al.*, 2008; Karatan *et al.*, 2004; Koide *et al.*, 2002, 2007;

Wojcik *et al.*, 2010). In these cases, ^{10}Fn3-based binding proteins serve as an autonomous module providing target-binding function and fusion partner provides additional functionality. The fact that ^{10}Fn3 scaffold folds irrespective of redox potential is particularly advantageous because some fusion partners need to be produced in a reduced environment (e.g., GFP) or in an oxidizing environment (e.g., alkaline phosphatase).

^{10}Fn3 has been used to create fusion proteins in which two fused domains synergistically function as a single binding unit (Huang *et al.*, 2008, 2009). In these so-called affinity clamps, ^{10}Fn3 scaffold is fused to a natural peptide-binding domain, such as the PDZ domain, in such a configuration that the target-binding loops of the ^{10}Fn3 scaffold would recognize a peptide/binding domain complex. The newly created binding interface located at the junction of the two linked domains is then optimized through directed evolution. Optical sensors have been designed by further fusing affinity clamps to a pair of fluorescent proteins (Huang and Koide, 2010).

Modularity of ^{10}Fn3 variants has particular promise for their therapeutic applications. For example, a bispecific, tandem therapeutic candidate currently under development by Adnexus and Bristol-Myers Squibb is a fusion of a ^{10}Fn3 variant that binds to epidermal growth factor receptor and a second ^{10}Fn3 variant that binds insulin-like growth factor I receptor, resulting in antiproliferative effects through two separate pathways (Emanuel *et al.*, 2011). ^{10}Fn3 variants with possible therapeutic activity are also being fused to modules designed to extend its half-life in the bloodstream, including chemical conjugation polyethylene glycol, as in CT-322, the candidate currently in clinical studies (Tolcher *et al.*, 2011), and genetic fusion with a second ^{10}Fn3 variant that binds human serum albumin (Lipovsek, 2011).

5.4. Expression inside cells

The lack of disulfides, high stability, and rapid folding of the ^{10}Fn3 make it straightforward to express ^{10}Fn3-based binding proteins in different cellular compartments, regardless of their redox potential. The gene encoding the binder of interest needs to be placed under an appropriate promotor and, if applicable, with a flanking amino-acid sequence for proper localization. ^{10}Fn3-based binding proteins have been expressed in the functional form in the yeast nucleus (Koide *et al.*, 2002) and the cytoplasm of mammalian cells (Wojcik *et al.*, 2010) using standard techniques.

6. Conclusion

We described the rationale behind the engineering of the ^{10}Fn3-based family of binding proteins, as well as the general guidelines for the design of ^{10}Fn3-based libraries, selection of high-affinity and specific target-binding proteins from such libraries, and production and evaluation of selected proteins. Whereas the fundamentals of this system are simple, its robust nature allows the use of a wide range of library designs and selection protocols, and thus an infinite number of possible outcomes. The adoption of the ^{10}Fn3-based system by a large number of different research groups (as reviewed in Lipovsek, 2011) demonstrates the versatility of ^{10}Fn3-based binding proteins. We hope that this collection of guidelines will encourage further use and study of the ^{10}Fn3-based family of target-binding proteins.

ACKNOWLEDGMENTS

This work was supported in part by the National Institutes of Health grants R01-GM072688, R01-GM090324 and U54-GM087519.

REFERENCES

Baggio, R., Burgstaller, P., Hale, S. P., Putney, A. R., Lane, M., Lipovsek, D., Wright, M. C., Roberts, R. W., Liu, R., Szostak, J. W., and Wagner, R. W. (2002). Identification of epitope-like consensus motifs using mRNA display. *J. Mol. Recognit.* **15,** 126–134.

Binz, H. K., Amstutz, P., and Pluckthun, A. (2005). Engineering novel binding proteins from nonimmunoglobulin domains. *Nat. Biotechnol.* **23,** 1257–1268.

Bloom, L., and Calabro, V. (2009). FN3: A new protein scaffold reaches the clinic. *Drug Discov. Today* **14,** 949–955.

Chao, G., Lau, W. L., Hackel, B. J., Sazinsky, S. L., Lippow, S. M., and Wittrup, K. D. (2006). Isolating and engineering human antibodies using yeast surface display. *Nat. Protoc.* **1,** 755–768.

Emanuel, S. L., Engle, L. J., Chao, G., Zhu, R. R., Cao, C., Lin, Z., Yamniuk, A. P., Hosbach, J., Brown, J., Fitzpatrick, E., Gokemeijer, J., Morin, P., *et al.* (2011). A fibronectin scaffold approach to bispecific inhibitors of epidermal growth factor receptor and insulin-like growth factor-I receptor. *mAbs* **3,** 38–48.

Fellouse, F. A., Esaki, K., Birtalan, S., Raptis, D., Cancasci, V. J., Koide, A., Jhurani, P., Vasser, M., Wiesmann, C., Kossiakoff, A. A., Koide, S., and Sidhu, S. S. (2007). High-throughput generation of synthetic antibodies from highly functional minimalist phage-displayed libraries. *J. Mol. Biol.* **373,** 924–940.

Getmanova, E. V., Chen, Y., Bloom, L., Gokemeijer, J., Shamah, S., Warikoo, V., Wang, J., Ling, V., and Sun, L. (2006). Antagonists to human and mouse vascular endothelial growth factor receptor 2 generated by directed protein evolution in vitro. *Chem. Biol.* **13,** 549–556.

Gilbreth, R. N., Esaki, K., Koide, A., Sidhu, S. S., and Koide, S. (2008). A dominant conformational role for amino acid diversity in minimalist protein-protein interfaces. *J. Mol. Biol.* **381,** 407–418.

Hackel, B. J., Ackerman, M. E., Howland, S. W., and Wittrup, K. D. (2010). Stability and CDR composition biases enrich binder functionality landscapes. *J. Mol. Biol.* **401,** 84–96.

Hackel, B. J., Kapila, A., and Wittrup, K. D. (2008). Picomolar affinity fibronectin domains engineered utilizing loop length diversity, recursive mutagenesis, and loop shuffling. *J. Mol. Biol.* **381,** 1238–1252.

Huang, J., Koide, A., Makabe, K., and Koide, S. (2008). Design of protein function leaps by directed domain interface evolution. *Proc. Natl. Acad. Sci. USA* **105,** 6578–6583.

Huang, J., and Koide, S. (2010). Rational conversion of affinity reagents into label-free sensors for Peptide motifs by designed allostery. *ACS Chem. Biol.* **5,** 273–277.

Huang, J., Makabe, K., Biancalana, M., Koide, A., and Koide, S. (2009). Structural basis for exquisite specificity of affinity clamps, synthetic binding proteins generated through directed domain-interface evolution. *J. Mol. Biol.* **392,** 1221–1231.

Jespers, L., Schon, O., Famm, K., and Winter, G. (2004). Aggregation-resistant domain antibodies selected on phage by heat denaturation. *Nat. Biotechnol.* **22,** 1161–1165.

Karatan, E., Merguerian, M., Han, Z., Scholle, M. D., Koide, S., and Kay, B. K. (2004). Molecular recognition properties of FN3 monobodies that bind the Src SH3 domain. *Chem. Biol.* **11,** 835–844.

Koide, A., Abbatiello, S., Rothgery, L., and Koide, S. (2002). Probing protein conformational changes by using designer binding proteins: Application to the estrogen receptor. *Proc. Natl. Acad. Sci. USA* **99,** 1253–1258.

Koide, A., Bailey, C. W., Huang, X., and Koide, S. (1998). The fibronectin type III domain as a scaffold for novel binding proteins. *J. Mol. Biol.* **284,** 1141–1151.

Koide, A., Gilbreth, R. N., Esaki, K., Tereshko, V., and Koide, S. (2007). High-affinity single-domain binding proteins with a binary-code interface. *Proc. Natl. Acad. Sci. USA* **104,** 6632–6637.

Koide, A., and Koide, S. (2007). Monobodies: Antibody mimics based on the scaffold of the fibronectin type III domain. *Methods Mol. Biol.* **352,** 95–109.

Koide, A., Wojcik, J., Gilbreth, R. N., Reichel, A., Piehler, J., and Koide, S. (2009). Accelerating phage-display library selection by reversible and site-specific biotinylation. *Protein Eng. Des. Sel.* **22,** 685–690.

Koide, S. (2010). Design and engineering of synthetic binding proteins using nonantibody scaffolds. *In* "Protein Engineering and Design," (S. J. Park and J. R. Cochran, eds.), pp. 109–130. CRC Press, Boca Raton, FL.

Koide, S., and Sidhu, S. S. (2009). The importance of being tyrosine: Lessons in molecular recognition from minimalist synthetic binding proteins. *ACS Chem. Biol.* **4,** 325–334.

Kunkel, T. A., Roberts, J. D., and Zakour, R. A. (1987). Rapid and efficient site-directed mutagenesis without phenotypic selection. *Methods Enzymol.* **154,** 367–382.

Lipes, B. D., Chen, Y. H., Ma, H., Staats, H. F., Kenan, D. J., and Gunn, M. D. (2008). An entirely cell-based system to generate single-chain antibodies against cell surface receptors. *J. Mol. Biol.* **379,** 261–272.

Lipovsek, D. (2011). Adnectins: Engineered target-binding protein therapeutics. *Protein Eng. Des. Sel.* **24,** 3–9.

Lipovsek, D., Lippow, S. M., Hackel, B. J., Gregson, M. W., Cheng, P., Kapila, A., and Wittrup, K. D. (2007). Evolution of an interloop disulfide bond in high-affinity antibody mimics based on fibronectin type III domain and selected by yeast surface display: Molecular convergence with single-domain camelid and shark antibodies. *J. Mol. Biol.* **368,** 1024–1041.

Lo Conte, L., Chothia, C., and Janin, J. (1999). The atomic structure of protein-protein recognition sites. *J. Mol. Biol.* **285,** 2177–2198.

Muyldermans, S., Cambillau, C., and Wyns, L. (2001). Recognition of antigens by single-domain antibody fragments: The superfluous luxury of paired domains. *Trends Biochem. Sci.* **26,** 230–235.

Plaxco, K. W., Spitzfaden, C., Campbell, I. D., and Dobson, C. M. (1996). Rapid refolding of a proline-rich all-beta-sheet fibronectin type III module. *Proc. Natl. Acad. Sci. USA* **93,** 10703–10706.

Pluckthun, A., and Mayo, S. L. (2007). The design of evolution and the evolution of design. *Curr. Opin. Struct. Biol.* **17,** 451–453.

Rakestraw, J. A., Aird, D., Aha, P. M., Baynes, B. M., and Lipovsek, D. (2011). Secretion-and-capture cell-surface display for selection of target-binding proteins. *Protein Eng. Des. Sel.*(in press).

Roberts, R. W., and Szostak, J. W. (1997). RNA-peptide fusions for the in vitro selection of peptides and proteins. *Proc. Natl. Acad. Sci. USA* **94,** 12297–12302.

Sidhu, S. S., and Koide, S. (2007). Phage display for engineering and analyzing protein interaction interfaces. *Curr. Opin. Struct. Biol.* **17,** 481–487.

Sidhu, S. S., Lowman, H. B., Cunningham, B. C., and Wells, J. A. (2000). Phage display for selection of novel binding peptides. *Methods Enzymol.* **328,** 333–363.

Skerra, A. (2007). Alternative non-antibody scaffolds for molecular recognition. *Curr. Opin. Biotechnol.* **18,** 295–304.

Steiner, D., Forrer, P., Stumpp, M. T., and Pluckthun, A. (2006). Signal sequences directing cotranslational translocation expand the range of proteins amenable to phage display. *Nat. Biotechnol.* **24,** 823–831.

Tolcher, A. W., Sweeney, C. J., Papadopoulos, K., Patnaik, A., Chiorean, E. G., Mita, A. C., Sankhala, K., Furfine, E., Gokemeijer, J., Iacono, L., Eaton, C., Silver, B. A., *et al.* (2011). Phase I and pharmacokinetic study of CT-322 (BMS-844203), a targeted Adnectin inhibitor of VEGFR-2 based on a domain of human fibronectin. *Clin. Cancer Res.* **17,** 363–371.

Wojcik, J., Hantschel, O., Grebien, F., Kaupe, I., Bennett, K. L., Barkinge, J., Jones, R. B., Koide, A., Superti-Furga, G., and Koide, S. (2010). A potent and highly specific FN3 monobody inhibitor of the Abl SH2 domain. *Nat. Struct. Mol. Biol.* **17,** 519–527.

Xu, L., Aha, P., Gu, K., Kuimelis, R., Kurz, M., Lam, T., Lim, A., Liu, H., Lohse, P., Sun, L., Weng, S., Wagner, R., *et al.* (2002). Directed evolution of high-affinity antibody mimics using mRNA display. *Chem. Biol.* **9,** 933–942.

Yeung, Y. A., and Wittrup, K. D. (2002). Quantitative screening of yeast surface-displayed polypeptide libraries by magnetic bead capture. *Biotechnol. Prog.* **18,** 212–220.

CHAPTER SEVEN

Anticalins: Small Engineered Binding Proteins Based on the Lipocalin Scaffold

Michaela Gebauer *and* Arne Skerra

Contents

1. Introduction 158
2. Cloning and Expression of Lipocalins and Anticalins in *E. coli* 166
   2.1. General considerations 166
   2.2. Procedure 167
3. Construction of a Genetic Anticalin Library 170
   3.1. General considerations 170
   3.2. Procedure 170
4. Preparation and Selection of a Phage Display Library for Anticalins 172
   4.1. General considerations 172
   4.2. Procedure 173
5. Preparation and Selection of a Bacterial Surface Display Library for Anticalins 176
   5.1. General considerations 176
   5.2. Procedure 176
6. Colony Screening for Anticalins with Specific Target-Binding Activity 177
   6.1. General considerations 177
   6.2. Procedure 178
7. Screening for Anticalins with Specific Target-Binding Activity Using Microtiter Plate Expression in *E. coli* 179
   7.1. General considerations 179
   7.2. Procedure 180
8. Measuring Target Affinity of Anticalins in an ELISA 181
   8.1. General considerations 181

Lehrstuhl für Biologische Chemie, Technische Universität München, Freising-Weihenstephan, Germany

Methods in Enzymology, Volume 503
ISSN 0076-6879, DOI: 10.1016/B978-0-12-396962-0.00007-0

8.2. Procedure 182
9. Measuring Target Affinity of Anticalins via Surface Plasmon Resonance 182
9.1. General considerations 182
9.2. Procedure 183
10. Application of Anticalins in Biochemical Research and Drug Development 184
References 186

Abstract

Anticalins are a novel class of small, robust proteins with designed ligand-binding properties derived from the natural lipocalin scaffold. Due to their compact molecular architecture, comprising a single polypeptide chain, they provide several benefits as protein therapeutics, such as high target specificity, good tissue penetration, low immunogenicity, tunable plasma half-life, efficient *Escherichia coli* expression, and suitability for furnishing with additional effector functions via genetic fusion or chemical conjugation. The lipocalins are a widespread family of proteins that naturally serve in many organisms, including humans, for the transport, storage, or sequestration of small biological compounds like vitamins and hormones. Their fold is dominated by an eight-stranded antiparallel β-barrel, which is open to the solvent at one end. There, four loops connect the β-strands in a pairwise manner and, altogether, they form the entry to a ligand-binding site. This loop region can be engineered via site-directed random mutagenesis in combination with genetic library selection techniques to yield "Anticalins" with exquisite specificities—and down to picomolar affinities—for prescribed molecular targets of either hapten or antigen type. Several Anticalins directed against medically relevant disease targets have been successfully engineered and can be applied, for example, for the blocking of soluble signaling factors or cell surface receptors or for tissue-specific drug targeting. While natural lipocalins were already subject to clinical studies in the past, a first Anticalin has completed Phase I trials in 2011, thus paving the way for the broad application of Anticalins as a promising novel class of biopharmaceuticals.

1. Introduction

Antibodies still constitute the paradigm for a class of biochemical reagents that enable specific and sensitive molecular recognition in biomedical research, and they have been in use for human therapy since more than one century. By now, immunoglobulins (Igs) provide an accepted class of biopharmaceuticals with considerable growth rate. Humanized or human recombinant (monoclonal) antibodies have become particularly successful due to their high target specificity, therapeutic efficacy, safety, and, hence, low failure rate during preclinical and clinical development. Nevertheless, they also show practical disadvantages due to the high cost of manufacturing, limitations in formulation and delivery, immunological side effects arising from the intrinsic effector region, and their naturally very long plasma half-life. As a promising alternative, non-Ig protein scaffolds with engineered binding properties have gathered attention to address areas of medical need (Gebauer and Skerra, 2009; Skerra, 2000a).

Among those scaffolds, endogenous human proteins such as the lipocalins (Åkerström *et al.*, 2006; Flower, 1996; Skerra, 2000b), which are abundant in plasma, secretions, and tissue fluids, are particularly attractive for clinical applications as they promise low immunogenicity, similar to the Igs (or their antigen-binding fragments). Lipocalins with predefined ligand/target-binding properties can be generated by applying methods of combinatorial protein design to yield so-called Anticalins (Beste *et al.*, 1999; Kim *et al.*, 2009; Schönfeld *et al.*, 2009). If directed against medically relevant disease targets, such Anticalins generally offer three mechanisms for therapeutic intervention: (i) as antagonists, for example, by binding to cellular receptors and blocking them from interaction with natural signaling molecules (or vice versa); (ii) as tissue-targeting vehicles, by addressing toxic molecules, radionuclides, or enzymes to disease-related cell surface receptors; (iii) as antidotes, by quickly scavenging toxic or otherwise irritating compounds from circulation.

Members of the natural lipocalin protein family share a structurally conserved β-barrel, whose eight antiparallel β-strands wind around a central axis, with an α-helix attached to its side (Fig. 7.1). At one end, the cup-shaped β-barrel opens to the solvent and supports four loops that form the entrance to the ligand pocket. The opposite end is closed by short loops, and densely packed amino acid side chains form a hydrophobic core inside the cylindrical structure. Despite extremely low mutual sequence homology, the β-barrel is structurally well conserved among the 10–12 different human lipocalins (Breustedt *et al.*, 2006), including family members from various other species whose three-dimensional structures are known. In contrast, the loop region that forms the ligand-binding site exhibits large differences in amino acid sequence, length, and conformation of the four polypeptide segments, which is in line with the diverse ligand specificities observed within this family (Skerra, 2000b).

In this regard, the protein architecture of the lipocalins shows resemblance with that of Igs, whose structurally conserved framework supports the hypervariable loop region (also known as complementarity-determining region, CDR) in the pair of variable domains, thus giving rise to the huge variety of possible antigen specificities (Skerra, 2003). However, antibodies have a large molecular size and complex composition, which relies on altogether four subunits (two light and two heavy chains). In contrast, lipocalins are composed of a single polypeptide chain, comprising merely 160–180 residues, and their binding site is composed of four loops instead of

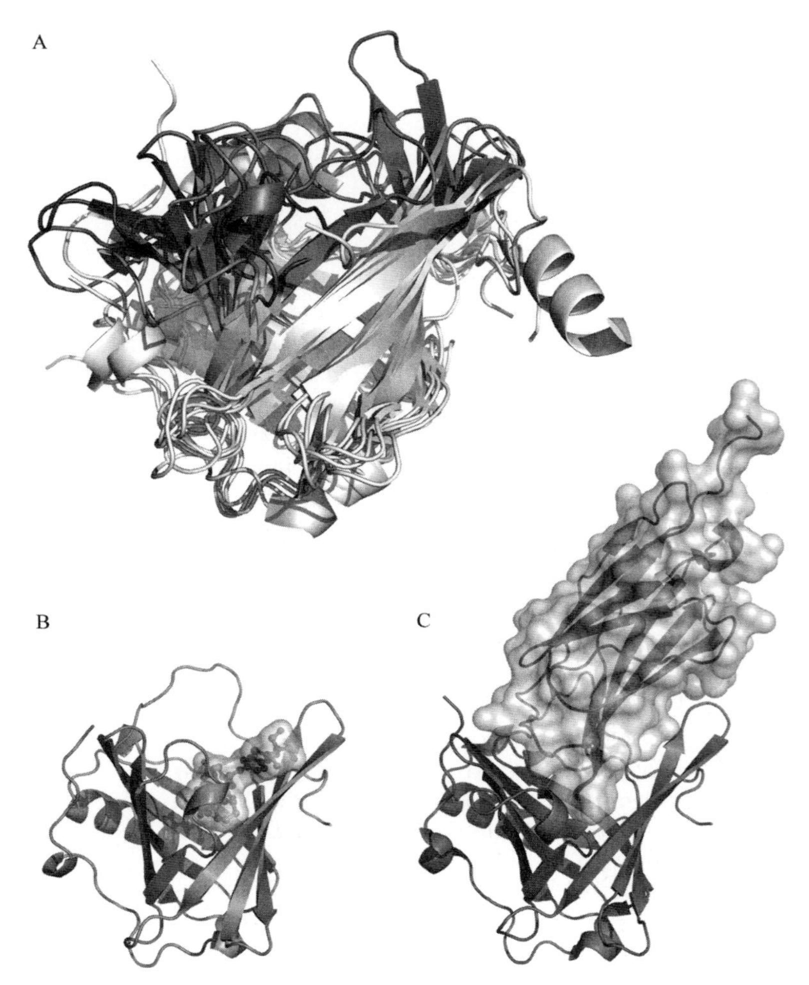

Figure 7.1 Structural basis for the design of Anticalin libraries. (A) Conformational plasticity in the loop region of natural lipocalins. The crystal structures of six human members of the lipocalin family—RBP, Tlc, ApoD, AGP, NGAL (Lcn2), c8γ (PDB entries 1RBP, 3EYC, 2HZQ, 3KQ0, 1L6M, 2QOS, respectively)—were superimposed via their β-barrel Cα positions (Skerra, 2000b). While the β-barrel itself as well as the loops at its bottom show a conserved three-dimensional structure, the four loops at the open end (top, colored dark) exhibit high conformational variability, not unlike the hypervariable region of immunoglobulins. (B) Crystal structure of an Anticalin based on Lcn2 recognizing a small molecule (PDB entry 3DSZ). Ribbon diagram of the Anticalin selected against an Y(III)–DTPA derivative, which is depicted as ball and sticks with a translucent surface (Kim *et al.*, 2009). Charged with suitable radionuclides, this Anticalin offers applications in radioimmunotherapy (RIT) and diagnostics (RID). (C) Crystal structure of an Lcn2-based Anticalin

six while exhibiting similar dimensions. Furthermore, similar to antibodies, human lipocalins occur as soluble proteins in the plasma and other tissue fluids, with concentrations up to 1.0mg/mL. Consequently, lipocalins provide a small and robust scaffold for the engineering of medically useful binding proteins, offering several benefits as they do not require disulfide bonds and can be easily manufactured in *Escherichia coli* or other microbial host cells (Schlehuber and Skerra, 2005; Skerra, 2007).

During the initial development of the Anticalin technology, the structurally and biochemically well-characterized bilin-binding protein (BBP) from the butterfly *Pieris brassicae* was employed as a prototypic lipocalin scaffold in order to tailor its binding site for ligands such as fluorescein, digoxigenin (Beste *et al.*, 1999; Schlehuber *et al.*, 2000), and other low molecular weight molecules including peptides. To this end, a genetic library with altogether 16 randomized amino acid positions covering the natural binding site of the BBP for biliverdin was prepared and used for phage display selection against the immobilized target compounds (Skerra, 2001). Sequence analysis of the engineered lipocalins revealed that the loop regions that had been subjected to targeted randomization essentially tolerated the entire set of possible side chains (Skerra, 2000b). Furthermore, X-ray structural analysis of the fluorescein-binding and digoxigenin-binding BBP variants showed that the extensive alteration of side chains, affecting 10% of all residues in the protein scaffold, did not impair the β-barrel fold, whereas the variegated loops adopted dramatically altered conformations compared with the wild-type lipocalin (Korndörfer *et al.*, 2003a,b).

Hence, it was demonstrated that the high structural plasticity of the binding site provided by the lipocalin architecture in general enables the generation of novel ligand-binding proteins with antibody-like properties, dubbed "Anticalins." Therapeutic potential of such engineered lipocalins became already apparent during these early studies. An affinity-improved variant of the digoxigenin-binding Anticalin with a sub-nanomolar dissociation constant proved to completely reverse intoxication by digitalis—a drug widely in use for the treatment of ventricular tachyarrhythmias and congestive heart failure—in an established guinea pig model, thus demonstrating *in vivo* efficacy as an antidote (Schlehuber and Skerra, 2005).

Beyond that, other natural lipocalins from insects have prompted biopharmaceutical development based on the notion that blood-sucking ticks

---

recognizing a macromolecular target (PDB entry 3BX7). Ribbon diagram of the Anticalin selected against the extracellular domain of the deactivating T-cell coreceptor CTLA-4, which is depicted in darker color with a translucent surface (Schönfeld *et al.*, 2009). This Anticalin efficiently blocks the negative regulatory activity of CTLA-4 and thus shows potential for stimulating the cellular immune response to treat infectious diseases or for the immunotherapy of cancer. (See Color Insert.)

have evolved proteins, in particular lipocalins, with sophisticated functions in order to undermine the host's immunological defense mechanisms. For example, the recombinant histamine-binding protein (HBP) from the saliva of *Rhipicephalus appendiculatus* was subject to clinical studies (rEV131; Evolutec Group plc) for the treatment of allergic disorders, especially conjunctivitis and rhinitis (Couillin *et al.*, 2004). rEV131 showed *in vivo* efficacy in a mouse model of asthma and proved beneficial for the treatment of human allergies in a Phase I/II clinical trial. Another insect lipocalin, the complement inhibitor of C5 activation (OmCI) from the soft tick *Ornithodoros moubata* (Nunn *et al.*, 2005), was under preclinical development for myasthenia gravis and acute myocardial infarction (rEV576; Evolutec Group plc).

Nevertheless, for therapeutic use, protein drugs based on an endogenous human counterpart generally are preferred, mostly to prevent an immune response leading to anti-drug antibodies by avoiding novel T- and B-cell epitopes. Consequently, the next generation of Anticalins was engineered on the basis of human lipocalins, in particular the tear lipocalin (Tlc, von Ebner's gland protein, Lcn1; Breustedt *et al.*, 2005, 2009) and the neutrophil gelatinase-associated lipocalin (NGAL, siderocalin, Lcn2; Goetz *et al.*, 2002). Moreover, during the design of corresponding advanced random libraries, considerations with regard to the shape and size of accessible molecular target structures came into play. For example, low molecular weight substances of the "hapten type" can be fully accommodated within the cup-shaped ligand pocket of the lipocalin scaffold. In this case, ideally a set of side chains pointing inward at the upper end of the β-barrel as well as adjoining parts of the loops are chosen for randomization. On the other hand, protein targets of the "antigen type" cannot penetrate as deeply into the ligand-binding site as a small molecule. Hence, a set of residues located at more exposed positions, primarily at the tips of the four hypervariable loops, is preferentially subjected to random mutagenesis in order to allow formation of an extended interface, allowing tight contacts with the macromolecular binding partner.

This concept can be illustrated with two Anticalins derived from the Lcn2 scaffold, one directed against an Y(III)–DTPA hapten (Kim *et al.*, 2009) and one recognizing the extracellular domain of cytotoxic T lymphocyte antigen 4 (CTLA-4) as antigen (Schönfeld *et al.*, 2009). In both cases, single-digit nanomolar to picomolar affinities were achieved, and the crystal structures of the Anticalins in complex with their different targets were solved (Fig. 7.1B and C). Again, the β-barrel architecture of the human lipocalin scaffold was fully preserved, while the loop segments at the entrance to the binding site showed high conformational variability. Notably, in the case of the Anticalin directed against CTLA-4, a pronounced induced fit was observed upon comparison with the uncomplexed crystal structure, thus confirming the mechanistic relationship between engineered lipocalins and Igs. This unique feature distinguishes lipocalins with their flexible loop region from other non-Ig scaffolds described so far whose binding sites reside on rigid elements of secondary structure (Gebauer and Skerra, 2009).

When comparing the three-dimensional structures of these exemplary Anticalins, it becomes clear that distinct sets of side chains dominate the interactions with two different types of target structures. However, apart from the overall location of variegated side chains within the lipocalin binding site, their total number and the applied genetic randomization strategy are also crucial to optimize the chances for successful selection of high-affinity Anticalins against a given target. Based on our experience and also on theoretical considerations (Skerra, 2003), a range of 16–24 randomized amino acid residues appears optimal for the design of Anticalin libraries. With a smaller number, there tends to be insufficient structural variation in order to initially achieve a good shape complementarity with the target molecule—and hence selectable binding activity—whereas with a larger number, the physical sampling of the drastically increasing combinatorial space becomes inadequate.

The combinatorial complexity arising from a given number of randomized residues can, to some extent, be narrowed by applying an elaborate mutagenesis strategy. For the first Anticalin libraries, synthetic oligodeoxynucleotides carrying degenerate NNS/NNK codons at the randomized amino acid positions were used in conjunction with polymerase chain reaction (PCR) assembly to create the gene libraries (Beste *et al.*, 1999; Skerra, 2001). These codon mixtures represent all 20 amino acids, yet with differing degeneracy, as well as one stop codon (amber, TAG). More advanced Anticalin libraries may be prepared using defined mixtures of nucleotide triplets as they have become available via the Slonomics® triplet overhang-based ligation approach (Van den Brulle *et al.*, 2008) or using triplet phosphoramidite DNA synthon chemistry (Kayushin *et al.*, 1996), for example. In this way, a fairly equal distribution of amino acid residues is ensured, and at the same time, stop codons are fully avoided. Furthermore, certain amino acids can be excluded from the libraries, e.g., cysteine, which is often considered deleterious as it can lead to disulfide cross-linking or isomer formation under conditions of periplasmic secretion (including phage display). Using this methodology, the effective number of combinations at each randomized position is reduced from 32 to 19, which leads to a significantly smaller number of theoretical combinations, i.e., 3.8×10^{25} versus 1.3×10^{30} when considering a library with, for example, 20 variegated positions. In principle, the number of codons used in the mixture at each amino acid position may be further reduced in an individual manner, but this would also restrict the chemistry of the allowed side chains, while corresponding effects on the functional quality of the resulting molecular library are difficult to predict.

State-of-the-art phage display libraries have physical complexities between 10^{10} and 10^{11} and, thus, permit sampling of naive libraries with complexities mentioned only in a limited way. Nevertheless, practical experience with selection from Anticalin libraries—and also from antibody or other scaffold libraries (cf. other chapters in this issue)—demonstrates that high-affinity binding proteins can be selected under those circumstances. Often, during early stages of selection from a naive random library, broad sampling is intended to achieve coverage of different binding modes and/or epitopes on the target. With increasing number of selection cycles and, likewise, also during affinity maturation approaches (see below), lower complexities are faced while better control of the selection process is desired.

Hence, for the combinatorial engineering of binding proteins, phage display selection, which essentially yields an enriched population of target-specific candidates, is usually followed by enzyme-linked immunosorbent assay (ELISA) screening (Bradbury *et al.*, 2011). While being clearly more laborious, the latter step allows ranking of candidates according to target affinity, cross-reactivity with control targets, and if performed with soluble protein extracts (as opposed to phage ELISA), expression efficiencies of the engineered scaffold proteins. Unfortunately, due to the complicated procedure, ELISAs are only feasible for the screening of around 100–1000 clones, even if performed using high-throughput instrumentation. To overcome these limitations and in order to allow controlled screening of much higher numbers of Anticalins, we have developed further methods such as the filter-sandwich colony screening assay (Schlehuber *et al.*, 2000) and bacterial cell surface display in conjunction with fluorescence-activated cell sorting (FACS) (Binder *et al.*, 2010), thus offering a comprehensive tool set for the quick selection as well as optimization of therapeutic Anticalin lead candidates (Fig. 7.2).

Using these methods, Anticalins have been successfully selected against a variety of disease-related protein antigens, including the immunological receptor CTLA-4 and soluble growth factors such as VEGF (Hohlbaum and Skerra, 2007). Typically, high selectivity and affinities in the subnanomolar range have been achieved. For example, CTLA-4 (CD152) is an activation-induced, transmembrane T-cell coreceptor with negative regulatory effect on T-cell-mediated immune responses. CTLA-4 antagonizes CD28-dependent co-stimulation of T cells, whereby CTLA-4 and CD28 share the same counter-receptors on antigen-presenting cells, B7.1 and B7.2. CTLA-4 constitutes a clinically relevant target for cancer immunotherapy (Peggs *et al.*, 2006), and the cognate antibody Ipilimumab (Yervoy®; Bristol-Myers Squibb Co) with blocking activity was recently approved by the US Food and Drug Administration. The Anticalin selected against the same target shields the relevant CTLA-4 epitope that is involved in the interaction with the counter-receptors B7.1 and B7.2 (Schönfeld *et al.*, 2009). Accordingly, antagonistic activity of the Anticalin toward CTLA-4

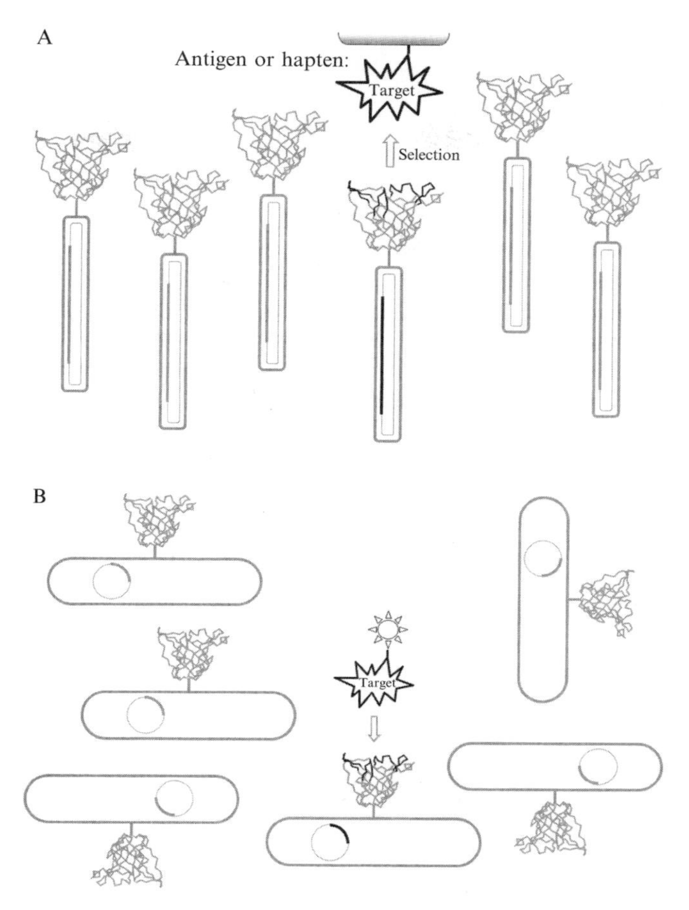

Figure 7.2 Strategies for selection of Anticalins from large genetic libraries. (A) Filamentous phage display. Anticalins are displayed on the surface of phagemids which carry the single-stranded vector DNA including the coding region packaged inside, thus providing a physical link between phenotype and genotype (Skerra, 2001). Individual phagemids differ by the mutation pattern in the variable loop regions of the encoded lipocalin variants, prepared using established library synthesis techniques to yield complexities of around 10^{10} different members (see text). For panning of the phagemids, the target compound, which can be a protein, a hapten, or a hapten–protein conjugate, is presented either in an immobilized form (e.g., coupled to a solid surface such as microtiter plate, immunostick, paramagnetic particle) or functionalized with a marker group, for example, biotin, followed by capturing of the target–phagemid complex using (strept)avidin reagents. Cognate Anticalins mediate binding of the entire phagemid particle to the target, whereas non-binding phagemids are washed away, leading to an enrichment by several orders of magnitude per panning cycle. After a couple of cycles, interspersed with phagemid amplification steps, the resulting enriched phagemid population is used to prepare single infected bacterial colonies, from which the plasmid DNA can be prepared for DNA sequencing of the Anticalin insert or

was confirmed in several *in vitro* cell culture tests and in a mouse model of *Leishmania* infection, where a T-cell-mediated immune response plays a role. Thus, the CTLA-4-specific Anticalin (PRS-010; Pieris AG) provides a promising drug candidate for the immunotherapy of infectious diseases and cancer. Its lack of immunological effector functions should limit off target toxicity as only antagonistic binding activity is needed. In fact, this is also the case for many other medical targets involved in immune regulation, inflammation, and neoangiogenesis.

Another therapeutic drug candidate is an Anticalin with strong antagonistic activity toward vascular endothelial growth factor (VEGF). VEGF is a well-characterized mediator of tumor angiogenesis and other neovascular diseases such as age-related macular degeneration (Ferrara and Kerbel, 2005). The cognate antibody Bevacizumab (Avastin®; F. Hoffmann-La Roche Ltd) constitutes an approved and highly successful cancer drug. The selected Anticalin (PRS-050; Pieris AG) exhibits a favorable binding and activity profile in direct comparison with established VEGF antagonists such as Avastin. A half-life extended version of the Anticalin has shown excellent efficacy in several animal models assessing VEGF-induced enhanced vascular permeability, angiogenesis, and anti-xenograft tumor activity. PRS-050 has completed Phase I clinical trials in 2011 (Pieris AG), demonstrating safety and tolerability of Anticalins in patients as a novel class of drugs.

2. Cloning and Expression of Lipocalins and Anticalins in *E. coli*

2.1. General considerations

A series of plasmid vectors have been developed for the cloning, bacterial expression—in various formats—and engineering of Anticalins. These vectors are based on the generic *E. coli* expression vector pASK75, which employs the tightly regulated tetracycline promoter/operator for chemically

subcloning on another expression vector for functional analysis. (B) Bacterial surface display. The Anticalins are anchored and functionally displayed on the surface of the outer *E. coli* membrane. Incubation with a target coupled to a bright fluorophore leads to the specific fluorescent labeling of bacterial cells that harbor a cognate Anticalin. These cells can be separated from a huge excess of cells encoding non-reactive Anticalins using automated fluorescence-activated cell sorting (FACS) (Binder *et al.*, 2010). This method is particularly attractive for selection from mid-size libraries, containing 10^6 to 10^8 different members, for example, for the purpose of affinity maturation, or at later stages of phage display selection from a naive library. FACS allows *in situ* assessment of the library composition and adjustment of sorting parameters, leading to better selection efficiency.

inducible gene expression (Skerra, 1994). Most of these vectors direct the recombinant lipocalin into the bacterial periplasm, where efficient disulfide bond formation can occur. This is achieved by N-terminal fusion with the OmpA signal peptide. While periplasmic secretion is required for phage display and bacterial surface display, expression of the engineered lipocalin itself can alternatively be achieved in the bacterial cytoplasm, albeit protein liberation and purification from the periplasmic cell fraction is often easier. In the case of a lipocalin scaffold that carries disulfide bonds, the use of an *E. coli* strain with oxidizing cytoplasm is recommended (Venturi *et al.*, 2002) if refolding from inclusion bodies is undesired. At the C-terminus, Anticalins are typically equipped with an affinity tag, preferentially the *Strep*-tag II (Schmidt and Skerra, 2007), which permits efficient one-step purification. Of course, while helpful during the research stage, such a tag can be omitted for later clinical development. Fusion partners, such as the minor phage coat protein pIII, an autotransporter β-domain, reporter enzymes, or other effector domains, are also usually fused to the C-terminus of the engineered lipocalin as their exchange is easily possible via a conserved unique restriction site (*Eco*47III). Furthermore, the central Anticalin-encoding cassette, which encompasses the entire variable loop region, can be exchanged between the various vectors using a pair of mutually distinct type IIa restriction sites (*Bst*XI). While many vectors for the engineering of different lipocalin scaffolds have been described up to now (for references see Section 1), this review focuses at the Lcn2 (NGAL), and a set of relevant vectors for the procedures described here is illustrated in Fig. 7.3. The Lcn2 scaffold carries one intramolecular disulfide bridge, whereas an unpaired Cys side chain characteristic for the wild-type protein (Goetz *et al.*, 2002) has been replaced by Ser in order to avoid the formation of non-physiological disulfide isomers or covalent protein dimers.

2.2. Procedure

The recombinant Lcn2 and its derived Anticalins can be functionally produced by periplasmic secretion in *E. coli* W3110 (Bachmann, 1972), *E. coli* JM83 (Yanisch-Perron *et al.*, 1985), *E. coli* BL21 (Studier and Moffatt, 1986), or the *E. coli supE* strain TG1/F$^-$ (a derivative of *E. coli* K12 TG1 that was cured from its episome using acridinium orange; Kim *et al.*, 2009), amongst others. For expression of wild-type Lcn2, *E. coli* BL21 is preferred as this strain lacks endogenous enterobactin, the natural ligand of this lipocalin (Goetz *et al.*, 2002). *E. coli* W3110, which can grow in glucose minimal medium, is recommended for laboratory fermenter production (Breustedt *et al.*, 2006).

For expression at the shake flask scale, the appropriate *E. coli* strain is transformed with pNGAL98—or its corresponding derivative encoding an Lcn2 variant—using standard techniques (Sambrook *et al.*, 1989). A single

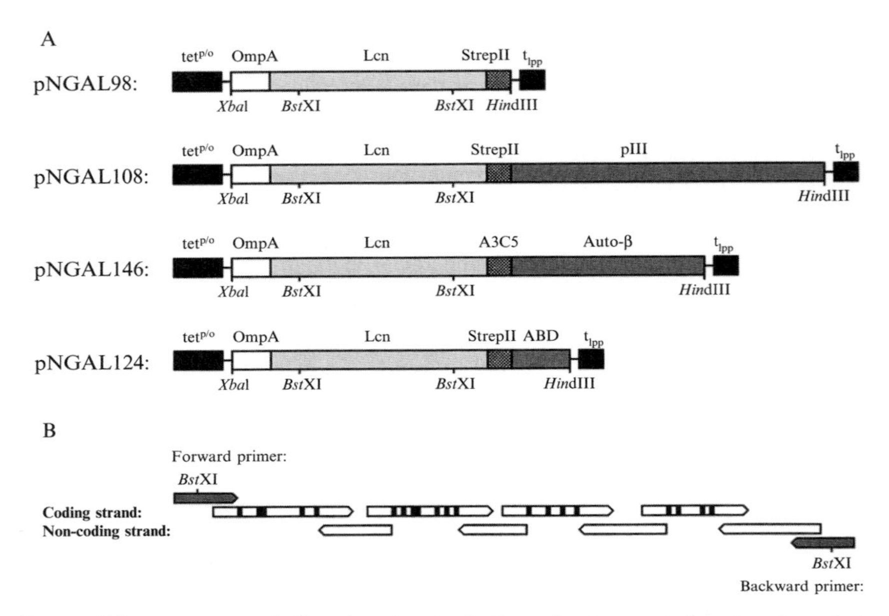

Figure 7.3 Vectors and libraries for Anticalin selection and functional analysis. (A) Design of vectors for Anticalin expression and selection. All plasmids are based on the generic *E. coli* expression vector pASK75, which harbors the tetracycline promoter/operator for tightly regulated transcriptional control (Skerra, 1994). The *tet*$^{p/o}$ is chemically inducible with anhydrotetracycline. On these vectors, the Lcn2 expression cassette encompasses an N-terminal OmpA signal peptide for periplasmic secretion, the mature part of the engineered lipocalin, an affinity tag (usually, the *Strep*-tag II (Schmidt and Skerra, 2007), or the A3C5 tag (Binder *et al.*, 2010) in case of pNGAL146), and, as required, a C-terminal effector domain. pNGAL98 encodes the plain Anticalin, suitable for soluble periplasmic expression followed by one-step affinity purification and biochemical analysis. pNGAL108 encodes a fusion protein with the gene III minor coat protein of filamentous bacteriophage M13 for phage display selection. pNGAL124 codes for an albumin-binding domain (ABD) from *streptococcal* protein G as effector region, which is useful both for functional bacterial colony screening (Schlehuber *et al.*, 2000) and for extending the plasma half-life in animal experiments (Schlapschy *et al.*, 2007). pNGAL146 encodes a fusion with the β-domain of the autotransporter EspP for bacterial cell surface display (Binder *et al.*, 2010). All vectors of this series share a pair of conserved *Bst*XI restrictions sites that generate two different non-palindromic sticky ends, allowing facile exchange of the central Anticalin-encoding DNA region in a directed manner. The expression cassette ends with the strong rho-independent *lpp* terminator. Further vector elements (not shown) include an ampicillin resistance gene (β-lactamase), a truncated ColEI origin of replication, and the intergenic region of the filamentous phage f1 for the biosynthesis of phagemid particles upon coinfection of *E. coli* with a helper phage. (B) Construction of an Lcn2-based Anticalin library via PCR assembly from a set of synthetic oligodeoxynucleotides. In the illustrated strategy, altogether 20 amino acid positions randomized in the four structurally variable loop segments are divided into four sequence subsets. For randomization of the amino acids in each subset, an oligodeoxynucleotide is synthesized wherein NNS mixtures of the nucleotides are employed at the mutated codons. N means a mixture of all four bases

colony is used to inoculate 50 mL of Luria–Bertani (LB) medium containing 100 µg/mL ampicillin (Amp), followed by culturing at 30 °C and 200 rpm overnight. Forty milliliters of this culture is transferred to 2 L LB/Amp medium in a 5-L Erlenmeyer flask, and cell growth at 22 °C and 200 rpm is monitored by measuring the optical density at 550 nm. Once OD_{550}=0.5 is reached, heterologous protein production is induced by adding 200 µL of 2 mg/mL anhydrotetracycline (aTc; ACROS Organics, Geel, Belgium) dissolved in dimethylformamide. After a 3-h induction period, the bacteria are harvested by centrifugation at 4400×*g* for 15 min at 4 °C in a prechilled centrifuge rotor (e.g., Sorvall SLA3000). The supernatant is thoroughly removed, and the bacterial pellet is placed on ice. Spheroplasts are prepared by carefully resuspending the cells from a 2-L culture in 20 mL ice-cold periplasmic extraction buffer (0.5 *M* sucrose, 1 m*M* EDTA, 0.1 *M* Tris/HCl, pH 8.0). The cell suspension is immediately transferred to a 50-mL centrifuge tube and incubated on ice for 30 min prior to centrifugation at 5100×*g* for 15 min. The supernatant containing the periplasmic cell fraction is carefully recovered from the sedimented spheroplasts, transferred to a fresh centrifuge tube, and once again centrifuged at 27,000×*g* for 15 min at 4 °C (e.g., Sorvall SS34).

The cleared extract is dialyzed against 2 L of chromatography buffer (150 m*M* NaCl, 1 m*M* EDTA, 0.1 *M* Tris/HCl, pH 8.0) overnight, passed through a 0.45-µm sterile filter, and applied to a 2 mL bed volume *Strep*-Tactin chromatography column (IBA, Göttingen, Germany) equilibrated with at least 10 bed volumes of chromatography buffer. Unbound protein is removed by washing the column with chromatography buffer. Once the base line is reached, the recombinant lipocalin is eluted by applying 2.5 m*M* D-desthiobiotin dissolved in chromatography buffer. Fractions containing the recombinant protein are dialyzed as appropriate or further purified by size exclusion chromatography on a Superdex 75 HR 10/30 column (Amersham-Pharmacia, Freiburg, Germany) using PBS (4 m*M* KH_2PO_4, 16 m*M* Na_2HPO_4, 115 m*M* NaCl, pH 7.4) or another suitable buffer. Protein purity is checked by SDS-PAGE (Fling and Gregerson, 1986), and the

A, C, G, and T while S means a mixture of only the two bases G and C; thus, such a triplet encodes all 20 natural amino acids as well as the amber stop codon (TAG), which is translated as glutamine in the *E. coli supE*-strains XL1-Blue or TG1 used here for phagemid production and gene expression. Four additional oligodeoxynucleotides with invariant nucleotide sequences, corresponding to the non-coding strand, serve for bridging the gaps in the assembly reaction. Hybridization overlaps typically comprise 18–19 base pairs. Two shorter flanking oligodeoxynucleotides, carrying biotin groups and applied in excess amount, are used as primers for the one-pot PCR amplification of the assembled, fully synthetic gene cassette. The two flanking primers each carry a *Bst*XI restriction site that is compatible with the one on the vectors described above, giving rise to mutually non-compatible DNA overhangs upon enzyme digest. This special arrangement of restriction sites enables a particularly efficient ligation and cloning of the synthetic gene.

resulting protein is quantified by absorption measurement at 280 nm using a calculated extinction coefficient (Gill and von Hippel, 1989).

3. Construction of a Genetic Anticalin Library

3.1. General considerations

Here we describe the preparation of an Anticalin library based on the Lcn2 scaffold via synthetic gene assembly using PCR with commercially available oligodeoxynucleotides that carry NNS degenerate codons at the variegated positions. Suitable sets of residues to be randomized for the recognition of haptens or antigens can be taken from the literature (see, e.g., Kim *et al.*, 2009; Schönfeld *et al.*, 2009). For the preparation of more elaborate libraries by employing codon-based synthesis strategies, suitable methods have been described in Section 1. Generally, it is sufficient to synthesize a gene fragment corresponding to the central DNA cassette flanked by the pair of mutually non-compatible *Bst*XI restriction sites mentioned below, which covers the relevant coding regions for all four structurally variable lipocalin loops at the open end of the β-barrel as well as adjoining β-strand segments (Fig. 7.3).

3.2. Procedure

A combinatorial library of Lcn2 variants can be prepared on the basis of the cloned cDNA (Breustedt *et al.*, 2006). Alternatively, a fully synthetic gene design with optimized codons for *E. coli* expression may be used. Importantly, we have introduced two amino acid substitutions: Cys87Ser to remove the single unpaired thiol side chain of Lcn2/NGAL and Gln28His to introduce an additional *Bst*XI restriction site. Another *Bst*XI restriction site naturally occurs downstream at amino acid position Ser158. To generate an Anticalin library, the central lipocalin gene cassette flanked by these two *Bst*XI restriction sites is prepared using synthetic DNA. In the simplest fashion, oligodeoxynucleotides carrying NNS mixtures at the randomized codons are employed. The gene cassette can be assembled via PCR (Beste *et al.*, 1999; Skerra, 2001) in a process also known as gene splicing by overlap extension (Heckman and Pease, 2007). In advancement of the original strategy, this gene assembly can even be performed in a one-pot reaction from eight long oligodeoxynucleotides—four of which each cover one randomized loop segment on the coding DNA strand—together with two shorter flanking primers that are applied in excess for the PCR amplification as schematically illustrated in Fig. 7.3B.

Typically, PCR steps can be performed using a conventional thermostable DNA polymerase as well as commercially available oligodeoxynucleotides in HPLC grade purity. Further purification by urea PAGE is highly recommended in order to avoid overt contamination with gene fragments carrying base pair deletions. The flanking PCR primers may be synthesized with a biotin group at their 5′-ends, thus allowing separation of the doubly cut DNA fragment after *Bst*XI cleavage from incompletely digested product via streptavidin-coated paramagnetic beads. To achieve maximal yields of PCR product, restriction digest, and subsequent ligation, the corresponding enzymatic reactions usually require careful optimization. In the case of the PCR, the number of amplification cycles should be kept low, for example, 15–20, because with increasing exponential amplification, the molecular diversity of the DNA pool diminishes, eventually narrowing the resulting combinatorial complexity of the Anticalin library.

A typical one-pot PCR assembly reaction can be set up in a 1000 μL mixture containing 10–50 pmol of each long oligodeoxynucleotide template and each 500 pmol of the two flanking primers. In addition, the reaction mix contains 100 μL of 10× *Taq* buffer, 20 μL dNTP-Mix (10 m*M* dATP, dCTP, dGTP, dTTP), and 50 U *Taq* DNA polymerase (5 U/μL; Promega, Mannheim, Germany). The mixture is divided into 100-μL aliquots, and PCR is performed under hot start conditions with 15–20 cycles of 1 min at 94 °C, 1 min at 60 °C, 1.5 min at 72 °C, followed by a final incubation for 5 min at 60 °C. The PCR product is purified using the E.Z.N.A. Cycle-Pure Kit (PeqLab Biotechnologie, Erlangen, Germany).

For subsequent cloning, this DNA fragment is cut with the restriction enzyme *Bst*XI (Promega) according to the manufacturer's instructions and then purified by preparative agarose gel electrophoresis. Subsequently, DNA fragments not or incompletely digested can be removed via their 5′-biotin tags using streptavidin-coated paramagnetic beads (Merck, Darmstadt, Germany). To this end, 100 μL of the commercially available suspension of the streptavidin-coated paramagnetic particles (at a concentration of 10 mg/mL) is washed three times with 100 μL TE buffer (10 m*M* Tris/HCl, pH 8.0, 1 m*M* EDTA). The particles are then drained with the help of a magnet and mixed with 50 pmol of the digested DNA fragment in 100 μL TE buffer for 15 min at room temperature. After collecting the paramagnetic particles at the wall of an Eppendorf tube, the supernatant containing the purified, fully digested DNA fragment is recovered for use in the following ligation reaction.

For a typical ligation reaction, approximately 2.5 μg (12.5 pmol) of the purified PCR fragment and a similar molar amount of the vector fragment (pNGAL108) are incubated in the presence of 250 Weiss units of T4 DNA ligase (Promega) in a total volume of 4300 μL (50 m*M* Tris/HCl, pH 7.8, 10 m*M* $MgCl_2$, 10 m*M* DTT, 1 m*M* ATP, 50 μg/mL BSA) for 48 h at 16 °C. The DNA is then precipitated by adding 100 μL yeast tRNA (10 mg/mL solution in H_2O; Roche Applied Science, Mannheim, Germany), 4300 μL of 5 *M* ammonium acetate, and 17.1 mL ethanol, followed by incubation at −20°C for at least 1.5 h. The DNA pellet is washed with cold 70% ethanol, dried, and dissolved in 400 μL of water.

The preparation of electrocompetent bacterial cells of the *E. coli* K-12 strain XL1-Blue (Bullock *et al.*, 1987) is carried out according to published methods (Hengen, 1996; Tung and Chow, 1995). One liter LB medium (10 g/L Bacto

Tryptone, 5g/L Bacto Yeast Extract, 5g/L NaCl, pH 7.5) is adjusted to an optical density at 550nm of OD_{550}=0.08 by addition of an appropriate volume of a stationary overnight culture of XL1-Blue and incubated at 26°C (with 140 rpm) in a 2-L Erlenmeyer flask. After reaching OD_{600}=0.6, the culture is chilled for 30min on ice and then centrifuged for 15min at 4000×*g* and 4°C. The cells are washed twice with 500mL of ice-cold 10% v/v glycerol and finally resuspended in 2mL ice-cold GYT-medium (10% v/v glycerol, 0.125% w/v Bacto Yeast Extract, 0.25% w/v Bacto Tryptone). The cells are finally aliquoted (200μL), shock-frozen in liquid nitrogen, and stored at −80°C.

Electroporation is carried out in the cold room at 4°C with a Micro Pulser system (e.g., Bio-Rad, München, Germany) using cuvettes from the same vendor (cuvette gap 2mm). An aliquot of 10μL of the ligated DNA solution (containing approximately 1μg salt-free DNA) is mixed with 100μL of the competent cell suspension, first incubated for 1min on ice and then transferred to the bottom of the prechilled cuvette, avoiding air bubbles between the electrodes. Electroporation is performed using a time constant of 5ms and 12.5kV/cm field strength. Immediately afterward, the cell suspension is diluted in 2mL ice-cold SOC medium (20g/L Bacto Tryptone, 5g/L Bacto Yeast Extract, 10m*M* NaCl, 2.5m*M* KCl, adjusted to pH 7.5 with NaOH and sterilized by autoclaving; 10mL/L of 1*M* $MgCl_2$, 10mL/L of 1*M* $MgSO_4$, 20mL/L of 1*M* glucose added as sterile solutions prior to use).

The electroporated cells are pooled, transferred to a sterile Erlenmeyer flask, and incubated without antibiotic selection at 37°C at 160rpm for 1h. To determine the library diversity, a series of 10-fold dilutions from a small aliquot (e.g., 50μL) is prepared and about 100-μL aliquots of the 10^{-4} to 10^{-6} dilutions are plated on LB/Amp agar to select for colonies that harbor the library plasmid pNGAL108. The entire bacterial culture is transferred to a 5-L Erlenmeyer flask containing 1L of 2xYT/Amp medium, and the initial optical density at 550nm is determined. If OD_{550} is ≥0.3, the volume of the culture is adjusted with 2xYT/Amp to an initial OD_{550}=0.2–0.3. Using the same conditions as above, the cells are grown until OD_{550} has risen by 0.6, which resembles the mid-log phase and will usually take about 6–8h. By employing a total of 40μg ligated DNA in 40 electroporation runs, up to approximately 10^{10} transformants can be obtained according to this procedure.

4. Preparation and Selection of a Phage Display Library for Anticalins

4.1. General considerations

For the selection of Anticalins, a phagemid selection system has been developed that relies on a plasmid vector which encodes the engineered lipocalin as fusion with the gene III minor coat protein of the filamentous bacteriophage M13 and also carries its intergenic region to allow single-stranded replication after superinfection with a helper phage. Here we use the Lcn2 scaffold fused to the full-length pIII, whereas fusions with only the C-terminal domain of pIII have been successfully used for Anticalin selection in the past. In principle, for panning of phagemids displaying Anticalins with a desired binding specificity, several strategies are possible. For solid-phase panning, protein antigens or protein–hapten conjugates can be (i) directly adsorbed to the wells of a microtiter plate or to immunosticks or, for example, after biotinylation, (ii) indirectly immobilized to streptavidin/avidin-coated plastic vessels or paramagnetic beads. For solution-phase selection, which is usually more stringent as there are no avidity effects and less secondary interactions with the phagemid particle, the phagemid solution is first incubated with the dissolved biotinylated target and the resulting complex is then conveniently trapped e.g., by means of streptavidin-coated paramagnetic beads. If the lipocalin–pIII fusion protein also contains the *Strep*-tag II, care should be taken that excess streptavidin-binding sites are blocked with desthiobiotin to abolish direct binding via the affinity tag (Schmidt and Skerra, 2007). For elution of bound phagemids from the complex with the target—attached to the plastic surface or beads—various strategies are possible: acidic or basic buffer conditions, denaturant solutions, competition with excess (unconjugated) target, proteolysis, reductive cleavage of disulfide linkers, or via direct bacterial infection. While in the following example we describe the method of acid elution, numerous alternative protocols have been published in conjunction with antibody phage display technology (Bradbury *et al.*, 2011) and can be applied accordingly. In general, for initial selection from a naive library, competitive elution with a huge excess of non-functionalized target compound may be advantageous as this usually favors the selection of target-specific Anticalins versus non-specific binding proteins. This effect is of less importance during advanced panning cycles or during the affinity maturation of Anticalins that do already show a desired epitope specificity. As with increasing affinity the kinetics of dissociation becomes relevant, enforced elution under extreme pH or denaturing conditions can then even provide a benefit.

4.2. Procedure

For preparation of the phagemid library, 400 mL of the entire culture from the electroporation of XL1-Blue cells with a freshly prepared lipocalin gene library described above—based on the plasmid vector pNGAL108—is infected with 1.3×10^{12} pfu VCS-M13 helper phage (Stratagene/Agilent Technologies, Waldbronn, Germany) at a multiplicity of infection of approximately 10:1. After gentle agitation at 37 °C for 30 min, kanamycin (Kan) is added at a concentration of 70 mg/L, and the incubation temperature is lowered to 26 °C. After temperature equilibration for 10 min, 25 µg/L aTc is added in order to induce gene expression for the lipocalin–pIII fusion protein. Phagemid production is allowed for 7h at 26°C.

The supernatant containing the phagemid library is separated from the cells by centrifugation for 30min, 18,000×*g* at 4°C, followed by passage through a 0.45-μm filter. The phagemid particles are precipitated by mixing the clear supernatant with one-fourth volume of ice-cold 20% w/v PEG 8000, 15% w/v NaCl, transferred into clean centrifuge tubes, and incubated on ice for at least 2h. After centrifugation (30min, 18,000×*g*, 4°C), the precipitated phagemid particles from the initial 400mL culture are dissolved in 16mL cold BBS/E (200m*M* Na-borate, pH 8.0, 160m*M* NaCl, 1m*M* EDTA) supplemented with 50m*M* benzamidine (Sigma-Aldrich, Munich, Germany) and 1μg/mL Pefabloc (Roth, Karlsruhe, Germany) to prevent proteolytic degradation of the phage coat fusion protein. Following incubation on ice for 1h and centrifugation of undissolved components (10min, 43,000×*g*, 4°C), each supernatant is transferred to a new plastic tube. Phagemids are then precipitated again by adding one-fourth volume of 20% w/v PEG 8000, 15% w/v NaCl, and incubating on ice for 60min. The entire suspension is aliquoted and frozen at −80°C for storage or directly used.

For the first panning cycle, phagemids are thawed (if necessary) and centrifuged (30min, 34,000×*g*, 4°C). The supernatant is thoroughly removed, and the precipitated phagemid particles are dissolved in 400μL PBS containing 50m*M* benzamidine. After incubation on ice for 30min, the solution is finally cleared by centrifugation (5min, 18,500×*g*, 4°C) and used for the phage display selection. A sample of the supernatant is titered by infection of XL1-Blue cells (see below).

Panning cycles are carried out using the Lcn2 random phagemid library by applying about 10^{12} recombinant phagemids. First, the phagemids dissolved in 300μL PBS, supplemented with 1m*M* EDTA (if applicable) and 50 m*M* benzamidine, are blocked for 30min by adding 100μL of 8% w/v BSA (Sigma-Aldrich) in PBS containing 0.4% v/v Tween 20 (polyoxyethylene sorbitan monolaurate; AppliChem, Darmstadt, Germany). Then, this solution is incubated for ~1h with streptavidin-coated magnetic beads (Merck or Invitrogen, Darmstadt, Germany) that were blocked with 2% w/v BSA in PBS/T (PBS containing 0.1% v/v Tween 20) for 1h and charged with 400μ L of a 100n*M* solution of the biotin-labeled target protein (labeling ratio: approximately 2 mol of biotin-X-NHS ester per 1 mol of protein).

After collecting the beads on a magnet and discarding the supernatant, 10 washing steps with PBS/T are performed. Then, remaining bound phagemids are eluted under acidic conditions with 400μL of 0.1*M* glycine/HCl, pH 2.2 for 8min, immediately followed by neutralization with an appropriate volume (~60μL) of 0.5*M* Tris–base. Some of the washing fractions and the elution fraction are used to infect exponentially growing *E. coli* XL1-Blue in order to follow the phagemid titer in the course of the panning cycle (Schlehuber *et al.*, 2000). In brief, 20 μL serial dilutions of the phagemid solution are mixed with 180 μL culture of *E. coli* XL1-Blue and incubated for 30 min at 37 °C. Aliquots of the infected cells (100 μL) are plated on LB/Amp agar and incubated overnight at 37 °C. On the next day, the colonies are counted and the titers of the phagemid solutions (cfu/mL) are calculated.

For phagemid reamplification, the remaining phagemid solution (440 μL) is used to infect 4 mL exponentially growing culture of *E. coli* XL1-Blue, followed by incubation at 37 °C for 30 min under agitation. After centrifugation at 4100×*g* for 5 min, the cells are resuspended in 600 μL 2xYT medium, plated on three large LB/Amp agar plates (Ø13.5 cm), and incubated at 32 °C overnight. The infected cells are recovered by scraping the colonies from the agar plates with addition of 50 mL 2xYT medium. The resulting cell suspension is used to inoculate 50 mL 2xYT/Amp medium to an initial OD_{550} of approximately 0.08. The culture is incubated at 37 °C until $OD_{550}=0.5$ and infected with 1.5×10^{11} pfu VCS-M13 helper phages according to the procedure described above.

For the next panning cycles, about 10^{11} to 10^{12} of the amplified phagemids are used, and elution of bound phagemids is performed under similar conditions as before. The selection process is monitored by plotting the total number of phagemids titered for input, washing, and elution fractions and comparing these figures between subsequent cycles. In case of successful enrichment of Anticalins against the target, both the relative number of phagemids in the washing cycles and, in particular, the number of phagemids in the elution fraction tend to rise (see, e.g., Schlehuber *et al.*, 2000). Notably, there will be discontinuities if the stringency of selection is increased in the course of the selection campaign (e.g., by lowering the concentration of target, increasing the number of washing steps or by switching from solid to solution panning) or if the method of elution is changed (i.e., acidic, basic, denaturing or by competition).

Typically, after three to five panning cycles the phage display selection is finished, and the colonies resulting from the last cycle are then scraped from the agar plate and collectively used to inoculate an overnight culture for plasmid preparation. Subsequently, the *Bst*XI DNA fragment is prepared and used for subcloning on a vector for soluble lipocalin expression (e.g., pNGAL98) or, alternatively, on a vector for selection by colony screening or bacterial surface display (see below). Also, it is advisable to recover a few single colonies, individually prepare the corresponding plasmids, subject the central coding region to DNA sequencing with the help of suitable primers, and investigate the mutated lipocalin sequence. Generally, the number of panning cycles should not be chosen too high because there is a certain risk that one selects gene products showing enhanced expression or even phagemids with gene deletions, which often lead to better replication, rather than high-affinity binding proteins, as desired.

5. Preparation and Selection of a Bacterial Surface Display Library for Anticalins

5.1. General considerations

Anticalins can be functionally displayed on the Gram-negative bacterial cell envelope via fusion to the β-domain of the bacterial autotransporter EspP (Binder *et al.*, 2010): first, the fusion protein is secreted into the *E. coli* periplasm by means of the N-terminal OmpA signal peptide, then the β-domain inserts into the outer membrane, catalyzes translocation of the attached lipocalin, and finally anchors it on the bacterial cell surface. After incubation with a fluorescence-labeled target, binding activity can be individually probed for each bacterial cell as part of a mixed population by quantitative flow cytofluorometry (FACS). Also, the target protein may be equipped with the *Strep*-tag II, for example, and detected with fluorescently labeled secondary antibodies.

When applying a combinatorial library, repeated cycles of cell sorting and propagation can be performed, hence allowing the selection of Anticalins with defined binding properties. Bacterial cell surface display combined with FACS offers the advantage of providing valuable quantitative information about the phenotypic status of the library during the sorting process. Thus, the gating parameters, which are crucial for the efficiency of selection, can be optimally adjusted in real time. In addition, FACS allows the simultaneous analysis of multiple features such as display level, target specificity, and affinity. To determine the display level *in situ*, we employ the A3C5 tag and a fluorescently labeled high-affinity anti-A3C5 Fab fragment to normalize the fluorescence signals arising from the Anticalin-target interaction.

5.2. Procedure

The mutagenized gene cassette of the pooled plasmid preparation from the last phage display panning step—or from a freshly prepared genetic library—is subcloned via the pair of *Bst*XI restriction sites on the expression plasmid pNGAL146, a derivative of pNGAL97, which encodes a fusion of the lipocalin with the β-domain of the EspP autotransporter (Fig. 7.3A) (Binder *et al.*, 2010). Chemically or electrocompetent cells of the *E. coli* K-12 strain JK321 (Jose *et al.*, 1996) whose *dsb*A$^-$ phenotype assists the export of those Anticalins carrying a disulfide bond are transformed with the corresponding ligation product and used to inoculate 200 mL LB/Amp medium. After growth to the exponential phase at 30 °C, gene expression is induced at $OD_{550}=0.5$ by addition of 2.5–5.0 ng/mL aTc for 2 h.

Approximately 2×10^8 cells (corresponding to 200 μL bacterial culture having an $OD_{550}=1.8$) are spun down in a bench-top centrifuge

(Eppendorf, Hamburg, Germany) for 3min (10,000rpm, 4°C), resuspended in 500μL ice-cold PBS/BSA (PBS containing 0.5% w/v BSA), centrifuged again and resuspended in PBS containing an appropriate concentration (0.1–2μ*M*) of the recombinant target protein carrying the *Strep*-tag II, followed by incubation for 1h on ice. After washing with 0.5mL PBS, cells are resuspended in 50μL PBS containing 100μg/mL *Strep*MAB-Immo (IBA) and incubated for 30min on ice. Following addition of 500μL ice-cold PBS, cells are pelleted once again and resuspended in a solution containing 3μ*M* DY634-labeled A3C5 Fab fragment (Binder *et al.*, 2010) as well as 200μg/mL FITC-labeled polyclonal anti-IgG H&L chain (Acris Antibodies, Herford, Germany). After incubation for 30min on ice, cells are finally centrifuged, washed once with PBS, and directly used for flow cytofluorometric sorting on a FACSAria instrument (BD Biosciences, Heidelberg, Germany), operated with filter-sterilized PBS as sheath fluid, into 0.5mL LB medium.

The DY634 and FITC fluorophores are excited at 633 and 488nm, respectively, while the emitted light is detected using 660/20- and 530/30-nm bandpass filters. To prevent loss of promising clones, a non-stringent sorting mode ("yield mode") is used in the first cycle, whereas in subsequent cycles, the "purity mode" is chosen to increase stringency. After each sorting cycle, cells are propagated by overnight growth at 37°C in 250mL LB/Amp in a shaker flask or by plating on a large Petri dish as described above. Finally, the sorted bacteria are plated on LB/Amp agar to grow colonies, and individual clones are subjected to DNA sequence analysis or subcloning for soluble expression.

6. Colony Screening for Anticalins with Specific Target-Binding Activity

6.1. General considerations

In this filter-sandwich colony screening assay (Skerra *et al.*, 1991), bacterial colonies harboring plasmids for the periplasmic expression of the engineered lipocalins, pre-selected via phage display, are grown on a hydrophilic membrane on an agar plate. Upon induction, the Anticalins are secreted into the periplasm and become partially released from the bacterial colonies. After diffusion through the hydrophilic membrane, the recombinant proteins are finally immobilized on a second, hydrophobic membrane that is placed underneath and coated with a capturing reagent. Thus, the Anticalins are functionally immobilized and subsequently can be probed for binding to a labeled target. Binding activity is directly visualized on this second membrane, for example, by using an enzyme conjugate followed by a chromogenic reaction. Colonies that give rise to a strong signal are then identified on the first membrane and propagated for further analysis, preparative expression, and/or subcloning of the coding regions.

6.2. Procedure

The mutagenized gene cassette isolated from preceding selection experiments is subcloned via the pair of *Bst*XI restriction sites on pNGAL124, a derivative of pNGAL98 which encodes a C-terminal fusion with the 46-amino acid bacterial albumin-binding domain (ABD). The high-affinity interaction of this tag with human serum albumin (HSA) used for coating of the hydrophobic filter membrane provides an efficient means for the directed functional immobilization of the Anticalin that becomes secreted and released from the individual colonies (Schlehuber *et al.*, 2000). About 100–200 μL of the cell suspension typically obtained from transformation of chemically competent *E. coli* TG1/F^- cells with the corresponding ligation mixture is uniformly plated on a hydrophilic PVDF membrane (GVWP, pore size 0.22 μm; Millipore, Schwalbach/Ts, Germany), which was cut to size and placed (avoiding formation of air bubbles) on an LB/Amp agar plate. Given the fact that transformation efficiency may vary from experiment to experiment, it is recommended to first estimate the appropriate volume of cell suspension needed to obtain about 500–1000 colonies on a normal LB/Amp agar plate. To evenly spread the cells on the filter, the appropriate volume of the suspension is then adjusted to 100 or 200 μL and the liquid is quickly but carefully plated with the help of a Drigalski spatula. The agar plate is incubated at 37 °C for 7–8 h until the first bacterial colonies (~ 0.5 mm diameter) become visible.

In the meantime, a hydrophobic PVDF membrane (Immobilon P, pore size 0.45 μm; Millipore) is cut to size, moisturized with methanol according to the instructions of the manufacturer, and rinsed with water and PBS. The membrane is coated with 10 mg/mL HSA in PBS for 4 h at room temperature and blocked by incubation with 3% w/v BSA, 0.5% v/v Tween 20 in PBS for 2h. Then, the membrane is washed with PBS twice for 10 min, followed by a single wash for 10 min in 20 mL LB/Amp supplemented with 200 μg/L aTc. The well-drained membrane is transferred onto an LB/Amp agar plate containing 200 μg/L aTc and stored at 4 °C until use. Once the bacterial colonies on the hydrophilic membrane are ready, the first membrane is placed on the HSA-coated second membrane with the colonies facing upward, again avoiding air bubbles.

The stack of two membranes is marked at several positions (e.g., by piercing with a needle) and incubated on the agar plate at 22 °C for 11–15 h. During this period, the lipocalin–ABD fusion proteins are secreted from the colonies on the upper membrane and become subsequently immobilized via complex formation with the bound HSA on the lower membrane. After that, the upper membrane is transferred onto a new LB/Amp agar plate, and the colonies are stored at 4 °C for later recovery. The hydrophobic lower membrane is removed and immediately washed three times in 20 mL PBS/T for 5 min.

Then, the immobilized Anticalins, which form an invisible replica of the colonies originally present on the upper membrane, are probed for binding activity by incubation with the appropriately labeled target protein, e.g., by using a biotin reagent (see, for example, Pierce Protein Research Products from Thermo Fisher Scientific, Bonn, Germany), in PBS/T for 1 h. The membrane is washed three times with PBS/T for 5 min and incubated with a suitable enzyme conjugate (e.g., avidin–alkaline phosphatase) diluted in PBS/T for 1 h. For Anticalin fusion proteins containing the *Strep*-tag II, the use of streptavidin reagents should be avoided; instead avidin or NeutrAvidin conjugates (Thermo Scientific) can be used for the specific detection of biotin groups without binding to the peptide tag (Schmidt and Skerra, 2007).

The membrane is finally washed twice with PBS/T and twice with PBS for 5 min before staining with suitable chromogenic reagents, followed by immersion in water. As result, a map of colonies producing Anticalins with target-binding activity is obtained, and colonies giving rise to the strongest signals can be identified on the first membrane by superimposing. To avoid drying, colonies should be quickly recovered with a sterile toothpick and used for inoculating a fresh LB/Amp agar plate or small liquid culture. It is recommended to check these colonies once again both for binding to the target protein and to dummy proteins in a secondary filter-sandwich screening assay by directly inoculating colonies in several copies (using a sterile toothpick) onto a fresh hydrophilic membrane.

7. Screening for Anticalins with Specific Target-Binding Activity Using Microtiter Plate Expression in *E. coli*

7.1. General considerations

This protocol was developed in order to allow assessment of Anticalin candidates selected from large libraries after individual expression in a soluble format with regard to target-binding activity and specificity. To this end, the engineered lipocalins are expressed in the periplasm of *E. coli* using microtiter plate cultures followed by periplasmic extraction and application of the cleared supernatant to an ELISA. In principle, two strategies are possible: (i) the protein target or hapten–protein conjugate can be adsorbed to the microtiter plate, the protein extract with the engineered lipocalin is applied and, after washing, bound Anticalin is detected, e.g., via the C-terminal *Strep*-tag II; (ii) using a suitable capturing antibody, for example, *Strep*MAB-Immo (Schmidt and Skerra, 2007), first the engineered lipocalin is functionally immobilized from the bacterial extract, then the target protein—if necessary, labeled with biotin or digoxigenin groups, for example—is applied and, after washing, bound target is detected with a secondary antibody.

As the protein extracted from a single clone can easily be subjected to several assays in parallel, this procedure also allows to test for cross-reactivity with related or dummy "antigens." Ovalbumin is recommended as a sensitive negative control to sort out mutant lipocalins with non-specific binding activity. Also, competition schemes are possible, using known binding partners such as antibodies or other Anticalins available for the target, in order to screen for defined epitope recognition or to ensure reversible and target-specific binding behavior. Finally, screening ELISA can be performed to quantify expression levels in the bacterial protein extract when using the capturing strategy mentioned above and detecting the recombinant lipocalin with a cognate antibody (directed against the conserved parts of the scaffold) or by using a second affinity tag, which may be fused to its N-terminus to ensure sterical accessibility.

7.2. Procedure

The mutagenized gene cassette of the pooled plasmid preparation from the last phage display panning step—or one of the other selection methods—is subcloned via the pair of *Bst*XI restriction sites on the expression plasmid pNGAL98, which encodes a fusion of the lipocalin with the OmpA signal peptide for periplasmic production in *E. coli* at the N-terminus and the *Strep*-tag II at the C-terminus (Fig. 7.3A). Chemically competent cells of the amber suppressor strain TG1/F^- are transformed with the corresponding ligation product, and individual colonies are picked from the LB/Amp agar plate into a sterile 96-well microtiter plate (e.g., PS Microtest plate with lid; Sarstedt, Nürnbrecht, Germany) containing 100 μL of Terrific Broth (TB)/Amp medium. This master plate is closed with the lid, and the small cultures are grown with shaking (500–800 rpm) in a thermomixer (e.g., Thermomixer comfort; Eppendorf) at 37 °C for 5–6 h in a humidified atmosphere.

About 10 μL cell suspension from each well is removed to inoculate a new 96-well plate containing 100 μL TB/Amp using a 96-well sterile transfer device or a multi-pipette. Growth is started at 37 °C (with 500–800 rpm) for 1 h and continued at 22 °C until the cells reach OD_{550}=0.2–0.4, which corresponds to the mid-log phase at this microtiter plate scale. Protein expression is then induced by adding 20 μL TB/Amp containing 1.2 μg/mL aTc to each well, and incubation is continued overnight at 22 °C (500–800 rpm). To extract the Anticalins from the bacterial periplasm, each well is mixed with 40 μL 4xBBS (640 m*M* NaCl, 8 m*M* EDTA, 0.8 *M* Na-borate, pH 8.0) containing 1 mg/mL lysozyme, followed by incubation for 1 h at room temperature or 4 °C. The lysates are supplemented with 40 μ

L of 10% w/v BSA in TBS (115 m*M* NaCl, 20 m*M* Tris/HCl, pH 7.4) containing 0.5% v/v Tween 20, mixed and incubated for 1 h. Then, the crude extract is cleared by centrifugation in a table-top centrifuge equipped with a swing-out microtiter plate rotor (e.g., SIGMA, Osterode, Germany) at 2000×*g* for 10 min.

A 96-well MaxiSorp plate (Nunc, Langenselbold, Germany) is coated with 50 μL solution of the target protein (10–100 μg/mL) in TBS (or PBS) per well and then blocked with 100 μL of 2% w/v BSA in TBS/T (TBS containing 0.1% v/v Tween 20), followed by washing three times with TBS/T. Fifty microliters of the cleared *E. coli* extract from above is carefully transferred into each well and incubated for 1 h. The plate is washed three times with TBS/T, and bound Anticalins are detected via the *Strep*-tag II by applying 50 μL of streptavidin–alkaline phosphatase conjugate (GE Healthcare, Munich, Germany), diluted 1:1500 in TBS/T. After 1 h, the plate is washed twice with TBS/T and once with TBS. Signals are developed by adding 50 μL of 0.5 mg/mL p-nitrophenyl phosphate in 100 m*M* Tris/HCl, pH 8.8, 100 m*M* NaCl, 5 m*M* $MgCl_2$ and quantified by measuring absorption at 405 nm.

Alternatively, the recombinant lipocalins can be captured from the bacterial extract via their C-terminal *Strep*-tag II. Therefore, a Nunc MaxiSorp microtiter plate is coated with 50 μL of 5 μg/mL *Strep*MAB-Immo in TBS (or PBS) at 4 °C overnight, washed three times and blocked with 100 μL TBS/T containing 2% w/v BSA for 1 h at room temperature. After washing, 50 μL of the cleared *E. coli* lysate is applied to each well and incubated for 1 h. The wells are washed three times with buffer, and then 50 μL of a 0.1–1 μ*M* solution of the target protein, e.g., labeled with digoxigenin (Roche), is added and incubated for 1 h. After washing three times, the bound target protein is detected in the presence of 50 μL anti-digoxigenin–Fab-alkaline phosphatase conjugate (Roche), diluted 1:4000 in TBS/T, for 1 h. After washing twice with TBS/T and once with TBS, signals are developed as above.

8. Measuring Target Affinity of Anticalins in an ELISA

8.1. General considerations

To determine target affinity of a selected Anticalin in an ELISA, the target is adsorbed to a series of wells in a microtiter plate, and the purified Anticalin (prepared as described in the protocol further above) is applied in a dilution series. Using curve fitting, a half maximal effective concentration can be determined, which provides an estimate for the complex dissociation constant (K_D). More elaborate strategies are also possible, for example, by indirect immobilization of the Anticalin itself—as described above—or by using various competition schemes as they are well known from the literature (Azimzadeh *et al.*, 1992). Notably, the direct adsorption of Anticalins to the plastic surface of a microtiter plate for ELISA measurement of target-binding activity is not recommended because, due to their small size, chances are high that the binding site becomes obstructed; indirect immobilization via a capturing reagent usually gives more reliable results.

8.2. Procedure

The target protein is adsorbed to the surface of a 96-well Nunc MaxiSorp plate by applying 50 μL of a 10–100 μg/mL protein solution per well in the presence of PBS at 4 °C overnight. The optimal temperature and incubation period as well as the suitable buffer should be determined individually. After three washing steps, wells are blocked with 100 μL of 2% w/v BSA in PBS for 2 h at room temperature, followed by washing. To prepare a series of twofold dilutions, 50 μL PBS is dispensed into each well of a row. In the first well, 50 μL of the purified Anticalin solution (see further above) is added, mixed with the buffer by repeated pipetting, and 50 μL of this solution is transferred to the next well, followed by mixing and so on, thereby generating 8 or 12 dilutions. The Anticalin is allowed to bind for 1 h at room temperature; then the wells are washed, and the Anticalin is detected via the C-terminal *Strep*-tag II by adding 50 μL of streptavidin–alkaline phosphatase conjugate, diluted 1:1500 in PBS/T. After 1-h incubation and washing twice with PBS/T and once with PBS, the signals are developed in the presence of 50 μL of 0.5 mg/mL *p*-nitrophenyl phosphate as described above. The time course of absorption $\Delta A/\Delta t$ at 405 nm is immediately measured for 15 min in a microtiter plate reader (e.g., SpectraMAX 250; Molecular Devices, Sunnyvale, CA). Alternatively, an end-point measurement is possible. The data are fitted, for example, using KaleidaGraph software (Synergy Software, Reading, PA), to the equation $\Delta A/\Delta t = (\Delta A/\Delta t)_{max} \times [Acn]_{tot}/(K_D + [Acn]_{tot})$, whereby $[Acn]_{tot}$ represents the concentration of the applied Anticalin in each well and K_D is the resulting dissociation constant (Voss and Skerra, 1997).

9. Measuring Target Affinity of Anticalins via Surface Plasmon Resonance

9.1. General considerations

To determine affinity of a selected Anticalin via surface plasmon resonance analysis, the target is covalently coupled to a hydrogel carbohydrate matrix commercially available, for example, as a Biacore or XanTec chip. The purified Anticalin (prepared as described in the protocol further above) is then applied at various concentrations in the mobile phase, and its association and dissociation is measured in real time. Using appropriate curve–fitting procedures, target binding can be evaluated both by means of kinetic and equilibrium methods. Ideally, these two procedures lead to the same K_D value. Care should be taken that measurements are carried out in the absence of mass transfer limitations and that control signals are properly subtracted (Myszka, 1999).

9.2. Procedure

For real-time analysis on a Biacore instrument (GE Healthcare), CM5 (GE Healthcare) or CMD200m (XanTec bioanalytics, Düsseldorf, Germany) sensor chips, both carrying a carboxymethyldextran matrix on a gold surface, are recommended. The sensor chip is equilibrated with HBS/ET (150 m*M* NaCl, 1 m*M* EDTA, 0.005% v/v Tween 20, 20 m*M* Hepes/NaOH, pH 7.5), and the target protein is covalently immobilized using standard amine coupling chemistry according to the manufacturer's instructions. The reference surface on the second flow channel is separately activated and blocked with ethanolamine. Alternatively, the target protein can be captured by an immobilized high-affinity antibody, especially if it is available as a recombinant Ig–Fc fusion protein. The amount of target protein charged to the chip largely depends on the expected affinities and the experimental setup, but should result in a maximum analyte response of 100–300 resonance units (RU) on a Biacore X instrument. For example, a target protein of 35 kDa directly immobilized at a density of 500 RU on the surface of a CMD200m chip yields a maximum response of around 250 RU for a 21 kDa Anticalin upon saturation. Nonetheless, the active ligand density on the chip surface should be kept as low as possible to avoid rebinding effects and mass transfer (Myszka, 1999).

Using HBS/ET as running buffer, up to 100 μL of the purified, sterile-filtered engineered lipocalin is applied to both channels simultaneously at a typical flow rate of 25 μL/min using six different concentrations, ideally in the range from $0.1 \times K_D$ to $20 \times K_D$. Serial dilutions should be performed immediately prior to use in low-protein-binding tubes (Eppendorf) using the running buffer as diluent. If the rate constant of dissociation (k_{off}) of the investigated Anticalin is sufficiently fast, the chip with the immobilized target is preferentially regenerated by simply washing with running buffer until the base line is reached. In the case of slow k_{off} values, the surface is regenerated with a short pulse of a denaturing buffer, for example, with high or low pH, elevated salt concentration, or containing urea. To avoid irreversible inactivation of the immobilized target, the mildest reagent applicable should be used.

To correct for bulk responses, the corresponding signals measured for the control channel are subtracted from the resulting sensorgrams, and baseline corrections are applied as necessary (Myszka, 1999). Kinetic data evaluation is performed by global fitting with BIAevaluation software provided by the instrument manufacturer (Karlsson *et al.*, 1991).

10. Application of Anticalins in Biochemical Research and Drug Development

Anticalins selected by means of the procedures described here can directly serve as affinity reagents in biochemical assays and even in animal experiments. Typical target affinities are in the low n*M* to p*M* range, which is sufficient for many practical applications. If higher binding activities are needed, methods of *in vitro* affinity maturation can be employed, similarly as they are well known in the field of antibody engineering. To this end, first random mutations are introduced at a low dose into a gene encoding a promising Anticalin candidate, and from the resulting library, new variants are selected using one of the methods described above by applying higher stringency.

Several strategies for introducing further random mutations into an engineered lipocalin gene are possible: for example, (i) error-prone PCR leading to scattered mutations over the entire *Bst*XI cassette, (ii) site-directed random mutagenesis of specific amino acid codons using degenerate oligodeoxynucleotides and synthetic gene assembly procedures similar to those described above (this time employing the gene encoding the Anticalin candidate as template), and (iii) re-randomization of one loop segment in the context of the other three loops at the open end of the calyx fixed ("loop walking"). For corresponding practical examples, the reader is referred to the literature (Kim *et al.*, 2009; Schönfeld *et al.*, 2009). Similarly, if Anticalins with a higher folding stability are desired, selection conditions can be chosen in the presence of denaturing agents or at elevated temperature. Usually, natural as well as engineered lipocalins are rather stable proteins with melting temperatures—measured via circular dichroism spectrometry, for example (Schlehuber and Skerra, 2002)—in the range of 60 to 70 °C, at least.

Applications in biochemical assays often require a label for detection—unless the Anticalin is used in a competition assay. The *Strep*-tag II is an excellent tool for this purpose, and a series of secondary reagents for the sensitive detection or stable immobilization of corresponding fusion proteins is available (Schmidt and Skerra, 2007). In this way, Anticalins can be utilized in conventional preclinical assays such as immunofluorescence microscopy of cells or tissues, detection of cells expressing disease markers in flow cytofluorometry (FACS), and the in immunohistochemical staining of tissue sections. The *Strep*-tag II can even be used to determine the plasma concentration of an applied Anticalin in pharmacokinetic animal studies, for example, by using a sandwich ELISA with the purified target protein adsorbed to the microtiter plate.

For some therapeutic applications, Anticalins having antagonistic binding activity for a disease-related receptor or soluble signaling factor do not need additional effector functions. For other mechanisms of treatment, Anticalins can be coupled to various proteins or reagents. One option for cancer therapy is the fusion to protein toxins or enzymes that activate chemical prodrugs. Furthermore, Anticalins of identical or even different specificities can be fused to yield so-called Duocalins, thus leading to signal triggering by clustering of cell surface receptors or to immune activation by cross-linking tumor markers with cytotoxic cells, for example. Alternatively, Anticalins can be chemically coupled to a variety of chemical labels such as fluorescent groups or radionuclides—either in a random manner via one or more of their Lys side chains at the protein surface or via specifically introduced free Cys side chains.

Similar strategies can be employed in order to extend the plasma half-life of Anticalins. Natural as well as engineered lipocalins are quickly cleared by renal filtration due to their small size of ca. 20kDa if they circulate as monomeric proteins. When conjugated with radioactive isotopes for *in vivo* diagnostics, these properties should lead to images of high contrast soon after administration. However, for medical indications that require prolonged treatment, the simple architecture and robustness of the lipocalin scaffold facilitate the preparation of fusion proteins or of site-directed conjugates to modulate clearance using established methodologies. Several techniques have proven to extend the plasma half-life of Anticalins, for example, the preparation of fusion proteins with the ABD or chemical PEGylation. A novel alternative technology to equip Anticalins with prolonged plasma half-life is PASylation, that is, the genetic fusion with a conformationally disordered, hydrophilic polypeptide comprising the small amino acids Pro, Ala, and Ser.

In conclusion, Anticalins provide binding sites with high structural plasticity and an extended molecular interface with their prescribed targets, comparable in size to that of antibodies. Thus, Anticalins with high specificity and affinity can be readily generated against haptens, peptides, and proteins. As Anticalins are readily derived from human lipocalin scaffolds, further reformatting such as by CDR-grafting during the humanization of antibodies is not required. Furthermore, the intrinsic monovalent binding activity decreases the risk of unwanted intermolecular cross-linking of cellular receptor targets. Available structural and functional data suggest that Anticalins are able to recognize a diverse set of epitopes on different target proteins and, therefore, offer considerable potential as antagonistic reagents in general. Compared with antibodies, Anticalins provide several practical advantages because they are much smaller, consist of a single polypeptide chain, do not require glycosylation or disulfide bonds, exhibit robust biophysical properties, provide a deep binding pocket, and can easily be produced in microbial expression systems such as *E. coli* or yeast. Since their structure–function relationships are well understood, rational engineering of additional functions such as site-directed PEGylation or conjugation with effector domains, dimerization modules, or even with a second Anticalin can be easily achieved, hence opening a wide range of therapeutic modes of action.

REFERENCES

Åkerström, B., Borregaard, N., Flower, D. A., and Salier, J.-S. (2006). Lipocalins. Landes Bioscience, Georgetown, TX.

Azimzadeh, A., Pellequer, J. L., and Van Regenmortel, M. H. (1992). Operational aspects of antibody affinity constants measured by liquid-phase and solid-phase assays. *J. Mol. Recognit.* **5,** 9–18.

Bachmann, B. J. (1972). Pedigrees of some mutant strains of *Escherichia coli* K-12. *Bacteriol. Rev.* **36,** 525–557.

Beste, G., Schmidt, F. S., Stibora, T., and Skerra, A. (1999). Small antibody-like proteins with prescribed ligand specificities derived from the lipocalin fold. *Proc. Natl. Acad. Sci. USA* **96,** 1898–1903.

Binder, U., Matschiner, G., Theobald, I., and Skerra, A. (2010). High-throughput sorting of an Anticalin library via EspP-mediated functional display on the *Escherichia coli* cell surface. *J. Mol. Biol.* **400,** 783–802.

Bradbury, A. R., Sidhu, S., Dübel, S., and McCafferty, J. (2011). Beyond natural antibodies: The power of *in vitro* display technologies. *Nat. Biotechnol.* **29,** 245–254.

Breustedt, D. A., Korndörfer, I. P., Redl, B., and Skerra, A. (2005). The 1.8-Å crystal structure of human tear lipocalin reveals an extended branched cavity with capacity for multiple ligands. *J. Biol. Chem.* **280,** 484–493.

Breustedt, D. A., Schönfeld, D. L., and Skerra, A. (2006). Comparative ligand-binding analysis of ten human lipocalins. *Biochim. Biophys. Acta* **1764,** 161–173.

Breustedt, D. A., Chatwell, L., and Skerra, A. (2009). A new crystal form of human tear lipocalin reveals high flexibility in the loop region and induced fit in the ligand cavity. *Acta Crystallogr. D Biol. Crystallogr.* **65,** 1118–1125.

Bullock, W. O., Fernandez, J. M., and Short, J. M. (1987). XL1-Blue: A high efficiency plasmid transforming recA *Escherichia coli* strain with beta-galactosidase selection. *Biotechniques* **5,** 376–378.

Couillin, I., Maillet, I., Vargaftig, B. B., Jacobs, M., Paesen, G. C., Nuttall, P. A., Lefort, J., Moser, R., Weston-Davies, W., and Ryffel, B. (2004). Arthropod-derived histamine-binding protein prevents murine allergic asthma. *J. Immunol.* **173,** 3281–3286.

Ferrara, N., and Kerbel, R. S. (2005). Angiogenesis as a therapeutic target. *Nature* **438,** 967–974.

Fling, S. P., and Gregerson, D. S. (1986). Peptide and protein molecular weight determination by electrophoresis using a high-molarity tris buffer system without urea. *Anal. Biochem.* **155,** 83–88.

Flower, D. R. (1996). The lipocalin protein family: Structure and function. *Biochem. J.* **318,** 1–14.

Gebauer, M., and Skerra, A. (2009). Engineered protein scaffolds as next-generation antibody therapeutics. *Curr. Opin. Chem. Biol.* **13,** 245–255.

Gill, S. C., and von Hippel, P. H. (1989). Calculation of protein extinction coefficients from amino acid sequence data. *Anal. Biochem.* **182,** 319–326.

Goetz, D. H., Holmes, M. A., Borregaard, N., Bluhm, M. E., Raymond, K. N., and Strong, R. K. (2002). The neutrophil lipocalin NGAL is a bacteriostatic agent that interferes with siderophore-mediated iron acquisition. *Mol. Cell* **10,** 1033–1043.

Heckman, K. L., and Pease, L. R. (2007). Gene splicing and mutagenesis by PCR-driven overlap extension. *Nat. Protoc.* **2,** 924–932.

Hengen, P. N. (1996). Methods and reagents. Preparing ultra-competent *Escherichia coli*. *Trends Biochem. Sci.* **21,** 75–76.

Hohlbaum, A. M., and Skerra, A. (2007). Anticalins: The lipocalin family as a novel protein scaffold for the development of next-generation immunotherapies. *Expert Rev. Clin. Immunol.* **3,** 491–501.

Jose, J., Kramer, J., Klauser, T., Pohlner, J., and Meyer, T. F. (1996). Absence of periplasmic DsbA oxidoreductase facilitates export of cysteine-containing passenger proteins to the *Escherichia coli* cell surface via the IgA_{β} autotransporter pathway. *Gene* **178,** 107–110.

Karlsson, R., Michaelsson, A., and Mattsson, L. (1991). Kinetic analysis of monoclonal antibody-antigen interactions with a new biosensor based analytical system. *J. Immunol. Methods* **145,** 229–240.

Kayushin, A. L., Korosteleva, M. D., Miroshnikov, A. I., Kosch, W., Zubov, D., and Piel, N. (1996). A convenient approach to the synthesis of trinucleotide phosphoramidites—Synthons for the generation of oligonucleotide/peptide libraries. *Nucleic Acids Res.* **24,** 3748–3755.

Kim, H. J., Eichinger, A., and Skerra, A. (2009). High-affinity recognition of lanthanide(III) chelate complexes by a reprogrammed human lipocalin 2. *J. Am. Chem. Soc.* **131,** 3565–3576.

Korndörfer, I. P., Beste, G., and Skerra, A. (2003a). Crystallographic analysis of an "anticalin" with tailored specificity for fluorescein reveals high structural plasticity of the lipocalin loop region. *Proteins Struct. Funct. Genet.* **53,** 121–129.

Korndörfer, I. P., Schlehuber, S., and Skerra, A. (2003b). Structural mechanism of specific ligand recognition by a lipocalin tailored for the complexation of digoxigenin. *J. Mol. Biol.* **330,** 385–396.

Myszka, D. G. (1999). Improving biosensor analysis. *J. Mol. Recognit.* **12,** 279–284.

Nunn, M. A., Sharma, A., Paesen, G. C., Adamson, S., Lissina, O., Willis, A. C., and Nuttall, P. A. (2005). Complement inhibitor of C5 activation from the soft tick *Ornithodoros moubata*. *J. Immunol.* **174,** 2084–2091.

Peggs, K. S., Quezada, S. A., Korman, A. J., and Allison, J. P. (2006). Principles and use of anti-CTLA4 antibody in human cancer immunotherapy. *Curr. Opin. Immunol.* **18,** 206–213.

Sambrook, J., Fritsch, E. F., and Maniatis, T. (1989). Molecular Cloning: A Laboratory Manual. Cold Spring Harbor Laboratory Press, Cold Spring Harbor, NY.

Schlapschy, M., Theobald, I., Mack, H., Schottelius, M., Wester, H. J., and Skerra, A. (2007). Fusion of a recombinant antibody fragment with a homo-amino-acid polymer: Effects on biophysical properties and prolonged plasma half-life. *Protein Eng. Des. Sel.* **20,** 273–284.

Schlehuber, S., and Skerra, A. (2002). Tuning ligand affinity, specificity, and folding stability of an engineered lipocalin variant—a so-called 'anticalin'—using a molecular random approach. *Biophys. Chem.* **96,** 213–228.

Schlehuber, S., and Skerra, A. (2005). Lipocalins in drug discovery: From natural ligand-binding proteins to "anticalins" *Drug Discov. Today* **10,** 23–33.

Schlehuber, S., Beste, G., and Skerra, A. (2000). A novel type of receptor protein, based on the lipocalin scaffold, with specificity for digoxigenin. *J. Mol. Biol.* **297,** 1105–1120.

Schmidt, T. G., and Skerra, A. (2007). The *Strep*-tag system for one-step purification and high-affinity detection or capturing of proteins. *Nat. Protoc.* **2,** 1528–1535.

Schönfeld, D., Matschiner, G., Chatwell, L., Trentmann, S., Gille, H., Hülsmeyer, M., Brown, N., Kaye, P. M., Schlehuber, S., Hohlbaum, A. M., and Skerra, A. (2009). An engineered lipocalin specific for CTLA-4 reveals a combining site with structural and conformational features similar to antibodies. *Proc. Natl. Acad. Sci. USA* **106,** 8198–8203.

Skerra, A. (1994). Use of the tetracycline promoter for the tightly regulated production of a murine antibody fragment in *Escherichia coli*. *Gene* **151,** 131–135.

Skerra, A. (2000a). Engineered protein scaffolds for molecular recognition. *J. Mol. Recognit.* **13,** 167–187.

Skerra, A. (2000b). Lipocalins as a scaffold. *Biochim. Biophys. Acta* **1482,** 337–350.

Skerra, A. (2001). 'Anticalins': A new class of engineered ligand-binding proteins with antibody-like properties. *J. Biotechnol.* **74,** 257–275.

Skerra, A. (2003). Imitating the humoral immune response. *Curr. Opin. Chem. Biol.* **7,** 683–693.

Skerra, A. (2007). Anticalins as alternative binding proteins for therapeutic use. *Curr. Opin. Mol. Ther.* **9,** 336–344.

Skerra, A., Dreher, M. L., and Winter, G. (1991). Filter screening of antibody Fab fragments secreted from individual bacterial colonies: Specific detection of antigen binding with a two-membrane system. *Anal. Biochem.* **196,** 151–155.

Studier, F. W., and Moffatt, B. A. (1986). Use of bacteriophage T7 RNA polymerase to direct selective high-level expression of cloned genes. *J. Mol. Biol.* **189,** 113–130.

Tung, W. L., and Chow, K. C. (1995). A modified medium for efficient electrotransformation of *E. coli*. *Trends Genet.* **11,** 128–129.

Van den Brulle, J., Fischer, M., Langmann, T., Horn, G., Waldmann, T., Arnold, S., Fuhrmann, M., Schatz, O., O'Connell, T., O'Connell, D., Auckenthaler, A., and Schwer, H. (2008). A novel solid phase technology for high-throughput gene synthesis. *Biotechniques* **45,** 340–343.

Venturi, M., Seifert, C., and Hunte, C. (2002). High level production of functional antibody Fab fragments in an oxidizing bacterial cytoplasm. *J. Mol. Biol.* **315,** 1–8.

Voss, S., and Skerra, A. (1997). Mutagenesis of a flexible loop in streptavidin leads to higher affinity for the *Strep*-tag II peptide and improved performance in recombinant protein purification. *Protein Eng.* **10,** 975–982.

Yanisch-Perron, C., Vieira, J., and Messing, J. (1985). Improved M13 phage cloning vectors and host strains: Nucleotide sequences of the M13mp18 and pUC19 vectors. *Gene* **33,** 103–119.

CHAPTER EIGHT

T Cell Receptor Engineering

Jennifer D. Stone, Adam S. Chervin, David H. Aggen, *and* David M. Kranz

Contents

1. Introduction 190
2. Stability and Affinity Engineering of T Cell Receptors by Yeast Surface Display 192
   2.1. Design and cloning for expression on yeast surface as single-chain variable fragment 192
   2.2. Stability engineering and selection 195
   2.3. Affinity engineering and selection 198
3. Affinity Engineering and Selection of T Cell Receptors by T Cell Display 200
   3.1. Design and construction of libraries for expression on T cell surface 201
   3.2. Packaging, transduction, and characterization of libraries in T cells 203
   3.3. Selection of higher affinity receptors from a T cell library 205
   3.4. Isolation of TCR sequences from selected T cell hybridomas 206
4. Expression, Purification, and Applications of Soluble scTv Proteins 207
   4.1. Design and cloning for scTv expression in *E. coli* as inclusion bodies 209
   4.2. Growth and induction of *E. coli* 209
   4.3. Isolation of inclusion bodies 210
   4.4. Solubilization and refolding of scTv proteins 211
   4.5. Purification of scTv proteins 211
   4.6. Detection of specific pepMHC on a cell surface using soluble, purified scTvs 212
   4.7. Monitoring of soluble protein levels using an scTv competition ELISA 214
5. Recipes for Media and Buffers 215
Acknowledgments 218
References 219

Department of Biochemistry, University of Illinois, Urbana, Illinois, USA

Methods in Enzymology, Volume 503
ISSN 0076-6879, DOI: 10.1016/B978-0-12-396962-0.00008-2

Abstract

T lymphocytes express on their surface a heterodimeric αβ receptor, called the T cell receptor (TCR), which recognizes foreign antigens. Unlike antibodies, the recognition requires both an antigenic peptide epitope and a protein encoded by the major histocompatibility complex (MHC). In contrast to conventional antibody-directed target antigens, antigens recognized by the TCR can include the entire array of potential intracellular proteins, which are processed and delivered to the cell surface as a peptide/MHC complex. In the past 10 years, there have been significant efforts to engineer TCRs in various formats, which would allow improved recognition and destruction of virus-infected cells or cancer. The proposed therapeutic approaches involve either the use of engineered, high-affinity TCRs in soluble forms, analogous to antibody-directed therapies, or the use of engineered TCRs whose genes are reintroduced into autologous T cells and transferred back into patients (T cell adoptive therapies). This chapter describes three methods associated with the engineering of TCRs for these therapeutic purposes: (1) use of a yeast display system to engineer higher affinity single-chain VαVβ TCRs, called scTv; (2) use of a T cell display system to engineer higher affinity full-length TCRs; and (3) expression, purification, and characterization of soluble TCRs in an *Escherichia coli* system.

1. Introduction

Proteins encoded by the major histocompatibility complex (MHC) bind to peptides derived from intracellular sources and deliver the peptide to the cell surface. The T cell system evolved to recognize these peptide/MHC (pepMHC) ligands, thereby triggering a response to "foreign" antigens (Zinkernagel and Doherty, 1974). Thus, aberrant expression of a protein (i.e., yielding peptides that are "nonself," in sequence or in quantity) elicits T cell activation upon binding of the heterodimeric αβ T cell receptor (TCR) to the peptide/MHC ligand (Davis *et al.*, 2007). The structural and biochemical features of the TCR:pepMHC interaction have been studied extensively over the past 15 years (Garcia *et al.*, 2009; Rudolph *et al.*, 2006). It has been found that the TCR docks onto the pepMHC in a conserved diagonal orientation, and that the specificity for the peptide is conferred largely by interactions between the complementarity-determining region (CDR) three loops of each TCR chain (CDR3α, CDR3β) and the peptide. The other four CDR loops (CDR1α, CDR2α, CDR1β, CDR2β) are generally more involved in MHC contact, although CDR1 loops can in some cases contact peptide (Armstrong *et al.*, 2008).

The affinities of wild-type TCR:pepMHC interactions, excluding perhaps allogeneic reactions, have been shown to be relatively low, with K_D values in the range of 10–300 μ*M* (Bowerman *et al.*, 2009; Davis *et al.*, 1998). Recent studies have also suggested an importance of the two-dimensional confinement of TCR and pepMHC in the opposing cell-to-cell environment (Aleksic *et al.*, 2010; Huang *et al.*, 2010; Huppa *et al.*, 2010). The observed low affinities are consistent with the inability of T cells to undergo extensive somatic mutation of the V regions, unlike antibodies.

TCR affinities in this range are adequate in large part because of the contribution of the coreceptors CD4 and CD8 on T helper cells and cytotoxic T cells, respectively. The CD4 coreceptor binds to the class II MHC ligand, and the CD8 coreceptor binds to the class I MHC ligand, increasing the sensitivity of the T cell such that it can be stimulated by just a few specific pepMHC ligands per target cell (Li *et al.*, 2004; Purbhoo *et al.*, 2004). The increase in sensitivity is also a consequence of the ability of the CD4/CD8 associated kinase, Lck, to be recruited to the TCR/CD3 complex (Artyomov *et al.*, 2010).

Importantly, the necessity for the coreceptor appears to be bypassed by expression of a TCR with adequate affinity. The affinity threshold for this coreceptor independence is a K_D of about 1 μM (Chervin *et al.*, 2009). Affinities above this level (i.e., K_D values less than 1 μM) allow a class I-restricted TCR to mediate activity of $CD4^+$ T cells (Zhao *et al.*, 2007). This finding has prompted the notion that it may be possible to recruit class I-restricted T helper cell activity, in an adoptive T cell setting, against tumors or virus-infected cells. However, the affinities of TCRs in this approach most likely do not require the low nanomolar K_D values observed with many antibodies, but rather only require modest improvements from the wild-type TCR range to approximately 1 μM range (Chervin *et al.*, 2009; Holler and Kranz, 2003; Robbins *et al.*, 2008; Zhao *et al.*, 2007). For these purposes, it is possible to use strategies that involve lower throughput directed evolution or screening. One method that is described in this chapter involves display of libraries of the TCR on a T cell line, followed by high-speed sorting with fluorescent pepMHC ligands (Chervin *et al.*, 2008; Kessels *et al.*, 2000).

As with antibodies, there has also emerged an interest in using the TCR antigen recognition system for developing soluble, antigen-specific targeting agents as therapeutics. Conceptually, this would greatly increase the number of targets available for specific therapies, since intracellular epitopes could be accessible through their processing and delivery to the surface by MHC. However, the soluble approach requires that the TCR be engineered for considerably higher affinity (e.g., low nanomolar K_D values) than the wild-type affinities found naturally, and it also requires that the TCR be fused to an effector molecule. The latter could include cytokines or, given recent success with scFv bispecific molecules (Bargou *et al.*, 2008), an anti-CD3 single-chain Fv for recruitment of T cell activity. The potency of the effector will be especially important as the number of pepMHC antigens on a target cell can be several orders of magnitude lower than the number of antigens on a target cell typically recognized by antibody-based therapeutics.

Despite considerable effort, the expression and engineering of soluble, extracellular TCR fragments (VαCα/VβCβ, or the single-chain VαlinkerVβ, called scTv here) have remained more difficult than antibody fragments (Fab or scFv) (Richman and Kranz, 2007). Although various strategies to overcome the challenges inherent with wild-type TCRs have been reported (Boulter *et al.*, 2003; Li *et al.*, 2005; Varela-Rohena *et al.*, 2008), we focus here on several methods that have been developed in our laboratory, including yeast display of human scTv (Aggen *et al.*, 2011), T cell display of full-length TCRs (Chervin *et al.*, 2008), and production of stabilized soluble scTv in *Escherichia coli* (Aggen *et al.*, 2011; Holler *et al.*, 2000; Jones *et al.*, 2006; Kieke *et al.*, 1999; Weber *et al.*, 2005).

2. Stability and Affinity Engineering of T Cell Receptors by Yeast Surface Display

2.1. Design and cloning for expression on yeast surface as single-chain variable fragment

Yeast display allows for cell surface expression of recombinant proteins as fusions to the C-terminus of the mating agglutinin protein AGA-2 (Fig. 8.1A). From 10,000 to 1,000,000 copies of the fusion protein can be expressed on the surface of each yeast cell. For display of TCRs, a recombinant scTv consisting of TCR variable domains attached by a flexible linker can be introduced into the yeast display vector pCT302 using an in-frame *Nhe*I restriction site and downstream *Bgl*II or *Xho*I restriction site (Boder and Wittrup, 1997, 2000). The pCT302 vector uses a galactose-inducible promoter, resulting in induction of the introduced protein when yeast are cultured in media containing galactose. The fusion contains an N-terminal hemagglutinin epitope (HA, sequence: YPYDVPDYA) upstream of the *Nhe*I site that can be used as a marker of yeast protein expression. A C-terminal tag (c-myc, sequence: EQKLISEEDL) can also be incorporated to monitor expression of the full construct. Upon fusion-protein induction, cells can be stained with antibodies specific for the epitope tags and flow cytometry used to monitor surface protein expression. The gene fusion should incorporate a stop codon, upstream of the *Bgl*II or *Xho*I restriction site (Fig. 8.1B), as no stop codon is included in the vector.

An scTv molecule includes the variable domains of the T cell receptor beta (Vβ) and alpha (Vα) chains connected by a linker (e.g., GSADDAKK-DAAKKDGKS) (Richman *et al.*, 2009a; Weber *et al.*, 2005). TCR genes derived from human or murine T cell clones have been constructed in the Vβ-L-Vα or Vα-L-Vβ orientation. Recently, we have shown that for human TCRs, the Vα2 region is particularly amenable for engineering scTv of improved stability (Aggen *et al.*, 2011). The scTv construct can be

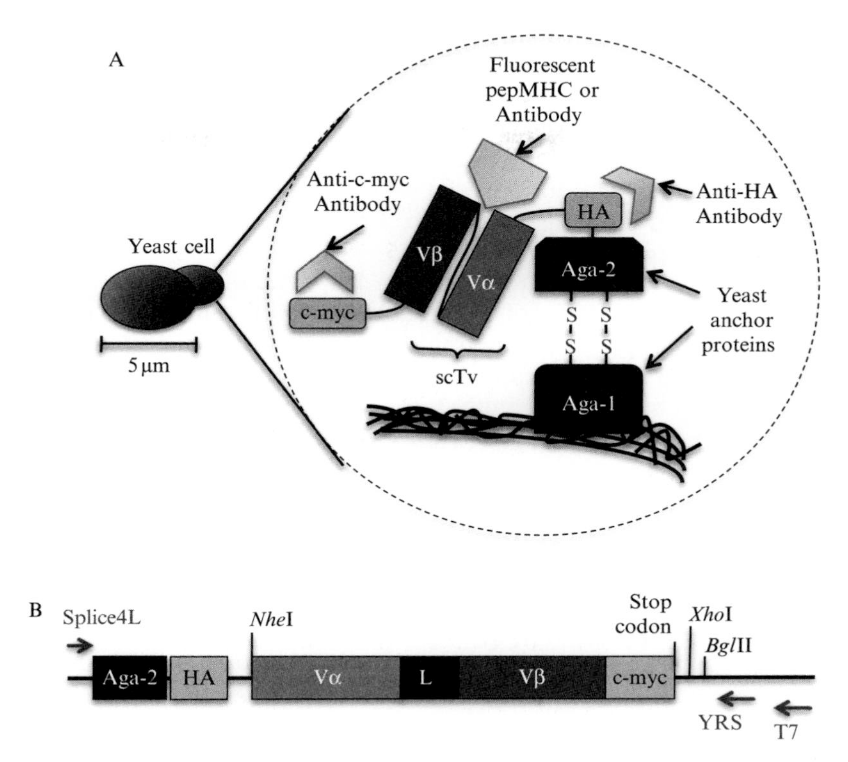

Figure 8.1 Schematic diagrams for yeast surface display. (A) Cartoon showing the display of an scTv construct, including N- and C-terminal expression tags (HA and c-myc, respectively), by fusion to the yeast mating protein Aga-2, which is, in turn, linked by disulfide bonds to the integral yeast cell wall protein Aga-1. Expression of the scTv may be monitored by antibodies against either expression tag, the particular TCR variable domains included, or with the specific pepMHC ligand. (B) An example of cloning an scTv construct into the yeast display vector pCT302. The variable domains are connected by a flexible linker and are cloned in-frame via an *Nhe*I restriction site with the gene for Aga-2 and HA on the 5′-end. A c-myc tag is included at the C-terminus before the stop codon and the downstream restriction sites *Xho*I and *Bgl*II. Approximate locations of gene amplification and sequencing primers Splice4L, YRS, and T7 are indicated. (For color version of this figure, the reader is referred to the Web version of this chapter.)

cloned into *Nhe*I and *Xho*I restriction sites of pCT302 by conventional cloning techniques and use of a subcloning competent strain of *E. coli*, such as DH5α or DH10B. The pCT302 vector contains the ampicillin resistance gene, allowing for ampicillin (100 μg/mL) selection of transformed *E. coli*. Sequencing primers, including "Splice4L" (5′, Fwd: GGCAGCCCCA-TAAACACACAGTAT), "T7" (3′, Rev: TAATACGACTCACTA-TAG), and "YRS" (3′, Rev: CGAGCTAAAAGTACAGTGGG), can be used to confirm proper scTv gene cloning in pCT302 (Fig. 8.1B).

The scTv fusion-containing plasmid can be introduced into the EBY100 yeast cell strain by LiOAc-mediated transformation (Gietz and Woods, 2002) or by electroporation (e.g., Biorad gene pulser, 0.2cm gap cuvette). After transformation, yeast are plated on SD-CAA plates and incubated at 30°C for 36–48h. Yeast colonies are cultured in 3mL SD-CAA liquid media, shaking at 200rpm at 30°C, until an OD_{600} of ~1.0–1.5 is reached (typically, 36–48 h). Cells are centrifuged at 1800×*g* for 3min, and suspended in SG-CAA media to induce protein expression for 36–48h at 20°C, shaking at 200rpm. The cells will continue to grow while in SG-CAA media, and should be resuspended at a final OD_{600} of 0.6–0.8. Optimal protein induction time and temperature will vary depending on the protein of interest. Liquid cultures can be stored at 4°C or frozen in 10% DMSO for storage.

To assess expression of the introduced scTv on the surface of yeast, cells can be analyzed by flow cytometry using commercially available antibodies against the HA and c-myc expression epitopes, the specific Vβ domain of the TCR, and in some cases the specific Vα domain of the TCR. While antibodies against Vα domains are not as readily available compared to Vβ domains, recently our lab has isolated a monoclonal antibody (mAb) specific for the human Vα2 domain that specifically detects Vα2-containing scTv proteins on the yeast (Fig. 8.2). Note that this yeast display system always

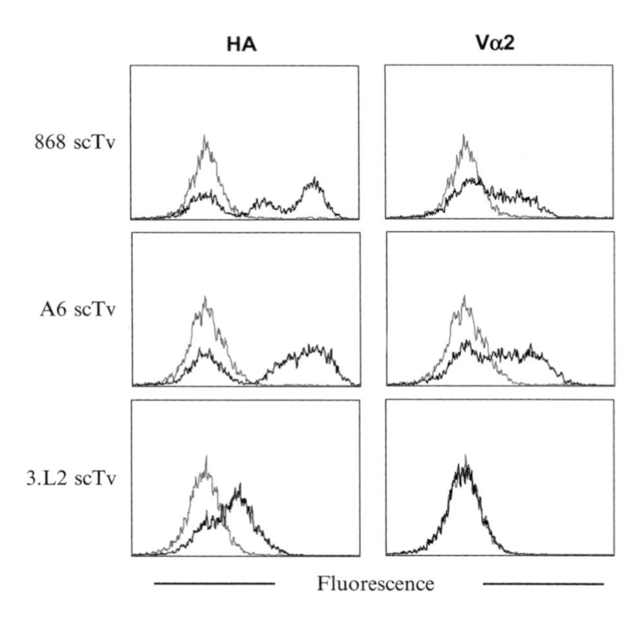

Figure 8.2 Detection of properly folded scTv variable domain on yeast. Staining was performed for the N-terminal expression tag HA (left column), and using a monoclonal, human Vα2-specific antibody (right column) for yeast expressing 868 scTv (top row, Vα2^{+}), A6 scTv (middle row, Vα2^{+}), and 3.L2 scTv (bottom row, Vα2^{-}).

yields a population of "negative" cells that have lost plasmid during propagation and induction. This negative population is useful, as it serves as an internal control for each antibody stain analyzed by flow cytometry. In addition, depending on growth conditions and the fusion protein, it is not uncommon to observe two "positive" populations, differing in the surface levels of the fusion (Fig. 8.2).

2.2. Stability engineering and selection

In most cases, a wild-type scTv is not optimally expressed on the surface of yeast in the absence of constant domains; if one is displayed, it may denature irreversibly at increased temperature or extreme pH (Orr *et al.*, 2003). It is possible to identify more stable, displayed scTv mutants by random mutagenesis followed by rounds of selection by fluorescence-activated cell sorting (FACS) with fluorescent-labeled antibodies against Vβ and/or Vα regions (Richman *et al.*, 2009a; Shusta *et al.*, 2000; Weber *et al.*, 2005). Random mutagenesis can be achieved by amplifying the gene encoding the scTv protein using error-prone PCR, resulting in a library of scTv variants. A protocol to achieve an approximately 0.5% error rate has been described previously (Richman *et al.*, 2009b). The library is introduced into pCT302 by homologous recombination in yeast, taking advantage of the extensive overlap between the *Nhe*I, *Bgl*II, and *Xho*I digested vector, and the error-prone PCR product of the scTv gene in pCT302, generated by using the "Splice4L" and "T7" primers (Fig. 8.1B). Digested vector and mutagenized scTv PCR products are coelectroporated into electrocompetent EBY100 yeast cells. Up to 10^7–10^8 variants can be generated with as few as four electroporations. Figure 8.3 shows a flowchart summarizing the protocol for engineering scTvs by yeast display, whether for improved stability or affinity.

Electrocompetent EBY100 cells into which the scTv library can be introduced are made with the following solutions, which are kept at 4 °C (selected recipes are provided at the end of the chapter): YPD media (500 mL), SD-CAA media (1 L), 1 *M* sorbitol (200 mL), 1 *M* sorbitol with 1 m*M* $CaCl_2$ (1 L), 0.1 *M* lithium acetate with 10 m*M* DTT (100 mL), autoclaved ddH_2O (1 L), and a 1:1 (volume:volume) mixture of YPD and 1 *M* sorbitol. Then, follow these steps:

1. Prepare 50 and 100 mL of YPD media in sterile 250 and 500 mL Erlenmeyer flasks, respectively. Inoculate the 250-mL flask with EBY100 from a colony or small volume (5–10 μL) of liquid culture and shake overnight (12–16 h) at 30 °C, 185 rpm. Prewarm the 500-mL flask to 30 °C.
2. The next morning, inoculate the 500-mL flask to a final OD_{600}=0.3 from the overnight EBY100 culture. Shake the flask at 185 rpm and 30 °C.

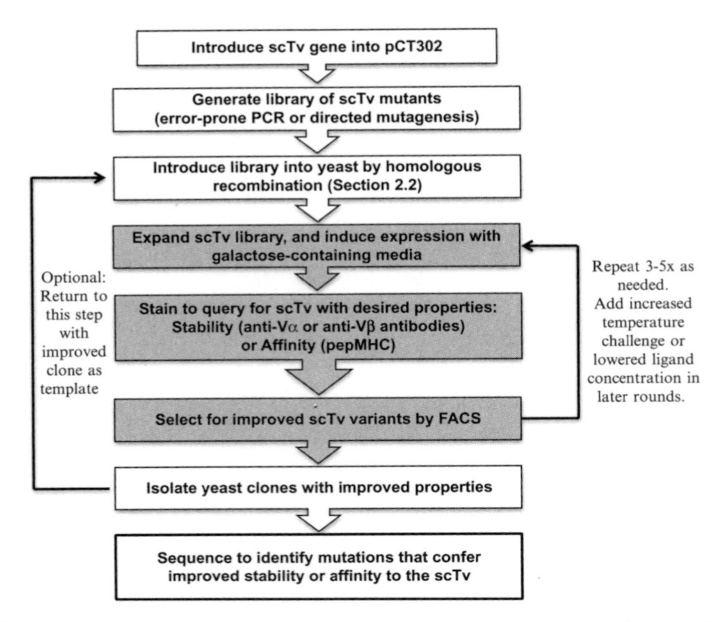

Figure 8.3 Flowchart outlining scTv engineering by yeast surface display.

3. Allow culture to grow to an OD_{600}=1.3–1.6, monitoring hourly.
4. Harvest yeast cells by centrifugation at 1800×*g* for 3min in 2×50mL conical tubes. Cells should be kept on ice or at 4°C during wash and centrifugation steps.
5. Suspend the yeast cell pellet in 50mL cold ddH_2O, and harvest cells by centrifugation at 1800×*g* and 4°C for 3min. Repeat for a total of two washes.
6. Suspend the yeast cell pellet in 50mL 1*M* sorbitol/1m*M* $CaCl_2$ solution. Harvest by centrifugation at 1800×*g* for 3min.
7. Suspend each cell pellet in 20mL of 0.1*M* lithium acetate/10m*M* DTT in ddH_2O. The lithium acetate/DTT solution should be made fresh the same day as library creation for optimal efficiency. Incubate the suspended cells in a sterile Erlenmeyer culture flask for 30min at 30°C on a shaker at 200–250rpm.
8. Harvest cells by centrifugation at 1800×*g* for 3min.
9. Wash by suspending the yeast cell pellet in 50mL cold 1*M* sorbitol/1 m*M* $CaCl_2$, and harvest cells by centrifugation at 1800×*g* and 4°C for 3 min. Repeat for a total of two washes.
10. Suspend cells in 50mL of cold 1*M* sorbitol solution, and harvest cells as in previous step.
11. Suspend the yeast cell pellet in 200–400μL of 1*M* sorbitol. *Note*: At this step, electrocompetent cells can be flash-frozen and stored at −80°C

for future use, but for maximal transformation efficiency, they should be used on the same day.

12. For library generation, mix 1 μg of pCT302 vector (digested with *Nhe*I, *Xho*I, and *Bgl*II) with 5 μg of error-prone scTv PCR product, pretreated with *Dpn*I to remove template. This mixture is divided among at least four electroporations to achieve maximal efficiency, and each aliquot is added to 100 μL electrocompetent EBY100 per electroporation. Using a BioRad GenePulser, electroporate at 2.5 kV, 25 μF capacitance with a 0.2 cm gap cuvette. Cells are allowed to recover in a sterile, 50 mL Erlenmeyer flask containing 2 mL per electroporation of a 1:1 mixture of 1 *M* sorbitol/YPD for 1 h at 30 °C. *Note*: PCR products encoding libraries with directed mutations, rather than error-prone PCR products, generated by splicing by overlap extension (SOE) with primers introducing targeted degeneracy, may be used.
13. Control electroporations should include (A) yeast only, (B) yeast with scTv PCR product only, and (3) yeast with digested pCT302 vector only. Each control electroporation is allowed to recover for 1 h at 30 °C in separate culture tubes in 2 mL of 1:1 1 *M* sorbitol/YPD.
14. Following recovery, 10 μL of each control culture are spread on an SD-CAA plate. From the library culture, 10 μL each of 10-fold serial dilutions from 1:10 to 1:10,000 is spread on SD-CAA plates to determine library size.
15. The library culture is harvested by centrifugation at 1800×*g* for 3 min. Cells are suspended in 10 mL 1 *M* sorbitol and added to 500 mL of SD-CAA media in a sterile, 2-L Erlenmeyer flask. The library is expanded by shaking at 200 rpm, 30 °C for 36–48 h. *Note*: After expansion of the library culture in SD-CAA, aliquots should be frozen in 10% DMSO.

To select for stabilized scTv mutants, a small volume (<3 mL) of the expanded library is centrifuged at 1800 rpm for 3 min and suspended in SG-CAA to a final OD_{600}~0.6–0.8. This corresponds to a sufficient number of yeast (~3×10^7) to fully cover the diversity of most error-prone PCR libraries. The library is induced in a sterile culture tube by shaking at 200 rpm, 20 °C for 36–48 h. After induction, yeast can be stained with an antibody, preferably a conformation-specific mAb directed against the specific Vβ or Vα domain of the scTv (an example of yeast expressing scTvs, stained with a mAb against the human Vα2 domain is shown in Fig. 8.2). Yeast cells that express the most stable scTv fusions can be isolated by high-speed FACS. *Note*: The number of cells to be sorted should sufficiently sample the diversity contained in the library (generally, a 10-fold excess over expected total diversity is recommended). Also, it is useful to collect at least 50,000 cells in order to have adequate numbers for growth and expansion. Often, we sort the top 1% of cells with the highest fluorescence (e.g., from 3×10^7 cells, collecting about 3×10^5 cells). Multiple rounds of expansion in culture, staining, and sorting may be required before a population of yeast expressing the stable scTv can be detected by flow cytometry (generally, up to four or five sorts may be required). Depending on each scTv, it also may require multiple rounds of mutagenesis, or DNA shuffling, in order to identify an optimally stable scTv.

Once an scTv-positive yeast population has been collected, further stability may be possible by temperature-based selections. In this case, induced yeast cells can be incubated for 30 min at increased temperature. Previous studies have shown that temperatures up to 46 °C do not result in significantly reduced viability (Shusta *et al.*, 2000) prior to staining on ice with an anti-Vα or -Vβ antibody, and FACS. This procedure, when carried out with a conformation-sensitive antibody, can isolate scTv with improved resistance to denaturation and enhanced expression levels in soluble form (Shusta *et al.*, 1999; Weber *et al.*, 2005). Cells isolated by FACS after several rounds of temperature-based selections are expanded and spread on SD-CAA plates to isolate single yeast clones expressing scTv with improved thermostability. Individual colonies are expanded in SD-CAA liquid media, and then each clone is assessed by analytical flow cytometry. An expanded range of temperatures (up to 80 °C) can be used to assess resistance to thermal denaturation for the isolated clones by flow cytometry, where the yeast will only be analyzed, and not collected and reexpanded. Yeast clones expressing the highest surface levels of scTv, with the most improved resistance to thermal denaturation, are determined based on the mean fluorescence of the Vβ- and/or Vα-positive peak.

Sequences of the most stable clones can be obtained by isolating plasmids from yeast clones, following the instructions of the Zymoprep II kit (Zymo Research). Plasmid DNA is transformed into DH10B or DH5α (Invitrogen) and colonies are cultured and used for minipreps to isolate DNA for sequence analysis. Alternatively, PCR may be carried out directly from the Zymoprep-derived DNA, or even from a boiled yeast colony, using "Splice4L" and "T7" primers. The sequences of the optimal, thermostable scTv clones can then be determined by sequencing with the "YRS" primer (Fig. 8.1B).

2.3. Affinity engineering and selection

Once a more stable scTv template has been generated by error-prone mutagenesis, site-directed libraries can be generated within specific CDRs of the scTv to improve binding affinities for the specific pepMHC ligand that the wild-type TCR recognized (Holler *et al.*, 2000; Weber *et al.*, 2005). The CDR loops are the regions that most closely contact the pepMHC (reviewed in Rudolph *et al.*, 2006). Affinities of most naturally occurring (wild-type) TCRs for a pepMHC antigen are in the range of K_D values from 1 to 100 μ*M* (reviewed in Stone *et al.*, 2009). In an effort to retain peptide specificity, mutagenesis of the CDR3 loops, which generally dock over the peptide and not the MHC, can serve as the target for mutagenesis. Selection with fluorescent-labeled pepMHC tetramers (Altman *et al.*, 1996) can yield TCRs with >1000-fold improvement in affinity using yeast display (Holler *et al.*, 2003). More recently, pepMHC IgG dimers (commercially available as "DimerX," BD Biosciences) consisting of pepMHC molecules attached to each arm of a murine IgG molecule (O'Herrin *et al.*, 1997) have been used for selection. This dimeric reagent can stain at comparable levels to pepMHC tetramers refolded *in vitro* for detection of high-affinity scTv on the surface of yeast (Fig. 8.4). To make a specific

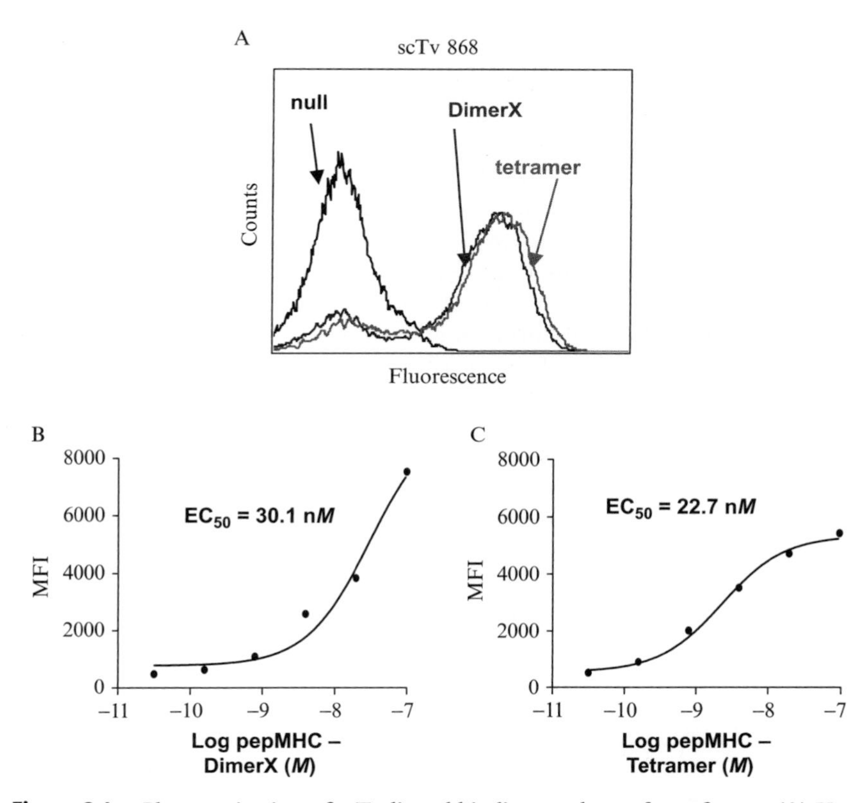

Figure 8.4 Characterization of scTv ligand binding on the surface of yeast. (A) Yeast expressing the high-affinity human scTv 868, specific for HLA-A2 in complex with the SL9 peptide derived from HIV, are stained with SL9/A2 Ig-dimer (DimerX, dark gray), SL9/A2 streptavidin-linked tetramer (light gray), or the null complex WT1/A2 streptavidin tetramer (black). (B, C) Quantification of the binding to scTv 868 yeast cells by various concentrations of (B) SL9/A2 Ig-dimer (DimerX) or (C) SL9/A2 streptavidin-linked tetramer. PepMHC ligand binding is detected by fluorescence of yeast cells by flow cytometry. (For color version of this figure, the reader is referred to the Web version of this chapter.)

pepMHC DimerX reagent, the peptide of interest is added to the MHC-Ig dimer at approximately 160-fold molar excess and incubated overnight at 37 °C. This reagent can be used directly for FACS to isolate scTv mutants with improved affinities.

Mutagenesis of CDR loops can be achieved using SOE PCR (Warrens *et al.*, 1997), where one of the primers can contain the codon degeneracy (e.g., we generally generate libraries with degeneracy at three to five codons). A detailed description of PCR product generation with CDR loop degeneracies for insertion into the pCT302 vector has been described (Richman *et al.*, 2009b). Once degenerate PCR products are generated, a library of mutants can be produced using the same protocol as described in Section 2.2. The library is sorted with biotin-labeled pepMHC tetramer or pepMHC Ig-dimer (DimerX, BD Biosciences). Secondary reagents in these cases can be streptavidin-phycoerythrin or a fluorescent-labeled secondary antibody, respectively. A similar, iterative selection process is carried out as described in Section 2.2 (Fig. 8.3). Individual yeast colonies are isolated and expanded, and binding to soluble pepMHC is assessed for each clone by flow cytometry. The sequences of the higher affinity scTv mutants are determined by isolation of the plasmid (Section 2.2). The scTv genes that encode mutants with the highest affinity (as judged by positive staining at the lowest pepMHC concentrations, see below) can be introduced into *E. coli* expression vectors, expressed, and refolded *in vitro* (Section 4).

A significant advantage of yeast surface display for engineering of scTvs is that binding properties can be analyzed directly on the yeast, without additional cloning, expression, and purification. High-affinity variants of the scTv can be expressed on the yeast surface, stained with soluble pepMHC ligands, and analyzed by flow cytometry for equilibrium binding or off rates. Monomeric, biotinylated pepMHC can be used as reagents in flow cytometry to approximate equilibrium binding constants, which have been reported to be in close agreement with a variety of other measurements (Gai and Wittrup, 2007).

3. Affinity Engineering and Selection of T Cell Receptors by T Cell Display

A promising treatment for eradicating cancer, called adoptive T cell therapy, involves the transfer of *ex vivo* amplified, tumor-specific T cells into a patient. A modification of this strategy includes the introduction of high functional avidity, tumor-specific TCR genes into T cells, thereby "reprogramming" the patient's T cells to efficiently recognize and kill tumors that express antigen specifically recognized by the introduced TCR (Engels and Uckert, 2007; Rosenberg, 2010; Schmitt *et al.*, 2009). This approach relies on efficient TCR transgene expression. An important potential advantage of this approach is that the TCR introduced into T cells can be engineered to have sufficient affinity in order to redirect the activity of CD4+ T helper cells against a class I MHC tumor (or viral) antigen. This approach relies on the finding that class I-restricted TCRs with affinities above the threshold of about 1 μM K_D value do not depend on the coreceptor CD8 for mediating activity (Chervin *et al.*, 2009).

As indicated above, some TCRs are not amenable to expression and engineering by yeast display. An alternative method of *in vitro* directed evolution, called T cell display, has been employed to isolate TCRs with higher affinity against a class I pepMHC (Chervin *et al.*, 2008). This method is based on the premise that the TCR/CD3 machinery allows expression of full-length TCRs on normal T cells, and that it is possible to generate libraries of mutated TCRs using a retrovirus-based transduction system (Kessels *et al.*, 2000). In this section, we describe the method of engineering TCRs by T cell display.

3.1. Design and construction of libraries for expression on T cell surface

Stable surface expression of introduced TCRs into T cells can be achieved through the use of lentiviral or retroviral vectors. These systems allow for integration of the transgene into the host's genome, permitting long-term expression on the surface of T cells without the requirement for continuous selection (e.g., with an antibiotic). While lentiviral vectors have become a common delivery system for human clinical trials (Deichmann *et al.*, 2007), others and we have used a retroviral system for *in vitro* engineering of murine TCRs in a murine T cell line (Chervin *et al.*, 2008; Kessels *et al.*, 2000). Initially, we employed the murine stem cell virus for delivery of TCR genes (Chervin *et al.*, 2008) into the $58^{-/-}$ T cell hybridoma (Letourneur and Malissen, 1989). More recently, we have used the pMP71 plasmid (Engels *et al.*, 2003) utilizing 5′ *Not*I and the 3′ *Bsp*EI restriction sites, with the TCR α and β chain genes linked by a picornavirus 2A-like (P2A) sequence (Fig. 8.5).

Since the TCR displayed on the surface of T cells includes full-length α and β chains, in complex with CD3 subunits for assembly and expression as they would occur naturally, there is no need for stability engineering. The first step for engineering a TCR by T cell display is to introduce mutations into CDRs of the TCR. This is accomplished with a primer that contains degeneracy at the desired codons, together with other primers that are used to generate a library by SOE PCR (see Fig. 8.5), similar to the strategy for targeted mutations in yeast display (Section 2.3). The major consideration at the stage of primer/library design that distinguishes yeast and T cell display is potential maximum library size. Since the maximum size using current technology is about 10^5 for T cell display, one must consider whether to generate degeneracy in fewer codons (e.g., three for T cell display,

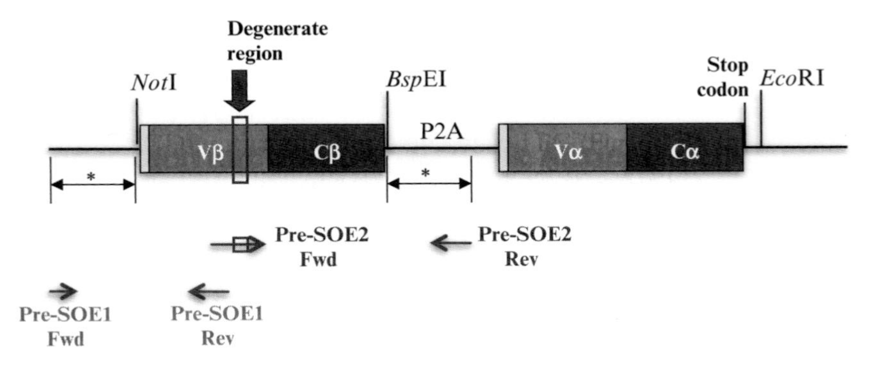

Figure 8.5 Schematic diagram of full-length TCR genes in retroviral vector. Arrangement of a TCR in the pMP71 retroviral vector, including restriction sites used for cloning. As an example, the locations and directionalities of primers designed to introduce variability into a stretch of the TCR β variable domain are shown. The amplified insert sequence includes a 15-base overlap on each end (labeled with an asterisk) with the pMP71 vector containing wild-type TCR, after removal of the sequence to be engineered (in this case, the β chain) by digesting with *Not*I and *Bsp*EI. This overlap facilitates use of the In-Fusion kit for homologous recombination in *E. coli*. The location of a sequencing primer that would be used to screen PCR amplified from *E. coli* colonies with selected library elements (Section 3.4) is also indicated. (For color version of this figure, the reader is referred to the Web version of this chapter.)

compared to five for yeast display) or in more codons, but sacrificing coverage of the sequence landscape at each position. This is typically approached on an empirical basis, although it is possible that better design criteria, such as libraries with overrepresented wild-type residues, might be desirable (Hackel and Wittrup, 2010).

To generate CDR diversity, two sets of primers are ordered (IDT Integrated DNA Technologies) for two separate pre-SOE PCR reactions (Fig. 8.5). The first reaction (pre-SOE1) includes a forward primer with the 5′-restriction site and a reverse primer with complementarity to the primer that contains the codon degeneracy. The second reaction (pre-SOE2) includes this forward primer that contains the codon degeneracy and a reverse primer that includes the 3′-restriction site. These two PCR products contain a 20- to 30-base pair (bp) overlap required for the SOE reaction with the flanking primers. The resulting SOE product (about 1 kbp in size) contains the diverse region of interest flanked by two restriction sites for incorporation into the retroviral vector (Fig. 8.5) via Clontech's In-Fusion kit.

At this point, the SOE PCR product can be sequenced directly, using an internal primer, to verify diversity in the region of interest. The PCR product containing the degenerate codons is then cloned into the pMP71 retroviral vector. This step is important, as low ligation efficiency limits library size and prevents adequate sampling of possible diversity. To obtain a sufficient library size, we take advantage of homologous recombination in *E. coli* using the In-Fusion kit (Clontech). The In-fusion reaction requires homology (e.g., 15 bp) between the ends of the PCR product and the linearized vector derived from digestion of pMP71 with *Not*I and *Bsp*EI. The PCR product and the vector are mixed at a 2:1 molar ratio, as recommended by the manufacturer. To create a library of approximately 10^4 mutants, we use a reaction scale of 55 ng of insert and 200 ng of linearized vector. The In-Fusion reaction mixture is transformed into a high-efficiency, high-copy *E. coli* cell line such as NEB Turbo Competent cells (New England Biolabs). A typical transformation includes 2.5 μL of the In-Fusion reaction and 50 μL of *E. coli* cells; this transformation is diluted to 1 mL with SOC media, incubated at 37 °C for a recovery period of 1 h, and then 25 μL of the culture is spread onto an LB plate containing 50 μg/mL ampicillin to determine library size. The remaining culture is stored at 4 °C until the library is analyzed.

Twenty-four hours after transformation and plating of *E. coli*, about 10 colonies are subjected to sequencing (either by isolation of plasmid DNA or by the rapid PCR method described above) to verify diversity in the desired region for the different clones. If the sequences show diversity, the remaining transformed culture is expanded and a larger scale plasmid preparation is purified (Maxi-prep, Qiagen) for use in packaging and transduction of T cells.

3.2. Packaging, transduction, and characterization of libraries in T cells

To perform *in vitro* engineering and selection of TCR genes, we expressed TCR libraries in the $58^{-/-}$ T cell hybridoma (Letourneur and Malissen, 1989) rather than in primary mouse T cells. The $58^{-/-}$ line has two major advantages for *in vitro* engineering purposes. First, in order for a cell to be transduced by a retrovirus, it must be actively dividing. Because $58^{-/-}$ is an immortalized hybridoma cell line, it is continuously passing through cell cycle, while primary lymphocytes must be induced to enter the cell cycle by providing an activation stimulus, and many of these cells will die by apoptosis. Second, since there is no endogenous TCR expressed on the surface of $58^{-/-}$, the detection and isolation of the introduced, engineered TCR is uncomplicated by competition with or chain mis-pairing with endogenous TCR α and β chains as are present in primary T cells.

Once the library plasmid mixture has been verified by sequencing, amplified, and purified, it can be packaged for introduction into T cells. Since the retroviral vector used for cloning the TCR does not contain all of the gene elements for producing an infectious virion (for safety reasons), a packaging cell line that contains the missing elements, *gag-pol* and *env* genes, must be used. Typically, packaging lines are produced in human embryonic kidney cells. We use the Plat-E cell line, generated from 293T cells (Morita *et al.*, 2000), which contain the *gag-pol* and *env* genes on selectable plasmids. As such, they are grown in DMEM-based media supplemented 10% fetal calf serum, penicillin, streptomycin, glutamine, HEPES, 10 μg/mL of blasticidin, and 2 μg/mL of puromycin. Plat-E cells adhere to plastic surfaces, and they require trypsinization for passaging.

To produce supernatant that contains infection-competent retroviruses encoding the TCR library elements, 4×10^6 Plat-E cells in 6 mL complete DMEM media are cultured at 37 °C, 5% CO_2, in a fresh, lysine-coated, 10 cm diameter plastic tissue culture plates. After 24 h, the pMP71 plasmid mixture containing the TCR library is transfected into the Plat-E cells using Lipofectamine 2000 (Invitrogen) as follows:

1. Prepare two separate 15 mL conical tubes
   a. Tube 1: Add 60 μL of Lipofectamine 2000 (Invitrogen) to 1.5 mL Opti-mem (Gibco) serum-free media. Incubate at room temperature for 5 min
   b. Tube 2: Add 40 μg of retroviral library DNA to 1.5 mL Opti-mem serum-free media. Incubate at room temperature for 5 min
2. Combine Tubes 1 and 2 and mix gently. Incubate at room temperature for 10 min. This is the transfection mixture.
3. Carefully aspirate 6 mL of complete DMEM media from the 10 cm Plat-E dish.
4. Add, dropwise and gently, 3 mL of transfection mixture to the Plat-E dish.
5. Incubate the Plat-E dish in the transfection mixture at 37 °C, 5% CO_2 for 4 h.
6. Aspirate transfection mixture. Wash plate by adding 6 mL of complete RPMI media (RPMI 1640 supplemented with 10% fetal calf serum and "K" supplement), dropwise and gently.
7. Aspirate wash media. Add 6.5 mL of RPMI-1640, dropwise and gently. Incubate the Plat-E dish in the transfection mixture at 37 °C, 5% CO_2 for 48 h.
8. After the transfection, collect and pass supernatant through a 0.45-μm syringe filter. Add 50 μL of Lipofectamine 2000 to filtered supernatant.

After the retroviral supernatant encoding the TCR library has been collected from the Plat-E culture, it can be used to transduce the $58^{-/-}$ T cell hybridoma as follows. *Note*: for sterile solutions, generally pass through a filter with a 0.22-μm diameter pore size unless otherwise indicated.

1. Coat wells of a 24-well tissue culture plate with 0.5 mL each sterile Retronectin (Takara) at 15 μg/mL in PBS for 3 h at room temperature.
2. Aspirate the Retronectin solution, and block the wells with 0.5% sterile PBS–BSA for 30 min at room temperature.

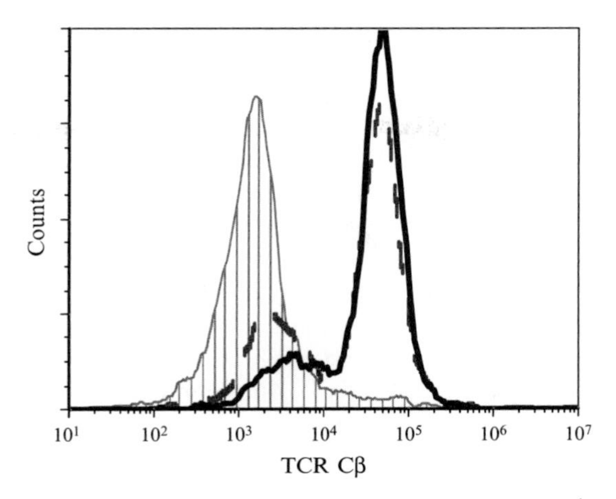

Figure 8.6 Transduced TCRs are expressed on the surface of $58^{-/-}$ T cell hybridomas. TCR β constant domain expression (as detected by the H57-597 antibody, BD Pharmingen) in $58^{-/-}$ cells 24 h posttransduction with mock (gray, shaded), 2C (gray, dashed), or m33 (dark gray, solid line) gene-optimized TCR sequences in the pMP71 vector, packaged into retrovirus particles by Plat-E. Transduction efficiencies were 69% for the 2C TCR and 84% for the m33 TCR. (For color version of this figure, the reader is referred to the Web version of this chapter.)

3. While the wells are incubating with PBS–BSA, suspend $58^{-/-}$ T cells at 1×10^6 cells/mL in fresh complete RPMI media.
4. Aspirate the PBS–BSA from the wells and add 1 mL of $58^{-/-}$ culture per well. To each $58^{-/-}$ well, add 1 mL of retroviral library supernatant.
5. Centrifuge the 24-well plate for 1 h at $2000\times g$ and 30 °C.
6. Move the plate to a 37 °C, 5% CO_2 incubator.

Transgene expression can be detected 24 h posttransduction. Since $58^{-/-}$ do not express endogenous TCR, detection of introduced TCR can be achieved by staining with a fluorescently labeled anti-mouse Cβ antibody (such as H57-597, BD Pharmingen) and analyzing by flow cytometry to assess transduction efficiency (Fig. 8.6).

3.3. Selection of higher affinity receptors from a T cell library

If the level of TCR expression is suitable, the T cells that express the library of TCRs can be stained with the original pepMHC antigen and subjected to FACS for selection of those TCRs with higher affinity. Almost all wild-type, class I restricted TCRs require the coreceptor CD8 in order to bind to pepMHC oligmers. Since $58^{-/-}$ does not express the MHC coreceptor CD8 on its surface, T cells that bind to soluble pepMHC are likely to be of higher affinity and to respond in a coreceptor-independent manner. We have used the DimerX (BD Bioscience) recombinant, soluble, dimeric MHC reagents for staining of libraries on the surface of T cells, as on yeast cells (see Section 2.3). T cells are stained on ice with DimerX MHC-Ig dimer pre-loaded with appropriate peptide, and the top binders (aim for at least 50,000–100,000 cells) are collected by FACS. The specific percentage collected varies with the requirements of a particular experiment; however, collecting 1–2% of positive binders is typical. Additional rounds of sorting, up to three or four, may be required to establish a strong, positive population. The staining concentration of pepMHC may be varied (e.g., reduce the DimerX concentration to isolate the tightest binders). *Note*: It is important to carry out all stains and sorts under sterile conditions to prevent contamination of the library. Typically, doubling time for $58^{-/-}$ is ~24h, so typical numbers of sorted cells (e.g., 50,000) will expand in a few days for further sorting or downstream use.

3.4. Isolation of TCR sequences from selected T cell hybridomas

Previously, $58^{-/-}$ clones were isolated from a positive-selected library population by limiting dilution (Chervin *et al.*, 2008). Instead, we currently use the following process of total RNA isolation from the mixed $58^{-/-}$ population, combined with direct sequencing of the resulting cDNA clones grown in an *E. coli* vector. There are significant technical challenges to performing limiting dilution on a T cell hybridoma population and the time it may take for a population to grow from a single cell (>3weeks). In addition, the same process involving isolating RNA, cloning, and sequencing remains a requirement to characterize the mutations that have been selected after limiting dilution procedure. Hence, we have adopted the more rapid protocol below.

Once the sorted library of cells have grown to a sufficient number (~10^6–10^7), RNA can be isolated and reverse transcribed into cDNA for PCR and direct sequencing or cloning and sequencing. *Note*: since RNA is more labile than DNA, it is important throughout this procedure to use RNAse-free, autoclaved tubes, pipet tips, and especially water. The Quantitect Reverse Transcriptase kit (Qiagen) is used to generate cDNA from $58^{-/-}$ libraries. To prepare the RNA for reverse transcription, cells are pelleted and lysed in 1mL TRIzol (Invitrogen) at room temperature for 5min. After lysis, 200μL of chloroform is added to the TRIzol mixture and mixed by shaking in a microcentrifuge tube. The mixture is then centrifuged at 12,000×*g* for 15min at 4°C, separating into a clear layer containing the RNA (top) and a pink layer containing the TRIzol (bottom). The top layer (~400μL) is transferred to a microcentrifuge tube and mixed by vortexing with an equal volume of 70% ethanol. The RNA/ethanol mixture is transferred to an Ambion PureLink RNA Mini kit by Life Technologies (Invitrogen) for purification, washing, and elution in RNAse-free water.

After recovery of RNA, it is reverse transcribed into cDNA using the Quantitect Reverse Transcriptase kit (Qiagen), following the manufacturer's instructions. The cDNA mixture is then subjected to PCR with Taq polymerase that selectively amplifies the region of interest for the engineered TCR (e.g., containing the mutated CDR region). For ease of further amplification and isolation of individual sequences, this PCR product is cloned into the linearized pCR4-TOPO vector (Invitrogen) by following the manufacturer's instructions. This cloning is template independent and ligation occurs through pairing of the single deoxyadenosine (A) added at the 3′ ends of PCR products by Taq polymerase and a single, overhanging deoxythymidine (T) at the 3′ ends of the linearized pCR4-TOPO vector. The ligation mixture is transformed into a high-competent, high-copy *E. coli* strain, such as TOP10 One Shot® Chemically Competent cells (Invitrogen). Spread 50 μL of the transformation reaction on LB agar plates containing 50 μg/mL ampicillin for selection and grow at 37 °C overnight, or until single colonies may be distinguished.

The sequence of the optimized TCR gene from a single *E. coli* colony may be obtained by performing colony PCR and sequencing of the PCR product. A single colony is removed from the agar plate, suspended in 7.5 μL ddH_2O in a sealed tube, and subjected to boiling for 5 min at 100 °C. One microliter from the boiled colony solution is used as the template for PCR amplification, using the same primers that were used to amplify the target TCR gene from cDNA. It is possible to screen many colonies in a single PCR by using a 96-well PCR plate. An internal primer may be used for sequencing the PCR product. *Note*: It is worthwhile to verify amplification of the gene from the *E. coli* colonies by agarose gel electrophoresis of PCR product from some colonies before sequencing.

The results from sequencing will identify those residues from the library that were selected from sorting, based on the selection criteria (e.g., stringency of pepMHC binding). Those residues identified by sequencing can be introduced back into the TCR retroviral vector and transduced into $58^{-/-}$ for further characterization of activity, as a pure, homogenous population of T cells. Alternatively, if the population of selected $58^{-/-}$ appears to be monoclonal based on sequencing, the FACS-selected $58^{-/-}$ may be used directly for T cell activity assays.

4. Expression, Purification, and Applications of Soluble scTv Proteins

A significant benefit of engineering TCRs by yeast and/or T cell display, in addition to the possibility of transducing high-affinity TCRs into $CD4^+$ T cells to recruit class I tumor-antigen-specific T cell help, is that the enhanced affinity can allow use of soluble versions of these proteins. The soluble TCRs could be used as probes for pepMHC (e.g., in diagnostic or vaccine monitoring setting) or to direct therapeutic molecules to the site of pathology. In principle, applications for which antibodies or antibody fragments (e.g., Fab or scFv) are frequently used could be substituted with engineered TCRs. This allows binding to epitopes derived from intracellular proteins, presented as pepMHC on the cell surface. Soluble TCRs have been constructed from the extracellular portions (VαCα and VβCβ) of the TCR (Boulter *et al.*, 2003; Davis *et al.*, 1998; Garboczi *et al.*, 1996; Willcox *et al.*, 1999), and there have been single-chain, three-domain constructs (Vα-VβCβ) produced as well (Card *et al.*, 2004; Chung *et al.*, 1994). By analogy with scFv fragments, we have produced stabilized, single-chain TCR variable domain (scTv) constructs (Aggen *et al.*, 2011; Holler *et al.*, 2000; Jones *et al.*, 2006; Soo Hoo *et al.*, 1992; Weber *et al.*, 2005) that are convenient to produce in *E. coli* or yeast, stable, and adaptable to different applications (Aggen *et al.*, 2011; Colf *et al.*, 2007; Zhang *et al.*, 2007). Diagrams of the configurations of different TCR constructs are shown in Fig. 8.7.

In this section, we describe the production, purification, and applications of high-affinity scTvs produced in *E. coli*.

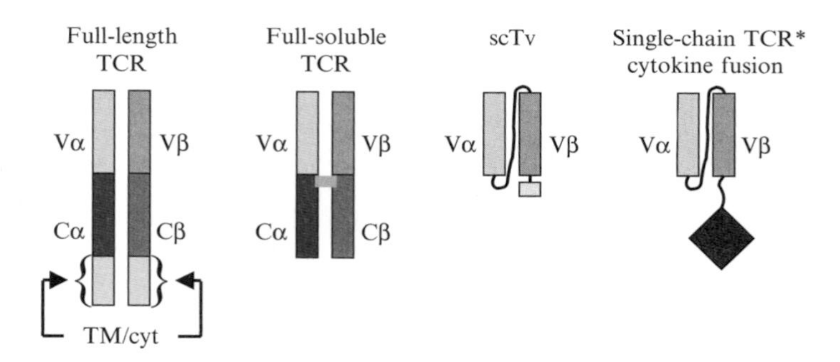

Figure 8.7 Schematic representations of T cell receptor expression formats. Full-length TCR contains Vα-Cα and Vβ-Cβ domains as well as transmembrane and cytoplasmic domains; full-length TCR is generally expressed on the surface of T cells in complex with CD3 proteins as a larger TCR complex. Full soluble TCR contains Vα-Cα and Vβ-Cβ domains, but the chains are truncated before the transmembrane domain and will often contain an introduced, nonnative disulphide bond between the constant domains to enhance chain pairing. Soluble three-domain scTCR includes the Vα and Vβ antigen-binding domains connected by a flexible linker, along with one constant domain, usually the Cβ. Soluble scTv, similar to scFv constructs derived from antibodies, contain only the Vα and Vβ antigen-binding domains connected by a flexible linker. These constructs may contain point mutations to allow for high expression and stable protein. (For color version of this figure, the reader is referred to the Web version of this chapter.)

4.1. Design and cloning for scTv expression in *E. coli* as inclusion bodies

Similar to scFv proteins, scTv proteins consist of only the variable domains of the TCR, joined by a flexible linker. While scTv proteins have been produced in *E. coli* as soluble proteins within the periplasm using the pET22b vector (Jones *et al.*, 2008a), the method described here uses scTv proteins produced in insoluble inclusion bodies using the pET28a vector from Novagen (Chervin *et al.*, 2009; Jones *et al.*, 2008b). The vector includes an N-terminal or C-terminal His_6 tag, a T7 tag, and a thrombin site that may be of use. In general, we include a His_6 tag in-frame on either the N- or C-terminus of the scTv for ease of purification; if downstream applications require a lack of His_6 sequence, the thrombin site may be employed between the gene and the His_6 tag to remove it. Protein expression in the pET28 vector uses the lac operon, with isopropyl-β-D-thiogalactopyranoside (IPTG) as inducer.

The binding domains included in an scTv are the Ig-like Vα and Vβ domains of a TCR, joined in-frame by a polypeptide linker. The linker can be designed in various ways but is generally a predicted soluble or flexible peptide chain of about 15–20 amino acids. One common soluble, flexible linker design is a repeating chain of glycine and serine, such as $(GGGGS)_4$ (Card *et al.*, 2004; Chung *et al.*, 1994; Jones *et al.*, 2008a). This has the advantage of being unstructured and fairly protease-resistant. Another commonly used linker in scTvs is lysine-rich, such as GSADDAKK-DAAKKDGKS (Aggen *et al.*, 2011; Weber *et al.*, 2005) or SSADDAKK-DAAKKDDAKKDDAKKDA (Jones *et al.*, 2008b). Although this linker may be more labile, the number of lysines allows multiple amine-reactive groups (such as fluorophores or biotin groups) to be added to the small protein minimizing affects on pepMHC binding function.

Once the scTv has been cloned into pET28a, it can be transfected into an *E. coli* strain optimized for protein production, such as BL21(DE3), BL21 Star (Invitrogen), or BL21 Codon Plus (Agilent Technologies). Growth of the bacteria can be carried out in the presence of 100 μg/mL kanamycin for selection.

4.2. Growth and induction of *E. coli*

Methods for the production of insoluble inclusion bodies of recombinant protein in *E. coli* have been described elsewhere for various systems (Fahnert *et al.*, 2004; Garboczi *et al.*, 1992; Nagai and Thogersen, 1987). The method used to induce scTv protein production is similar to previously described methods and will be outlined here briefly, using an scTv gene in pET28a as an example.

Expression of scTv proteins is frequently initiated by fresh transformation of the vector containing the scTv gene into a competent, protein production-optimized strain of *E. coli*, expanded under selection (100 μg/mL kanamycin for pET28a vector) at 37 °C in LB medium. Cultures should be rotated in test tubes or shaken in Erlenmeyer flasks rapidly to ensure adequate aeration. It is important not to allow the cells to reach past log phase during expansion, or protein production levels may be reduced. For typical expression batches, cultures are expanded up to 9–10 L, for example, 1.5 L of culture in each of 6, 6 L flasks. The cultures should be grown to a density of roughly $OD_{600}=0.7$ before induction. It is convenient to take a 1-mL sample of the culture at this stage for characterization of the *E. coli* "preinduction" (see below).

To induce protein expression, IPTG is added to each flask to a final concentration of 0.75 m*M*. The culture is shaken at 37 °C for an additional 2–4 h, depending on optimal expression of the construct. At the end of the induction period, take a second 1-mL sample of the induced culture. The two samples are centrifuged (microfuge, 5 min), the cell pellet suspended in SDS sample loading buffer, and run on an SDS-PAGE gel to verify that the protein of interest has been produced as expected (i.e., a band at about 30,000 Da that is prominent in the induced sample, compared to the "preinduced" sample). The remaining *E. coli* cultures are harvested by spinning at 4200 × *g* in a 4 °C centrifuge, keeping the cell pellets ice cold at all times from this point onward.

4.3. Isolation of inclusion bodies

To release inclusion bodies from *E. coli*, various methods can be used, including French press, sonication, homogenization, and microfluidization (reviewed in Grabski, 2009). As we use microfluidization to disrupt the cells, it is described here. The cell pellet is suspended in up to 100 mL ice-cold lysis buffer until the solution is smooth. The solution is passed through a coarse filter such as a tea strainer to ensure there are no clumps, which could block the microfluidizer (Microfluidics). The solution is passed through the microfluidizer several (~4) times until it is the consistency of water, flushing with osmotic shock buffer (no additional detergent). The solution is centrifuged for 30 min at 11,000 × *g* in a 4 °C centrifuge to pellet inclusion bodies; the supernatant can be discarded. At this stage, the inclusion bodies may be stored at −20 °C before processing is continued.

The inclusion body pellet is suspended in approximately 40 mL osmotic shock buffer+triton to wash it, thoroughly breaking up the pellet by blending, pipetting, or vortexing. The solution is centrifuged for 30 min at 11,000 × *g* at 4 °C, and this pellet washing step is repeated at least four times total. The pellet should look clean and white after these washes. Next, the pellet is washed by suspending in approximately 40 mL osmotic shock buffer (no additional detergent), again making sure to thoroughly break up the pellet. The solution is centrifuged for 30 min at 11,000 × *g* at 4 °C, and this detergent-free wash is repeated for a total of three washes.

At this stage, the pellet is suspended with 6–7 mL osmotic shock buffer without detergent, breaking up the pellet as thoroughly as possible. To each of a set of preweighed, 1.5 mL microcentrifuge tubes, 1 mL of inclusion body slurry is added. The tubes are spun in a tabletop centrifuge for 2 min at 16,100 × *g*, and the supernatant is aspirated. The tubes are each weighed, noting the difference, which is made up of inclusion bodies. There should be about 500–800 mg per tube at this scale of expression batch. Inclusion body aliquots can be stored frozen at −80 °C.

4.4. Solubilization and refolding of scTv proteins

A typical scale of refolding is described here, which would use one aliquot (500–800 mg) of inclusion bodies. First, 400 mL of refolding buffer is prepared and equilibrated to 4 °C. To the inclusion body aliquot, osmotic shock buffer (without detergent) is added to a final volume of 1 mL, and then 1.1 g guanidine hydrochloride and 2.5 μL β-mercaptoethanol are added. The mixture is incubated at 37 °C to solubilize the inclusion bodies, vortexing frequently (it may take about an hour to dissolve the pellet).

Once the inclusion bodies are dissolved in the reduced guanidine solution, 2 m*M* (246 mg) reduced glutathione and 0.2 m*M* (49 mg) oxidized glutathione are added to the 400 mL refolding buffer, keeping at 4 °C, and stirring to dissolve. The solubilized inclusion bodies are added dropwise to the stirring refolding buffer. The mixture is incubated, continuously stirring, at 4 °C for 4 h. After this period, the refolding buffer and inclusion body mixture is transferred into a large container (≥3 L). Over the next 24 h, dilution buffer is slowly added to the gently stirring mixture using a low-flow siphon to a final volume of 2.5 L.

4.5. Purification of scTv proteins

The presence of a His_6 sequence at the N- or C-terminus of the scTv is convenient for purification. After the dilution is completed, 2 mL of Ni–NTA agarose bead slurry (Qiagen) is added to the refolding mixture. The mixture is stirred slowly at 4 °C overnight (≥12 h). The next morning, the beads are collected by vacuum filtration through a sintered glass funnel, allowing the volume to get down to approximately 10 mL. The remaining slurry is transferred into a 15-mL conical tube and centrifuged at 800 × *g* for 5 min. The supernatant is discarded and the beads are washed one time in 10 mL PBS, and centrifuged at 800 × *g* for 5 min. To the washed beads, 1 mL of freshly prepared 0.5 *M* imidazole in PBS is added and the mixture is allowed to incubate briefly (>1 min), is transferred to a 0.22 μm Spin-X Centrifuge Tube Filter (Corning Inc.) and is centrifuged for 1 min at 16,100 × *g* in a tabletop centrifuge.

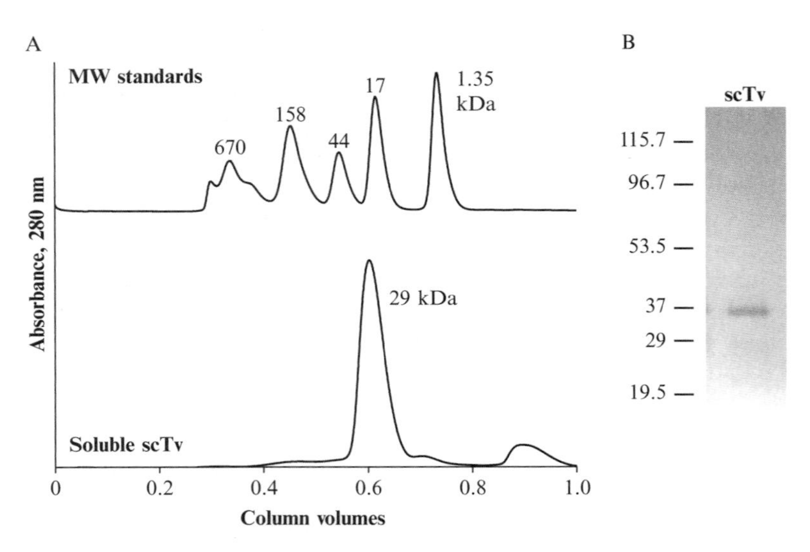

Figure 8.8 Purification and characterization of scTv proteins. (A) Size exclusion chromatography purification trace of scTv m33 after Ni–NTA bead elution (bottom trace) compared to molecular weight standards (top trace). (B) SDS-PAGE gel of purified, non-reduced scTv m33.

The eluted protein is further purified by size exclusion chromatography by injecting onto a Superdex 200 10/300 column (GE Biosciences) preequilibrated in filtered, degassed PBS. A typical size exclusion trace for an scTv purification can be seen in Fig. 8.8. Fractions that contain purified scTv (as determined by SDS-PAGE gel) are pooled and the concentration of the scTv can be measured by UV–Vis spectroscopy using the calculated extinction coefficient at 280nm; for example, the scTv m33 has an extinction coefficient of ε_{280}=32,100 and an expected molecular weight of 29,119Da.

4.6. Detection of specific pepMHC on a cell surface using soluble, purified scTvs

One application for high-affinity scTv proteins is to detect specific pepMHC epitopes on the surface of antigen-presenting cells. This information could be useful diagnostically in determining the level of a particular complex that is present on a cell to be targeted by a TCR-based therapeutic, or to assess if a specific vaccine containing the antigenic epitope might be of value. The level of pepMHC complex presentation, combined with other considerations of receptor affinity, will determine the feasibility of targeting the cells.

In Fig. 8.9, sample data are shown for detection of the HTLV-derived peptide Tax_{11-19} (LLFGYPVYV) in complex with HLA-A2 on the surface of T2, a TAP-deficient cell line that predominantly presents exogenously loaded peptides. The Tax/A2 complex is specifically recognized by the A6 TCR (Utz *et al.*, 1996). An engineered higher affinity scTv A6 (K_D for Tax/A2=54 n*M* (Aggen *et al.*, 2011) containing the lysine-rich interdomain linker (GSADDAKKDAAKKDGKS) was purified and biotinylated using *N*-hydroxysulfosuccinimide (Sulfo-NHS) ester biotin (Pierce). The biotinylated scTv A6 was purified to remove excess biotin and then incubated at various concentrations with T2 cells loaded with either a cognate (Tax)

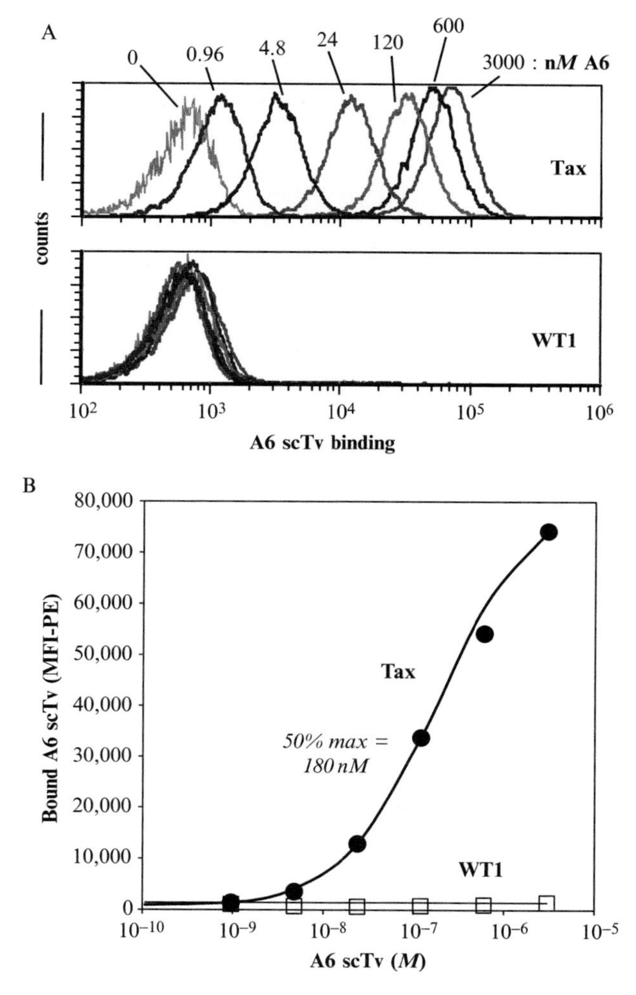

Figure 8.9 Detection of specific pepMHC levels on antigen-presenting cells with scTv. (A) Various dilutions of biotinylated scTv A6 specifically detect Tax/A2 complexes on T2 cells loaded with Tax peptide, but not WT1 peptide. (B) Graph of fluorescence detected on peptide-loaded T2 cells stained with biotinylated scTv A6 and detected with fluorescent streptavidin. (For color version of this figure, the reader is referred to the Web version of this chapter.)

peptide or a null (WT1) peptide. The cells are washed and then incubated with fluorescent-labeled streptavidin to detect bound complex. As can be seen in Fig. 8.9, only cells presenting the specific peptide are stained with the scTv. The staining is sensitive, as a significant shift in the flow histogram can still be seen at 1 n*M* A6. Soluble, recombinant scTv or full-length TCR proteins can also be used to detect endogenously processed pepMHC complexes presented by cells, as described previously (Purbhoo *et al.*, 2006; Zhang *et al.*, 2007). Whether a particular high-affinity scTv will detect endogenous pepMHC complexes will depend largely on their cell surface levels. While the lower limit of standard flow cytometry approaches may be about 1000 molecules per cell, it is possible that advanced confocal microscopy techniques, or improved labeling of the scTv, might allow lower levels of detection.

4.7. Monitoring of soluble protein levels using an scTv competition ELISA

To quantitate soluble scTv, it is convenient to be able to use a plate-based ELISA assay. For example, for pharmacokinetic studies, serum levels of scTv in animals could be monitored under various conditions and over time. Here, we describe such an assay to detect unlabeled, soluble scTv containing the human Vα2 domain using a competition ELISA format with biotinylated Vα2-containing scTv and an immobilized mouse mAb against the human Vα2 domain. In this example, the scTv construct of the human A6 TCR was used (Aggen *et al.*, 2011).

The mAb, clone 2B7, was produced in serum-free hybridoma culture and immobilized on high-binding Immulon 2HB 96-well plates (Thermo Fisher Scientific) by diluting the supernatant 1:100 in PBS and adding 50 μL/well. After coating for 3h at room temperature, wells were washed three times with 200mL PBS–T and then blocked with 200 μL/well PBS–BSA overnight at 4 °C. The plate was washed one time with PBS–T, and then either (1) incubated for 1 h at room temperature with various concentrations of biotinylated scTv A6 in PBS–BSA (Fig. 8.10A) or (2) preincubated with unmodified scTv A6 in PBS–BSA, followed by addition of biotinylated scTv A6 to a final concentration of 1.5 n*M* for an additional 1 h incubation, all at room temperature (Fig. 8.10B). The plate was then washed three times with PBS–T and then incubated for 30 min at room temperature with 50 μL per well streptavidin conjugated to horseradish peroxidase (BD Pharmingen), diluted 1:10,000 in PBS–BSA. The wells were washed three times with PBS–T and then developed with 50 μL per well TMB substrate solution (KPL) until blue color developed. The reaction was stopped with 50 μL per well 1 N sulfuric acid, and absorbance in each well at 450 nm was read using an EL_X-800 Universal Plate Reader (BioTek Instruments).

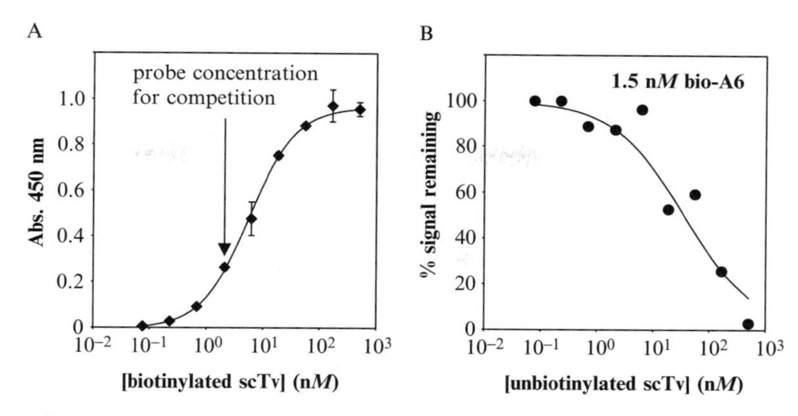

Figure 8.10 Quantitation of soluble scTv A6 by ELISA. (A) Various concentrations of biotinylated scTv A6 captured by immobilized monoclonal antibody against the Vα2 TCR domain, and then detected with streptavidin linked to horseradish peroxidase, developed with TMB substrate. (B) Competition of 1.5n*M* biotinylated scTv A6 binding to the immobilized monoclonal anti-Vα2 antibody by unbiotinylated scTv A6 in the same well.

Biotinylated scTv A6 was detected at concentrations down to 1n*M* by the anti-Vα2 capture ELISA (Fig. 8.10A). This allowed selection of a concentration of the biotinylated probe to detect unlabeled scTv A6 in a competition ELISA, showing that unlabled scTv could be detected at concentrations as low as 1–10n*M* (Fig. 8.10B). The competition ELISA could be employed to monitor scTv persistence (e.g., in serum or tumor samples) or localization in situations where biotinylation or other labels would be undesirable such as some therapeutic applications.

5. Recipes for Media and Buffers

SD-CAA (500mL)	
7.4g	Sodium citrate
2.1g	Citric acid monohydrate
2.5g	Casamino acids
3.35g	Yeast nitrogen base
10g	Dextrose

Dissolve in 500mL dH_2O, and filter through 0.22μ*M* filter to sterilize. For SG-CAA, substitute equal mass galactose for dextrose.

SD-CAA plates (500 mL total)	
Autoclave the below reagents dissolved in 400 mL dH_2O:	
91.085 g	Sorbitol
7.5 g	Bacto-agar
7.4 g	Sodium citrate
2.1 g	Citric acid monohydrate
When above solution is cooled, add the following filter-sterilized solution and mix:	
2.5 g	Casamino acids
10 g	Dextrose
3.35 g	Yeast nitrogen base
100 mL	dH_2O

Pour ~20 mL of media per plate.

YPD media (1 L)	
10 g	Yeast extract
20 g	Bacto-peptone
20 g	Dextrose

Dissolve in 1 L dH_2O and autoclave. For plates, add 15 g bactoagar to media before placing in autoclave.

1 *M* sorbitol (1 L)	
182.17 g	Sorbitol

Dissolve in 1 L dH_2O. For 1 M sorbitol with 1 mM $CaCl_2$, add 0.11 g $CaCl_2$ to 1 L solution.

0.1 *M* LiOAc/10 m*M* DTT	
Dissolve the following components in 100 mL dH_2O and filter sterilize:	
0.659 g	LiOAc
0.154 g	DTT

Note: Solution should always be made fresh when making library.

LB medium (1 L)	
5 g	Yeast extract
10 g	Sodium chloride
10 g	Bacto-tryptone

Dissolve in 1 L dH_2O, and autoclave to sterilize.

SOC medium (500 mL total)	
Autoclave the below reagents dissolved in 400 mL dH_2O:	
2.5 g	Yeast extract
10 g	Bacto-tryptone
0.292 g	Sodium chloride
0.932 g	Potassium chloride
When above solution is cooled, add the following filter-sterilized solution and mix:	
1.015 g	Magnesium chloride
1.233 g	Magnesium sulfate
3.6 g	Dextrose
100 mL	dH_2O

K supplement (675 mL)	
20.23 g	4-(2-Hydroxyethyl)-1-piperazineethanesulfonic acid (HEPES, 106 m*M*)
3.4 g	L-Glutamine (35 m*M*)
170 mL	Penicillin/streptomycin, liquid (5000 units/mL penicillin and 5000 μg/mL streptomycin in 0.85% saline, Invitrogen)
59.5 μL	2-Mercaptoethanol liquid (Sigma-Aldrich); bring volume to 675 mL with ddH_2O, pass through a 0.22 μm filter to sterilize, and store 25 mL aliquots at −20 °C. Add one aliquot for 500 mL media.

Lysis buffer (500 mL)	
50 m*M*	Tris base
100 m*M*	Sodium chloride
0.1%	Sodium azide
1%	Triton X-100
10 m*M*	Dithiothreitol (DTT, add fresh)
1 m*M*	Phenylmethylsulfonyl fluoride (PMSF, dissolve first in 1 mL isopropanol and add fresh); bring volume to 1 L with ddH_2O and adjust pH to 8.0 with hydrochloric acid

Osmotic shock buffer (1 L)	
20 m*M*	Tris base
2.5 m*M*	Ethylenediaminetetraacetic acid (EDTA); bring volume to 1 L with ddH_2O and adjust pH to 8.0 with hydrochloric acid

Osmotic shock buffer+triton (1 L)	
To 1 L osmotic shock buffer, add	
0.5%	Triton X-100

Refolding buffer (400 mL)	
3 *M*	Urea
50 m*M*	Tris base; bring volume to 400 mL with ddH_2O, and adjust pH to 8.0 with hydrochloric acid

Dilution buffer (3.0 L)	
200 m*M*	Sodium chloride
50 m*M*	Tris base; bring volume to 3 L with ddH_2O and adjust pH to 8.0 with hydrochloric acid

PBS (phosphate-buffered saline, 1 L)	
8 g	Sodium chloride (137 m*M*)
0.2 g	Potassium chloride (2.7 m*M*)
1.15 g	Sodium phosphate heptahydrate, dibasic ($Na_2HPO_4 \cdot 7H_2O$, 4.3 m*M*)
0.2 g	Potassium phosphate, monobasic (KH_2PO_4 anhydrous, 1.47 m*M*); bring volume to 1 L with ddH_2O and adjust pH to 7.4

PBS–T
0.05% Tween-20 in PBS

PBS–BSA
1% bovine serum albumin in PBS

ACKNOWLEDGMENTS

We thank members of the lab, past and present, for valuable contributions to the development of these methods. Research in the lab has been funded by grants from the National Institutes of Health R01 GM55767, R01 AI064611, P01 CA97296, and U54 AI57153 (from the NIH-supported Great Lakes Regional Center for Excellence) to D. M. K. and a Samuel and Ruth Engelberg/Irvington Institute Postdoctoral Fellowship of the Cancer Research Institute to J. D. S.

REFERENCES

Aggen, D. H., Chervin, A. S., Insaidoo, F. K., Piepenbrink, K. H., Baker, B. M., and Kranz, D. M. (2011). Identification and engineering of human variable regions that allow expression of stable single-chain T cell receptors. *Protein Eng. Des. Sel.* **24,** 361–372.

Aleksic, M., Dushek, O., Zhang, H., Shenderov, E., Chen, J. L., Cerundolo, V., Coombs, D., and van der Merwe, P. A. (2010). Dependence of T cell antigen recognition on T cell receptor-peptide MHC confinement time. *Immunity* **32,** 163–174.

Altman, J. D., Moss, P. A. H., Goulder, P. J. R., Barouch, D. H., McHeyzer-Williams, M. G., Bell, J. I., McMicheal, A. J., and Davis, M. M. (1996). Phenotypic analysis of antigen-specific T lymphocytes. *Science* **274,** 94–96.

Armstrong, K. M., Piepenbrink, K. H., and Baker, B. M. (2008). Conformational changes and flexibility in T-cell receptor recognition of peptide-MHC complexes. *Biochem. J.* **415,** 183–196.

Artyomov, M. N., Lis, M., Devadas, S., Davis, M. M., and Chakraborty, A. K. (2010). CD4 and CD8 binding to MHC molecules primarily acts to enhance Lck delivery. *Proc. Natl. Acad. Sci. USA* **107,** 16916–16921.

Bargou, R., Leo, E., Zugmaier, G., Klinger, M., Goebeler, M., Knop, S., Noppeney, R., Viardot, A., Hess, G., Schuler, M., *et al.* (2008). Tumor regression in cancer patients by very low doses of a T cell-engaging antibody. *Science* **321,** 974–977.

Boder, E. T., and Wittrup, K. D. (1997). Yeast surface display for screening combinatorial polypeptide libraries. *Nat. Biotechnol.* **15,** 553–557.

Boder, E. T., and Wittrup, K. D. (2000). Yeast surface display for directed evolution of protein expression, affinity, and stability. *Methods Enzymol.* **328,** 430–444.

Boulter, J. M., Glick, M., Todorov, P. T., Baston, E., Sami, M., Rizkallah, P., and Jakobsen, B. K. (2003). Stable, soluble T-cell receptor molecules for crystallization and therapeutics. *Protein Eng.* **16,** 707–711.

Bowerman, N. A., Crofts, T. S., Chlewicki, L., Do, P., Baker, B. M., Christopher Garcia, K., and Kranz, D. M. (2009). Engineering the binding properties of the T cell receptor:peptide:MHC ternary complex that governs T cell activity. *Mol. Immunol.* **46,** 3000–3008.

Card, K. F., Price-Schiavi, S. A., Liu, B., Thomson, E., Nieves, E., Belmont, H., Builes, J., Jiao, J. A., Hernandez, J., Weidanz, J., *et al.* (2004). A soluble single-chain T-cell receptor IL-2 fusion protein retains MHC-restricted peptide specificity and IL-2 bioactivity. *Cancer Immunol. Immunother.* **53,** 345–357.

Chervin, A. S., Aggen, D. H., Raseman, J. M., and Kranz, D. M. (2008). Engineering higher affinity T cell receptors using a T cell display system. *J. Immunol. Methods* **339,** 175–184.

Chervin, A. S., Stone, J. D., Holler, P. D., Bai, A., Chen, J., Eisen, H. N., and Kranz, D. M. (2009). The impact of TCR-binding properties and antigen presentation format on T cell responsiveness. *J. Immunol.* **183,** 1166–1178.

Chung, S., Wucherpfenning, K. A., Friedman, S. M., Hafler, D. A., and Strominger, J. L. (1994). Functional three-domain single chain T-cell receptors. *Proc. Natl. Acad. Sci. USA* **91,** 12654–12658.

Colf, L. A., Bankovich, A. J., Hanick, N. A., Bowerman, N. A., Jones, L. L., Kranz, D. M., and Garcia, K. C. (2007). How a single T cell receptor recognizes both self and foreign MHC. *Cell* **129,** 135–146.

Davis, M. M., Boniface, J. J., Reich, Z., Lyons, D., Hampl, J., Arden, B., and Chien, Y. (1998). Ligand recognition by alpha beta T cell receptors. *Annu. Rev. Immunol.* **16,** 523–544.

Davis, M. M., Krogsgaard, M., Huse, M., Huppa, J., Lillemeier, B. F., and Li, Q. J. (2007). T cells as a self-referential, sensory organ. *Annu. Rev. Immunol.* **25,** 681–695.

Deichmann, A., Hacein-Bey-Abina, S., Schmidt, M., Garrigue, A., Brugman, M. H., Hu, J., Glimm, H., Gyapay, G., Prum, B., Fraser, C. C., *et al.* (2007). Vector integration is nonrandom and clustered and influences the fate of lymphopoiesis in SCID-X1 gene therapy. *J. Clin. Invest.* **117,** 2225–2232.

Engels, B., and Uckert, W. (2007). Redirecting T lymphocyte specificity by T cell receptor gene transfer—A new era for immunotherapy. *Mol. Aspects Med.* **28,** 115–142.

Engels, B., Cam, H., Schuler, T., Indraccolo, S., Gladow, M., Baum, C., Blankenstein, T., and Uckert, W. (2003). Retroviral vectors for high-level transgene expression in T lymphocytes. *Hum. Gene Ther.* **14,** 1155–1168.

Fahnert, B., Lilie, H., and Neubauer, P. (2004). Inclusion bodies: Formation and utilisation. *Adv. Biochem. Eng. Biotechnol.* **89,** 93–142.

Gai, S. A., and Wittrup, K. D. (2007). Yeast surface display for protein engineering and characterization. *Curr. Opin. Struct. Biol.* **17,** 467–473.

Garboczi, D. N., Hung, D. T., and Wiley, D. C. (1992). HLA-A2-peptide complexes: Refolding and crystallization of molecules expressed in Escherichia coli and complexed with single antigenic peptides. *Proc. Natl. Acad. Sci. USA* **89,** 3429–3433.

Garboczi, D. N., Utz, U., Ghosh, P., Seth, A., Kim, J., VanTienhoven, E. A., Biddison, W. E., and Wiley, D. C. (1996). Assembly, specific binding, and crystallization of a human TCR-alphabeta with an antigenic Tax peptide from human T lymphotropic virus type 1 and the class I MHC molecule HLA-A2. *J. Immunol.* **157,** 5403–5410.

Garcia, K. C., Adams, J. J., Feng, D., and Ely, L. K. (2009). The molecular basis of TCR germline bias for MHC is surprisingly simple. *Nat. Immunol.* **10,** 143–147.

Gietz, R. D., and Woods, R. A. (2002). Transformation of yeast by lithium acetate/single-stranded carrier DNA/polyethylene glycol method. *Methods Enzymol.* **350,** 87–96.

Grabski, A. C. (2009). Advances in preparation of biological extracts for protein purification. *Methods Enzymol.* **463,** 285–303.

Hackel, B. J., and Wittrup, K. D. (2010). The full amino acid repertoire is superior to serine/tyrosine for selection of high affinity immunoglobulin G binders from the fibronectin scaffold. *Protein Eng. Des. Sel.* **23,** 211–219.

Holler, P. D., and Kranz, D. M. (2003). Quantitative analysis of the contribution of TCR/pepMHC affinity and CD8 to T cell activation. *Immunity* **18,** 255–264.

Holler, P. D., Holman, P. O., Shusta, E. V., O'Herrin, S., Wittrup, K. D., and Kranz, D. M. (2000). *In vitro* evolution of a T cell receptor with high affinity for peptide/MHC. *Proc. Natl. Acad. Sci. USA* **97,** 5387–5392.

Holler, P. D., Chlewicki, L. K., and Kranz, D. M. (2003). TCRs with high affinity for foreign pMHC show self-reactivity. *Nat. Immunol.* **4,** 55–62.

Huang, J., Zarnitsyna, V. I., Liu, B., Edwards, L. J., Jiang, N., Evavold, B. D., and Zhu, C. (2010). The kinetics of two-dimensional TCR and pMHC interactions determine T-cell responsiveness. *Nature* **464,** 932–936.

Huppa, J. B., Axmann, M., Mortelmaier, M. A., Lillemeier, B. F., Newell, E. W., Brameshuber, M., Klein, L. O., Schutz, G. J., and Davis, M. M. (2010). TCR-peptide-MHC interactions in situ show accelerated kinetics and increased affinity. *Nature* **463,** 963–967.

Jones, L. L., Brophy, S. E., Bankovich, A. J., Colf, L. A., Hanick, N. A., Garcia, K. C., and Kranz, D. M. (2006). Engineering and characterization of a stabilized alpha1/alpha2 module of the class I major histocompatibility complex product Ld. *J. Biol. Chem.* **281,** 25734–25744.

Jones, L. L., Colf, L. A., Bankovich, A. J., Stone, J. D., Gao, Y. G., Chan, C. M., Huang, R. H., Garcia, K. C., and Kranz, D. M. (2008a). Different thermodynamic binding mechanisms and peptide fine specificities associated with a panel of structurally similar high-affinity T cell receptors. *Biochemistry* **47,** 12398–12408.

Jones, L. L., Colf, L. A., Stone, J. D., Garcia, K. C., and Kranz, D. M. (2008b). Distinct CDR3 conformations in T cell receptors determine the level of cross-reactivity for diverse antigens, but not the docking orientation. *J. Immunol.* **181,** 6255–6264.

Kessels, H. W., van Den Boom, M. D., Spits, H., Hooijberg, E., and Schumacher, T. N. (2000). Changing T cell specificity by retroviral T cell receptor display. *Proc. Natl. Acad. Sci. USA* **97,** 14578–14583.

Kieke, M. C., Shusta, E. V., Boder, E. T., Teyton, L., Wittrup, K. D., and Kranz, D. M. (1999). Selection of functional T cell receptor mutants from a yeast surface- display library. *Proc. Natl. Acad. Sci. USA* **96,** 5651–5656.

Letourneur, F., and Malissen, B. (1989). Derivation of a T cell hybridoma variant deprived of functional T cell receptor alpha and beta chain transcripts reveals a nonfunctional alpha-mRNA of BW5147 origin. *Eur. J. Immunol.* **19,** 2269–2274.

Li, Q. J., Dinner, A. R., Qi, S., Irvine, D. J., Huppa, J. B., Davis, M. M., and Chakraborty, A. K. (2004). CD4 enhances T cell sensitivity to antigen by coordinating Lck accumulation at the immunological synapse. *Nat. Immunol.* **5,** 791–799.

Li, Y., Moysey, R., Molloy, P. E., Vuidepot, A. L., Mahon, T., Baston, E., Dunn, S., Liddy, N., Jacob, J., Jakobsen, B. K., *et al.* (2005). Directed evolution of human T-cell receptors with picomolar affinities by phage display. *Nat. Biotechnol.* **23,** 349–354.

Morita, S., Kojima, T., and Kitamura, T. (2000). Plat-E: An efficient and stable system for transient packaging of retroviruses. *Gene Ther.* **7,** 1063–1066.

Nagai, K., and Thogersen, H. C. (1987). Synthesis and sequence-specific proteolysis of hybrid proteins produced in Escherichia coli. *Methods Enzymol.* **153,** 461–481.

O'Herrin, S. M., Lebowitz, M. S., Bieler, J. G., al-Ramadi, B. K., Utz, U., Bothwell, A. L., and Schneck, J. P. (1997). Analysis of the expression of peptide-major histocompatibility complexes using high affinity soluble divalent T cell receptors. *J. Exp. Med.* **186,** 1333–1345.

Orr, B. A., Carr, L. M., Wittrup, K. D., Roy, E. J., and Kranz, D. M. (2003). Rapid method for measuring ScFv thermal stability by yeast surface display. *Biotechnol. Prog.* **19,** 631–638.

Purbhoo, M. A., Irvine, D. J., Huppa, J. B., and Davis, M. M. (2004). T cell killing does not require the formation of a stable mature immunological synapse. *Nat. Immunol.* **5,** 524–530.

Purbhoo, M. A., Sutton, D. H., Brewer, J. E., Mullings, R. E., Hill, M. E., Mahon, T. M., Karbach, J., Jager, E., Cameron, B. J., Lissin, N., *et al.* (2006). Quantifying and imaging NY-ESO-1/LAGE-1-derived epitopes on tumor cells using high affinity T cell receptors. *J. Immunol.* **176,** 7308–7316.

Richman, S. A., and Kranz, D. M. (2007). Display, engineering, and applications of antigen-specific T cell receptors. *Biomol. Eng.* **24,** 361–373.

Richman, S. A., Aggen, D. H., Dossett, M. L., Donermeyer, D. L., Allen, P. M., Greenberg, P. D., and Kranz, D. M. (2009a). Structural features of T cell receptor variable regions that enhance domain stability and enable expression as single-chain ValphaVbeta fragments. *Mol. Immunol.* **46,** 902–916.

Richman, S. A., Kranz, D. M., and Stone, J. D. (2009b). Biosensor detection systems: Engineering stable, high-affinity bioreceptors by yeast surface display. *Methods Mol. Biol.* **504,** 323–350.

Robbins, P. F., Li, Y. F., El-Gamil, M., Zhao, Y., Wargo, J. A., Zheng, Z., Xu, H., Morgan, R. A., Feldman, S. A., Johnson, L. A., *et al.* (2008). Single and dual amino acid substitutions in TCR CDRs can enhance antigen-specific T cell functions. *J. Immunol.* **180,** 6116–6131.

Rosenberg, S. A. (2010). Of mice, not men: No evidence for graft-versus-host disease in humans receiving T-cell receptor-transduced autologous T cells. *Mol. Ther.* **18,** 1744–1745.

Rudolph, M. G., Stanfield, R. L., and Wilson, I. A. (2006). How TCRs bind MHCs, peptides, and coreceptors. *Annu. Rev. Immunol.* **24,** 419–466.

Schmitt, T. M., Ragnarsson, G. B., and Greenberg, P. D. (2009). T cell receptor gene therapy for cancer. *Hum. Gene Ther.* **20,** 1240–1248.

Shusta, E. V., Kieke, M. C., Parke, E., Kranz, D. M., and Wittrup, K. D. (1999). Yeast polypeptide fusion surface display levels predict thermal stability and soluble secretion efficiency. *J. Mol. Biol.* **292,** 949–956.

Shusta, E. V., Holler, P. D., Kieke, M. C., Kranz, D. M., and Wittrup, K. D. (2000). Directed evolution of a stable scaffold for T-cell receptor engineering. *Nat. Biotechnol.* **18,** 754–759.

Soo Hoo, W. F., Lacy, M. J., Denzin, L. K., Voss, E. W. J., Hardman, K. D., and Kranz, D. M. (1992). Characterization of a single-chain T cell receptor expressed in *E. coli*. *Proc. Natl. Acad. Sci. USA* **89,** 4759–4763.

Stone, J. D., Chervin, A. S., and Kranz, D. M. (2009). T-cell receptor binding affinities and kinetics: Impact on T-cell activity and specificity. *Immunology* **126,** 165–176.

Utz, U., Banks, D., Jacobson, S., and Biddison, W. E. (1996). Analysis of the T-cell receptor repertoire of human T-cell leukemia virus type 1 (HTLV-1) Tax-specific CD8+ cytotoxic T lymphocytes from patients with HTLV-1-associated disease: Evidence for oligoclonal expansion. *J. Virol.* **70,** 843–851.

Varela-Rohena, A., Molloy, P. E., Dunn, S. M., Li, Y., Suhoski, M. M., Carroll, R. G., Milicic, A., Mahon, T., Sutton, D. H., Laugel, B., *et al.* (2008). Control of HIV-1 immune escape by CD8 T cells expressing enhanced T-cell receptor. *Nat. Med.* **14,** 1390–1395.

Warrens, A. N., Jones, M. D., and Lechler, R. I. (1997). Splicing by overlap extension by PCR using asymmetric amplification: An improved technique for the generation of hybrid proteins of immunological interest. *Gene* **186,** 29–35.

Weber, K. S., Donermeyer, D. L., Allen, P. M., and Kranz, D. M. (2005). Class II-restricted T cell receptor engineered *in vitro* for higher affinity retains peptide specificity and function. *Proc. Natl. Acad. Sci. USA* **102,** 19033–19038.

Willcox, B. E., Gao, G. F., Wyer, J. R., Ladbury, J. E., Bell, J. I., Jakobsen, B. K., and Anton van der Merwe, P. (1999). TCR binding to peptide-MHC stabilizes a flexible recognition interface. *Immunity* **10,** 357–365.

Zhang, B., Bowerman, N. A., Salama, J. K., Schmidt, H., Spiotto, M. T., Schietinger, A., Yu, P., Fu, Y. X., Weichselbaum, R. R., Rowley, D. A., *et al.* (2007). Induced sensitization of tumor stroma leads to eradication of established cancer by T cells. *J. Exp. Med.* **204,** 49–55.

Zhao, Y., Bennett, A. D., Zheng, Z., Wang, Q. J., Robbins, P. F., Yu, L. Y., Li, Y., Molloy, P. E., Dunn, S. M., Jakobsen, B. K., *et al.* (2007). High-affinity TCRs generated by phage display provide CD4+ T cells with the ability to recognize and kill tumor cell lines. *J. Immunol.* **179,** 5845–5854.

Zinkernagel, R. M., and Doherty, P. C. (1974). Restriction of *in vitro* T cell-mediated cytotoxicity in lymphocytic choriomeningitis within a syngeneic or semiallogeneic system. *Nature* **248,** 701–702.

CHAPTER NINE

Engineering Knottins as Novel Binding Agents

Sarah J. Moore *and* Jennifer R. Cochran

Contents

1. Introduction 224
2. Knottins as Scaffolds for Engineering Molecular Recognition 225
3. Engineering Knottins by Yeast Surface Display 227
4. Knottin Library Construction 230
 4.1. Protocol 1: Generation of a yeast-displayed knottin library 232
5. Screening Yeast-Displayed Knottin Libraries 234
6. Knottin production by chemical synthesis or recombinant expression 237
 6.1. Protocol 2: Synthetic production of knottins 238
 6.2. Protocol 3: Recombinant expression of knottins in *Pichia pastoris* 241
7. Cell binding assays 243
 7.1. Protocol 4: Direct and competition binding assays 244
8. Summary 247
Acknowledgments 247
References 247

Abstract

Cystine-knot miniproteins, also known as knottins, contain a conserved core of three tightly woven disulfide bonds which impart extraordinary thermal and proteolytic stability. Interspersed between their conserved cysteine residues are constrained loops that possess high levels of sequence diversity among knottin family members. Together these attributes make knottins promising molecular scaffolds for protein engineering and translational applications. While naturally occurring knottins have shown potential as both diagnostic agents and therapeutics, protein engineering is playing an important and increasing role in creating designer molecules that bind to a myriad of biomedical targets. Toward this goal, rational and combinatorial approaches have been used to engineer knottins with novel molecular recognition properties. Here, methods are described for creating

Department of Bioengineering, Cancer Institute, and Bio-X Program, Stanford University, Stanford, California, USA

Methods in Enzymology, Volume 503
ISSN 0076-6879, DOI: 10.1016/B978-0-12-396962-0.00009-4

and screening knottin libraries using yeast surface display and fluorescence-activated cell sorting. Protocols are also provided for producing knottins by synthetic and recombinant methods, and for measuring the binding affinity of knottins to target proteins expressed on the cell surface.

1. Introduction

Cystine-knot miniproteins, also known as knottins, have emerged as an important class of molecules with applications as therapeutic and diagnostic agents (Daly and Craik, 2011; Kolmar, 2010). Knottins are a structural family that share a common fold, in which one disulfide bond threads though a macrocycle created by two other disulfide bonds and the peptide backbone (Le Nguyen *et al.*, 1990) (Fig. 9.1). Highly diverse loop regions are anchored to a core of anti-parallel β strands and constrained by this molecular "knot" topology. Knottins are likely an example of convergent evolution (Carugo *et al.*, 2001), in which many species have evolved a similar solution to the problem of creating small, stable folds. Polypeptides containing cystine-knot motifs have been found in fungi, plants, and animals, and perform many

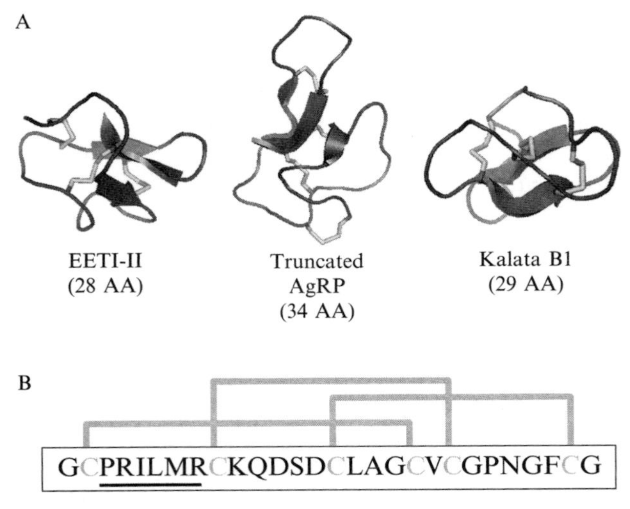

Figure 9.1 (A) Three-dimensional structures of knottins used as molecular scaffolds for protein engineering. The *E. elaterium* trypsin inhibitor-II (EETI-II, PDB 2ETI), a truncated form of the human Agouti Related Protein (AgRP, PDB 1MR0), and the Kalata B1 cyclotide (PDB 2F2J). (B) The sequence of wild-type EETI-II is shown, including connectivities of the three disulfide bonds that form the cystine-knot core. The trypsin binding loop is underlined. (For color version of this figure, the reader is referred to the Web version of this chapter.)

different functions including ion channel blockade, antimicrobial activity, and protease inhibition (Pallaghy *et al.*, 1994). The KNOTTIN database (http://knottin.cbs.cnrs.fr/) provides standardized information on knottin family members (Gracy *et al.*, 2008).

Knottins possess several natural properties that make them attractive for *in vivo* applications. First, their disulfide-constrained core confers high thermal and proteolytic stability (Werle *et al.*, 2006). Consequently, knottins have been shown to retain their three-dimensional structure after boiling or exposure to acid, base, or serum and have been rationally engineered for increased bioavailability when administered orally (Kolmar, 2009). Second, due to their high stability, knottins are believed to be nonimmunogenic through inefficient peptide-MHC presentation to the immune system (Craik *et al.*, 2007; Maillere *et al.*, 1995). Third, their small size (typically 30–50 amino acids) and stability has translated into desirable pharmacokinetic and biodistribution properties for molecular imaging applications, namely high tumor uptake and rapid clearance from nontarget tissues (Kimura *et al.*, 2009a). Further, knottins are amenable to production by chemical synthesis or recombinant expression (Jackson *et al.*, 2002; Le Nguyen *et al.*, 1989; Nishio *et al.*, 1993; Schmoldt *et al.*, 2005; Silverman *et al.*, 2009).

Naturally occurring knottins have been developed for a variety of biomedical applications. The most notable example of a knottin in clinical use is the ω-conotoxin MVIIa, which blocks ion channels in nerves that transmit pain signals (Williams *et al.*, 2008). This knottin is FDA-approved to be administered intrathecally for severe chronic pain, under the brand name Prialt from Elan Pharmaceuticals. In another example, the naturally occurring chlorotoxin knottin from scorpion venom was conjugated to a near-infrared dye and used to discriminate between neuroectodermal tumors and healthy tissue in murine models (Lyons *et al.*, 2002; Veiseh *et al.*, 2007). A structurally related family of polypeptides known as cyclotides have also been used for a variety of medicinal applications (Craik *et al.*, 2007). Craik and colleagues pioneered the study of these molecules, which contain a head-to-tail cyclized peptide backbone in addition to a cystine-knot motif (Craik *et al.*, 1999; http://www.cyclotide.com). As cyclotides are discussed in detail in a separate chapter in this volume, they will only be briefly mentioned here.

2. Knottins as Scaffolds for Engineering Molecular Recognition

Although naturally occurring knottins have found interesting clinical applications, protein engineering can be used to create tailor-made molecules against a variety of biomedical targets. There is considerable potential to engineer knottins with novel molecular recognition properties, as evidenced by the sequence plasticity and loop length variability observed in knottin family members, and the high stability afforded by their disulfide-bonded core. In general, researchers have utilized only a select subset of available knottins as molecular scaffolds for protein engineering applications, with several common examples shown in Fig. 9.1. First, the *Ecballium elaterium* trypsin inhibitor II (EETI-II) (Favel *et al.*, 1989) is a representative member of the knottin family whose structure and folding pathway have been well studied. EETI-II, which is found in the seeds of the squirting cucumber, is composed of 28 amino acids and contains several prominent disulfide-constrained loops. Second, the agouti-related protein (AgRP) is a 132-amino acid neuropeptide that binds to melanocortin receptors in the human brain and is involved in regulating metabolism and appetite (Ollmann *et al.*, 1997). The biological activity of AgRP is mediated by its C-terminal cystine-knot domain, which contains five disulfide bonds (residues 87–132) (McNulty *et al.*, 2001), but a fully active 34-amino acid truncated AgRP sequence was developed by Millhauser and colleagues that contains only four disulfide bonds (Jackson *et al.*, 2002). Third, the cyclotide Kalata B1 from the African plant *Oldenlandia*, which naturally has uterotonic activities, has been extensively studied for tolerance to mutation (Simonsen *et al.*, 2008) and was recently used as a scaffold for engineering peptides with new molecular recognition properties (Gunasekera *et al.*, 2008).

Kolmar and colleagues have pioneered the study and development of knottins as molecular scaffolds for over a decade. In one early example, a key structural loop of EETI-II was examined for tolerance to mutation, and residues conducive to forming β-turns were found to be critical for folding (Wentzel *et al.*, 1999). In addition, 13- or 17-amino acid functional binding epitopes were grafted in place of the 6-amino acid EETI-II trypsin inhibitor loop, and these epitope-containing knottins were isolated from a pool of wild-type knottin using bacterial surface display (Christmann *et al.*, 1999). In later studies, biologically active loops from disintegrin proteins were grafted into EETI-II and a modified version of the AgRP knottin scaffold and were shown to inhibit platelet aggregation (Reiss *et al.*, 2006). Additionally, monomeric knottins with grafted peptides that bound the thrombopoietin receptor acted as antagonists, while chemically linked dimers of these engineered knottins functioned as potent agonists with sub-nanomolar EC_{50} values (Krause *et al.*, 2007). Most recently, modeling and sequence analysis were used to introduce mutations to knottins that inhibit trypsin to enable them to also inhibit human mast cell tryptase β, a potential clinical target for treating allergic and inflammatory disorders (Sommerhoff *et al.*, 2010). Others have also grafted functional loops into EETI-II to create knottins that inhibit porcine pancreatic elastase (Hilpert *et al.*, 2003).

Examples using bioinformatics and combinatorial methods to study the tolerance of knottin scaffolds to mutation and examples of engineering knottins with high-affinity binding against clinically relevant targets have recently emerged from our research group. First, yeast-displayed libraries of EETI-II variants with randomized amino acid sequences and varied loop lengths were screened to identify mutants that retained the ability to bind trypsin. These EETI-II variants were then analyzed for sequence covariance, and while there was broad tolerance to mutation in general, sequence trends were found that facilitated folding (Lahti *et al.*, 2009). Second, yeast surface display was used to engineer EETI-II and AgRP knottin variants with low- to sub-nanomolar binding affinity to integrin receptors (Kimura *et al.*, 2009b; Silverman *et al.*, 2009, 2011). Knottins engineered for specificity to the platelet integrin receptor ($\alpha_{iib}\beta_3$) inhibited platelet aggregation as well as or slightly better than FDA-approved peptidomimetic eptifibatide (Silverman *et al.*, 2011). In addition, knottins engineered against integrin receptors expressed on tumors and the tumor vasculature ($\alpha_5\beta_1$, $\alpha_v\beta_3$, and $\alpha_v\beta_5$ integrins) have shown promise in radiotherapy applications (Jiang *et al.*, 2010b) and as diagnostic agents for noninvasive imaging of integrin expression in living subjects (Jiang *et al.*, 2010a; Kimura *et al.*, 2009a, 2011, 2010; Miao *et al.*, 2009; Nielsen *et al.*, 2010; Willmann *et al.*, 2010). While initial knottin engineering studies have focused on introducing functionality into a single loop region, multiple loops of EETI-II (Kimura *et al.*, 2011) or AgRP (Silverman *et al.*, 2009) can be simultaneously mutated to potentially increase the surface area of the binding interface.

3. Engineering Knottins by Yeast Surface Display

Yeast surface display is a powerful combinatorial technology that has been used to engineer proteins with novel molecular recognition properties, increased target binding affinity, proper folding, and improved stability (Gai and Wittrup, 2007; Pepper *et al.*, 2008). In this platform, libraries of protein variants are generated and screened in a high-throughput manner to isolate mutants with desired biochemical and biophysical properties. To date, yeast surface display has been the most successful combinatorial method used to engineer knottins with altered molecular recognition. Alternatively, EETI-II variants have been screened using bacterial display (Christmann *et al.*, 1999; Wentzel *et al.*, 1999) or mRNA display (Baggio *et al.*, 2002); however, in one report phage display was not successful in enriching functional EETI-II mutants against trypsin (Wentzel *et al.*, 1999). Compared to these other display formats, yeast surface display benefits from quality control mechanisms of the eukaryotic secretory pathway, chaperone-assisted folding, and efficient disulfide bond formation (Ellgaard and Helenius, 2003).

General methods and applications of yeast surface display using *Saccharomyces cerevesiae*, pioneered by Wittrup and colleagues, have been extensively described (Boder and Wittrup, 2000; Chao *et al.*, 2006; Colby *et al.*, 2004; Moore *et al.*, 2009). Thus, this chapter provides only a brief overview of the yeast display platform and highlights points to consider when applying this technology to knottin engineering. In Fig. 9.2, schematics of the yeast surface display system are presented at both the gene and protein level. The knottin to be engineered is genetically fused to the yeast mating agglutinin protein Aga2p, which is attached by two disulfide binds to the yeast cell wall protein Aga1p (Boder and Wittrup, 1997). Expression of this

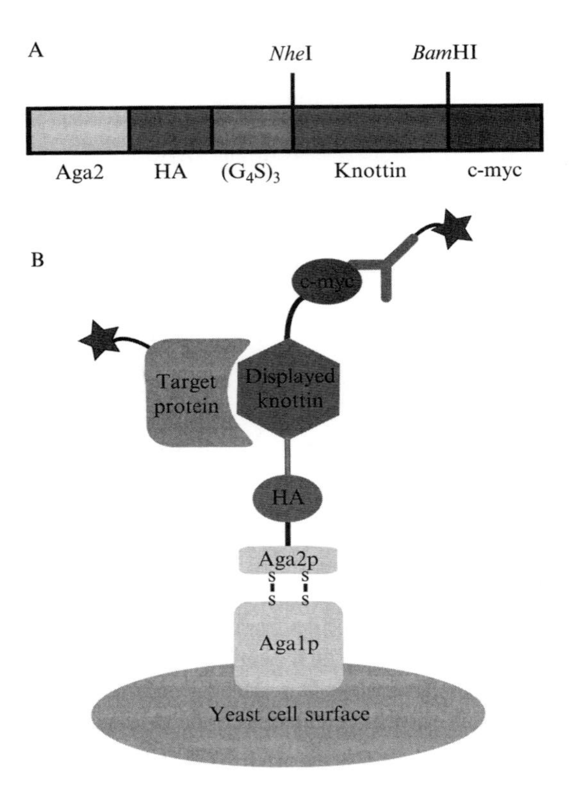

Figure 9.2 The pCT yeast surface display construct. (A) Open reading frame of the pCT genetic construct. $(G_4S)_3$=$(Gly_4Ser)_3$ linker. Relevant restriction enzyme sites are indicated. (B) Knottin variants are genetically fused to the yeast mating agglutinin protein Aga2p, which is disulfide-bonded to the yeast cell wall protein Aga1p. Hemagglutinin (HA) and c-myc epitope tags are include for detection of displayed protein by flow cytometry using fluorescently labeled antibodies. In addition, knottin binding can be measured against a fluorescently labeled target protein. Figure adapted with permission from Lahti and Cochran (2009). (For color version of this figure, the reader is referred to the Web version of this chapter.)

Aga2p-fusion construct, and of a chromosomally integrated Aga1p expression cassette, is under the control of a galactose-inducible promoter. N- or C-terminal epitope tags are included to measure cell surface expression levels by flow cytometry using fluorescently labeled primary or secondary antibodies. The pCT construct represents the most widely used display format, where the N-terminus of the knottin (or other protein to be engineered) is fused to Aga2, but several alternative variations of the yeast surface display plasmid have been described (Moore *et al.*, 2009). High-throughput fluorescence-activated cell sorting (FACS) has been the predominant method used to screen libraries of knottin variants against a target of interest. One of the benefits of this screening platform over panning-based methods used with phage or mRNA display is that two-color FACS (described below) can be used to quantitatively discriminate clones that differ by as little as twofold in binding affinity to the desired target (VanAntwerp and Wittrup, 2000).

Library sizes of 10^7–10^8 are typically created by homologous recombination (Swers *et al.*, 2004), with each yeast cell displaying tens of thousands of copies of a single mutant on its surface. However, libraries of this size only allow exploration of roughly 6 amino acid residues (e.g., $20^6 = 6.4 \times 10^7$), much lower than the theoretical diversity needed to fully sample mutations within a complete knottin scaffold. Thus, multiple rounds of mutagenesis and screening are often required to obtain high-affinity binders, which mimics natural evolution (Boder *et al.*, 2000; Hackel *et al.*, 2008; Hanes *et al.*, 1998; Kimura *et al.*, 2009b). Larger library sizes of 10^9 or greater could be created by combining multiple yeast transformations (Feldhaus *et al.*, 2003), but screening is limited by the throughput of current commercial flow cytometers (typically 10^7–10^8 cells/h). Alternatively, initial rounds of library screening could be performed by magnetic-activated cell sorting (MACS) (Ackerman *et al.*, 2009; Siegel *et al.*, 2004), where magnetic particles either conjugated to or specific for the target of interest are used to isolate yeast-displayed knottin binders. With MACS, 10^9–10^{10} clones can be screened at one time, significantly increasing the number of library members that can be analyzed. In addition, magnetic bead-based enrichment can afford avidity effects that allow isolation of weak (micromolar) binders, which will often not survive the wash steps needed for FACS sample preparation and analysis. Although smaller library sizes are generated with yeast surface display compared to other protein display technologies, the quality of a library is also critical. For example, when an identical antibody library was presented on the surface of yeast (library size $\sim 10^7$) or phage (library size $\sim 10^9$), yeast surface display identified all of the antibodies selected by phage display, in addition to twice as many unique antibodies (Bowley *et al.*, 2007).

Finally, a genotype–phenotype linkage provides a connection between a protein variant and its corresponding DNA sequence, allowing amino acid mutations to be identified following library screening. For yeast surface display, this involves recovering plasmid DNA from individual yeast clones and analyzing its genetic code through sequencing methods.

4. Knottin Library Construction

The general strategy for engineering knottins by yeast surface display begins with creating a library of mutants, typically by varying the disulfide-constrained knottin loop regions.

When engineering protein and peptide loops, it is important to vary the loop length in addition to sequence diversity (Hackel *et al.*, 2008; Lee *et al.*, 2004), as a larger surface contact area may be required to obtain high-affinity binding to a target of interest. However, altered loop sizes could decrease stability or not be tolerated by a particular knottin scaffold. Loops can be entirely randomized (Lahti *et al.*, 2009; Wentzel *et al.*, 1999), or a functional peptide motif can be included as a starting point for target binding. For an example of the latter, knottins with high affinity to particular integrin receptors were isolated by optimizing the position, loop length, and identity of residues flanking an arginine–glycine–aspartic acid (RGD) binding motif (Kimura *et al.*, 2009b).

The overall process of generating a yeast-displayed knottin library is depicted in Fig. 9.3. First, to selectively mutate knottin loop regions at the DNA level, degenerate codons can be introduced by oligonucleotide assembly using overlap extension PCR. For this purpose, "NNS" or "NNK" codons have been used, where N=A, C, T, or G; S=equal mixture of G or C; and K=equal mixture of G or T (Kimura *et al.*, 2011, 2009b; Lahti *et al.*, 2009; Silverman *et al.*, 2011, 2009). These combinations capture the diversity of all 20 amino acids while only including one of three possible stop codons, minimizing undesired truncations. When designing primers for gene assembly, the degenerate portion must be flanked by sufficient sequence homology with neighboring oligonucleotides (typically ~15–30 nucleotides) to yield melting temperatures of 40–60 °C that allow for specific annealing. In addition, long degenerate oligonucleotides are prone to forming secondary structure which can inhibit PCR reactions, but the addition of betaine and/or dimethylsulfoxide (DMSO) can help mitigate these effects. The gene assembly process can be performed in a stepwise manner, essentially building the full-length gene two primers at a time. Alternatively, one can combine all of the overlapping oligonucleotides together into one PCR assembly reaction. Next, in the second step of creating the DNA library, the genetic material is amplified using flanking primers with sufficient overlap with the yeast display vector for homologous recombination in yeast (Swers *et al.*, 2004).

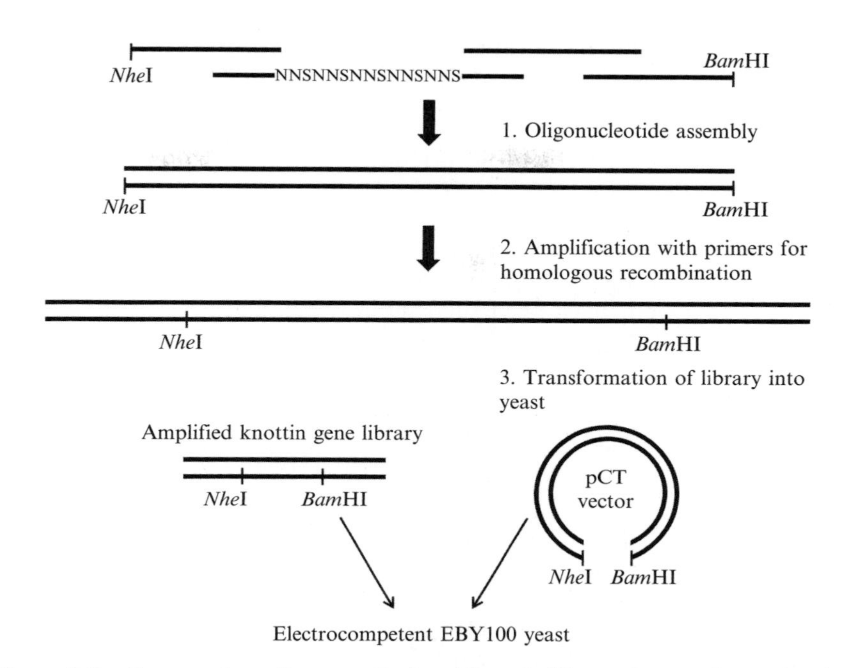

Figure 9.3 Construction of a yeast-displayed knottin library. In step 1, overlapping oligonucleotides containing degenerate codons for knottin loop mutation are assembled by PCR. In step 2, this product is amplified using primers with overlap amenable for homologous recombination with the yeast surface display plasmid. In step 3, the knottin gene library and the restriction enzyme-digested yeast surface display plasmid are transformed by electroporation into the yeast strain EBY100, which contains Aga1p on a chromosomally integrated galactose-inducible expression cassette. *Nhe*I and *Bam*HI restriction sites are shown as points of reference.

Although assembly and amplification methods allow knottin libraries to be created at relatively low cost and effort, drawbacks of this strategy include frameshifts and truncations that result from stop codons, incorrect annealing, or errors incorporated when synthesizing long oligonucleotide sequences. For example, when analyzing clones from unsorted libraries at the genetic level, we found that 40–60% of sequences contained stop codons or frameshifts (Silverman *et al.*, 2009). In addition, there are also inherent biases present in the genetic code. To address these limitations, commercial suppliers offer synthetic oligonucleotide libraries and new methods have been developed that allow defined control over library composition. Some of these technologies have been recently reviewed in detail (Lipovsek *et al.*, 2009).

Here, a detailed protocol is provided for generating a yeast-displayed knottin library by oligonucleotide assembly and amplification. However, given the recent decline in gene synthesis costs, commercially gene-synthesized knottin libraries are becoming a viable alternative that should be considered if economically feasible.

4.1. Protocol 1: Generation of a yeast-displayed knottin library

DNA assembly reaction:

1. Design oligonucleotides corresponding to the open reading frame of the knottin gene of interest for use in overlap extension PCR. Primers of 70 bp in length have been successfully used (Silverman *et al.*, 2009). In addition, primers should be composed of yeast-optimized codons and contain overlap regions (typically ~15–30 nucleotides) corresponding to melting temperatures of 40–60 °C. It is important for primers to have similar melting temperatures so that PCR will result in specific annealing for all oligonucleotides in the reaction. Degenerate NNS or NNK codons can be used for residue positions that sample all 20 amino acids. Alternative codons can be used if a more confined set of amino acids are desired (Kimura *et al.*, 2009b).
2. Assemble oligonucleotides by PCR using KOD polymerase (Novagen). The addition of betaine and DMSO is optional (Lahti *et al.*, 2009), and concentrations can be optimized for a particular PCR assembly reaction if necessary. In addition, if no product is obtained, alternative polymerases such as PfuUltra (Stratagene) or Phusion (New England Biolabs) could be tried, following manufacturer's recommended PCR conditions and temperatures. Low-fidelity Taq polymerase (Invitrogen) has also been used (Silverman *et al.*, 2009). Sequential assembly steps may be required for multiple overlap extension primers, although including all primers in a single assembly reaction may also result in the desired product.
3. An example PCR reaction (50 μl) will contain 1 μ*M* of each overlapping oligonucleotide, 1 unit of polymerase, 200 μ*M* dNTPs, 1× concentration PCR buffer, 1 m*M* $MgCl_2$, 1 *M* betaine, and 3% DMSO. Use standard PCR conditions for 10–30 cycles with extension temperature recommended for the chosen polymerase, and annealing temperature slightly below the melting temperature of the primers.
4. Analyze PCR product by agarose gel electrophoresis to confirm the presence of an assembled gene of the correct size. Purify PCR product by gel excision and extraction using a commercially available kit.

DNA amplification reaction:

5. Amplify library DNA using Pfx50 polymerase (Invitrogen). Library amplification primers should each have 30–50 bp or more of overlap with the pCT yeast surface display plasmid for homologous recombination, as described previously (Chao *et al.*, 2006; Colby *et al.*, 2004).

6. An example PCR reaction (50 μl) will contain purified template from the assembly reaction, 0.5 μ*M* of both the forward and reverse homologous recombination primers with overlap with the assembled gene, 1 unit of polymerase, 200 μ*M* dNTPs, 1× concentration PCR buffer, and 1 m*M* $MgCl_2$. To ensure high sequence diversity and sufficient quantities of DNA for yeast transformation below, all of the template from the assembly reaction should be used, typically by dividing the purified assembly reaction over 10 parallel amplification reactions. Use standard PCR conditions for 30–40 cycles with extension temperatures recommended for the chosen polymerase.
7. Analyze PCR product by agarose gel electrophoresis to confirm the presence of an amplified product of the correct size. Purify PCR product by gel excision and extraction using a commercially available kit. Alternatively, if a single product is obtained after PCR, purification columns can be used instead to increase DNA yield for library transformations.

DNA transformation into yeast:

8. Prepare the yeast surface display vector for homologous recombination by restriction enzyme digestion, and electroporate the vector and library insert into the *S. cerevisiae* strain EBY100 as described previously (Chao *et al.*, 2006; Colby *et al.*, 2004). Multiple transformation reactions can be performed in parallel to increase the library size. A molecular weight ratio of insert to vector is typically advised for yeast transformations (Chao *et al.*, 2006); however, due to the relatively short length of the knottin gene these ratios may be difficult to obtain. In one example of a randomized knottin loop library, four separate transformation reactions were performed with 10–15 μg of insert and 1 μg of vector per transformation. This process was repeated for five different loop lengths to yield a final library of $\sim 5 \times 10^6$–10^7 transformants (Lahti *et al.*, 2009). Alternatively, in our experience, a 5–10-fold molar excess of insert to 1 μg of vector per transformation can be used along with parallel transformations to yield a knottin library with an expected size of $\sim 10^7$ transformants.
9. Estimate the library size (i.e., the number of yeast transformants) by serial dilution plating and colony counting. Propagate the library in selective media as previously described (Chao *et al.*, 2006; Colby *et al.*, 2004).
10. Following transformation of the mutant DNA into yeast, it is important to verify the diversity of the library and the percentage of full-length knottin sequences before screening. First, recover plasmids from an aliquot of yeast culture using a Zymoprep Yeast Plasmid Miniprep Kit (Zymo Research), and transform this DNA into supercompetent *E. coli* cells (Stratagene). Second, isolate DNA from individual bacterial colonies

using a commercial plasmid isolation kit, and sequence to confirm the diversity in knottin loop length and amino acid composition.

11. Induce the knottin library for expression on the yeast cell surface and quantify expression levels using methods described elsewhere (Boder *et al.*, 2000; Chao *et al.*, 2006; Colby *et al.*, 2004). Typically, yeast surface expression levels are measured by flow cytometry using a primary antibody directed against the C-terminal c-myc epitope tag (clone 9E10, Covance) and a fluorescently labeled secondary antibody. Stop codons and frameshifts that occur in degenerate libraries prepared using the above methods will result in the expression of truncated peptides on the yeast cell surface. In addition, with yeast surface display there is always a population of yeast cells that do not express the fusion protein on their surface, which has generally been attributed to plasmid loss (Chao *et al.*, 2006), but a role for protein stability also has been proposed (Boder and Wittrup, 2000). Thus, in our experience anywhere from 20% to 50% of clones in the unsorted libraries can be expected to express the c-myc epitope tag.

5. Screening Yeast-Displayed Knottin Libraries

A description is provided for screening yeast-displayed knottin libraries by FACS; however, as described above, magnetic-based library sorting methods could also be used (Ackerman *et al.*, 2009; Siegel *et al.*, 2004). When screening knottin libraries by FACS, an enriched pool of binders generally emerges in 4–7 rounds of sorting, although this is highly variable and depends on many factors including: (1) the instrument used, (2) the concentration of target, (3) the stringency of the sort gate, and (4) the starting affinity of the binding interaction.

Two-color FACS is typically used for library screening, where one fluorescent label can be used to detect the c-myc epitope tag and the other to measure interaction of the knottin mutant against the binding target of interest. Different instrument lasers and/or filter sets can be used to measure excitation and emission properties of the two fluorophores at single-cell resolution. This enables yeast expression levels to be normalized with binding; in other words, a knottin that exhibits poor yeast expression but binds a high amount of a target can be distinguished from a knottin that is expressed at high levels but binds weakly to a target. Accordingly, a two-dimensional flow cytometry plot of expression versus binding will result in a diagonal population of yeast cells that bind to target antigen. High-affinity binders can be isolated using library sort gates similar to those depicted in Fig. 9.4 (sort rounds 2–6). Alternatively, in an initial sort round it could be useful to clear the library of undesired clones that do not express full-length

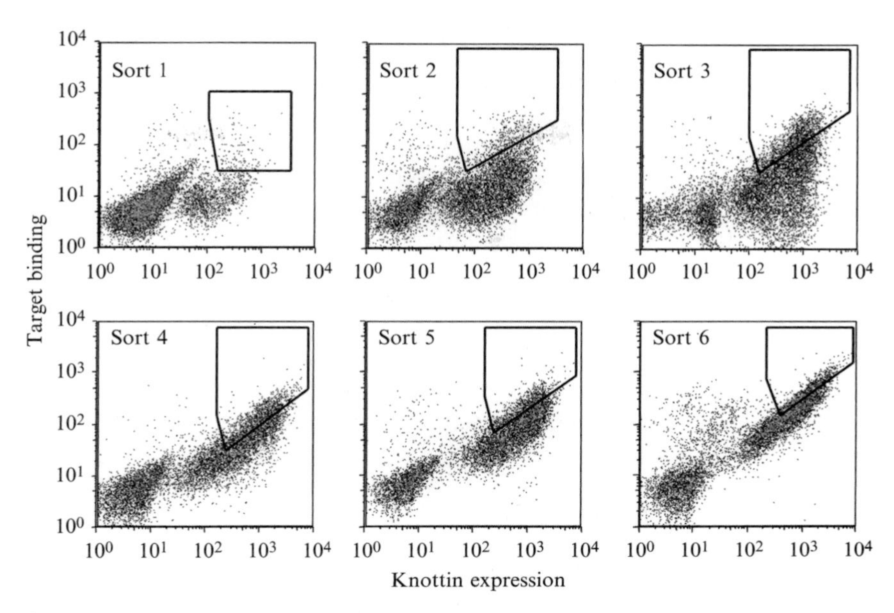

Figure 9.4 Example of knottin library sort progression using two-color FACS. Density dot plots, indicating knottin expression (*x*-axis, Alexa 555), as measured through antibody detection of the C-terminal c-myc epitope tag, and target binding (*y*-axis, fluorescein). The initial knottin library contained very few of these "double positive" yeast cells. Polygons indicate collection gates used for library sorting. Collected yeast cells were propagated in culture and surface expression was induced for the next sort round. In this example, six rounds of FACS were used to isolate an enriched population of yeast with high target binding affinity. A diagonal gate was used in sort rounds 2–6 to isolate knottin mutants that bound the highest level of target for a given amount of surface expression. To increase stringency, target concentrations were reduced in successive sort rounds from 500 n*M* (rounds 1–2), 100n*M* (round 3), and 50 n*M* (rounds 4–6). Data are used with permission from Kimura *et al.* (2009b). (For color version of this figure, the reader is referred to the Web version of this chapter.)

knottin by screening for and isolating yeast that bind to antibodies directed against the C-terminal c-myc epitope tag.

Because knottins are significantly smaller than most other proteins engineered using yeast surface display, one concern is that the antibodies used to detect the terminal c-myc tag may block binding to the desired target antigen. If the engineered knottin library cannot explicitly be shown to bind simultaneously to c-myc antibodies and the target, as is often the case with a naïve library, then library screening should be performed with single-color FACS against the binding target, with subsequent confirmation that the enriched pool of mutants expresses the C-terminal c-myc epitope tag. As an alternative strategy, single-color FACS (or MACS) can first be used to rid the starting library of truncated peptides that do not express the c-myc tag as described above. In early rounds of sorting, it is advisable to start with a relatively high concentration of target antigen (e.g., 0.5–1 μ*M*) and conservative sort gates, collecting as many as 2–3% of the clones with the highest binding signal. As binding populations emerge, more stringent screening conditions can be used, decreasing concentration of target and drawing more restrictive sort gates containing ~1% of the total cell population. The original yeast library as well as products from each sort round should be frozen and stored as glycerol stocks (Chao *et al.*, 2006) should it become necessary to try alternate sort strategies or analyze these cell populations at a later time.

The library diversity must be oversampled by at least 10- to 25-fold in each sort round to ensure a ≥99% theoretical probability that every clone will be sampled at least once (Bosley and Ostermeier, 2005). Thus, for each round of sorting, the number of cells analyzed and collected should be noted to determine the theoretical maximum diversity of the remaining library. These numbers will dictate how many cells should be sorted in the next round to achieve at least 10-fold oversampling. Two library screening strategies have been previously described in detail, namely equilibrium binding sorts and kinetic sorts (Boder and Wittrup, 1997, 1998, 2000; Boder *et al.*, 2000). Briefly, to isolate mutants with low or moderate affinity interactions, target antigen is incubated with the yeast library and allowed to come to equilibrium before sorting by FACS. The time needed to reach equilibrium, at which the binding signal no longer increases at longer incubation times, should be empirically determined. To further enhance already high-affinity interactions ($K_D < \sim 10$ n*M*), kinetic sorts are used to isolate mutants with the slowest kinetic off-rates. Here, a saturating amount of labeled target antigen is first allowed to come to equilibrium with the yeast library, so that all displayed proteins that can bind target are engaged in a binding interaction. After a wash step, unlabeled competitor is then incubated with the library, and the original labeled antigen is allowed to dissociate over time, from hours to days, before sorting for yeast that retain the highest amount of labeled antigen. Using this strategy, a single-chain antibody fragment with femtomolar binding affinity and a 10,000-fold decrease in kinetic off-rate to fluorescein was isolated from a yeast-displayed library (Boder *et al.*, 2000).

The quality of the fluorescently labeled target antigen is crucial to successful knottin engineering by yeast surface display. Whether commercially available or produced by the researcher, an antigen should be structurally and functionally relevant for the final application. In addition, if secondary reagents such as streptavidin or antibodies will be used for detection purposes, then steps must be taken to ensure that the library sorts are not enriching for knottins that bind to these molecules instead of the desired target. This can be carried out by alternating the detection reagents used in each sort round, or by sorting the library against detection reagent alone (i.e., a negative sort) and collecting the yeast that do not bind.

Following enrichment of knottin libraries for binders against a desired target, the yeast plasmids are recovered as described above and 10–30 clones should initially be sequenced to identify any sequence patterns that exist and assess the diversity of the isolated mutants. Additional rounds of FACS can be performed under increased sorting stringency if the sequence diversity is high or consensus sequences have not emerged. The binding affinities or kinetic off-rates of individual yeast-displayed knottin clones are then measured to determine which mutants are most promising to produce by recombinant or synthetic approaches. Detailed methods for performing binding assays with yeast-displayed proteins have been extensively described (Boder *et al.*, 2000; Chao *et al.*, 2006; Colby *et al.*, 2004).

When one considers the limitations of library size compared to the theoretical diversity, it is possible that weak binders will be isolated from a naïve starting library composed of randomized knottin sequences. If desired binding properties are not achieved with mutants isolated from a single library, additional rounds of mutagenesis and screening can be performed. A second-generation knottin library could be designed, for example, by fixing specific residues and varying others to a limited set of amino acids informed by the sequences recovered from the initial library (Kimura *et al.*, 2009b). DNA shuffling is a useful recombination tool for long genes (Stemmer, 1994); however, for knottin loop engineering, repeating the gene assembly process with oligonucleotides encoding for a more narrowly defined library will better sample combinations of mutations (Kimura *et al.*, 2009b). This second-generation library can then be screened against target antigen as described above. Alternatively, a library-isolated clone could be used as a starting point for creating a second-generation library, where a neighboring knottin loop could be randomized and screened to identify mutants with improved binding affinity or specificity. For example, an integrin-binding AgRP mutant was mutated to create second-generation knottin libraries containing three separate randomized loop regions (Silverman *et al.*, 2009). These studies highlighted the tolerance of the AgRP scaffold to mutations in multiple loop regions and suggested that such mutations could potentially influence the target binding interaction.

6. Knottin Production by Chemical Synthesis or Recombinant Expression

Once knottin mutants have been identified by yeast surface display, it is important to produce them in soluble form for further characterization. The small size of knottins makes them amenable to production by both chemical synthesis and recombinant expression. In our experience, the folding behavior of engineered knottins is unpredictable and often requires optimization and troubleshooting for both synthetic and recombinant approaches.

6.1. Protocol 2: Synthetic production of knottins

Solid phase peptide synthesis followed by *in vitro* folding is a useful strategy for producing knottins for biophysical and biological studies (Jackson *et al.*, 2002; Kimura *et al.*, 2009b; Le Nguyen *et al.*, 1989; Nishio *et al.*, 1993). Compared to standard recombinant expression methods, chemical synthesis permits facile incorporation of nonnatural amino acids or other chemical handles into knottins for a variety of applications (Avrutina *et al.*, 2005). Recombinant production strategies often incorporate peptide epitope tags or fusion partners to facilitate purification from a high background of endogenous host proteins; however, these modifications can potentially interfere with binding and biological function. To address these concerns, methods for proteolytic cleavage of fusion partners following knottin purification have been described (Schmoldt *et al.*, 2005). Alternatively, knottins can be produced in a "tag-free" format using chemical synthesis and folding. The structural stability afforded by the disulfide-bonded knottin core is compatible with organic solvents used for purification by reversed-phase HPLC (RP-HPLC), which would otherwise denature typical globular proteins.

Knottin peptide sequences are readily synthesized using standard 9-fluorenylmethyloxycarbonyl (Fmoc)-based solid phase peptide chemistry on an automated synthesizer (Chan and White, 2000). The linear peptide is then folded under conditions that promote oxidation of cysteine side chain thiols to form disulfide bonds, followed by purification with RP-HPLC (Fig. 9.5). While some knottins, including EETI-II (Le Nguyen *et al.*, 1989), form the native disulfide-bonded pattern readily, extensive screening and optimization of knottin folding conditions may be necessary (Daly *et al.*,

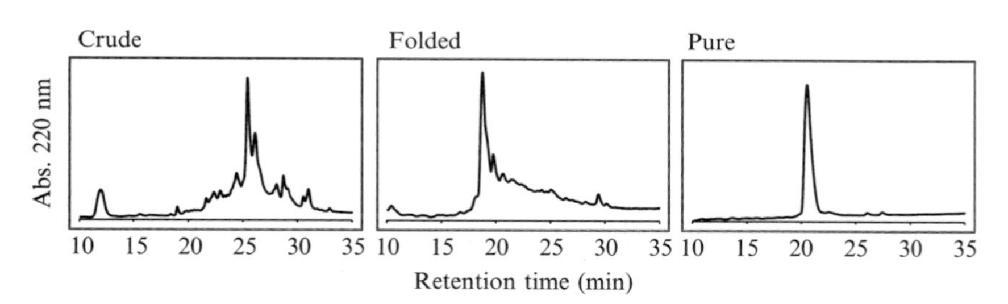

Figure 9.5 RP-HPLC traces of knottin synthesis and folding. An example of an engineered EETI-II knottin is shown. Representative traces of crude (unfolded, unpurified), folded (unpurified), or pure (folded, purified) knottin analyzed on a Vydac C18 analytic scale HPLC column at a flow rate of 1 ml/min. Gradient used was 10–50% Solvent B over 30 min, following 4 min at 10% Solvent B. Solvent A is water with 0.1% TFA and Solvent B is 90% acetonitrile in water with 0.1%TFA. Absorbance was monitored at 220 nm. In this example, the knottin was folded without intermediate purification of the crude linear peptide.

1999; Nishio *et al.*, 1993). Alternatively, combinations of different cysteine thiol protecting groups can be incorporated into the peptide sequence and selectively removed stepwise in pairs to facilitate folding (Cuthbertson and Indrevoll, 2003). In our experience, folded knottins can be stored in PBS at 4 °C, frozen at −20 or −80 °C, or lyophilized, but specific conditions will need to be determined empirically for each new variant.

An overview of knottin synthesis, folding, and purification is outlined below. Appropriate safety procedures for chemical handling and waste disposal must be followed when performing these experiments.

Automated solid phase peptide synthesis:

1. Weigh the required amount of resin for performing a 0.25 mmol scale synthesis based on the loading of reactive sites. A Rink resin is suggested, which provides a C-terminal amide upon peptide cleavage that enhances peptide stability compared to a carboxylate group.
2. Use amino acid precursors that contain standard acid-labile side chain protecting groups (Novabiochem/EMD Chemicals Inc.). For chain elongation, use a fourfold excess of Fmoc-protected amino acids to help drive coupling reactions to completion.
3. Use *N,N*-dimethylformamide (DMF) to initially swell the resin and for all subsequent wash steps. Remove Fmoc groups on the growing peptide chain with 20% piperidine in DMF for 20 min.
4. Simultaneously dissolve individually weighed 1 mmol portions of protected amino acids and 1 mmol of hydroxybenzotriazole (HOBt) in DMF, and activate for coupling with a stoichiometric amount of diisopropylcarbodiimide (DIC). This chemistry avoids activating agents that require tertiary amine bases, such as the commonly used reagent *N,N*-diisopropylethlyamine, which can cause cysteine racemization (Angell *et al.*, 2002). While we have found DIC/HOBt to be effective and convenient, alternative reagents and conditions can be applied at particular steps where inefficient coupling is observed.
5. After the synthesis is complete, remove the final Fmoc group and wash with DMF. Eliminate all traces of DMF prior to acid-mediated peptide cleavage and side chain deprotection by washing multiple times with dichloromethane (DCM) followed by methanol. Fully dry the peptidyl-resin under vacuum for several minutes, record the weight, and store in a sealed vial.

Peptide cleavage and deprotection:

6. Weigh approximately half of the peptidyl-resin into a 20 ml glass vial containing a Teflon-coated stir bar. In a well-ventilated fume hood, add 10 ml of a cleavage cocktail composed of trifluoroacetic acid (TFA),

1,2-ethanedithiol (EDT), triisopropylsilane (TIS), and water in a ratio of 94:2.5:1:2.5. After a few minutes, loosely cover the reaction vial with a Teflon-lined screw cap and stir for at least 2 h at room temperature.

7. Collect the resin on a fritted-glass disc and remove volatile components in the filtrate with either a steady stream of nitrogen gas or a rotary evaporator. Add 150ml of cold, anhydrous diethyl ether (Et_2O) to the oily substance that forms and collect the resulting white powder by centrifugation.
8. Wash the material an additional two to three times with fresh Et_2O to extract any remaining cleavage cocktail. Resuspend the white precipitate in 20–30ml of a 1:1 mixture of water and acetonitrile containing 0.1%TFA, freeze in a bath of dry ice and acetone, and lyophilize.

Analysis and purification of crude peptide:

9. Fully dissolve a small portion of the crude material and examine by RP-HPLC. Typically, a C_{18} preparative column is used with varying gradients composed of Solvent A (water with 0.1% TFA) and Solvent B (90% acetonitrile/10% water/0.1%TFA).
10. Determine which peaks in the chromatogram are the desired peptide product using either electrospray ionization (ESI) or matrix-assisted laser desorption/ionization time of flight (MALDI-TOF) mass spectrometry.
11. Scale up the purification, and pool and lyophilize collected fractions from multiple RP-HPLC runs. Assess the degree of peptide purity using an analytical-scale C_{18} column.

Knottin folding and analysis:

12. Perform folding reactions on a small scale under different conditions and monitor progress over time by RP-HPLC using an analytic scale C_{18} column. A single peak corresponding to the folded knottin should appear at a different retention time compared to the unfolded form and other misfolded by-products. A good starting point for folding conditions for engineered knottin variants are those previously reported for the wild-type knottin. Note that folding efficiency is sequence specific and may need to be optimized for a particular knottin variant.
    Examples of folding conditions for engineered integrin-binding knottins:
    a. EETI-II-based knottin (33 amino acids): 2.5m*M* reduced glutathione and 20% (v/v) DMSO in 0.1*M* ammonium bicarbonate, pH 8 at room temperature with gentle rocking overnight (Kimura *et al.*, 2009b).

b. Truncated AgRP-based knottin (38 amino acids): 4 *M* guanidinium chloride, 10 m*M* reduced glutathione, 2 m*M* oxidized glutathione, and 3.5% (v/v) DMSO at pH 7.5 at room temperature for 3 days (Jiang *et al.*, 2010a).

Alternatively, depending on the knottin scaffold and the level of purity, efficient folding may be achieved with the crude material without an initial RP-HPLC purification step.

13. Once conditions are determined, scale up the folding reaction and use RP-HPLC as described above to purify the folded product and assess final purity. The mass of the folded knottin should be confirmed with either ESI or MALDI-TOF mass spectrometry and will be 2 Da less for each disulfide bond formed.
14. Determine the concentration of folded knottin stock solutions using UV–vis spectrophotometry (Pace *et al.*, 1995) or amino acid analysis.

6.2. Protocol 3: Recombinant expression of knottins in *Pichia pastoris*

Multiple strategies have been developed for producing knottins using recombinant methods. For example, functional knottins have been produced with barnase as a genetic fusion partner, which promotes folding in the *E. coli* periplasmic space and serves as a useful purification handle (Schmoldt *et al.*, 2005). Here, recombinant expression of knottins in the yeast strain *Pichia pastoris* is described. This approach has been used to produce 2–10 mg/l of purified engineered knottins, particularly those based on a truncated version of AgRP (Silverman *et al.*, 2011, 2009). Recombinant expression of knottins in *S. cerevisiae* might also be possible, although has not been thoroughly explored. The yeast expression construct contains an N-terminal FLAG epitope tag for knottin binding detection and a C-terminal hexahistadine tag for purification by metal chelating chromatography (Ni-NTA). Size exclusion chromatography is then used to remove aggregates or misfolded multimers. While epitope tags are convenient for purification and biochemical assays, their inclusion or location could affect knottin folding and expression yield, or interfere with binding or biological function. These potential complications should be considered when developing a knottin production and purification strategy.

Protocols are essentially as described by the manufacturer in the Invitrogen "Multi-Copy *Pichia* Expression Kit," with the following modifications for knottins. All media, reagents, and strains are described in the product manual, which can also be consulted for information on optimization of protein expression levels as necessary.

1. Clone the gene encoding for the open reading frame of the engineered knottin into the pPIK9K plasmid, following the manufacturer's

instructions, except include an N-terminal FLAG epitope tag (DYKDDDDK) for binding detection and a C-terminal hexahistadine (HHHHHH) tag for purification. The final construct is therefore:

(*Sna*BI)-DYKDDDDK-(*Avr*II)-knottin-(*Mlu*I)-HHHHHH-(two stop codons)-(*Not*I)

The FLAG epitope tag can be inserted between the *Sna*BI and *Avr*II restriction sites in the original pPIK9K plasmid, and the hexahistadine tag can be inserted between an additional *Mlu*I restriction site and the existing *Not*I restriction site.

2. Purify plasmid DNA from a 50 ml culture of *E. coli* containing the cloned knottin sequence using standard protocols. Cut DNA with a restriction enzyme to create a linear construct as described in the product manual, then ethanol precipitate the DNA and resuspend in 10 μl of Tris–EDTA buffer.
3. Transform 5 μl of the DNA suspension into the GS115 *P. pastoris* yeast strain using the electrotransformation method described in the manual. Propagate on selection plates as described.
4. Select 3–6 individual yeast colonies from YPD-Geneticin plates to test knottin expression. Grow colonies at 30 °C in 5 ml BMGY media overnight to an OD_{600} between 3 and 6, and induce expression in 5 ml BMMY+1% casamino acids at an initial $OD_{600}=1$. Induce separate cultures at both 20 and 30 °C, as some knottins are better expressed at lower temperature. The addition of casamino acids helps to reduce potential proteolysis. Since methanol is metabolized and also evaporates, it should be replenished by daily addition to the cultures at a final concentration of 0.5%, usually by addition of 250 μl of 10% methanol every 24 h.
5. After three days of induction, take 100 μl of culture and pellet cells for 1 min at maximum speed in a microcentrifuge. Measure knottin expression by analyzing yeast supernatant by SDS-PAGE followed by Western blot, using an anti-FLAG epitope tag antibody for detection (Sigma). Select highest expressing yeast cultures for scale-up. *Note*: some knottin sequences may be expressed as a combination of monomers and disulfide-bonded multimers; this can be determined by comparing reduced and non-reduced samples by SDS-PAGE or Western blot. Monomeric knottins can usually be purified from higher-order multimers by size exclusion chromatography as described in step 8.
6. Scale-up yeast cultures as described in the product manual, except include 1% casamino acids in the BMMY media, and induce for expression at the optimal temperature as determined for 5 ml cultures.
7. Purify hexahistidine-tagged knottin peptides with Ni-NTA resin using standard methods and analyze elution fractions using SDS-PAGE.
8. Isolate monomeric knottin species by size exclusion chromatography using a high resolution preparative-scale column, such as a Sephadex 75

or Sephadex Peptide column (GE Healthcare). Concentrate elution fractions and store in PBS at 4 °C. As described above, properly folded knottins will appear as a single peak when analyzed by RP-HPLC. Thus, RP-HPLC should be used to determine if recombinantly expressed knottins consist of multiple folded states.

9. Confirm knottin mass using either ESI or MALDI-TOF mass spectrometry. Determine the concentration of folded knottin stock solutions using UV–Vis spectrophotometry (Pace *et al.*, 1995) or amino acid analysis.

7. Cell Binding Assays

Here, methods are provided to measure the affinity of knottins to receptors expressed on the surface of mammalian cells using direct binding or competition binding assays (Fig. 9.6). In a direct binding assay, an equilibrium binding constant (K_D) is measured using a knottin conjugated to a fluorophore or radioisotope, or a knottin that contains an N- or C-terminal epitope tag for detection by a labeled antibody. If labels or tags are not feasible or desired, a competition binding assay can be used to determine

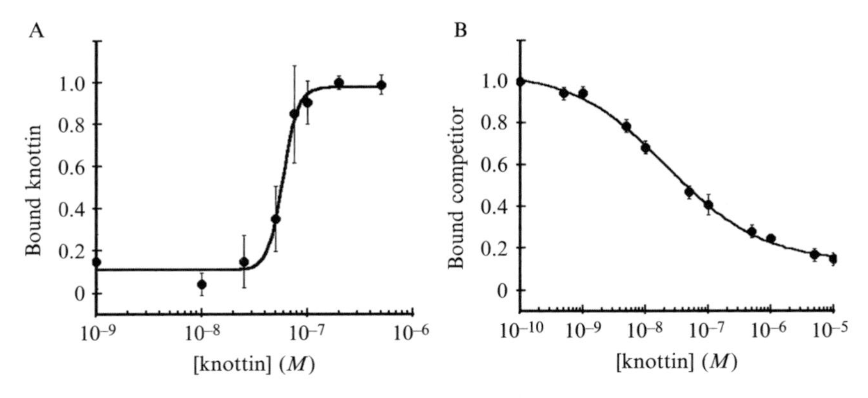

Figure 9.6 Binding of engineered knottins to receptors expressed on mammalian cells. (A) Direct binding assay of a recombinantly expressed AgRP knottin to cells that have been transfected to express integrin receptors. Binding was detected by flow cytometry using a fluorescently labeled antibody directed against the knottin C-terminal hexahistadine tag. Adapted with permission from Silverman *et al.* (2011). (B) Competition binding assay of a synthetically produced AgRP knottin and radiolabeled protein (^{125}I-labeled echistatin) to integrin-expressing tumor cells. Binding was detected using a gamma counter; however, a fluorescently labeled protein could alternatively be used as the competitor for detection by flow cytometry. Adapted with permission from Jiang *et al.* (2010a). Error bars represent standard deviation of triplicate data.

the half-maximal inhibitory concentration (IC_{50}), the amount of unlabeled knottin at which 50% of the maximal signal of the labeled competitor is detectable. A K_D value can then be calculated from the measured IC_{50} value (Eq. (9.1)) (Cheng and Prusoff, 1973).

$$K_D \text{ knottin} = \frac{IC_{50} \text{ knottin}}{1 + \frac{[\text{competitor}]}{K_D \text{ competitor}}} \tag{9.1}$$

For this equation to be valid, several assumptions must be made. First, the binding interaction is 1:1 and is not cooperative. Second, the binding reaction has reached equilibrium, and the binding is reversible and follows the law of mass action. Third, the K_D value of the labeled competitor is measured under similar conditions on the same cell line. Fourth, the amount of ligand in solution remains constant over the course of the experiment and is not depleted through binding to receptor. Ligand depletion will be more pronounced when measuring high-affinity interactions over a lower concentration range (Colby *et al.*, 2004), and can be avoided or minimized by decreasing the number of cells added in the experiment or by increasing the binding reaction volumes as described below.

7.1. Protocol 4: Direct and competition binding assays

Methods are provided for fluorescent detection of knottins using flow cytometry. Protocols can be adjusted as needed for radiolabeled knottins or competitors (Jiang *et al.*, 2010a; Kimura *et al.*, 2009b).

Direct binding assay:

1. Aliquot 5×10^4 mammalian cells expressing target protein into 1.5–2ml polypropylene tubes. In general, mammalian cells may be pelleted for 5 min at 4°C at 800×*g* in a microcentrifuge, but cell viability should be evaluated after this treatment.
2. Resuspend the cells in an appropriate binding buffer with a range of knottin concentrations. The composition of the binding buffer will depend upon the target; for example, include any salt or metal cofactors needed for the binding interaction. Cell culture media in which fetal bovine serum is replaced with 0.1–1% BSA (to prevent nonspecific binding) may also be used if appropriate. Knottin concentration ranges of at least 100-fold above and below the expected K_D should be tested. To avoid the effects of knottin (ligand) depletion from solution, binding should be measured at several volumes, ranging from around 50µl to up to 5ml. An example is shown in Fig. 9.7, where a single concentration of a high-affinity ligand was tested at multiple concentrations, illustrating the artificial decrease in binding signal observed with small reaction volumes.

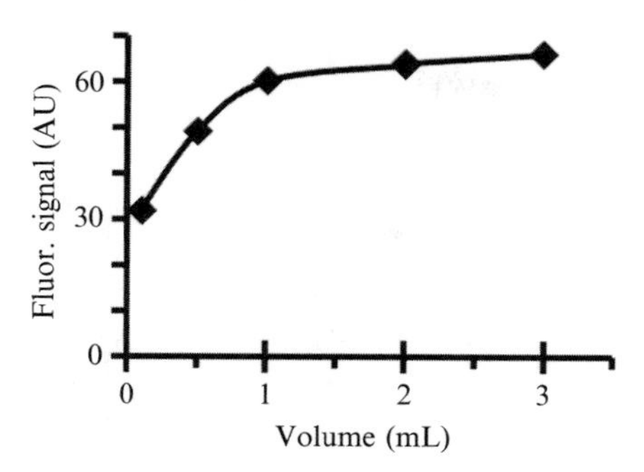

Figure 9.7 Example of reduced receptor binding signal observed under ligand depleting conditions. A fluorescently labeled ligand (30 n*M*) was incubated with 5×10^4 mammalian cells in several different volumes. At low labeling volumes, the receptor binding signal measured by flow cytometry was significantly reduced compared with larger volumes, likely due to depletion of ligand from the solution during the binding reaction.

3. Incubate cells with knottin until equilibrium is reached. This will typically take several hours but should be determined experimentally by measuring how long it takes for the binding signal to saturate. In general, mammalian cell binding assays should be performed at 4 °C to prevent receptor internalization that may occur. Alternatively, room temperature or 37 °C could be used to measure binding events that require receptor clustering or rearrangement that cannot occur at lower temperatures; in this case, a phosphatase inhibitor such as phenylarsine oxide can be used to prevent receptor internalization (Lazari *et al.*, 1997). Use caution, as longer incubation times and higher temperatures can cause mammalian cell death.
4. If the knottin is directly labeled with a fluorophore or other detection handle, then skip this step and proceed to step 5. Alternatively, if knottin contains an N- or C-terminal epitope tag, wash samples by adding 1 ml of PBS containing 0.1% BSA (PBSA) and pellet as described above. Aspirate the supernatant and resuspend cells in appropriate concentration of fluorescently labeled anti-epitope tag antibody in PBSA. The antibody will need to be titrated for each lot and cell line, so that a saturating concentration is used. An example is a fluorescein-labeled conjugated goat anti-His antibody (Bethyl Laboratories) that was used to measure binding of knottins containing a C-terminal hexahistidine tag (Silverman *et al.*, 2011). Incubate on ice for 20–30 min.
5. Wash cells in PBSA as described above and analyze cell fluorescence using a flow cytometer. Determine the mean fluorescence signal at each knottin concentration using data analysis software such as FlowJo (TreeStar, Inc.) or CellQuest (Becton Dickinson). Plot concentration versus mean fluorescence signal on a semilog plot and fit to a sigmoidal

curve for a 1:1 binding interaction; the half-maximal signal will be equivalent to the K_D value (Fig. 9.6A). Curve shapes other than sigmoidal are indicative of alternate binding stoichiometries or the involvement of nonspecific binding.

6. Additional considerations: Proper controls should be included, such as cells alone, and cells labeled with antibody alone, when appropriate. If possible, use mammalian cells that do not express the target receptor as a negative control. In addition, confirm expression levels of target receptor using commercially available antibodies. Finally, it is important to differentiate live cells using instrument scatter parameters or a DNA intercalating agent dye such as propidium iodide, as dead cells will often bind nonspecifically to ligand and interfere with analysis.

Competition binding assay:

Follow protocol for a direct binding assay, with these modifications:

1. In the first binding step, incubate cells with a range of knottin concentrations and a constant concentration of competitor. The competitor can be a protein or small molecule labeled with a fluorophore or radioisotope, or a protein containing an epitope tag for antibody detection. The concentration of competitor used should be below its K_D value: high enough to obtain a detectable signal but avoiding an excess that cannot be effectively competed. As in the direct binding assay, aim for knottin concentrations that are 100-fold above and below the expected IC_{50} value. In addition, it is important to avoid ligand depleting conditions and to incubate binding reactions until equilibrium is reached.
2. For fluorescently labeled competitor, proceed directly to analysis by flow cytometry. Alternatively, for a competitor containing an epitope tag, follow step 4 above for labeling with an anti-epitope tag antibody, and then analyze by flow cytometry. Plot concentration versus mean fluorescence signal on a semilog plot and fit to a sigmoidal curve for a 1:1 binding interaction; the half-maximal signal will be equivalent to the IC_{50} value (Fig. 9.6B), and a theoretical K_D can be calculated as given in Eq. (9.1).
3. Additional considerations: IC_{50} values are influenced by several factors including: (1) The affinity of the knottin for the target receptor. If the K_D value is small (high affinity), the IC_{50} will also be low. (2) The affinity of the competitor for the receptor. It will take more unlabeled knottin to compete for a tightly bound competitor compared to a weakly bound competitor. (3) The concentration of competitor. If the concentration of competitor used greatly exceeds its K_D value, the binding curves will be ambiguous, resulting in similar IC_{50} values.

8. Summary

The approaches and protocols described here have been used for knottin library creation and screening, and for knottin production, purification, and binding characterization. Despite successful application of rational and combinatorial engineering methods that have resulted in knottins with unique molecular recognition properties, there are still many open questions to be explored. For example, one current area of research is determining how broadly knottins can be engineered, from both the standpoints of expanding which knottins can be used as scaffolds and of investigating the diversity of targets that will yield successful results. In addition, it will be important to compare various native knottin scaffolds and how introduced mutations alter biophysical properties such as folding and stability. Applications where engineering can also play a role include using knottins to deliver therapeutic payloads to the cells they target, and for development of orally available peptide therapeutics. Computational methods for rational knottin design and engineering and for the study of knottin folding and structure also hold exciting promise.

ACKNOWLEDGMENTS

Work relevant to this chapter was supported by NIH grants 5K01 CA104706, R21 CA143498, and P50 CA114747, the Center for Biomedical Imaging at Stanford (CBIS), the Edward Mallinckrodt Jr. Foundation, the V Foundation for Cancer Research, and the Sidney Kimmel Foundation for Cancer Research. S. J. M. has been supported by an NSF Graduate Research Fellowship, the Medtronic Stanford Graduate Fellowship, and a Siebel Scholars Fellowship.
The authors thank Frank Cochran, Adam Silverman, and Jennifer Lahti for helpful comments and feedback.

REFERENCES

Ackerman, M., Levary, D., Tobon, G., Hackel, B., Orcutt, K. D., and Wittrup, K. D. (2009). Highly avid magnetic bead capture: An efficient selection method for de novo protein engineering utilizing yeast surface display. *Biotechnol. Prog.* **25,** 774–783.

Angell, Y. M., Alsina, J., Albericio, F., and Barany, G. (2002). Practical protocols for stepwise solid-phase synthesis of cysteine-containing peptides. *J. Pept. Res.* **60,** 292–299.

Avrutina, O., Schmoldt, H. U., Gabrijelcic-Geiger, D., Le Nguyen, D., Sommerhoff, C. P., Diederichsen, U., and Kolmar, H. (2005). Trypsin inhibition by macrocyclic and open-chain variants of the squash inhibitor MCoTI-II. *Biol. Chem.* **386,** 1301–1306.

Baggio, R., Burgstaller, P., Hale, S. P., Putney, A. R., Lane, M., Lipovsek, D., Wright, M. C., Roberts, R. W., Liu, R., Szostak, J. W., and Wagner, R. W. (2002). Identification of epitope-like consensus motifs using mRNA display. *J. Mol. Recognit.* **15,** 126–134.

Boder, E. T., and Wittrup, K. D. (1997). Yeast surface display for screening combinatorial polypeptide libraries. *Nat. Biotechnol.* **15,** 553–557.
Boder, E. T., and Wittrup, K. D. (1998). Optimal screening of surface-displayed polypeptide libraries. *Biotechnol. Prog.* **14,** 55–62.
Boder, E. T., and Wittrup, K. D. (2000). Yeast surface display for directed evolution of protein expression, affinity, and stability. *Methods Enzymol.* **328,** 430–444.
Boder, E. T., Midelfort, K. S., and Wittrup, K. D. (2000). Directed evolution of antibody fragments with monovalent femtomolar antigen-binding affinity. *Proc. Natl. Acad. Sci. USA* **97,** 10701–10705.
Bosley, A. D., and Ostermeier, M. (2005). Mathematical expressions useful in the construction, description and evaluation of protein libraries. *Biomol. Eng.* **22,** 57–61.
Bowley, D. R., Labrijn, A. F., Zwick, M. B., and Burton, D. R. (2007). Antigen selection from an HIV-1 immune antibody library displayed on yeast yields many novel antibodies compared to selection from the same library displayed on phage. *Protein Eng. Des. Sel.* **20,** 81–90.
Carugo, O., Lu, S., Luo, J., Gu, X., Liang, S., Strobl, S., and Pongor, S. (2001). Structural analysis of free and enzyme-bound amaranth alpha-amylase inhibitor: Classification within the knottin fold superfamily and analysis of its functional flexibility. *Protein Eng.* **14,** 639–646.
Chan, W. C., and White, P. D. (2000). Fmoc Solid Phase Peptide Synthesis: A Practical Approach. Oxford University Press Inc., New York.
Chao, G., Lau, W. L., Hackel, B. J., Sazinsky, S. L., Lippow, S. M., and Wittrup, K. D. (2006). Isolating and engineering human antibodies using yeast surface display. *Nat. Protoc.* **1,** 755–768.
Cheng, Y., and Prusoff, W. H. (1973). Relationship between the inhibition constant (K_I) and the concentration of inhibitor which causes 50 per cent inhibition (I_{50}) of an enzymatic reaction. *Biochem. Pharmacol.* **22,** 3099–3108.
Christmann, A., Walter, K., Wentzel, A., Kratzner, R., and Kolmar, H. (1999). The cystine knot of a squash-type protease inhibitor as a structural scaffold for *Escherichia coli* cell surface display of conformationally constrained peptides. *Protein Eng.* **12,** 797–806.
Colby, D. W., Kellogg, B. A., Graff, C. P., Yeung, Y. A., Swers, J. S., and Wittrup, K. D. (2004). Engineering antibody affinity by yeast surface display. *Methods Enzymol.* **388,** 348–358.
Craik, D. J., Daly, N. L., Bond, T., and Waine, C. (1999). Plant cyclotides: A unique family of cyclic and knotted proteins that defines the cyclic cystine knot structural motif. *J. Mol. Biol.* **294,** 1327–1336.
Craik, D. J., Clark, R. J., and Daly, N. L. (2007). Potential therapeutic applications of the cyclotides and related cystine knot mini-proteins. *Expert Opin. Investig. Drugs* **16,** 595–604.
Cuthbertson, A., and Indrevoll, B. (2003). Regioselective formation, using orthogonal cysteine protection, of an alpha-conotoxin dimer peptide containing four disulfide bonds. *Org. Lett.* **5,** 2955–2957.
Daly, N. L., and Craik, D. J. (2011). Bioactive cystine knot proteins. *Curr. Opin. Chem. Biol.* **15,** 362–368.
Daly, N. L., Love, S., Alewood, P. F., and Craik, D. J. (1999). Chemical synthesis and folding pathways of large cyclic polypeptides: Studies of the cystine knot polypeptide kalata B1. *Biochemistry* **38,** 10606–10614.
Ellgaard, L., and Helenius, A. (2003). Quality control in the endoplasmic reticulum. *Nat. Rev. Mol. Cell Biol.* **4,** 181–191.
Favel, A., Mattras, H., Coletti-Previero, M. A., Zwilling, R., Robinson, E. A., and Castro, B. (1989). Protease inhibitors from *Ecballium elaterium* seeds. *Int. J. Pept. Protein Res.* **33,** 202–208.

Feldhaus, M. J., Siegel, R. W., Opresko, L. K., Coleman, J. R., Feldhaus, J. M., Yeung, Y. A., Cochran, J. R., Heinzelman, P., Colby, D., Swers, J., Graff, C., Wiley, H. S., *et al.* (2003). Flow-cytometric isolation of human antibodies from a nonimmune Saccharomyces cerevisiae surface display library. *Nat. Biotechnol.* **21,** 163–170.

Gai, S. A., and Wittrup, K. D. (2007). Yeast surface display for protein engineering and characterization. *Curr. Opin. Struct. Biol.* **17,** 467–473.

Gracy, J., Le Nguyen, D., Gelly, J. C., Kaas, Q., Heitz, A., and Chiche, L. (2008). KNOTTIN: The knottin or inhibitor cystine knot scaffold in 2007. *Nucleic Acids Res.* **36,** D314–D319.

Gunasekera, S., Foley, F. M., Clark, R. J., Sando, L., Fabri, L. J., Craik, D. J., and Daly, N. L. (2008). Engineering stabilized vascular endothelial growth factor-A antagonists: Synthesis, structural characterization, and bioactivity of grafted analogues of cyclotides. *J. Med. Chem.* **51,** 7697–7704.

Hackel, B. J., Kapila, A., and Wittrup, K. D. (2008). Picomolar affinity fibronectin domains engineered utilizing loop length diversity, recursive mutagenesis, and loop shuffling. *J. Mol. Biol.* **381,** 1238–1252.

Hanes, J., Jermutus, L., Weber-Bornhauser, S., Bosshard, H. R., and Pluckthun, A. (1998). Ribosome display efficiently selects and evolves high-affinity antibodies in vitro from immune libraries. *Proc. Natl. Acad. Sci. USA* **95,** 14130–14135.

Hilpert, K., Wessner, H., Schneider-Mergener, J., Welfle, K., Misselwitz, R., Welfle, H., Hocke, A. C., Hippenstiel, S., and Hohne, W. (2003). Design and characterization of a hybrid miniprotein that specifically inhibits porcine pancreatic elastase. *J. Biol. Chem.* **278,** 24986–24993.

Jackson, P. J., McNulty, J. C., Yang, Y. K., Thompson, D. A., Chai, B., Gantz, I., Barsh, G. S., and Millhauser, G. L. (2002). Design, pharmacology, and NMR structure of a minimized cystine knot with agouti-related protein activity. *Biochemistry* **41,** 7565–7572.

Jiang, L., Kimura, R. H., Miao, Z., Silverman, A. P., Ren, G., Liu, H., Li, P., Gambhir, S. S., Cochran, J. R., and Cheng, Z. (2010a). Evaluation of a ^{64}Cu-labeled cystine-knot peptide based on agouti-related protein for PET of tumors expressing alphav beta3 integrin. *J. Nucl. Med.* **51,** 251–258.

Jiang, L., Miao, Z., Kimura, R. H., Liu, H., Cochran, J. R., Culter, C. S., Bao, A., Li, P., and Cheng, Z. (2010b). Preliminary evaluation of ^{177}Lu-labeled knottin peptides for integrin receptor-targeted radionuclide therapy. *Eur. J. Nucl. Med. Mol. Imaging* **38,** 613–622.

Kimura, R. H., Cheng, Z., Gambhir, S. S., and Cochran, J. R. (2009a). Engineered knottin peptides: A new class of agents for imaging integrin expression in living subjects. *Cancer Res.* **69,** 2435–2442.

Kimura, R. H., Levin, A. M., Cochran, F. V., and Cochran, J. R. (2009b). Engineered cystine knot peptides that bind alphav beta3, alphav beta5, and alpha5 beta1 integrins with low-nanomolar affinity. *Proteins* **77,** 359–369.

Kimura, R. H., Miao, Z., Cheng, Z., Gambhir, S. S., and Cochran, J. R. (2010). A dual-labeled knottin peptide for PET and near-infrared fluorescence imaging of integrin expression in living subjects. *Bioconjug. Chem.* **21,** 436–444.

Kimura, R. H., Jones, D. S., Jiang, L., Miao, Z., Cheng, Z., and Cochran, J. R. (2011). Functional mutation of multiple solvent-exposed loops in the *Ecballium elaterium* trypsin inhibitor-II cystine knot miniprotein. *PLoS One* **6,** e16112.

Kolmar, H. (2009). Biological diversity and therapeutic potential of natural and engineered cystine knot miniproteins. *Curr. Opin. Pharmacol.* **9,** 608–614.

Kolmar, H. (2010). Engineered cystine-knot miniproteins for diagnostic applications. *Expert Rev. Mol. Diagn.* **10,** 361–368.

Krause, S., Schmoldt, H. U., Wentzel, A., Ballmaier, M., Friedrich, K., and Kolmar, H. (2007). Grafting of thrombopoietin-mimetic peptides into cystine knot miniproteins yields high-affinity thrombopoietin antagonists and agonists. *FEBS J.* **274,** 86–95.

Lahti, J. L., and Cochran, J. R. (2009). Yeast surface display. *In* "Therapeutic Monoclonal Antibodies: From Bench to Clinic," (Z. An and W. Strohl, eds.), pp. 213–237. John Wiley & Sons, Inc., Hoboken, NJ.

Lahti, J. L., Silverman, A. P., and Cochran, J. R. (2009). Interrogating and predicting tolerated sequence diversity in protein folds: Application to *E. elaterium* trypsin inhibitor-II cystine-knot miniprotein. *PLoS Comput. Biol.* **5,** e1000499.

Lazari, M. F., Porto, C. S., Freymuller, E., Abreu, L. C., and Picarelli, Z. P. (1997). Receptor-mediated endocytosis of angiotensin II in rat myometrial cells. *Biochem. Pharmacol.* **54,** 399–408.

Le Nguyen, D., Nalis, D., and Castro, B. (1989). Solid phase synthesis of a trypsin inhibitor isolated from the Cucurbitaceae *Ecballium elaterium. Int. J. Pept. Protein Res.* **34,** 492–497.

Le Nguyen, D., Heitz, A., Chiche, L., Castro, B., Boigegrain, R. A., Favel, A., and Coletti-Previero, M. A. (1990). Molecular recognition between serine proteases and new bioactive microproteins with a knotted structure. *Biochimie* **72,** 431–435.

Lee, C. V., Liang, W. C., Dennis, M. S., Eigenbrot, C., Sidhu, S. S., and Fuh, G. (2004). High-affinity human antibodies from phage-displayed synthetic Fab libraries with a single framework scaffold. *J. Mol. Biol.* **340,** 1073–1093.

Lipovsek, D., Mena, M., Lippow, S. M., Basu, S., and Baynes, B. M. (2009). Library construction for protein engineering. *In* "Protein Engineering and Design," (S. J. Park and J. R. Cochran, eds.), pp. 83–108. Taylor and Francis, Boca Raton.

Lyons, S. A., O'Neal, J., and Sontheimer, H. (2002). Chlorotoxin, a scorpion-derived peptide, specifically binds to gliomas and tumors of neuroectodermal origin. *Glia* **39,** 162–173.

Maillere, B., Mourier, G., Herve, M., Cotton, J., Leroy, S., and Menez, A. (1995). Immunogenicity of a disulphide-containing neurotoxin: Presentation to T-cells requires a reduction step. *Toxicon* **33,** 475–482.

McNulty, J. C., Thompson, D. A., Bolin, K. A., Wilken, J., Barsh, G. S., and Millhauser, G. L. (2001). High-resolution NMR structure of the chemically-synthesized melanocortin receptor binding domain AGRP(87-132) of the agouti-related protein. *Biochemistry* **40,** 15520–15527.

Miao, Z., Ren, G., Liu, H., Kimura, R. H., Jiang, L., Cochran, J. R., Gambhir, S. S., and Cheng, Z. (2009). An engineered knottin peptide labeled with ^{18}F for PET imaging of integrin expression. *Bioconjug. Chem.* **20,** 2342–2347.

Moore, S. J., Olsen, M. J., Cochran, J. R., and Cochran, F. V. (2009). Cell surface display systems for protein engineering. *In* "Protein Engineering and Design," (J. S. Park and J. R. Cochran, eds.), pp. 23–50. Taylor and Francis, Boca Raton.

Nielsen, C. H., Kimura, R. H., Withofs, N., Tran, P. T., Miao, Z., Cochran, J. R., Cheng, Z., Felsher, D., Kjaer, A., Willmann, J. K., and Gambhir, S. S. (2010). PET imaging of tumor neovascularization in a transgenic mouse model with a novel ^{64}Cu-DOTA-knottin peptide. *Cancer Res.* **70,** 9022–9030.

Nishio, H., Kumagaye, K. Y., Kubo, S., Chen, Y. N., Momiyama, A., Takahashi, T., Kimura, T., and Sakakibara, S. (1993). Synthesis of omega-agatoxin IVA and its related peptides. *Biochem. Biophys. Res. Commun.* **196,** 1447–1453.

Ollmann, M. M., Wilson, B. D., Yang, Y. K., Kerns, J. A., Chen, Y., Gantz, I., and Barsh, G. S. (1997). Antagonism of central melanocortin receptors in vitro and in vivo by agouti-related protein. *Science* **278,** 135–138.

Pace, C. N., Vajdos, F., Fee, L., Grimsley, G., and Gray, T. (1995). How to measure and predict the molar absorption coefficient of a protein. *Protein Sci.* **4,** 2411–2423.

Pallaghy, P. K., Nielsen, K. J., Craik, D. J., and Norton, R. S. (1994). A common structural motif incorporating a cystine knot and a triple-stranded beta-sheet in toxic and inhibitory polypeptides. *Protein Sci.* **3,** 1833–1839.

Pepper, L. R., Cho, Y. K., Boder, E. T., and Shusta, E. V. (2008). A decade of yeast surface display technology: Where are we now? *Comb. Chem. High Throughput Screen.* **11,** 127–134.
Reiss, S., Sieber, M., Oberle, V., Wentzel, A., Spangenberg, P., Claus, R., Kolmar, H., and Losche, W. (2006). Inhibition of platelet aggregation by grafting RGD and KGD sequences on the structural scaffold of small disulfide-rich proteins. *Platelets* **17,** 153–157.
Schmoldt, H. U., Wentzel, A., Becker, S., and Kolmar, H. (2005). A fusion protein system for the recombinant production of short disulfide bond rich cystine knot peptides using barnase as a purification handle. *Protein Expr. Purif.* **39,** 82–89.
Siegel, R. W., Coleman, J. R., Miller, K. D., and Feldhaus, M. J. (2004). High efficiency recovery and epitope-specific sorting of an scFv yeast display library. *J. Immunol. Methods* **286,** 141–153.
Silverman, A. P., Levin, A. M., Lahti, J. L., and Cochran, J. R. (2009). Engineered cystine-knot peptides that bind alphav beta3 integrin with antibody-like affinities. *J. Mol. Biol.* **385,** 1064–1075.
Silverman, A. P., Kariolis, M. S., and Cochran, J. R. (2011). Cystine-knot peptides engineered with specificities for alphaIIb beta3 or alphaIIb beta3 and alphav beta3 integrins are potent inhibitors of platelet aggregation. *J. Mol. Recognit.* **24,** 127–135.
Simonsen, S. M., Sando, L., Rosengren, K. J., Wang, C. K., Colgrave, M. L., Daly, N. L., and Craik, D. J. (2008). Alanine scanning mutagenesis of the prototypic cyclotide reveals a cluster of residues essential for bioactivity. *J. Biol. Chem.* **283,** 9805–9813.
Sommerhoff, C. P., Avrutina, O., Schmoldt, H. U., Gabrijelcic-Geiger, D., Diederichsen, U., and Kolmar, H. (2010). Engineered cystine knot miniproteins as potent inhibitors of human mast cell tryptase beta. *J. Mol. Biol.* **395,** 167–175.
Stemmer, W. P. (1994). Rapid evolution of a protein in vitro by DNA shuffling. *Nature* **370,** 389–391.
Swers, J. S., Kellogg, B. A., and Wittrup, K. D. (2004). Shuffled antibody libraries created by in vivo homologous recombination and yeast surface display. *Nucleic Acids Res.* **32,** e36.
VanAntwerp, J. J., and Wittrup, K. D. (2000). Fine affinity discrimination by yeast surface display and flow cytometry. *Biotechnol. Prog.* **16,** 31–37.
Veiseh, M., Gabikian, P., Bahrami, S. B., Veiseh, O., Zhang, M., Hackman, R. C., Ravanpay, A. C., Stroud, M. R., Kusuma, Y., Hansen, S. J., Kwok, D., Munoz, N. M., *et al.* (2007). Tumor paint: A chlorotoxin:Cy5.5 bioconjugate for intraoperative visualization of cancer foci. *Cancer Res.* **67,** 6882–6888.
Wentzel, A., Christmann, A., Kratzner, R., and Kolmar, H. (1999). Sequence requirements of the GPNG beta-turn of the *Ecballium elaterium* trypsin inhibitor II explored by combinatorial library screening. *J. Biol. Chem.* **274,** 21037–21043.
Werle, M., Schmitz, T., Huang, H. L., Wentzel, A., Kolmar, H., and Bernkop-Schnurch, A. (2006). The potential of cystine-knot microproteins as novel pharmacophoric scaffolds in oral peptide drug delivery. *J. Drug Target.* **14,** 137–146.
Williams, J. A., Day, M., and Heavner, J. E. (2008). Ziconotide: An update and review. *Expert Opin. Pharmacother.* **9,** 1575–1583.
Willmann, J. K., Kimura, R. H., Deshpande, N., Lutz, A. M., Cochran, J. R., and Gambhir, S. S. (2010). Targeted contrast-enhanced ultrasound imaging of tumor angiogenesis with contrast microbubbles conjugated to integrin-binding knottin peptides. *J. Nucl. Med.* **51,** 433–440.

SECTION SIX

PHARMACOKINETICS

CHAPTER TEN

Practical Theoretic Guidance for the Design of Tumor-Targeting Agents

K. Dane Wittrup,*,†,‡ Greg M. Thurber,* Michael M. Schmidt,† *and* John J. Rhoden*

Contents

1. Introduction 256
2. What Molecular Size Is Best for Tumor Uptake? 256
3. Will Targeting Increase Nanoparticle Accumulation in a Tumor? 259
4. How Does Affinity Affect Biodistribution? 261
5. What Dose Is Necessary in Order to Overcome the "Binding Site Barrier"? 263
6. Conclusions 264
References 265

Abstract

Theoretical analyses of targeting agent pharmacokinetics provides specific guidance with respect to desirable design objectives such as agent size, affinity, and target antigen. These analyses suggest that IgG-sized macromolecular constructs exhibit the most favorable balance between systemic clearance and vascular extravasation, resulting in maximal tumor uptake. Quantitative predictions of the effects of dose and binding affinity on tumor uptake and penetration are also provided. The single bolus dose required for saturation of xenografted tumors in mice can be predicted from knowledge of antigen expression level and metabolic half-life. The role of high binding affinity in tumor uptake can be summarized as: essential for small peptides, less important for antibodies, and negligible for nanoparticles.

* Department of Chemical Engineering, Massachusetts Institute of Technology, Cambridge, Massachusetts, USA
† Department of Biological Engineering, Massachusetts Institute of Technology, Cambridge, Massachusetts, USA
‡ Koch Institute for Integrative Cancer Research, Massachusetts Institute of Technology, Cambridge, Massachusetts, USA

Methods in Enzymology, Volume 503
ISSN 0076-6879, DOI: 10.1016/B978-0-12-396962-0.00010-0

1. Introduction

A powerful molecular toolbox is available to vary the size and binding affinity of tumor-targeting agents essentially at will. Such agents can vary in size from small peptidic scaffolds just a few nanometers across to nanoparticles hundreds of nanometers in diameter, and directed evolution can be used to engineer extremely tight binding to tumor-specific antigens. This impressive raw capability raises important questions with regard to the design criteria for a tumor-targeting agent. How does size variation affect the delivery of a drug payload to and throughout a tumor? Can ligand targeting alter nanoparticle biodistribution? How does binding affinity affect overall biodistribution and penetration throughout the tumor volume? What dosage is necessary to obtain uniform penetration of an antibody throughout a tumor? Recently completed analyses of the key rate processes and the quantitative balances among them in macromolecular pharmacokinetics lead to several recommendations and quantitative predictions.

2. What Molecular Size Is Best for Tumor Uptake?

We recently utilized a compartmental model to quantitatively analyze the effect of size on tumor uptake, using previously published data correlating macromolecular size with the key transport parameters (vascular permeability coefficient and the half-life for systemic clearance; Schmidt and Wittrup, 2009). The trend in tumor uptake with increasing size is shown in Fig. 10.1.

A local optimum in tumor uptake is predicted for targeting agents approximately the size of an IgG immunoglobulin. (It should be noted that this prediction accounts only for the effect of passive clearance (e.g., renal); in fact, FcRn-mediated lifetime extension substantially further improves tumor uptake for antibodies.) Mathematically, an optimum such as this results from the balance between two opposing trends. The crux of the behavior is a trade-off between systemic clearance and extravasation. It has been shown experimentally that the vascular extravasation rate drops precipitously with increasing macromolecular size (Dreher *et al.*, 2006)—however, this unfavorable effect is partially compensated for by the benefit of extended systemic circulation due to decreased renal filtration. It is remarkable to note that of the wide range of targeting agent sizes created by genetic and materials engineers, spanning two orders of magnitude, the best size for tumor uptake is in the range that natural selection already converged to for the primary targeting agents of the humoral immune system.

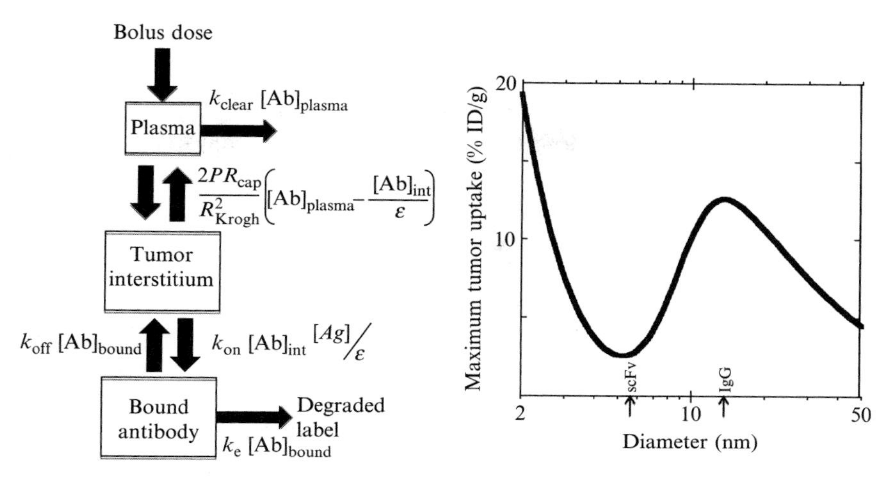

Figure 10.1 The relationship between tumor uptake and size of the molecular targeting agent (Schmidt and Wittrup, 2009). On the left, a schematic diagram of a compartmental model for targeting biodistribution is shown. On the right, the predicted maximum tumor uptake is plotted as a function of the size of the targeting agent, with scFv and IgG sizes indicated with arrows. The parameters used were appropriate for HER2 binding molecules with K_d=1 nM and labeled with ^{99m}Tc, and the size effects on the plasma clearance rate constant k_{clear}, tumor vascular permeability P, and tumor void fraction ε were correlated from published experimental measurements of these parameters.

Antibody fragments such as scFvs reside in an apparent "death valley" for uptake versus size. This prediction dovetails with the widespread experimental observation that scFvs and Fabs deliver relatively anemic tumor localization by comparison to whole antibodies. It has often been wishfully proposed that the several fold smaller size of antibody fragments might provide advantageous extravasation and intratumoral penetration characteristics. However, the rapid renal clearance of these small proteins collapses the circulating concentration too rapidly to reap the benefits of their favorable transport parameters (Thurber *et al.*, 2007). Consequently, scFvs, Fabs, diabodies, and the like essentially inhabit the worst of both worlds: too large for sufficiently rapid extravasation and too small to escape renal clearance.

Interestingly, agents with hydrodynamic diameters <5 nm are predicted to exhibit increased improvements in tumor uptake with further reductions in molecular size. This is because in this size range, renal clearance is essentially first-pass and so there is little incremental cost to further shrinkage in size. However, the extravasation rate rises rapidly with decreasing size, and so even if the tumor-targeting agent only circulates once through the bloodstream before being renally cleared, more agent is taken up into the tumor as size decreases. Experimental evidence for the benefits of decreasing size in this range have been provided for tumor-targeted scaffold proteins of the affibody (Orlova *et al.*, 2006) and DARPin (Zahnd *et al.*, 2010) types. However, binding scaffolds in this size range are occasionally pushed into the trough of the curve in Fig. 10.1 by fusion to additional binding domains, thereby obviating the key advantage of their small size.

Plückthun and coworkers' elegant and systematic study of the effect of size on DARPin uptake into tumors directly confirms the minimum in tumor uptake with respect to size shown in Fig. 10.1. A picomolar affinity anti-HER2 DARPin accumulated in xenografted mouse tumors to 8.1% ID/g at 24h. However, a larger heterodimer of this DARPin with a nonbinding DARPin accumulated to only 1.8% ID/g. Further increasing the size by addition of 20kDa PEG to a DARPin monomer raises the 24-h accumulation to 13% ID/g. Each of these three constructs had a single identical binding domain attached to varying amounts of additional macromolecular mass, allowing the effect of size alone to be determined. Tumor accumulation was therefore experimentally demonstrated to go through a minimum with respect to size, as predicted by the compartmental model (Fig. 10.1).

The overall recommendation arising from this analysis is that a tumor-targeting agent the size of an antibody, or slightly larger, should accumulate within tumors to the greatest extent. It is noteworthy that tumor uptake is predicted to continually decrease as a consequence of decreased extravasation as size increases above 20nm in radius—a significant dilemma for nanomedicine.

As an approximate prediction of the effects of size on tumor uptake, a relationship can be derived from the compartmental model of Schmidt and Wittrup (2009), for the high-affinity limit ($K_d \rightarrow 0$). This relationship is as follows:

$$[\mathrm{Ab}]_{\mathrm{tumor}} = [\mathrm{Ab}]_{\mathrm{plasma},t=0} \frac{2PR_{\mathrm{cap}}}{R_{\mathrm{Krogh}}^2} \frac{\left(\mathrm{e}^{-k_{\mathrm{clear}}t} - \mathrm{e}^{-k_{\mathrm{e}}t}\right)}{k_{\mathrm{e}} - k_{\mathrm{clear}}}$$

where $[\mathrm{Ab}]_{\mathrm{tumor}}$ is the antibody concentration in the tumor, $[\mathrm{Ab}]_{\mathrm{plasma},t=0}$ is the initial peak antibody concentration in the plasma, $\frac{2PR_{\mathrm{cap}}}{R_{\mathrm{Krogh}}^2}$ is the permeability coefficient times the vascular surface to tumor volume ratio (R_{cap} is the capillary radius; R_{Krogh} is the radius of the cylinder of tissue supplied by the capillary), k_{clear} is the effective first-order systemic clearance rate constant, and k_{e} is the rate constant for endocytic turnover of the tumor surface-bound antibody. To estimate the predicted effect of size variation on systemic clearance in the mouse, the following sigmoidal curve fit to published pharmacokinetic data can be used (Schmidt and Wittrup, 2009):

$$ln(k_{\mathrm{clear}}) = -3.3 + \frac{4.9}{1 + \mathrm{e}^{\left(\frac{ln(R)-1.4}{0.25}\right)}}$$

where k_{clear} is the clearance rate constant in units of h^{-1}, and R is the hydrodynamic radius of the targeting agent, in units of nm.

In the absence of a specific measurement of the metabolic half-life for surface-bound antibody, a reasonable approximation (Mattes *et al.*, 1994; Schmidt *et al.*, 2008) would be to use a constitutive half-life of 12h, providing $k_e=0.06\,h^{-1}$.

An estimate of the dependence of the vascular permeability coefficient on size can be obtained from the following curve fit to published data, assuming a capillary radius of 8 μm and $R_{Krogh}=75\,\mu m$ (Dreher *et al.*, 2006; Yuan *et al.*, 1994, 1995):

$$ln\left(\frac{2PR_{cap}}{R_{Krogh}^2}\right) = 1.1 - 4.3\,ln(R) + 1.3(ln(R))^2 - 0.16(ln(R))^3$$

in the range $1\,nm<R<100\,nm$, with $2PR_{cap}/R_{Krogh}^2$ in units of h^{-1}.

These relationships can be used to obtain a rough approximation of the expected tumor uptake time trajectory of a given targeting agent as a function of its size.

3. Will Targeting Increase Nanoparticle Accumulation in a Tumor?

It is a common strategy to add targeting ligands to the surface of nanoparticles, with the expressed intention of increasing tumor uptake. It is an unfortunate fact however that this approach does not succeed in that objective for agents 100 nm in diameter or larger. This is because the loss of unbound nanoparticles from the tumor by intravasation is slower than constitutive internalization and consumption of nanoparticles within the tumor interstitium (Schmidt and Wittrup, 2009). There are nevertheless benefits to attaching ligands to nanoparticles, as tumor cell endocytic uptake effectively delivers nanoparticle payloads.

The impact (or lack thereof) of molecular targeting on biodistribution is only discernable when negative, nontargeted controls are included in studies. A survey of such biodistribution studies of targeted and untargeted nanoparticles is shown in Fig. 10.2. There is, in general, an insignificant difference in tumor uptake between targeted and untargeted nanoparticles, indicating that accumulation is predominantly via the passive enhanced permeability and retention (EPR) effect (Bartlett *et al.*, 2007; Gabizon *et al.*, 2003; Gu *et al.*, 2008; Hussain *et al.*, 2007; Kirpotin *et al.*, 2006; Lopes de Menezes *et al.*, 1998; Mamot *et al.*, 2005). One might suspect that transport limitations might conceivably differ in qualitative respects in actual human tumors by

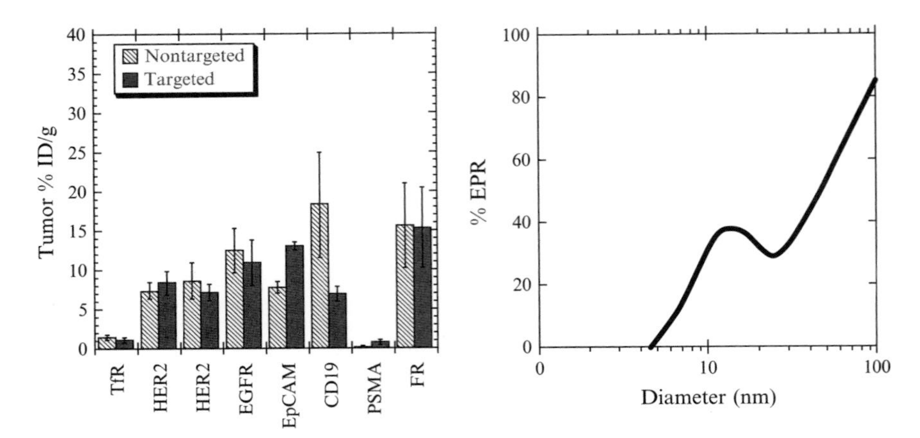

Figure 10.2 Biodistribution of targeted and nontargeted nanoparticles for the antigens transferrin receptor (TfR; Bartlett *et al.*, 2007), HER2 (Kirpotin *et al.*, 2006; Lub-de Hooge *et al.*, 2004), EGFR (Mamot *et al.*, 2005; Ping Li *et al.*, 2008), EpCAM (Goldrosen *et al.*, 1990; Hussain *et al.*, 2007), CD19 (Lopes de Menezes *et al.*, 1998), PSMA (Gu *et al.*, 2008; Smith-Jones *et al.*, 2003), and the folate receptor (FR; Coliva *et al.*, 2005; Gabizon *et al.*, 2003). Data are for 24-h postinjection in all cases, and all of the nanoparticles were ~100nm in diameter. Negative controls for the nanoparticle studies are generally irrelevant ligands. In the panel on the right, a compartmental model is used to predict what proportion of observed tumor uptake at 24h could be attained via the enhanced permeability and retention (EPR) effect without the use of ligand targeting (Schmidt and Wittrup, 2009). (For color version of this figure, the reader is referred to the Web version of this chapter.)

comparison to these model xenografts. However, the presence of passive targeting has been directly demonstrated in humans, as untargeted stealth liposomes accumulate at significant levels in tumors (Harrington *et al.*, 2001).

Nanoparticles that are sufficiently small (<50nm diameter) are predicted to distribute in a fashion more similar to proteins and can therefore exhibit targeting-mediated tumor accumulation (Fig. 10.2, right). This predicted size dependence is consistent with a number of published experimental observations. Dendrimers, which are generally <5nm in diameter, have been shown to accumulate in tumors to a greater extent when conjugated to a ligand (Kukowska-Latallo *et al.*, 2005). Similarly, iron oxide nanoparticles under 20nm in diameter accumulate in tumors to a greater extent when conjugated to a specific antibody (DeNardo *et al.*, 2005). Targeting peptide-conjugated cross-linked iron oxide particles under 40nm in diameter also exhibits improved tumor accumulation (Kelly *et al.*, 2008).

Despite the fact that targeting does not significantly increase tumor accumulation of nanoparticles above 50nm in diameter, targeting can enhance therapeutic efficacy. These advantages are discernable when negative, nontargeted controls are included in studies. Targeting has been shown to increase tumor cell internalization of nanoparticles, whereas untargeted nanoparticles are consumed by reticuloendothelial cells such as tumor-associated macrophages (Kirpotin *et al.*, 2006). Targeting to rapidly internalized antigens provides greater antitumor efficacy than targeting to a slowly internalized antigens (Sapra and Allen, 2002). Thus targeting-driven endocytosis within the tumor improves efficacy (Bartlett *et al.*, 2007; Kirpotin *et al.*, 2006), but targeting has a negligible effect at the whole organism biodistribution level where partitioning amongst clearance organs and the tumor occurs.

4. How Does Affinity Affect Biodistribution?

There is not a single answer to this question—the effect depends dramatically on the size of the targeting agent. As shown in Fig. 10.3, a compartmental model predicts that smaller targeting agents require higher affinity in order to be retained in the tumor, while the retention of larger agents does not depend on affinity to as great an extent. As emphasized in Section 1, intermediate-sized agents from ~10 to 100kDa in size reside in a "death valley" where tumor uptake is weak. Smaller peptide-sized targeting agents depend strongly on binding affinity to accumulate within the tumor, with subnanomolar affinity essentially required to obtain significant tumor retention.

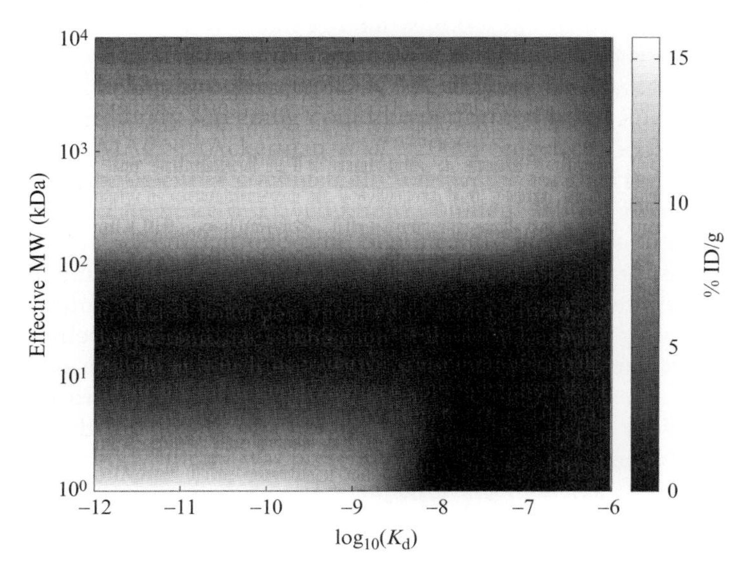

Figure 10.3 Topographical plot of the effect of size and binding affinity on tumor uptake 24h following a bolus dose of targeting agent (Schmidt and Wittrup, 2009).

The general features of this relationship were experimentally confirmed by Zahnd *et al.* (2010), with different size and affinity Darpins against HER2. Darpins, at 14.5 kDa molecular weight, are expected to lie on the steep gradient of improving tumor uptake versus affinity at the bottom of Fig. 10.3. A series of anti-HER2 Darpins were engineered with increasing affinity (Zahnd *et al.*, 2010): G3-HAVD (K_d=270 n*M*), G3-AVD (K_d=10 n*M*), G3-D (K_d=1.5 n*M*), and G3 (K_d=0.091 n*M*). The tumor uptake at 24 h of technetium-labeled Darpins were: 0.57% ID/g (G3-HAVD), 2.4% ID/g (G3-AVD), 3.7% ID/g (G3-D), and 8.1% ID/g (G3). Clearly, for this series of small binding scaffolds, improving affinity directly improves tumor uptake, consistent with the steep gradient of increasing uptake versus affinity at the bottom of Fig. 10.3. Fascinatingly, the story changes when these same binders are conjugated to PEG20, giving them an apparent molecular weight >300 kDa by size exclusion chromatography. This size places the PEGylated binders at the top of the plateau in uptake versus K_d and at sizes from 10^2 to 10^3 kDa. G3-HAVD-PEG20 accumulates to 3.0% ID/g, G3-AVD-PEG20 to 9.2% ID/g, G3-D-PEG20 to 8.6% ID/g, and G3-PEG20 to 13% ID/g. Fully consistent with the model prediction, for these larger targeting agents, the dependence of tumor uptake on binding affinity is weak once K_d<100 n*M*.

The affinity–uptake relationship predicted in Fig. 10.3 was further validated by a series of anti-HER2 antibodies of varying affinity (Rudnick *et al.*, 2011). These antibodies differed only by point mutations, bound at the same HER2 epitope, and had very similar plasma clearance kinetics. The lowest affinity antibody used in this study (K_d=270 n*M*) is predicted to be on the edge of the plateau in Fig. 10.3, while the three higher affinity antibodies (K_d=23, 7.3, and 0.56 n*M*) are predicted to lie squarely on the plateau and therefore have similar accumulation. When biodistribution studies were performed with all four antibodies using nonresidualizing radioisotopes, tumor uptake of the 270-n*M* affinity antibody was significantly lower than the uptake of all the higher affinity antibodies. By contrast, the accumulation of the higher affinity antibodies did not differ significantly from each other.

The relationship in Fig. 10.3 only captures the bulk uptake as % ID/g, and does not provide any information on the pharmacodynamic effect of the targeting agent, or the microdistribution within the tumor (considered in the following section). For antitumor agents designed to block receptor/ligand or receptor/receptor interactions, clearly affinity will be a dominant variable in achieving efficacious tumor control. Further, when antibody-directed cellular cytotoxicity (ADCC) is the objective, it has been shown that higher affinity antibodies lead to more effective NK-mediated cell killing (Tang *et al.*, 2007). Consequently, the plateau in Fig. 10.3 that extends above 10 n*M* K_d for agents the size of antibodies should not be interpreted as the absence of potential therapeutic benefit for higher, p*M* affinity antibodies—it is simply a prediction that for K_d<10 n*M*, bulk antibody uptake in tumors will not be a strong function of affinity.

5. What Dose Is Necessary in Order to Overcome the "Binding Site Barrier"?

It has been known since 1989 that high-affinity antibodies may accumulate around the vasculature and fail to distribute evenly throughout tumors, a phenomenon known by the term "binding site barrier" (Fujimori *et al.*, 1989). The word "barrier" implies a rigidity and permanence that fails to appropriately capture the phenomenon, which is actually a dynamic moving front balancing diffusion, binding, and endocytic consumption (Graff and Wittrup, 2003; Thurber and Wittrup, 2008). The critical role for tumor metabolism of antibodies in determining penetration has recently been confirmed experimentally in xenografted tumor models (Rudnick *et al.*, 2011). In this study, it was concluded that "high-density, rapidly internalizing antigens subject high-affinity antibodies to greater internalization and degradation, thereby limiting their penetration of tumors." In essence, the depth an antibody penetrates into tumor tissue is a dynamic balance between degradation and diffusion (Thurber *et al.*, 2008a,b), and since diffusion is driven by a concentration gradient, one can in principle "dose through" the internalization-driven limitation of penetration. In fact, the Weinstein group that coined the "barrier" nomenclature also demonstrated its dynamic and flexible nature by overcoming poor microdistribution by raising antibody doses (Blumenthal *et al.*, 1991; Saga *et al.*, 1995).

Quantitative comparison of key timescales can be gainfully employed to determine which rate processes dominate observed system behaviors. This type of scaling analysis is widely applied in engineering—for example, the Reynolds number $Re=Dv\rho/\mu$ provides a dimensionless ratio of inertial and viscous forces in a fluid flow (D is the pipe diameter, v the fluid velocity, ρ the density, and μ the viscosity). The units in the numerator and denominator cancel out, resulting in a fundamental dimensionless ratio. When $Re>2100$, the flow is turbulent; when $Re<2100$, the flow is laminar. We have applied this flavor of dimensional analysis to tumor-targeting agents and found that there are two such dimensionless numbers of particular value in understanding tumor penetration (Thurber *et al.*, 2007): a clearance modulus and the Thiele modulus.

The clearance modulus Γ compares systemic clearance time to diffusion time within the tumor. For rapidly clearing agents smaller than the renal filtration threshold, it is possible for systemic clearance to remove the agent from circulation before the moving diffusion front within the tumor has fully penetrated the tumor. The clearance modulus is defined as follows:

$$\Gamma = \frac{R_{\mathrm{Krogh}}^2[\mathrm{Ag}]}{2PR_{\mathrm{cap}}\mathrm{AUC}_{\mathrm{plasma}}}$$

where AUC_{plasma} is the area under curve in plasma for the agent, that is, the integral of plasma concentration versus time. When $\Gamma > 1$, the agent clears from plasma before substantial penetration can occur. Consequently, $\Gamma < 1$ is a necessary (but not sufficient) criterion to achieve penetration of the agent throughout the tumor volume. scFv-sized agents very often fail this criterion due to their rapid clearance rate. In contrast, IgGs are typically not limited by clearance due to their relatively long serum half-life.

For IgGs and other agents that are not limited by systemic clearance, saturation can be predicted from the balance between extravasation rate and endocytic consumption. The dimensionless ratio of these two rates is the Thiele modulus, named after Ernest Thiele who first derived it in order to understand why industrial catalysts varied in their activity with size (Thiele, 1939). Substituting endocytic metabolism for surface catalytic rate, and vascular extravasation rate for diffusion through ceramic pores, one obtains a Thiele modulus describing tumor penetration (Thurber *et al.*, 2007, 2008a,b):

$$\phi^2 = \frac{k_e R_{\text{Krogh}}^2 [\text{Ag}]/\varepsilon}{D\left(\frac{2PR_{\text{cap}}[\text{Ab}]_{\text{plasma}}}{D\varepsilon} + K_d\right)}$$

where k_e is the net endocytosis rate constant for degradation of bound antibody. For typical parameter values, this relationship predicts that for affinities of $K_d > 1\,nM$, poor spatial distribution is unlikely to occur. For p*M* affinities, one can predict the required dosage in a mouse xenograft experiment by assuming constitutive membrane turnover rates, as in Fig. 10.4. For example, for a tumor cell line expressing 10^5 antigens/cell, a bolus of at least 10–70 μg of IgG should be sufficient to permeate throughout a tumor. However, for a more highly expressed antigen at 10^6/cell, one would require at least 100–700 μg doses for saturation. Since the Thiele modulus is inversely proportional to the surface half-life of bound antibody, one could expect to encounter significant difficulty with permeation of antibody–drug conjugates targeted to rapidly internalized antigens with half-lives under an hour.

6. Conclusions

In this chapter, we have provided simple theoretic relationships that provide predictions of tumor uptake and penetration as a function of targeting agent size and binding affinity. The particular value of these relationships lies not in their quantitative accuracy, but rather in predicting trends in expected outcomes as these key design variables are adjusted.

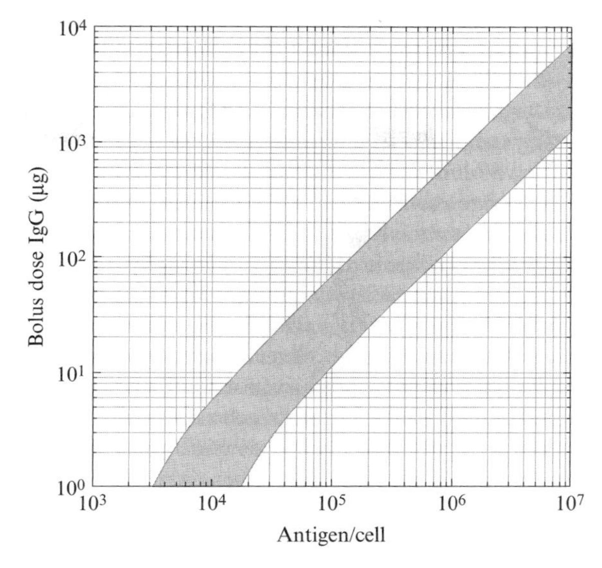

Figure 10.4 Predicted bolus dose required to achieve tumor saturation in a mouse. The gray band represents the combination of doses and expression levels for which $\phi^2=1$, the threshold at which degradative consumption equals extravasation, when the degradation half-life for antibody/antigen complexes is in the range 10–54 h, typical for constitutive turnover. The high-affinity limit for ($K_d=10\,\text{p}M$) is represented. Other parameters are (Schmidt and Wittrup, 2009): $\varepsilon=0.24$, $D=2.5\times10^{-7}\text{cm}^2/\text{s}$, $P=3.9\times10^{-7}\ \text{cm}^2/\text{s}$, $R_{cap}=8\mu\text{m}$, $R_{Krogh}=60\mu\text{m}$, mouse blood volume=2 ml, and tumor density=3×10^8 cells/ml (Schmidt *et al.*, 2008).

For example, greatest tumor accumulation is expected for proteins ~200 kDa in size—favoring IgG-like constructs. Nanoparticles 100 nm in diameter are not predicted (and have not been observed) to accumulate to a greater extent in tumors when targeting ligands are attached to their surfaces. The dynamically poor microdistribution termed the "binding site barrier" can be overcome by predictable increases in bolus dosing. These relationships may assume particular importance in the design of antibody–drug conjugates, which are often selected for rapid endocytosis, and must be dosed at lower levels due to toxicity—conditions particularly conducive to poor microdistribution.

REFERENCES

Bartlett, D. W., Su, H., Hildebrandt, I. J., Weber, W. A., and Davis, M. E. (2007). Impact of tumor-specific targeting on the biodistribution and efficacy of siRNA nanoparticles measured by multimodality in vivo imaging. *Proc. Natl. Acad. Sci. USA* **104**(39), 15549–15554.

Blumenthal, R. D., Fand, I., Sharkey, R. M., Boerman, O. C., Kashi, R., and Goldenberg, D. M. (1991). The effect of antibody protein dose on the uniformity of tumor distribution of radioantibodies: An autoradiographic study. *Cancer Immunol. Immunother.* **33**(6), 351–358.

Coliva, A., Zacchetti, A., Luison, E., Tomassetti, A., Bongarzone, I., Seregni, E., Bombardieri, E., Martin, F., Giussani, A., Figini, M., and Canevari, S. (2005). 90Y Labeling of monoclonal antibody MOv18 and preclinical validation for radioimmunotherapy of human ovarian carcinomas. *Cancer Immunol. Immunother.* **54**(12), 1200–1213.

DeNardo, S. J., DeNardo, G. L., Miers, L. A., Natarajan, A., Foreman, A. R., Gruettner, C., Adamson, G. N., and Ivkov, R. (2005). Development of tumor targeting bioprobes ((111)In-chimeric L6 monoclonal antibody nanoparticles) for alternating magnetic field cancer therapy. *Clin. Cancer Res.* **11**(19 Pt 2), 7087s–7092s.

Dreher, M. R., Liu, W., Michelich, C. R., Dewhirst, M. W., Yuan, F., and Chilkoti, A. (2006). Tumor vascular permeability, accumulation, and penetration of macromolecular drug carriers. *J. Natl. Cancer Inst.* **98**(5), 335–344.

Fujimori, K., Covell, D. G., Fletcher, J. E., and Weinstein, J. N. (1989). Modeling analysis of the global and microscopic distribution of immunoglobulin G, F(ab')2, and Fab in tumors. *Cancer Res.* **49**(20), 5656–5663.

Gabizon, A., Horowitz, A. T., Goren, D., Tzemach, D., Shmeeda, H., and Zalipsky, S. (2003). In vivo fate of folate-targeted polyethylene-glycol liposomes in tumor-bearing mice. *Clin. Cancer Res.* **9**(17), 6551–6559.

Goldrosen, M. H., Biddle, W. C., Pancook, J., Bakshi, S., Vanderheyden, J. L., Fritzberg, A. R., Morgan, A. C., Jr., and Foon, K. A. (1990). Biodistribution, pharmacokinetic, and imaging studies with 186Re-labeled NR-LU-10 whole antibody in LS174T colonic tumor-bearing mice. *Cancer Res.* **50**(24), 7973–7978.

Graff, C. P., and Wittrup, K. D. (2003). Theoretical analysis of antibody targeting of tumor spheroids: Importance of dosage for penetration, and affinity for retention. *Cancer Res.* **63**(6), 1288–1296.

Gu, F., Zhang, L., Teply, B. A., Mann, N., Wang, A., Radovic-Moreno, A. F., Langer, R., and Farokhzad, O. C. (2008). Precise engineering of targeted nanoparticles by using self-assembled biointegrated block copolymers. *Proc. Natl. Acad. Sci. USA* **105**(7), 2586–2591.

Harrington, K. J., Mohammadtaghi, S., Uster, P. S., Glass, D., Peters, A. M., Vile, R. G., and Stewart, J. S. (2001). Effective targeting of solid tumors in patients with locally advanced cancers by radiolabeled pegylated liposomes. *Clin. Cancer Res.* **7**(2), 243–254.

Hussain, S., Pluckthun, A., Allen, T. M., and Zangemeister-Wittke, U. (2007). Antitumor activity of an epithelial cell adhesion molecule targeted nanovesicular drug delivery system. *Mol. Cancer Ther.* **6**(11), 3019–3027.

Kelly, K. A., Setlur, S. R., Ross, R., Anbazhagan, R., Waterman, P., Rubin, M. A., and Weissleder, R. (2008). Detection of early prostate cancer using a hepsin-targeted imaging agent. *Cancer Res.* **68**(7), 2286–2291.

Kirpotin, D. B., Drummond, D. C., Shao, Y., Shalaby, M. R., Hong, K., Nielsen, U. B., Marks, J. D., Benz, C. C., and Park, J. W. (2006). Antibody targeting of long-circulating lipidic nanoparticles does not increase tumor localization but does increase internalization in animal models. *Cancer Res.* **66**(13), 6732–6740.

Kukowska-Latallo, J. F., Candido, K. A., Cao, Z., Nigavekar, S. S., Majoros, I. J., Thomas, T. P., Balogh, L. P., Khan, M. K., and Baker, J. R., Jr. (2005). Nanoparticle targeting of anticancer drug improves therapeutic response in animal model of human epithelial cancer. *Cancer Res.* **65**(12), 5317–5324.

Lopes de Menezes, D. E., Pilarski, L. M., and Allen, T. M. (1998). In vitro and in vivo targeting of immunoliposomal doxorubicin to human B-cell lymphoma. *Cancer Res.* **58** (15), 3320–3330.

Lub-de Hooge, M. N., Kosterink, J. G., Perik, P. J., Nijnuis, H., Tran, L., Bart, J., Suurmeijer, A. J., de Jong, S., Jager, P. L., and de Vries, E. G. (2004). Preclinical characterisation of 111In-DTPA-trastuzumab. *Br. J. Pharmacol.* **143**(1), 99–106.

Mamot, C., Drummond, D. C., Noble, C. O., Kallab, V., Guo, Z., Hong, K., Kirpotin, D. B., and Park, J. W. (2005). Epidermal growth factor receptor-targeted immunoliposomes significantly enhance the efficacy of multiple anticancer drugs in vivo. *Cancer Res.* **65**(24), 11631–11638.

Mattes, M. J., Griffiths, G. L., Diril, H., Goldenberg, D. M., Ong, G. L., and Shih, L. B. (1994). Processing of antibody-radioisotope conjugates after binding to the surface of tumor cells. *Cancer* **73**(Suppl. 3), 787–793.

Orlova, A., Magnusson, M., Eriksson, T. L., Nilsson, M., Larsson, B., Hoiden-Guthenberg, I., Widstrom, C., Carlsson, J., Tolmachev, V., Stahl, S., and Nilsson, F. Y. (2006). Tumor imaging using a picomolar affinity HER2 binding affibody molecule. *Cancer Res.* **66**(8), 4339–4348.

Ping Li, W., Meyer, L. A., Capretto, D. A., Sherman, C. D., and Anderson, C. J. (2008). Receptor-binding, biodistribution, and metabolism studies of 64Cu-DOTA-cetuximab, a PET-imaging agent for epidermal growth-factor receptor-positive tumors. *Cancer Biother. Radiopharm.* **23**(2), 158–171.

Rudnick, S. I., Lou, J., Shaller, C. C., Tang, Y., Klein-Szanto, A. J., Weiner, L. M., Marks, J. D., and Adams, G. P. (2011). Influence of affinity and antigen internalization on the uptake and penetration of anti-HER2 antibodies in solid tumors. *Cancer Res.* **71**(6), 2250–2259.

Saga, T., Neumann, R. D., Heya, T., Sato, J., Kinuya, S., Le, N., Paik, C. H., and Weinstein, J. N. (1995). Targeting cancer micrometastases with monoclonal antibodies: A binding-site barrier. *Proc. Natl. Acad. Sci. USA* **92**(19), 8999–9003.

Sapra, P., and Allen, T. M. (2002). Internalizing antibodies are necessary for improved therapeutic efficacy of antibody-targeted liposomal drugs. *Cancer Res.* **62**(24), 7190–7194.

Schmidt, M. M., and Wittrup, K. D. (2009). A modeling analysis of the effects of molecular size and binding affinity on tumor targeting. *Mol. Cancer Ther.* **8**(10), 2861–2871.

Schmidt, M. M., Thurber, G. M., and Wittrup, K. D. (2008). Kinetics of anti-carcinoembryonic antigen antibody internalization: Effects of affinity, bivalency, and stability. *Cancer Immunol. Immunother.* **57**(12), 1879–1890.

Smith-Jones, P. M., Vallabhajosula, S., Navarro, V., Bastidas, D., Goldsmith, S. J., and Bander, N. H. (2003). Radiolabeled monoclonal antibodies specific to the extracellular domain of prostate-specific membrane antigen: Preclinical studies in nude mice bearing LNCaP human prostate tumor. *J. Nucl. Med.* **44**(4), 610–617.

Tang, Y., Lou, J., Alpaugh, R. K., Robinson, M. K., Marks, J. D., and Weiner, L. M. (2007). Regulation of antibody-dependent cellular cytotoxicity by IgG intrinsic and apparent affinity for target antigen. *J. Immunol.* **179**(5), 2815–2823.

Thiele, E. W. (1939). Relation between catalytic activity and size of particle. *Ind. Eng. Chem.* **31**(7), 916–920.

Thurber, G. M., and Wittrup, K. D. (2008). Quantitative spatiotemporal analysis of antibody fragment diffusion and endocytic consumption in tumor spheroids. *Cancer Res.* **68**(9), 3334–3341.

Thurber, G. M., Zajic, S. C., and Wittrup, K. D. (2007). Theoretic criteria for antibody penetration into solid tumors and micrometastases. *J. Nucl. Med.* **48**(6), 995–999.

Thurber, G. M., Schmidt, M. M., and Wittrup, K. D. (2008a). Antibody tumor penetration: Transport opposed by systemic and antigen-mediated clearance. *Adv. Drug Deliv. Rev.* **60**(12), 1421–1434.

Thurber, G. M., Schmidt, M. M., and Wittrup, K. D. (2008b). Factors determining antibody distribution in tumors. *Trends Pharmacol. Sci.* **29**(2), 57–61.

Yuan, F., Leunig, M., Huang, S. K., Berk, D. A., Papahadjopoulos, D., and Jain, R. K. (1994). Microvascular permeability and interstitial penetration of sterically stabilized (stealth) liposomes in a human tumor xenograft. *Cancer Res.* **54**(13), 3352–3356.

Yuan, F., Dellian, M., Fukumura, D., Leunig, M., Berk, D. A., Torchilin, V. P., and Jain, R. K. (1995). Vascular permeability in a human tumor xenograft: Molecular size dependence and cutoff size. *Cancer Res.* **55**(17), 3752–3756.

Zahnd, C., Kawe, M., Stumpp, M. T., de Pasquale, C., Tamaskovic, R., Nagy-Davidescu, G., Dreier, B., Schibli, R., Binz, H. K., Waibel, R., and Pluckthun, A. (2010). Efficient tumor targeting with high-affinity designed ankyrin repeat proteins: Effects of affinity and molecular size. *Cancer Res.* **70**(4), 1595–1605.

CHAPTER ELEVEN

Reengineering Biopharmaceuticals for Targeted Delivery Across the Blood–Brain Barrier

William M. Pardridge* *and* Ruben J. Boado*,†

Contents

1. Introduction 270
2. Blood–Brain Barrier Receptor-Mediated Transport and Molecular Trojan Horses 271
3. Reengineering Recombinant Proteins for Targeted Brain Delivery 274
4. Genetic Engineering of Expression Plasmid DNA Encoding IgG Fusion Proteins 278
5. Pharmacokinetics and Brain Uptake of IgG Fusion Proteins 279
   5.1. Pharmacokinetics 279
   5.2. Brain uptake 281
   5.3. Brain targeting 283
6. CNS Pharmacological Effects of IgG Fusion Proteins 284
7. Immune Response Against IgG Fusion Proteins 287
8. Summary 288
References 289

Abstract

Recombinant protein therapeutics cannot enter brain drug development because these large molecule drugs do not cross the blood–brain barrier (BBB). However, recombinant proteins can be reengineered as BBB-penetrating IgG fusion proteins, where the IgG part is a genetically engineered monoclonal antibody (MAb) against an endogenous BBB receptor, such as the human insulin receptor (HIR) or the transferrin receptor (TfR). The IgG binds the endogenous insulin receptor or TfR to trigger transport across the BBB and acts as a molecular Trojan horse (MTH) to ferry into brain the fused protein therapeutic. The most potent MTH to date is a MAb against the HIR, designated the HIRMAb, which is active in humans and Old World primates, such as the Rhesus monkey. There is no known MAb against the mouse insulin

* Department of Medicine, UCLA, Los Angeles, California, USA
† ArmaGen Technologies, Inc., Santa Monica, California, USA

Methods in Enzymology, Volume 503
ISSN 0076-6879, DOI: 10.1016/B978-0-12-396962-0.00011-2

receptor. For drug delivery in the mouse, protein therapeutics are fused to a chimeric MAb against the mouse TfR, designated the cTfRMAb. The HIRMAb or cTfRMAb Trojan horses have been engineered and expressed as fusion proteins with multiple classes of protein therapeutics, including lysosomal enzymes, neurotrophins, decoy receptors, single chain Fv therapeutic antibodies, and avidin. The pharmacokinetic (PK) properties of the IgG fusion proteins differ from that of typical MAb drugs and resemble the PK profiles of small molecules due to rapid uptake by peripheral tissues, as well as brain. The brain uptake of the IgG fusion proteins, 2–3% of injected dose/brain, is comparable to the brain uptake of small molecules. The IgG fusion proteins have been administered chronically in mouse models, and the immune response is low titer and has no effect on the fusion protein clearance from blood or brain uptake *in vivo*. The BBB MTH technology enables the reengineering of a wide spectrum of recombinant protein therapeutics for targeted drug delivery to the brain.

1. Introduction

Treatment of the brain with the products of biotechnology, for example, recombinant proteins and monoclonal antibodies (MAbs), or even short interfering RNA (siRNA), is not possible without fundamental solutions to the problem of blood–brain barrier (BBB) delivery. All biotechnological pharmaceutics are large molecules that do not cross the BBB. In the absence of an effective BBB drug delivery technology, the brain drug developer is left with the traditional, yet ineffective, brain drug delivery strategies, including transcranial drug delivery to the brain, BBB disruption, or small molecules. Transcranial drug delivery, such as convection-enhanced diffusion, only delivers drug to the local injection site (Pardridge, 2010) and is ineffective as a brain drug delivery technology for the 1200g human brain. BBB disruption leads to chronic neuropathologic changes (Salahuddin *et al.*, 1988) and is too toxic to be widely used in humans. Small molecules are hardly an alternative strategy because 98% of small molecules tested do not cross the BBB. To be brain penetrating, the small molecule must be lipid soluble, form <8–10 hydrogen bonds with water, and have a molecular weight <400–500Daltons (Da) (Pardridge, 2005). Few small molecule pharmaceutical candidates have these molecular properties. Thus, even if a peptidomimetic small molecule were produced in lieu of drug development of a recombinant protein, the small molecule would most likely still need a BBB drug targeting technology to advance in clinical drug development.

The failure of biotechnology to treat the brain is illustrated with the biologic tumor necrosis factor (TNF)-α inhibitors (TNFI). Etanercept, the TNF receptor (TNFR); decoy receptor:Fc fusion protein, infliximab;

the chimeric anti-TNFα MAb; and adalimumab, the human anti-TNFα MAb, had combined revenues of $16 billion in 2008 (Tansey and Szymbowski, 2009). However, there was no penetration of the CNS markets with the biologic TNFIs, despite the primary role played by TNFα in chronic brain diseases such as Alzheimer's disease (AD) or Parkinson's disease (PD) (Park and Bowers, 2010). The biologic TNFIs have not been developed for the brain because these molecules do not cross the BBB, and no BBB drug targeting technology has been developed within the pharmaceutical industry. The purpose of this chapter is to review the BBB molecular Trojan horse (MTH) technology for BBB-targeted delivery of recombinant protein pharmaceuticals.

2. Blood–Brain Barrier Receptor-Mediated Transport and Molecular Trojan Horses

L-DOPA is an effective drug for PD because this water-soluble small molecule penetrates the BBB via carrier-mediated transport (CMT) on the BBB large neutral amino-acid transporter type 1, which is LAT1 (Boado *et al.*, 1999). Once in brain, the L-DOPA is decarboxylated to dopamine, the monoamine deficient in PD. Apart from LAT1, other BBB CMT systems include the GLUT1 glucose transporter, the CNT2 concentrative nucleoside transporter, and others (Pardridge, 2005). The CMT systems transport water-soluble small molecule nutrients and vitamins from blood to brain. Similarly, certain large molecules in blood, such as insulin or transferrin (Tf), are also transported across the BBB via receptor-mediated transport (RMT) systems, such as the insulin receptor or the transferrin receptor (TfR). These RMT systems are to be contrasted with BBB receptor-mediated endocytosis (RME) systems, such as the scavenger receptor (SR), which serves only to transport modified low-density lipoprotein (LDL) from blood to the endothelial compartment in brain (Triguero *et al.*, 1990). Similarly, LDL-related protein type 1 (LRP1) is also a RME system, not a transcytosis system (Nazer *et al.*, 2008). The BBB TfR mediates the bidirectional transport of holo-Tf from blood to brain, and apo-Tf, from brain back to blood (Zhang and Pardridge, 2001a). The BBB neonatal Fc receptor (FcRn) mediates the asymmetric efflux of IgG's from brain to blood but does not mediate the transport of IgGs from blood to brain (Zhang and Pardridge, 2001b). Transcytosis of RMT ligands from blood to brain has been demonstrated with emulsion autoradiography at the light microscopic level (Boado and Pardridge, 2009) by immunogold methods at the electron microscopic level (Bickel *et al.*, 1994) and by the capillary depletion method (Triguero *et al.*, 1990). The most definitive evidence of brain penetration via the BBB RMT systems is the demonstration of *in vivo* CNS pharmacologic effects following intravenous (IV) administration, as reviewed below.

The discovery of the BBB RMT systems suggested that RMT ligands, such as insulin or Tf, could act as a BBB MTH to ferry into brain drugs that were attached to the receptor ligand (Pardridge, 1986). Apart from the endogenous ligand, a receptor-specific peptidomimetic MAb may also function as a BBB Trojan horse. The MAb must be endocytosing and bind an exofacial epitope on the BBB receptor. The MAb should bind an epitope on the receptor that is spatially removed from the endogenous ligand binding site on the receptor, such that there is no interference with transport into brain of the endogenous ligand. The antibody-based MTHs are species specific. For brain drug transport in the rat, a mouse MAb against the rat TfR is used (Pardridge *et al.*, 1991); this MAb is not active in the mouse (Lee *et al.*, 2000). For brain transport in the mouse, a rat MAb against the mouse TfR is used (Lee *et al.*, 2000). These antibodies are not active in higher species. For brain drug delivery in the Rhesus monkey, a mouse MAb against the HIR is used (Pardridge *et al.*, 1995).

A potential BBB MTH must embody several important properties in order to be pharmacologically effective following systemic administration, and these properties are listed in Table 11.1. First, the BBB receptor that is targeted by the MTH must be a transcytosis system, not an endocytosis system. Second, the MTH must have a high affinity for the target BBB receptor, as represented by a receptor dissociation constant (K_D) in the low nanomolar range. If the K_D is $>10\,nM$, then brain uptake of the MTH, as represented by the BBB permeability–surface area (PS) product, may be too

Table 11.1 Properties of blood–brain barrier molecular Trojan horses

Number	Property
1.	Trojan horse targets a BBB receptor that is a transcytosis, not an endocytosis, system
2.	Trojan horse binding to BBB receptor is high affinity (low nM binding K_D)
3.	High-affinity binding of Trojan horse to BBB receptor (low nM binding K_D) is retained following fusion or conjugation of drug to Trojan horse
4.	High-affinity binding or high activity of drug or enzyme is retained following fusion or conjugation of the drug to the Trojan horse
5.	High brain uptake of Trojan horse-drug molecule by brain, for example, >2%ID/g brain in the mouse
6.	*In vivo* CNS pharmacologic effects are observed following intravenous administration of Trojan horse-drug molecule

From Pardridge (2010).

low to produce pharmacologic effects in brain for most pharmaceuticals. Moreover, if the putative MTH has a receptor $K_D > 10\,nM$, the capillary depletion method should not be used to confirm transcytosis in brain uptake models. A ligand with a low affinity for a target brain endothelial receptor will dissociate from the receptor during preparation of the brain homogenate (Triguero *et al.*, 1990). Third, the high affinity of the MTH for the BBB receptor must be retained following conjugation or fusion of the pharmaceutical to the MTH. If the affinity of the MTH for the BBB receptor is adversely affected following fusion to the pharmaceutical, then BBB transport will be minimal. Fourth, the affinity or activity of the pharmaceutical must be retained following fusion to the MTH. If the affinity of the pharmaceutical for its cognate receptor is impaired following fusion to the MTH, then there is little expectation for success at the preclinical animal model stage. Fifth, the brain uptake of the MTH-drug fusion or conjugate must be high so that pharmacologically active doses of the drug are delivered to brain. As a general rule, the brain uptake of the fusion protein in the mouse should be >2% of injected dose (ID)/g brain. The brain uptake, or % ID/g, is directly related to (a) the BBB PS product (mL/min/g) and (b) the plasma area under the concentration curve (AUC) (% ID min/mL), as shown in the following relationship:

$$\%\mathrm{ID/g} = (\mathrm{BBB\ PS\ product}) \times (\mathrm{plasma\ AUC})$$

The overall pharmacological potency of the fusion protein is directly related to the brain uptake, as reflected by the measured % ID/g, and this is directly related to the BBB PS product and the plasma AUC. Therefore, there are two causes for poor brain uptake of the fusion protein *in vivo*. First, if the BBB PS product is low, for example, because the MTH has a low affinity for the target BBB receptor, then the brain uptake will be proportionately reduced. However, the BBB PS product could be reasonably high, but if the plasma AUC is very low, then the brain uptake will also be low. A low plasma AUC is caused by the rapid removal of the MTH from blood via clearance by peripheral tissues such as liver or kidney. For example, cationic import peptides are cleared from blood within seconds of administration and have plasma AUCs that are prohibitively low (Lee and Pardridge, 2001). Similarly, receptor-associated protein (RAP), which is a ligand for LRP1, is rapidly removed from blood with a half-time of <2 min (Warshawsky *et al.*, 1993). A low plasma AUC causes a low brain uptake, irrespective of whether the MTH has a high BBB permeation constant. The role of the plasma AUC is discussed in subsequent sections of this chapter on pharmacokinetics (PK). Finally, a putative BBB MTH should be demonstrably effective in producing the desired *in vivo* pharmacologic effects in brain following IV administration, as discussed in subsequent sections of this chapter.

3. Reengineering Recombinant Proteins for Targeted Brain Delivery

The most active BBB MTH is a MAb against the HIR. The BBB expresses an insulin receptor, and this has been demonstrated with human brain capillaries used as an *in vitro* model of the human BBB (Pardridge *et al.*, 1985). A murine MAb against the HIR bound to the capillary endothelium in brain of the Rhesus monkey in an immunocytochemical study (Pardridge *et al.*, 1995). This HIRMAb rapidly penetrates the BBB in the Rhesus monkey following IV administration with a brain uptake of 2–3%ID/brain (Pardridge *et al.*, 1995). The HIRMAb has been genetically engineered, and both chimeric and humanized forms of the HIRMAb have been produced (Boado *et al.*, 2007a). The HIRMAb cross-reacts with the insulin receptor of Old World primates, such as the Rhesus monkey, but does not cross-react with the insulin receptor of lower animals, including New World primates, such as the squirrel monkey (Pardridge *et al.*, 1995). There is no known MAb against the rat or mouse insulin receptor that can be used as a BBB MTH in rodents. Therefore, a surrogate MTH is used in mouse studies, which is a genetically engineered chimeric MAb against the mouse TfR, designated the cTfRMAb (Boado *et al.*, 2009a).

IgG fusion proteins have been genetically engineered, expressed, and tested for both the HIRMAb, for the development of human therapeutics and preclinical testing in Rhesus monkeys, and for the cTfRMAb, for efficacy testing in mouse models of brain disease (Table 11.2). There are two general strategies for the engineering of IgG fusion proteins. First, the therapeutic protein may be fused to the carboxyl terminus of either the IgG heavy chain (HC) or light chain (LC). This is exemplified in Fig. 11.1A which shows the structure of a fusion protein formed by fusion of the human lysosomal enzyme, β-glucuronidase (GUSB), to the carboxyl terminus of the HC of the HIRMAb, and this fusion protein is designated HIRMAb-GUSB (Boado and Pardridge, 2010). In this case, the GUSB signal peptide was removed from the GUSB sequence. This approach is advantageous in that the site of fusion of the protein therapeutic to the IgG is far removed from the receptor binding sites within the complementarity-determining regions (CDRs) located within the amino terminal part of the HC and LC. However, any signal peptide or propeptide part of the therapeutic protein, which may play a crucial role in protein folding, cannot be incorporated in the protein sequence. If the prepeptide or propeptide sequence was incorporated in the protein sequence, then the protein therapeutic part of the fusion protein would be cleaved as the prepropeptide cleavage sites. This problem would be obviated by fusion of the prepro-protein to the amino terminus of the HC or LC of the IgG. GUSB was fused to the amino terminus of the HC of the HIRMAb, and this fusion protein is

Table 11.2 IgG fusion proteins engineered for targeted brain delivery

Category	Protein therapeutic	Reference
Neurotrophins	Brain-derived neurotrophic factor (BDNF)	Boado *et al.* (2007b)
	Glial-derived neurotrophic factor (GDNF)	Boado *et al.* (2008a)
	Erythropoietin (EPO)	Boado *et al.* (2010a)
Enzyme	α-L-Iduronidase (IDUA)	Boado *et al.* (2008b)
	Iduronate-2-sulfatase (IDS)	Lu *et al.* (2010)
	β-Glucuronidase (GUSB)	Boado and Pardridge (2010)
	Paraoxonase (PON)-1	Boado *et al.* (2008d)
Decoy receptor	Tumor necrosis factor receptor (TNFR) type II	Hui *et al.* (2009)
Monoclonal antibody	Antiamyloid antibody (AAA)	Boado *et al.* (2007c)
Other	Avidin	Boado *et al.* (2008c)

designated GUSB-HIRMAb (Fig. 11.1B). In this case, the GUSB signal peptide was included in the GUSB sequence fused to the IgG. However, the GUSB protein is now spatially close to the CDR sequences, which could impact on HIRMAb binding to the HIR. The HIRMAb-GUSB and the GUSB-HIRMAb fusion proteins were engineered, transiently expressed in COS cells, and purified with protein A affinity chromatography. The GUSB enzyme activity of the HIRMAb-GUSB fusion protein was >95% lost following fusion to the carboxyl terminus of the HIRMAb, although the high-affinity binding of this fusion protein to the HIR was retained (Boado and Pardridge, 2010). Conversely, when the GUSB enzyme was fused to the amino terminus of the HC of the HIRMAb, the GUSB enzyme activity was comparable to recombinant human GUSB, but the affinity of GUSB-HIRMAb fusion protein binding to the HIR was decreased >95%. Thus, there is a trade-off in retention of the bifunctionality of the IgG fusion protein depending on whether the therapeutic protein is fused to the amino or carboxyl terminus of the IgG chain.

The therapeutic protein is typically fused to the carboxyl terminus of the HC of the HIRMAb, so as to avoid steric interference with CDR binding. This design also places the therapeutic protein in a dimeric configuration (Fig. 11.1A), which is suitable for proteins that normally function as a dimeric structure. In most cases tested, the activity of the therapeutic protein is retained following fusion to the carboxyl terminus of the IgG. The HIRMAb or cTfRMAb fusion proteins that have been engineered and validated are listed in Table 11.2. Human GUSB is the only therapeutic

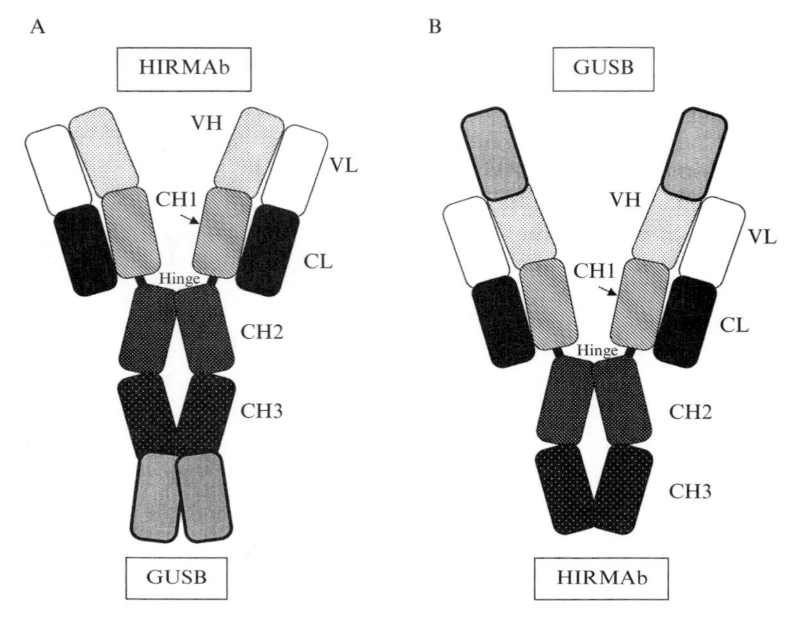

Figure 11.1 Structures of HIRMAb-GUSB fusion proteins. In the structure on the left, the human GUSB protein without the 22 amino-acid signal peptide, and without the 18 amino-acid carboxyl terminal propeptide, is fused to the carboxyl terminus of the heavy chain of the engineered HIRMAb. In the structure on the right, the human GUSB protein with the 22 amino-acid signal peptide, but without the 18 amino-acid carboxyl terminal propeptide, is fused to the amino terminus of the heavy chain of the engineered HIRMAb. From Boado and Pardridge (2010).

protein among these proteins that loses biological activity following fusion to the carboxyl terminus of the IgG chain. Other enzymes, such as human α-L-iduronidase (IDUA), iduronate-2-sulfatase (IDS), or paraoxonase (PON)-1, retain high enzyme activity following fusion to the carboxyl terminus of the HC of the HIRMAb (Table 11.2). Neurotrophins, such as brain-derived neurotrophic factor (BDNF), glial-derived neurotrophic factor (GDNF), or erythropoietin (EPO), retain high-affinity binding to the cognate neurotrophin receptor following fusion of the mature protein to the carboxyl terminus of the HIRMAb or the cTfRMAb (Table 11.2). Decoy receptor pharmaceuticals can also be fused to the carboxyl terminus of the HIRMAb or cTfRMAb, and fusion proteins have been engineered for the human type II TNF-α receptor (TNFR) (Hui *et al.*, 2009; Zhou *et al.*, 2011a). All other IgG decoy receptor fusion proteins involve the fusion of the carboxyl terminus of the decoy receptor to the amino terminus of the human IgG Fc region. However, the high-affinity binding of the

TNFR-II for TNFα is retained following fusion of the decoy receptor to the carboxyl terminus of the HIRMAb or the cTfRMAb (Hui *et al.*, 2009; Zhou *et al.*, 2011a).

The engineering of the MTH fusion protein is a special case when the therapeutic protein is also a MAb. Antiamyloid antibodies (AAAs) bind the Abeta amyloid peptide and disaggregate amyloid plaque in AD. However, the AAA must physically contact the amyloid plaque to cause disaggregation (Seubert *et al.*, 2008), and the plaque in brain is behind the BBB. The AAA, like other large molecules, does not cross the BBB (Boado *et al.*, 2007c). AAAs may enter brain when the BBB is disrupted, and chronic treatment of AD transgenic mice with high systemic doses of the AAA causes BBB disruption and cerebral microhemorrhage (Wilcock *et al.*, 2006). However, the preferred therapeutic is an AAA that penetrates the brain from blood in the absence of BBB disruption. RMT of an AAA across the BBB is possible following the reengineering of the AAA as a fusion protein with the BBB MTH. The AAA was first reengineered as a single chain Fv (ScFv) antibody, and the ScFv was then fused to the carboxyl terminus of the HC of the HIRMAb or cTfRMAb (Boado *et al.*, 2007c, 2010b). The structure of the ScFv fusion protein, which is designated the HIRMAb-ScFv protein, is shown in Fig. 11.2. The HIRMAb-ScFv and cTfRMAb-ScFv fusion

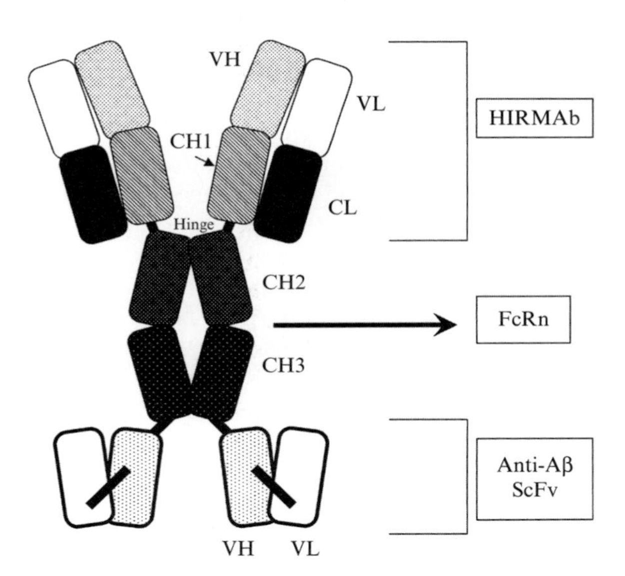

Figure 11.2 Antibody fusion protein with three functional domains. The first domain, the HIRMAb, binds the BBB HIR to trigger influx across the BBB. The second domain, the CH2/CH3 interface of the Fc region, binds to the BBB FcRn to trigger efflux from brain back to blood. The third domain, the anti-AβScFv fused to the CH3 region, binds to the Aβ amyloid peptide of AD to cause disaggregation of amyloid in brain. From Boado *et al.* (2007c).

proteins have three functional domains. The first domain is the HIRMAb or cTfRMAb part which binds the HIR or TfR on the BBB to trigger receptor-mediated influx across the BBB from blood to brain. The second domain is the antiamyloid ScFv, which binds and disaggregates amyloid plaque in brain behind the BBB. The third domain is the CH2–CH3 interface of the constant (C)-region, which binds the neonatal FcRn on the BBB to trigger receptor-mediated efflux across the BBB from brain to blood. The fusion protein is actively effluxed from brain via an Fc receptor (FcR) mechanism at the BBB (Boado *et al.*, 2007c). Confocal microscopy shows the FcRn type of FcR is abundantly expressed at the microvascular endothelium in brain (Schlachetzki *et al.*, 2002). The BBB FcR is an asymmetric transporter and mediates only the "reverse transcytosis" of IgG from brain to blood, but not the influx of IgG from blood to brain (Zhang and Pardridge, 2001b).

4. Genetic Engineering of Expression Plasmid DNA Encoding IgG Fusion Proteins

IgG fusion proteins such as those depicted in Figs. 11.1 and 11.2 are heterotetrameric proteins comprising two HCs and two LCs, which are glycosylated both at a single site on the C-region of the HC and at one or multiple sites on the therapeutic protein. Such proteins are generally expressed in eukaryotic host cells transfected with separate HC and LC genes. High host cell expression of the IgG fusion protein requires equally high expression of both the HC and the LC gene. In addition, the host cell must be transfected with a selectable marker gene, such as dihydrofolate reductase (DHFR). To insure high expression of all three genes in the host cell, a single expression plasmid DNA is engineered called a tandem vector (TV). The design of the TV is shown in Fig. 11.3, which shows the single plasmid DNA incorporates separate and tandem expression cassettes for the HC fusion gene, the LC gene, the DHFR gene, as well as the neomycin (neo) resistance gene. The HC and LC genes are 5′-flanked by the cytomegalovirus promoter (CMVp) and 3′-flanked by the bovine growth hormone (BGH) polyA (pA) sequence. The DHFR gene is 5′-flanked by the SV40 promoter (SV40p) and 3′-flanked by the hepatitis B virus (HBV) pA sequence. Following linearization of the TV, and electroporation of the host cell, a high producing line is produced by treatment with methotrexate (MTX). The high producing cell lines undergo two rounds of limited dilution cloning at 1 cell/well. The cDNAs encoding different parts of the IgG fusion protein may be produced with the polymerase chain reaction (PCR) using sequence specific oligodeoxynucleotide (ODN) primers. For engineering of the TV encoding the cTfRMAb, the cDNAs encoding the

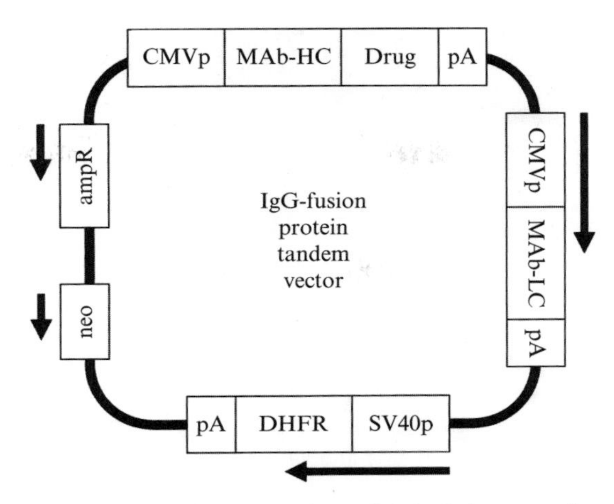

Figure 11.3 Tandem vector for expression of IgG fusion proteins. (A) A tandem vector contains the separate expression cassettes for the IgG heavy chain (HC) fusion gene, the IgG light chain (LC) gene, the dihydrofolate reductase (DHFR) gene, and the neomycin resistance gene (neo). The HC and LC genes are bordered by the cytomegalovirus promoter (CMVp) on the 5′-flank, and the bovine growth hormone polyA (pA) transcription termination sequence on the 3′-flank. The DHFR gene is bordered by the SV40 promoter (SV40p) on the 5′-flank and the hepatitis B virus pA sequence on the 3′-flank; ampR=ampicillin resistance gene.

variable region of the heavy chain (VH) and the variable region of the light chain (VL) were produced by PCR with RNA isolated from the rat hybridoma cell line expressing a rat MAb against the mouse TfR (Boado *et al.*, 2009a). The cDNAs encoding the C-region of the mouse IgG1 HC and the C-region of the mouse kappa LC were produced by PCR with RNA isolated from a mouse myeloma line. Following the PCR reaction, the four cDNAs were purified with agarose gel electrophoresis as shown in Fig. 11.4. Following isolation of the cDNAs, these TVs expressing the IgG fusion protein may be engineered.

5. Pharmacokinetics and Brain Uptake of IgG Fusion Proteins

5.1. Pharmacokinetics

The HIRMAb and cTfRMAb fusion proteins are purified by protein A and protein G affinity chromatography, respectively, following collection of host cell-conditioned serum-free medium. For fusion protein manufacturing for GLP toxicology studies, or GMP production of the fusion protein, the

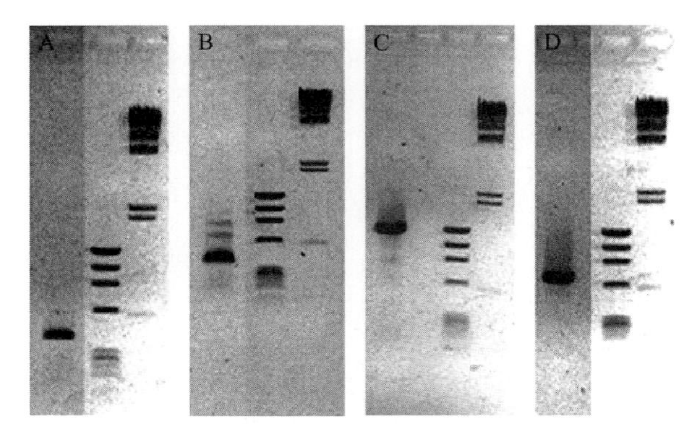

Figure 11.4 PCR cloning of fusion protein cDNAs. Agarose gel electrophoresis and ethidium bromide staining of PCR cloning of 0.4 kb TfRMAb VH (A), 0.4 kb TfRMAb VL (B), 1.4 kb mouse IgG1 C-region (C), and 0.7 kb mouse kappa C-region (D). The PCR generated cDNA is shown in lane 1, and DNA size standards are shown in lanes 2 and 3 for each panel. From Boado *et al.* (2009a).

protein A affinity column is followed by a cation exchange and/or anion exchange chromatographic step. Preclinical PK and brain uptake studies are performed in the Rhesus monkey for the HIRMAb fusion proteins and in the mouse for the cTfRMAb fusion proteins. The fusion protein concentration in blood is measured by radioisotope counting, if the fusion protein is radiolabeled, or by a sandwich ELISA when the fusion protein is not radiolabeled. The plasma concentration of the HIRMAb-IDUA fusion protein in the Rhesus monkey was measured over 2 h after IV injection at three doses of the fusion protein, 0.2, 2, and 20 mg/kg (Fig. 11.5; Boado *et al.*, 2009b). The plasma concentration, $A(t)$, is fit to the following equation:

$$A(t) = A_1 e^{-k1t} + A_2 e^{-k2t}$$

Based on the intercepts (A) and slopes (k) of the exponential decay in the plasma concentration, the PK parameters may be computed, including the area under the concentration curve (AUC), the median residence time (MRT), the central volume of distribution (Vc), the extravascular volume of distribution (Vss), and the systemic clearance (Cl). The plasma concentration profiles of the HIRMAb-IDUA fusion protein were analyzed with the PK model, and the plasma AUC was shown to be proportional to the fusion protein ID, which is typical of a linear PK profile (Boado *et al.*, 2009b). The PK properties of the HIRMAb or cTfRMAb fusion proteins are much closer to the PK profile of a small molecule, rather than a typical MAb therapeutic. The MRT of the lead antiamyloid MAb drug for AD, bapineuzumab, is 30 days in humans (Black *et al.*, 2010). In contrast, the MRT of the HIRMAb-IDUA fusion protein is 1–2 h in the Rhesus monkey (Boado *et al.*, 2009b). The rapid rate of removal from blood of

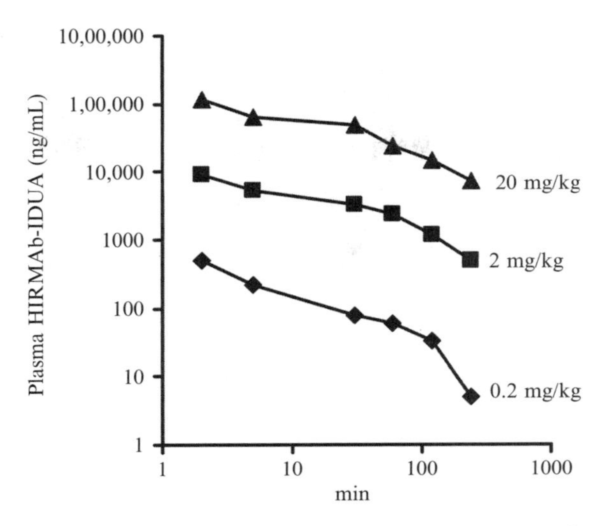

Figure 11.5 Plasma clearance of HIRMAb-IDUA fusion protein in the Rhesus monkey. Concentration of immunoreactive HIRMAb-IDUA at different times after IV injections of three doses, 0.2, 2, and 20 mg/kg, of the fusion protein. The plasma area under the concentration curve (AUC) is directly related to the injection dose, which is indicative of linear pharmacokinetics. From Boado *et al.* (2009b).

the fusion protein is shown by the data in Fig. 11.5, as the plasma concentration decreases by a log order of magnitude within the first 2 h after IV injection. The HIRMAb-IDUA fusion protein is rapidly removed from blood owing to uptake by insulin receptor-rich peripheral organs such as liver and spleen. The high uptake by liver and spleen is attributed to the fact that these organs are perfused by fenestrated vessels that are highly permeable to proteins. Therefore, the fusion protein rapidly distributes to the insulin receptor on parenchymal cells in liver and spleen. The high uptake by liver and spleen is not a phagocytotic process caused by aggregation of the protein, as the HIRMAb-IDUA fusion protein has <1% aggregates based on size exclusion chromatography (Boado *et al.*, 2009b). A typical MAb drug is confined to the plasma compartment and the systemic volume of distribution (Vss) is typically no greater than the central plasma volume (Vc). In contrast, the Vss of the HIRMAb-IDUA fusion protein is four- to fivefold greater than the Vc (Boado *et al.*, 2009b).

5.2. Brain uptake

The HIRMAb and cTfRMAb fusion proteins are rapidly taken up by brain, owing to binding of the HIRMAb or cTfRMAb to the BBB insulin receptor or BBB TfR. In the Rhesus monkey brain, the brain uptake of the HIRMAb-TNFR fusion protein is 3.0±0.1%ID/100g brain (Boado *et al.*, 2010c), which is comparable to the brain uptake of the murine HIRMAb in the Rhesus monkey (Pardridge *et al.*, 1995). The brain uptake in the monkey is expressed per 100g brain because the weight of the brain in the Rhesus monkey is 100g. This level of brain uptake of the HIRMAb fusion protein is comparable to the brain uptake of a lipid soluble small molecule. Fallypride, a lipid soluble small molecule with a molecular weight <400Da, is a ligand for the dopamine D2/D3 receptor and is used in positron emission tomography (PET) to image the level of dopamine receptor in the striatium. The uptake of fallypride in the cerebellum in the Rhesus monkey peaks at about 4%ID/100g at 5min after IV injection and then declines to 0.3%ID/100g at 2h after injection, owing to rapid washout from a region of brain lacking dopamine receptor (Christian *et al.*, 2009). Similarly, the brain uptake of the cTfRMAb fusion proteins in the mouse is comparable to the brain uptake of a lipid soluble small molecule. The brain uptake of the cTfRMAb fusion proteins in the mouse ranges from 2.0% to 3.5% ID/g (Table 11.3). The brain uptake in the mouse is expressed per gram brain because the weight of the brain in the mouse is about 1g. The uptake of fallypride in the dopamine transporter-rich striatum in the mouse is 0.58% ID/g (Honer *et al.*, 2004). The brain uptake of morphine, a neuroactive lipid soluble small molecule in the rat is 0.081±0.001% ID/g (Wu *et al.*, 1997). Therefore, the brain uptake of the cTfRMAb fusion protein actually exceeds that of a neuroactive small molecule. The BBB PS product of the small molecules is much greater than the BBB PS product of the HIRMAb or cTfRMAb fusion proteins. However, the plasma AUC of the fusion proteins is much greater than the plasma AUC for the small molecules, which is why the brain uptake, or % ID/g, of the fusion protein is comparable to the brain uptake of a small molecule. Although the plasma AUC of the fusion proteins is generally much greater than the plasma AUC for a small molecule, the plasma AUC of a typical MAb is much greater than the plasma AUC for the fusion proteins. The residence time of MAb therapeutics, which do not target the insulin receptor or TfR, is measured on the order of weeks (Black *et al.*, 2010).

Table 11.3 Brain uptake of IgG fusion proteins in the mouse

Fusion protein	Brain uptake (% ID/g)	Reference
cTfRMAb-ScFv	3.5±0.7	Boado *et al.* (2010b)
cTfRMAb-GDNF	3.1±0.2	Zhou *et al.* (2010a)
cTfRMAb-TNFR	2.8±0.5	Zhou *et al.* (2011a)
cTfRMAb-EPO	2.0±0.1	Zhou *et al.* (2010b)

5.3. Brain targeting

The brain uptake of the fusion protein, as reflected by the % ID/g measurement, divided by the terminal plasma AUC, is equal to the brain clearance, also called the permeability–surface area (PS) product. The PS product in brain and peripheral organs was measured for the HIRMAb-TNFR fusion protein in the Rhesus monkey. In parallel, the PS product in brain and peripheral organs was measured for the TNFR:Fc fusion protein (Boado *et al.*, 2010c). The commercial form of the TNFR:Fc fusion protein is called etanercept and is formed by fusion of the carboxyl terminus of the TNFR type II decoy receptor to the amino terminus of human IgG1 Fc region. Etanercept was not measurably transported across the BBB in the primate (Boado *et al.*, 2010c) because etanercept has no affinity to any BBB receptor transport system. However, etanercept is able to penetrate peripheral organs owing to transport across the porous walls of capillaries perfusing nonbrain organs. The ratio of the PS product for the HIRMAb-TNFR fusion protein, relative to the PS product for the TNFR:Fc fusion protein, is plotted for brain and six peripheral organs (Fig. 11.6). While the PS product ratio in brain was 30, the PS product ratio was only 5 in spleen and near unity for liver and other peripheral organs (Fig. 11.6). The PS product ratio for brain is an underestimate because the PS product for etanercept in brain is zero. These data demonstrate the selective targeting of the TNFR decoy receptor to brain following fusion to the HIRMAb (Boado *et al.*, 2010c).

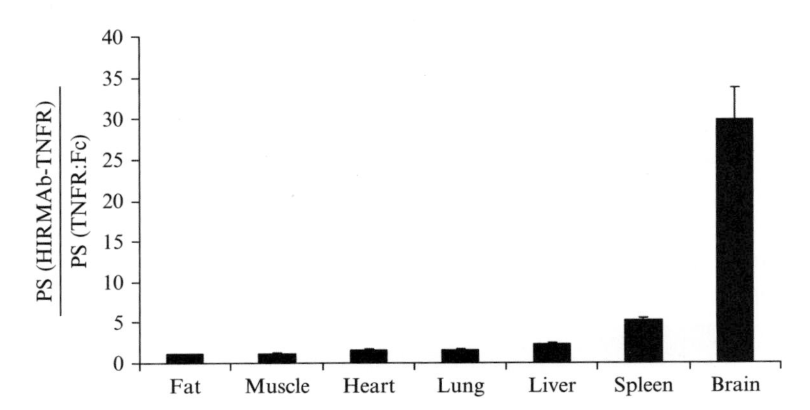

Figure 11.6 Brain targeting by molecular Trojan horse. Ratio of the organ permeability–surface area (PS) product, which is a measure of organ clearance, for the HIRMAb-TNFR fusion protein, relative to the organ PS product for the TNFR:Fc fusion protein, is plotted for brain and peripheral organs in the Rhesus monkey. The analysis shows selective targeting of the TNFR decoy receptor to brain following reengineering as the HIRMAb fusion protein. From Boado *et al.* (2010c).

6. CNS Pharmacological Effects of IgG Fusion Proteins

The BBB insulin receptor and TfR are actual transport systems and cause the net movement of the ligand across the endothelial barrier followed by distribution into brain parenchyma. The delivery to brain cells beyond the BBB was demonstrated by emulsion autoradiography at the light microscopic level for insulin (Duffy and Pardridge, 1987), for transferrin (Skarlatos *et al.*, 1995), and for HIRMAb fusion proteins (Boado and Pardridge, 2009). The capillary depletion method was developed to provide a quantitative measure of protein transport across the BBB (Triguero *et al.*, 1990) and can be used to corroborate morphologic methods such as emulsion autoradiography. The most direct morphologic method showing MTH distribution to neurons beyond the BBB is the finding of global expression of a plasmid DNA in the Rhesus monkey brain at 2days following the IV injection of the DNA encapsulated in liposomes, which were targeted with the HIRMAb (Zhang *et al.*, 2003). If the MTH is replaced by an IgG with no receptor specificity, then no gene expression is observed in brain.

The definitive test of brain penetration of MTH fusion proteins is the demonstration of *in vivo* CNS pharmacological effects in brain following the IV injection of the MTH fusion protein. Several recent pharmacologic studies have been reported, and these are summarized in Table 11.4. The neuroprotective properties of the HIRMAb-EPO fusion protein were examined in a stroke model in the rat (Fu *et al.*, 2010a). The HIRMAb-EPO fusion protein cannot be given IV in a rat stroke model because the HIRMAb part of the fusion protein does not recognize the rat insulin receptor. So as to prove that the HIRMAb-EPO fusion protein is therapeutic in stroke, following entry into the brain, the fusion protein was administered by intracerebral injection in rats subjected to a permanent middle cerebral artery occlusion (MCAO). The MCAO rats were treated with high, middle, and low doses of the HIRMAb-EPO fusion protein

Table 11.4 *In vivo* pharmacological effects in the brain following the administration of Trojan horse fusion proteins

Species	Route	Drug	Disease	Reference
Rat	Intracranial	EPO	Stroke	Fu *et al.* (2010a)
Mouse	Intravenous	EPO	Stroke	Fu *et al.* (2011)
	Intravenous	EPO	Parkinson's disease	Zhou *et al.* (2011b)
	Intravenous	GDNF	Parkinson's disease	Fu *et al.* (2010b)
	Intravenous	AAA	Alzheimer's disease	Zhou *et al.* (2011c)
	Intravenous	IDUA	Hurler's syndrome	Boado *et al.* (2011)

administered into the brain under stereotactic guidance 15min after the middle cerebral artery was occluded. The fusion protein concentration in each injection solution was quantified by sandwich ELISA. Rats were euthanized 24h after the MCAO for measurement of stroke volume. Coronal sections of brain were stained with 2% 2,3,5-triphenyltetrazolium chloride (TTC), which stains healthy brain red and leaves infarcted brain unstained. The permanent occlusion of the middle cerebral artery produced a large infarction in the nontreated rats. The intracerebral injection of the HIRMAb alone without the fused EPO had no effect on the stroke volume in the MCAO. The intracerebral injection of a high dose (64pmol) of the HIRMAb-EPO fusion protein reduced the hemispheric infarct volume by 98% (Fu *et al.*, 2010a). The intracerebral injection of a middle dose (4.7 pmol) of the HIRMAb-EPO fusion protein reduced the hemispheric infarct by 84%. In the middle dose-treated rats, there was no visible cortical infarct, and only residual subcortical infarct could be detected. Therefore, the cortical and subcortical infarct volumes were computed for both the control rats and the rats treated with the middle dose of HIRMAb-EPO fusion protein. This analysis shows the middle dose had no effect on the subcortical stroke volume but completely eliminated the cortical stroke. The intracerebral injection of the low dose of HIRMAb-EPO fusion protein (0.9pmol) had no effect on the stroke volume. The neurologic deficit was scored at 24h after the MCAO, and the neural deficit in the rats treated with either HIRMAb alone or low dose HIRMAb-EPO fusion protein was not significantly different from the neural deficit in the control rats (Fu *et al.*, 2010a). Treatment with the middle dose of HIRMAb-EPO fusion protein reduced the neural deficit by 35% ($p<0.005$), and treatment with the high dose of HIRMAb-EPO fusion protein reduced the neural deficit by 90% ($p<0.0005$). These studies show that the HIRMAb-EPO fusion protein is a potent neuroprotective agent in acute stroke, and the high-dose fusion protein virtually eliminated the infarct and neural deficit in the rats despite the permanent occlusion of the middle cerebral artery.

The efficacy of the cTfRMAb-EPO fusion protein in the MCAO model in the mouse could be tested with IV administration of the fusion protein because the cTfRMAb fusion proteins recognize the mouse TfR. Acute stroke in the mouse was produced by the permanent occlusion of the MCAO (Fu *et al.*, 2011). Permanent MCAO in untreated mice produced a large hemispheric infarct. The volume of the infarct in the untreated mouse was $103\pm17\,mm^3$. The IV treatment with recombinant EPO, at doses of 1000 or 10,000 units/kg, within 15min of the arterial occlusion produced no reduction in stroke volume. Conversely, the stroke volume was reduced 81% to $20\pm10\,mm^3$ following the IV injection of 1mg/kg of the cTfRMAb-EPO fusion protein (Fu *et al.*, 2011). The neurologic deficit was scored at 24h after the permanent MCAO, and the control mice exhibited high neural deficit scores of 3.7 ± 0.2. IV treatment with recombinant EPO produced no reduction in neural deficit scores. However, the IV treatment with 1.0mg/kg of the cTfRMAb-EPO fusion protein resulted in a 78% decrease in neural deficit score to 0.8±0.3 (Fu *et al.*, 2011). The IV administration of recombinant EPO to patients with acute stroke was shown to have no neuroprotective effect (Ehrenreich *et al.*, 2009). This lack of effect of EPO alone was confirmed in the mouse MCAO study (Fu *et al.*, 2011). However, EPO alone is not expected to have any neuroprotective effect in stroke following IV administration because (a) EPO alone does not cross the BBB (Boado *et al.*, 2010a) and (b) the BBB is intact in the early hours after stroke when neuroprotection is still possible (Belayev *et al.*, 1996).

EPO is also neuroprotective in experimental PD if the neurotrophin is administered by intracerebral injection (Xue *et al.*, 2007). The EPO must be given directly into the brain because systemic EPO is not effective in PD, owing to lack of transport of EPO across the BBB. Conversely, mice with experimental PD were successfully treated with repeat IV dosing with the cTfRMAb-EPO fusion protein (Zhou *et al.*, 2011b). Following 3weeks of treatment, PD mice were euthanized for measurement of striatal tyrosine hydroxylase (TH) enzyme activity. Mice treated with the cTfRMAb-EPO fusion protein showed a 306% increase in striatal TH enzyme activity, which correlated with improvement in three assays of neurobehavior. The blood hematocrit increased only 10% at 2weeks, with no further changes at 3weeks of treatment with the cTfRMAb-EPO fusion protein (Zhou *et al.*, 2011b). In contrast, treatment of mice with comparable doses of EPO causes a 52% and 75% increase in hematocrit at 2 and 4weeks of treatment (Grignaschi *et al.*, 2007). The minimal effect on erythropoiesis caused by chronic treatment with the cTfRMAb-EPO fusion protein is due to the different PK properties of the cTfRMAb-EPO fusion protein as opposed to EPO alone. The cTfRMAb-EPO fusion protein is cleared from blood via the TfR in peripheral tissues. Owing to the TfR-mediated clearance of the fusion protein, the plasma AUC of the cTfRMAb-EPO fusion protein is at least 10-fold reduced as compared to the plasma AUC of EPO in the mouse (Zhou *et al.*, 2010b). Similarly, the plasma AUC of the HIRMAb-EPO fusion protein is 13-fold lower than the plasma AUC of EPO in the Rhesus monkey (Boado *et al.*, 2010a). The hematopoietic effect of systemic EPO is directly proportional to the plasma AUC of the EPO analogue (Elliot *et al.*, 2008). Analogues of EPO with increased plasma AUC have enhanced effects on erythropoiesis. Conversely, analogues of EPO, such as the MTH-EPO fusion protein, with decreased plasma AUC, have diminished effects on erythropoiesis (Zhou *et al.*, 2011b).

The therapeutic effect of the cTfRMAb-ScFv fusion protein has been tested in a murine transgenic model of AD (Zhou *et al.*, 2011c). The cTfRMAb-ScFv fusion protein was administered by tail vein injection at a dose of 1mg/kg twice a week for 12weeks to double transgenic APPswe, PSEN1dE9 mice at 12months of age. The mice were shown to have extensive Abeta amyloid plaques in cerebral cortex based on immunocytochemistry. After 12 weeks of treatment, the brain $A\beta^{1-42}$ concentration was reduced 40% in the fusion protein-treated mice, and this was associated with no elevations in plasma $A\beta^{1-42}$ concentrations and no cerebral microhemorrhage (Zhou *et al.*, 2011c). Chronic treatment with the cTfRMAb-ScFv fusion protein caused no increase in plasma $A\beta^{1-42}$ peptide concentrations (Zhou *et al.*, 2011c) because the fusion protein is rapidly removed from blood. The MRT of the cTfRMAb-ScFv fusion protein in the mouse is 3 h (Boado *et al.*, 2010b). In contrast, the plasma MRT of bapineuzumab in humans is 30 days (Black *et al.*, 2010). The long MRT of the AAA causes a marked increase in the plasma concentration of the Abeta amyloid peptide in either AD transgenic mice (Wilcock *et al.*, 2006) or in humans with AD (Siemers *et al.*, 2010). The Abeta amyloid peptide is a toxic molecule that causes BBB disruption and cerebral microhemorrhage (Jansco *et al.*, 1998; Su *et al.*, 1999). In contrast, the chronic treatment of AD mice with the cTfRMAb-ScFv fusion protein was associated with no increase in plasma amyloid peptide and no cerebral microhemorrhage (Zhou *et al.*, 2011c).

Hurler mice express no IDUA and are a model of human mucopolysaccharidosis (MPS) type I, also called Hurler's syndrome. Hurler mice, like humans with MPSI, develop severe lysosomal inclusion bodies in cells of brain and peripheral organs. Patients with MPSI are treated with enzyme replacement therapy (ERT) and recombinant human IDUA. However, IDUA does not cross the BBB, and ERT does not treat the brain in MPSI (Tokic *et al.*, 2007). Hurler mice of advanced age, 6–8 months, were treated with twice weekly IV injections of the cTfRMAb-IDUA fusion protein for 8 weeks (Boado *et al.*, 2011). The mice were euthanized and lysosomal inclusion bodies in the brain were quantitated from semi-thin sections stained with *o*-toluidine blue and normalized per 100 nucleoli per brain section. Treatment of the MPSI mice with the cTfRMAb-IDUA reduced intracellular lysosomal inclusion bodies by 73% in brain, as compared to the MPSI mice treated with saline (Boado *et al.*, 2011). The glycosoaminoglycan levels in liver, spleen, heart, and kidney were reduced by >95%, 80%, 36%, and 20%, respectively, by cTfRMAb-IDUA fusion protein treatment. This study demonstrates the reversal of preexisting neural pathology in the brain of adult MPSI mice with receptor-mediated ERT of the brain.

7. Immune Response Against IgG Fusion Proteins

The chronic treatment of mice with cTfRMAb fusion proteins has not resulted in a preclinically significant immune response. Even in the MPSI mouse study, where the mice had never been exposed to the IDUA lysosomal enzyme, the immune response was of low titer (Boado *et al.*, 2011). In a chronic treatment study, mice were treated for 12weeks with twice weekly cTfRMAb-GDNF fusion protein at a dose of 2mg/kg/dose or 4mg/kg/week (Zhou *et al.*, 2011d). The immune response against the fusion protein was measured with a bridging ELISA, wherein the cTfRMAb-GDNF fusion protein was used as the capture reagent and biotinylated cTfRMAb-GDNF fusion protein was used as the detector reagent. The titer of the immune response is quantitated as the OD units per microliter undiluted serum recorded in the ELISA. A titer of <10 is considered evidence of tolerance to the biologic agent in a canine model of MPS (Dickson *et al.*, 2008). The titer against the cTfRMAb-GDNF fusion protein was <1 at the end of the 12-week treatment study (Zhou *et al.*, 2011d). There was minimal immune response directed against the GDNF part of the fusion protein. The low titer immune response was directed at the variable (V)-region of the cTfRMAb. However, these low titer antibodies had no functional consequence on fusion protein interaction with the endogenous TfR in the mouse. The PK profile and the brain uptake of the cTfRMAb-GDNF fusion protein were measured at the end of the 12-week treatment period, and there was no effect on either fusion protein clearance from blood or fusion protein transport across the mouse BBB *in vivo* (Zhou *et al.*, 2011d).

The reengineering of recombinant proteins as IgG fusion proteins may be a preferred formulation of the protein with respect to immune response. This is because the C-region of IgG contains certain amino-acid sequences, called Tregitopes, proposed to induce T-cell immune tolerance (De Groot *et al.*, 2008). These sequences may induce immune tolerance to the overall fusion protein.

8. Summary

Biotechnology has considerable potential for the development of new treatments of brain and spinal cord disorders. However, recombinant proteins must be reengineered to enable BBB transport and brain penetration. This is possible with the MTH technology, wherein recombinant proteins are fused to genetically engineered, receptor-specific MAb. The MAb, such as the HIRMAb, crosses the BBB via RMT and carries the fused therapeutic protein into brain. The IgG fusion proteins have PK profiles very different from typical MAb drugs and are rapidly cleared from blood similar to small molecules. The level of brain uptake of the MTH fusion proteins is comparable to the brain uptake of small molecules. The reengineering of recombinant proteins as IgG fusion proteins may result in improved immune tolerance to the therapeutic protein.

REFERENCES

Belayev, L., Busto, R., Zhao, W., and Ginsberg, M. D. (1996). Quantitative evaluation of blood-brain barrier permeability following middle cerebral artery occlusion in rats. *Brain Res.* **739,** 88–96.

Bickel, U., Kang, Y. S., Yoshikawa, T., and Pardridge, W. M. (1994). In vivo demonstration of subcellular localization of anti-transferrin receptor monoclonal antibody-colloidal gold conjugate within brain capillary endothelium. *J. Histochem. Cytochem.* **42,** 1493–1497.

Black, R. S., Sperling, R. A., Safirstein, B., Motter, R. N., Pallay, A., Nicols, A., and Grundman, M. (2010). A single ascending dose study of bapineuzumab in patients with Alzheimer disease. *Alzheimer Dis. Assoc. Disord.* **24,** 198–203.

Boado, R. J., and Pardridge, W. M. (2009). Comparison of blood-brain barrier transport of GDNF and an IgG-GDNF fusion protein in the Rhesus monkey. *Drug Metab. Dispos.* **37,** 2299–2304.

Boado, R. J., and Pardridge, W. M. (2010). Genetic engineering of IgG-glucuronidase fusion proteins. *J. Drug Target.* **18,** 205–211.

Boado, R. J., Li, J. Y., Nagaya, M., Zhang, C., and Pardridge, W. M. (1999). Selective expression of the large neutral amino acid transporter (LAT) at the blood-brain barrier. *Proc. Natl. Acad. Sci. USA.* **96,** 12079–12084.

Boado, R. J., Zhang, Y. F., Zhang, Y., and Pardridge, W. M. (2007a). Humanization of anti-human insulin receptor antibody for drug targeting across the human blood-brain barrier. *Biotechnol. Bioeng.* **96,** 381–391.

Boado, R. J., Zhang, Y., Zhang, Y., and Pardridge, W. M. (2007b). Genetic engineering, expression, and activity of a fusion protein of a human neurotrophin and a molecular Trojan horse for delivery across the human blood-brain barrier. *Biotechnol. Bioeng.* **97,** 1376–1386.

Boado, R. J., Zhang, Y. F., Zhang, Y., Xia, C. F., and Pardridge, W. M. (2007c). Fusion antibody for Alzheimer's disease with bi-directional transport across the blood-brain barrier transport and Abeta fibril disaggregation. *Bioconjug. Chem.* **18,** 447–455.

Boado, R. J., Zhang, Y., Zhang, Y., Wang, Y., and Pardridge, W. M. (2008a). GDNF fusion protein for targeted-drug delivery across the human blood brain barrier. *Biotechnol. Bioeng.* **100,** 387–396.

Boado, R. J., Zhang, Y., Zhang, Y., Xia, C. F., Wang, Y., and Pardridge, W. M. (2008b). Genetic engineering of a lysosomal enzyme fusion protein for targeted delivery across the human blood-brain barrier. *Biotechnol. Bioeng.* **99,** 475–484.

Boado, R. J., Zhang, Y., Zhang, Y., Xia, C. F., Wang, Y., and Pardridge, W. M. (2008c). Genetic engineering, expression and activity of a chimeric monoclonal antibody-avidin fusion protein for receptor-mediated delivery of biotinylated drugs in humans. *Bioconjug. Chem.* **19,** 731–739.

Boado, R. J., Zhang, Y., Zhang, Y. Wang, and Pardridge, W. M. (2008d). IgG-paraoxonase-1 fusion protein for targeted drug delivery across the human blood-brain barrier. *Mol. Pharm.* **5,** 1037–1043.

Boado, R. J., Zhang, Y., Wang, Y., and Pardridge, W. M. (2009a). Engineering and expression of a chimeric transferrin receptor monoclonal antibody for blood-brain barrier delivery in the mouse. *Biotechnol. Bioeng.* **102,** 1251–1258.

Boado, R. J., Hui, E. K. W., Lu, J. Z., and Pardridge, W. M. (2009b). AGT-181: Expression in CHO cells and pharmacokinetics, safety, and plasma iduronidase enzyme activity in Rhesus monkeys. *J. Biotechnol.* **144,** 135–141.

Boado, R. J., Hui, E. K., Lu, J. Z., and Pardridge, W. M. (2010a). Drug targeting of erythropoietin across the primate blood-brain barrier with an IgG molecular Trojan horse. *J. Pharmacol. Exp. Ther.* **33,** 961–969.

Boado, R. J., Zhou, Q. H., Lu, J. Z., Hui, E. K., and Pardridge, W. M. (2010b). Pharmacokinetics and brain uptake of a genetically engineered bifunctional fusion antibody targeting the mouse transferrin receptor. *Mol. Pharm.* **7,** 237–244.

Boado, R. J., Hui, E. K., Lu, J. Z., Zhou, Q. H., and Pardridge, W. M. (2010c). Selective targeting of a TNFR decoy receptor pharmaceutical to the primate brain as a receptor-specific IgG fusion protein. *J. Biotechnol.* **146,** 84–91.

Boado, R. J., Hui, E. K. W., Lu, J. Z., Zhou, Q. H., and Pardridge, W. M. (2011). Reversal of lysosomal storage in brain of adult MPS-I mice with intravenous Trojan horse-iduronidase fusion protein. *Mol. Pharm.* **8,** 1342–1350.

Christian, B. T., Vandehey, N. T., Fox, A. S., Murali, D., Oakes, T. R., Converse, A. K., Nickles, R. J., Shelton, S. E., Davidson, R. J., and Kalin, N. H. (2009). The distribution of D2/D3 receptor binding in the adolescent rhesus monkey using small animal PET imaging. *NeuroImage* **44,** 1334–1344.

De Groot, A. S., Moise, L., McMurry, J. A., Wambre, E., Van Overtvelt, L., Moingeon, P., Scott, D. W., and Martin, W. (2008). Activation of natural regulatory T cells by IgG Fc-derived peptide "Tregitopes" *Blood* **112,** 3303–3311.

Dickson, P., Peinovich, M., McEntee, M., Lester, T., Le, S., Krieger, A., Manuel, H., Jabagat, C., Passage, M., and Kakkis, E. D. (2008). Immune tolerance improves the efficacy of enzyme replacement therapy in canine mucopolysaccharidosis. *J. Clin. Invest.* **118,** 2868–2876.

Duffy, K. R., and Pardridge, W. M. (1987). Blood-brain barrier transcytosis of insulin in developing rabbits. *Brain Res.* **420,** 32–38.

Ehrenreich, H., Weissenborn, K., Prange, H., Schneider, D., Weimar, C., Wartenberg, K., Schellinger, P. D., Bohn, M., Becker, H., Wegrzyn, M., Jahnig, P., Herrman, M., *et al.* (2009). Recombinant human erythropoietin in the treatment of acute ischemic stroke. *Stroke* **40,** e647–e656.

Elliot, S., Pham, E., and Macdougall, I. C. (2008). Erythropoietins: A common mechanism of action. *Exp. Hematol.* **36,** 1573–1584.

Fu, A., Hui, E. K., Lu, J. Z., Boado, R. J., and Pardridge, W. M. (2010a). Neuroprotection in experimental stroke in the rat with an IgG-erythropoietin fusion protein. *Brain Res.* **1360,** 193–197.

Fu, A., Zhou, Q. H., Hui, E. K., Lu, J. Z., Boado, R. J., and Pardridge, W. M. (2010b). Intravenous treatment of experimental Parkinson's disease in the mouse with an IgG-GDNF fusion protein that penetrates the blood-brain barrier. *Brain Res.* **1352,** 208–213.

Fu, A., Hui, E. K., Lu, J. Z., Boado, R. J., and Pardridge, W. M. (2011). Neuroprotection in stroke in the mouse with intravenous erythropoietin-Trojan horse fusion protein. *Brain Res.* **1369,** 203–207.

Grignaschi, G., Zennaro, E., Tortarolo, M., Calvaresi, N., and Bendotti, C. (2007). Erythropoietin does not preserve motor neurons in a mouse model of familial ALS. *Amyotroph. Lateral Scler.* **8,** 31–35.

Honer, H., Bruhlmeier, M., Missimer, J., Schubiger, A. P., and Ametamey, S. M. (2004). Dynamic imaging of striatal D_2 receptors in mice using quad-HIDAC PET. *J. Nucl. Med.* **45,** 464–470.

Hui, E. K. W., Boado, R. J., and Pardridge, W. M. (2009). Tumor necrosis factor receptor-IgG fusion protein for targeted-drug delivery across the human blood brain barrier. *Mol. Pharm.* **6,** 1536–1543.

Jansco, G., Domoki, F., Santha, P., Varga, J., Fischer, J., Orosz, K., Penke, B., Becskei, A., Dux, M., and Toth, L. (1998). Beta-amyloid (1-42) peptide impairs blood-brain barrier function after intracarotid infusion in rats. *Neurosci. Lett.* **253,** 139–141.

Lee, H. J., and Pardridge, W. M. (2001). Pharmacokinetics and delivery of TAT and TAT-protein conjugate to tissues in vivo. *Bioconjug. Chem.* **12,** 995–999.

Lee, H. J., Engelhardt, B., Lesley, J., Bickel, U., and Pardridge, W. M. (2000). Targeting rat anti-mouse transferrin receptor monoclonal antibodies through the blood-brain barrier in the mouse. *J. Pharmacol. Exp. Ther.* **292,** 1048–1052.

Lu, J. Z., Hui, E. K. W., Boado, R. J., and Pardridge, W. M. (2010). Genetic engineering of a bi-functional IgG fusion protein with iduronate 2-sulfatase. *Bioconjug. Chem.* **21,** 151–156.

Nazer, B., Hong, S., and Selkoe, D. J. (2008). LRP promotes endocytosis and degradation, but not transcytosis, of the amyloid-beta peptide in a blood-brain barrier in vitro model. *Neurobiol. Dis.* **30,** 94–102.

Pardridge, W. M. (1986). Receptor-mediated peptide transport through the blood-brain barrier. *Endocr. Rev.* **7,** 314–330.

Pardridge, W. M. (2005). The blood-brain barrier: Bottleneck in brain drug development. *NeuroRx* **2,** 3–14.

Pardridge, W. M. (2010). Biopharmaceutical drug targeting to the brain. *J. Drug Target.* **18,** 157–167.

Pardridge, W. M., Eisenberg, J., and Yang, J. (1985). Human blood-brain barrier insulin receptor. *J. Neurochem.* **44,** 1771–1778.

Pardridge, W. M., Buciak, J. L., and Friden, P. M. (1991). Selective transport of anti-transferrin receptor antibody through the blood-brain barrier in vivo. *J. Pharmacol. Exp. Ther.* **259,** 66–70.

Pardridge, W. M., Kang, Y. S., Buciak, J. L., and Yang, J. (1995). Human insulin receptor monoclonal antibody undergoes high affinity binding to human capillaries in vitro and rapid transcytosis through the blood-brain barrier in vivo in the primate. *Pharm. Res.* **12,** 807–816.

Park, K. M., and Bowers, W. J. (2010). Tumor necrosis factor-alpha mediated signaling in neuronal homeostasis and dysfunction. *Cell. Signal.* **22,** 977–983.

Salahuddin, T. S., Johansson, B. B., Kalimo, H., and Olsson, Y. (1988). Structural changes in the rat brain after carotid infusions of hyperosmolar solutions. An electron microscopic study. *Acta Neuropathol.* **77,** 5–13.

Schlachetzki, F., Zhu, C., and Pardridge, W. M. (2002). Expression of the neonatal Fc receptor (FcRn) at the blood-brain barrier. *J. Neurochem.* **81,** 203–206.

Seubert, P., Barbour, R., Khan, K., Motter, R., Tang, P., Kholodenko, D., Kling, K., Schenk, D., Johnson-Wood, K., Schroeter, S., Gill, D., Jacobsen, J. S., *et al.* (2008). Antibody capture of soluble Abeta does not reduce cortical Abeta amyloidosis in the PDAPP mouse. *Neurodegener. Dis.* **5,** 65–71.

Siemers, E. R., Friedrich, S., Dean, R. A., Gonzales, C. R., Farlow, M. R., Paul, S. M., and Demattos, R. B. (2010). Safety and changes in plasma and cerebrospinal fluid amyloid beta after a single administration of an amyloid beta monoclonal antibody in subjects with Alzheimer disease. *Clin. Neuropharmacol.* **33,** 67–73.

Skarlatos, S., Yoshikawa, T., and Pardridge, W. M. (1995). Transport of [^{125}I]transferrin through the rat blood-brain barrier in vivo. *Brain Res.* **683,** 164–171.

Su, G. C., Arendash, G. W., Kalaria, R. N., Bjugstad, K. B., and Mullan, M. (1999). Intravascular infusions of soluble beta-amyloid compromise the blood-brain barrier, activate CNS glial cells and induce peripheral hemorrhage. *Brain Res.* **818,** 105–117.

Tansey, M. G., and Szymbowski, D. E. (2009). The TNF superfamily in 2009: New pathways, new indications, and new drugs. *Drug Discov. Today* **14,** 1082–1088.

Tokic, V., Barisic, I., Huzjak, N., Petkovic, G., Fumic, K., and Paschke, E. (2007). Enzyme replacement therapy in two patients with an advanced severe (Hurler) phenotype of mucopolysaccharidosis I. *Eur. J. Pediatr.* **166**(727), 732.

Triguero, D., Buciak, J. B., and Pardridge, W. M. (1990). Capillary depletion method for quantifying blood-brain barrier transcytosis of circulating peptides and plasma proteins. *J. Neurochem.* **54,** 1882–1888.

Warshawsky, I., Bu, G., and Schwartz, A. L. (1993). 39-kD protein inhibits tissue-type plasminogen activator clearance in vivo. *J. Clin. Invest.* **92,** 937–944.

Wilcock, D. M., Alamed, J., Gottschall, P. E., Grimm, J., Rosenthal, A., Pons, J., Ronan, V., Symmonds, K., Gordon, M. N., and Morgan, D. (2006). Deglycosylated anti-amyloid-beta antibodies eliminate cognitive deficits and reduce parenchymal amyloid with minimal vascular consequences in aged amyloid precursor protein transgenic mice. *J. Neurosci.* **26,** 5340–5346.

Wu, D., Kang, Y. S., Bickel, U., and Pardridge, W. M. (1997). Blood-brain barrier permeability to morphine-6-glucuronide is markedly reduced compared to morphine. *Drug Metab. Dispos.* **25,** 768–771.

Xue, Y. Q., Zhao, L. R., Guo, W. P., and Duan, W. M. (2007). Intrastriatal administration of erythropoietin protects dopaminergic neurons and improves neurobehavioral outcome in a rat model of Parkinson's disease. *Neuroscience* **146,** 1245–1258.

Zhang, Y., and Pardridge, W. M. (2001a). Rapid transferrin efflux from brain to blood across the blood-brain barrier. *J. Neurochem.* **76,** 1597–1600.

Zhang, Y., and Pardridge, W. M. (2001b). Mediated efflux of IgG molecules from brain to blood across the blood-brain barrier. *J. Neuroimmunol.* **114,** 168–172.

Zhang, Y., Schlachetzki, F., and Pardridge, W. M. (2003). Global non-viral gene transfer to the primate brain following intravenous administration. *Mol. Ther.* **7,** 11–18.

Zhou, Q. H., Boado, R. J., Lu, J. Z., Hui, E. K., and Pardridge, W. M. (2010a). Monoclonal antibody-glial derived neurotrophic factor fusion protein penetrates the blood-brain barrier in the mouse. *Drug Metab. Dispos.* **38,** 566–572.

Zhou, Q. H., Boado, R. J., Lu, J. Z., Hui, E. K., and Pardridge, W. M. (2010b). Re-engineering erythropoietin as an IgG fusion protein that penetrates the blood-brain barrier in the mouse. *Mol. Pharm.* **7,** 2148–2155.

Zhou, Q. H., Boado, R. J., Hui, E. K., Lu, J. Z., and Pardridge, W. M. (2011a). Brain-penetrating TNFR decoy receptor in the mouse. *Drug Metab. Dispos.* **39,** 71–76.

Zhou, Q. H., Hui, E. K. W., Lu, J. Z., Boado, R. J., and Pardridge, W. M. (2011b). Brain penetrating IgG-erythropoietin fusion protein is neuroprotective following intravenous treatment in Parkinson's disease in the mouse. *Brain Res.* **1382,** 315–320.

Zhou, Q. H., Fu, A., Boado, R. J., Hui, E. K. H., Lu, J. Z., and Pardridge, W. M. (2011c). Receptor-mediated Abeta amyloid antibody targeting to Alzheimer's disease mouse brain. *Mol. Pharm.* **8,** 280–285.

Zhou, Q. H., Boado, R. J., Lu, J. Z., Hui, E. K., and Pardridge, W. M. (2011d). Chronic dosing of mice with a transferrin receptor monoclonal antibody-GDNF fusion protein. *Drug Metab. Dispos.* **39,** 1149–1154.

CHAPTER TWELVE

Engineering and Identifying Supercharged Proteins for Macromolecule Delivery into Mammalian Cells

David B. Thompson,[1] James J. Cronican,[1] *and* David R. Liu

Contents

1. Introduction 294
   1.1. Discovery and basic properties of supercharged proteins 295
   1.2. Aggregation resistance of supercharged proteins 296
   1.3. Cell penetration of supercharged proteins 297
   1.4. Macromolecule delivery by supercharged proteins 300
2. Methods 301
   2.1. Theory underlying protein supercharging 301
   2.2. Engineering supercharged proteins 303
   2.3. Screening for functional supercharged proteins 305
   2.4. Identifying naturally supercharged proteins 307
   2.5. Protein expression 308
   2.6. Protein purification 309
   2.7. Cation exchange and endotoxin removal 310
   2.8. Construction of fusion proteins 311
   2.9. Nucleic acid delivery by supercharged proteins *in vitro* 312
   2.10. Protein delivery by supercharged proteins *in vitro* 313
   2.11. Protein delivery by supercharged proteins *in vivo* 315
3. Conclusion 317
References 318

Abstract

Supercharged proteins are a class of engineered or naturally occurring proteins with unusually high positive or negative net theoretical charge. Both super-negatively and superpositively charged proteins exhibit a remarkable ability to withstand thermally or chemically induced aggregation. Superpositively

Howard Hughes Medical Institute, Department of Chemistry and Chemical Biology, Harvard University, Cambridge, Massachusetts, USA
[1] These authors contributed equally to this work.

Methods in Enzymology, Volume 503
ISSN 0076-6879, DOI: 10.1016/B978-0-12-396962-0.00012-4

charged proteins are also able to penetrate mammalian cells. Associating cargo with these proteins, such as plasmid DNA, siRNA, or other proteins, can enable the functional delivery of these macromolecules into mammalian cells both *in vitro* and *in vivo*. The potency of functional delivery in some cases can exceed that of other current methods for macromolecule delivery, including the use of cell-penetrating peptides such as Tat and adenoviral delivery vectors. This chapter summarizes methods for engineering supercharged proteins, optimizing cell penetration, identifying naturally occurring supercharged proteins, and using these proteins for macromolecule delivery into mammalian cells.

1. Introduction

Most medicines are small molecules, chemical compounds containing less than ~100 atoms. For many diseases, however, small-molecule-based therapies have not been found. Recent efforts to discover bioactive molecules including human therapeutics have increasingly focused on macromolecules—for the purpose of this chapter, proteins or nucleic acids. Indeed, ~180 protein drugs are currently prescribed including insulin, erythropoietin, interferons, and a variety of antibodies (Overington *et al.*, 2006). Due to their significant folding energies, macromolecules are able to adopt large, stable three-dimensional conformations suitable for strong binding to targets even when they lack hydrophobic clefts commonly associated with small-molecule binding. Moreover, the strength of macromolecule-target binding can be sufficient to interfere with native protein–protein or protein–nucleic acid interfaces that have traditionally been difficult to address using small molecules (Arkin and Wells, 2004). The stability, size, and complexity of macromolecules can result in specificities that are not easily achievable using small molecules, as demonstrated by certain antibodies and RNAi agents.

Unfortunately, macromolecules are typically not able to diffuse into cells, and as a result, virtually all existing macromolecule therapeutics address extracellular targets. Perhaps the most challenging and widespread impediment to the broader use of proteins and nucleic acids as therapies is therefore the delivery of macromolecules into cells and into subcellular locations of interest.

To address this challenge, various macromolecule delivery approaches have been developed including electroporation (Jantsch *et al.*, 2008), ultrasound-mediated plasmid delivery (Shen *et al.*, 2008), viral delivery (Zhou *et al.*, 2008), nebulization (Durcan *et al.*, 2008), and direct chemical modification (Akdim *et al.*, 2010). In addition, other strategies associate a macromolecule with a nonviral delivery vehicle such as lipidoids (Akinc *et al.*, 2008), liposomes (Akinc *et al.*, 2010), dendrimers (Patil *et al.*, 2009), cationic polymers (Watanabe *et al.*, 2009), inorganic nanoparticles (Lee *et al.*, 2011), carbon nanotubes (Kam *et al.*, 2005), cell-penetrating peptides (Jo *et al.*, 2001), small molecules (Soutschek *et al.*, 2004), or receptor ligands (Nishina *et al.*, 2008). In 2009, we reported the use of cationic supercharged proteins as general and potent vectors for macromolecule delivery into mammalian cells.

In this chapter, we summarize the basic properties of supercharged proteins, as well as methods for engineering them, identifying naturally occurring supercharged proteins within a proteome database and applying them to deliver nucleic acids and proteins *in vitro* and *in vivo*.

1.1. Discovery and basic properties of supercharged proteins

We reported the creation and characterization of supercharged proteins in 2007 (Lawrence *et al.*, 2007). Engineered supercharged proteins are the product of extensive mutagenesis in which solvent-exposed residues throughout the protein's surface are substituted with either acidic or basic amino acids. In several cases tested, the resulting proteins retain much of their original activity. For example, we generated supercharged GFP proteins with a wide range of net theoretical charges (−30 to +48) that possess nearly identical excitation and emission spectra as starting GFP (Fig. 12.1A; Lawrence *et al.*, 2007). Mutations that alter the excitation and emission maxima of GFP, resulting in blue, cyan, and yellow fluorescent proteins, are also amenable to supercharging (Fig. 12.1B; Thompson *et al.*). Likewise, supercharged streptavidin (a tetramer with +52 net theoretical charge) also retains the ability to tetramerize and bind biotin, albeit at a reduced affinity (Lawrence *et al.*, 2007). Last, supercharged glutathione *S*-transferase (a dimer with −40 net theoretical charge) retains much of its catalytic activity (Lawrence *et al.*, 2007). The preservation of function across a diverse set of

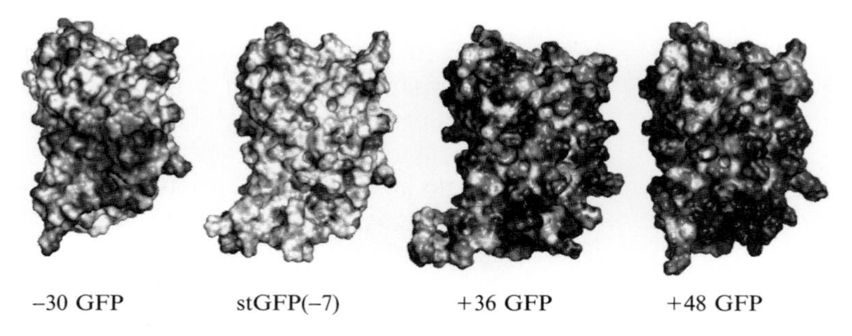

Figure 12.1 Electrostatic surface potentials of −30 GFP, stGFP, +36 GFP, and +48 GFP colored from −25kT/e (red) to +25kT/e (blue). (See Color Insert.)

proteins following extensive surface mutagenesis is a testament to the mutability of residues identified in the supercharging method (described below) and to the potential generality of protein supercharging for generating aggregation-resistant and cell-penetrating protein reagents.

1.2. Aggregation resistance of supercharged proteins

Supercharged proteins are remarkably resistant to both thermal- and chemical-induced aggregation and can efficiently refold to regain much of their original function. We demonstrated this aggregation resistance with supercharged GFP, streptavidin, and glutathione *S*-transferase (Lawrence *et al.*, 2007). Boiled supercharged proteins denature and lose their activity similar to their wild-type counterparts. However, their ability to avoid aggregation events including the association of exposed hydrophobic residues, even in the unfolded state, enables supercharged proteins to refold and regain much of their original activity after cooling (Fig. 12.2; Lawrence *et al.*, 2007). Further, extended exposure to 40% 2,2,2-trifluoroethanol (TFE), conditions that induce the denaturation and aggregation of typical proteins, does not result in any measurable aggregation of +36 GFP (Lawrence *et al.*, 2007). These observations suggest that supercharged proteins can avoid common aggregation pathways upon thermal or chemical denaturation and, in the absence of aggregation, can refold into functional proteins once conditions are restored that favor folding. Note that the net charge, rather than the number of charged residues, is necessary for these unusual properties; indeed, wild-type proteins and supercharged proteins can contain similar numbers of charged residues even though their total net charge differs dramatically (Lawrence *et al.*, 2007).

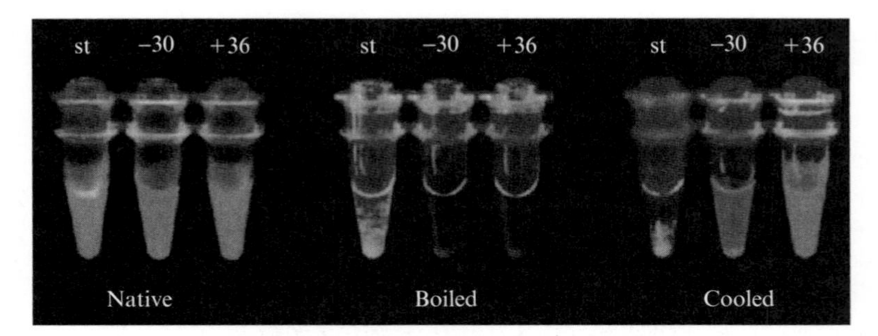

Figure 12.2 UV-illuminated samples of purified GFP variants (native), those samples heated 1 min at 100 °C (boiled), and those samples subsequently cooled for 2 h at 25 °C (cooled). (See Color Insert.)

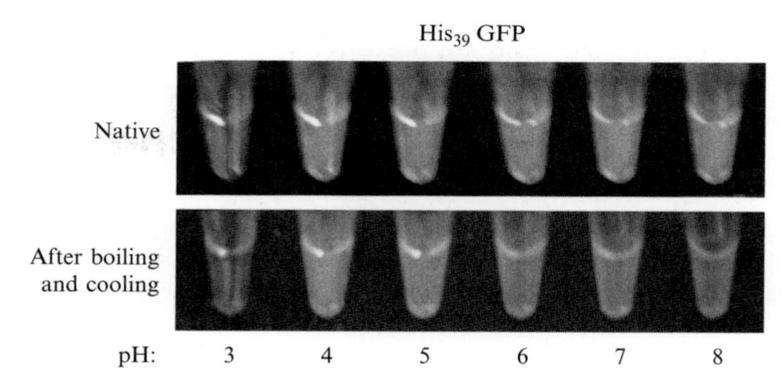

Figure 12.3 UV-illuminated samples of His_{39} GFP (native) at different pH values and after those samples were heated 1 min at 100 °C (boiled), and subsequently cooled for 2 h at 25 °C (cooled). (See Color Insert.)

In support of this model, we have also created a conditionally supercharged protein in the form of a GFP containing 39 histidine residues, primarily at the positions previously substituted with Lys/Arg in +36 GFP. This His_{39} GFP also folds and fluoresces and, like +36 GFP, recovers fluorescence when boiled and cooled at low pH (Fig. 12.3). As the pH increases past ~6, however, the histidine side chains become neutral, the protein decreases in charge, and the ability to refold after boiling is lost (Fig. 12.3).

The high charge of supercharged proteins enables them to reversibly complex with oppositely charged macromolecules, including nucleic acids. For example, upon mixing +36 GFP with nucleic acids, complexes form that sequester the nucleic acid out of solution and generate particles that can be isolated by centrifugation (Lawrence *et al.*, 2007). These complexes can be dissociated by the addition of high concentration of salt, presumably by competing the electrostatic interactions necessary for complex formation (Lawrence *et al.*, 2007).

1.3. Cell penetration of supercharged proteins

Many known nonviral macromolecule delivery vehicles are cationic. These vehicles typically bind to negatively charged components on the cell membrane and are endocytosed into cells, either by stimulating endocytosis or by simple membrane recycling (Wadia *et al.*, 2004). In some cases, a fraction of these endocytosed molecules are able to escape endosomes and access the intracellular environment (Wadia *et al.*, 2004). The HIV Tat cell-penetrating peptide is cationic and is one example of a delivery vehicle that is thought to use this macromolecule delivery mechanism (Wadia *et al.*, 2004). Engineered proteins with cationic regions have also been found to penetrate cells, including a pentamutant GFP containing a patch of five arginine residues (Fuchs and Raines, 2007).

We hypothesized and subsequently demonstrated that superpositively charged proteins such as +36 GFP can penetrate mammalian cells with potencies much greater than that of cationic peptides or modestly cationic engineered proteins (McNaughton *et al.*, 2009; Cronican *et al.* 2010; Cronican *et al.* 2011; Thompson *et al.*, submitted). Live-cell fluorescence microscopy images of mammalian cells reveal that the entire cell membrane becomes associated with +36 GFP within seconds of exposure to low nanomolar concentrations of the protein (McNaughton *et al.*, 2009). Within minutes, +36 GFP can be observed within the body of the cell as bright puncti, presumably contained within endosomes. Following uptake, +36 GFP-containing puncti can be found distributed throughout the extranuclear space (McNaughton *et al.*, 2009).

We have probed the requirements for +36 GFP uptake to better understand the mechanism by which it achieves such potent internalization. During internalization assays, the treated cells can be washed several times with heparin, a highly sulfated glycosaminoglycan, to remove surface-bound +36 GFP. This washing procedure effectively removes surface-bound +36 GFP as measured by both flow cytometry and confocal microscopy, while the intracellular puncti remain (Cronican *et al.*, 2010; McNaughton *et al.*, 2009; Veldhoen *et al.*, 2006). Examination of cells washed in this manner therefore enables measurement of total internalized protein as well as measurements of uptake and trafficking kinetics. The high negative charge of heparin also enables its use as a competitor with the cell surface for +36 GFP binding. For example, at 4° C endocytosis is known to be inhibited in mammalian cells (Harding *et al.*, 1983). At 4° C, +36 GFP will bind to the outside of the cell membrane but will not be internalized. Washing cells exposed to +36 GFP with heparin at 4° C will remove all +36 GFP fluorescence from the cell (McNaughton *et al.*, 2009). These results indicate that +36 GFP does not passively traverse the lipid bilayers of the cell's plasma membrane but instead requires endocytosis to drive internalization (McNaughton *et al.*, 2009).

The surface of mammalian cells is decorated with sulfated proteoglycans that contribute greatly to the anionic nature of the cell surface. Treatment with sodium chlorate, an inhibitor of proteoglycan sulfation, inhibits internalization of +36 GFP, as well as cell-surface association of +36 GFP (McNaughton *et al.*, 2009). Further, cells that have been genetically modified to produce only non-sulfated proteoglycans are incapable of internalizing +36 GFP (McNaughton *et al.*, 2009). Although these experiments were performed with +36 GFP, the results likely apply to all superpositively charged proteins in light of our observation that +36 GFP acts as a competitive inhibitor to the internalization of six unrelated naturally supercharged human proteins (Cronican *et al.*, 2011). Together, the above observations suggest an essential role for electrostatic interactions, derived in large part from sulfated proteoglycans (Gump *et al.*, 2010), and subsequent endocytosis in supercharged protein internalization.

While it is clear that positive charge promotes internalization of supercharged proteins, the relationship between charge and cell penetration is not so simple (Turcotte *et al.*, 2009). We have generated dozens of charged GFP variants from shuffling subsequences of starting GFP, +15 GFP, +25 GFP, +36 GFP, and +48 GFP (Thompson *et al.*, submitted). Using a single protein scaffold to display residues contributing to a variety of net charges allows examination of the role of charge magnitude and distribution without the complication of gross structural variation. We found scGFP cellular uptake to be highly charge dependent and strongly sigmoidal, exhibiting both low-potency and high-potency forms with the transition occurring near +21 charge units (Fig. 12.4) (Thompson *et al.*, submitted). Similarly, mammalian cells incubated with His_{39} GFP in media of pH values from 4 to

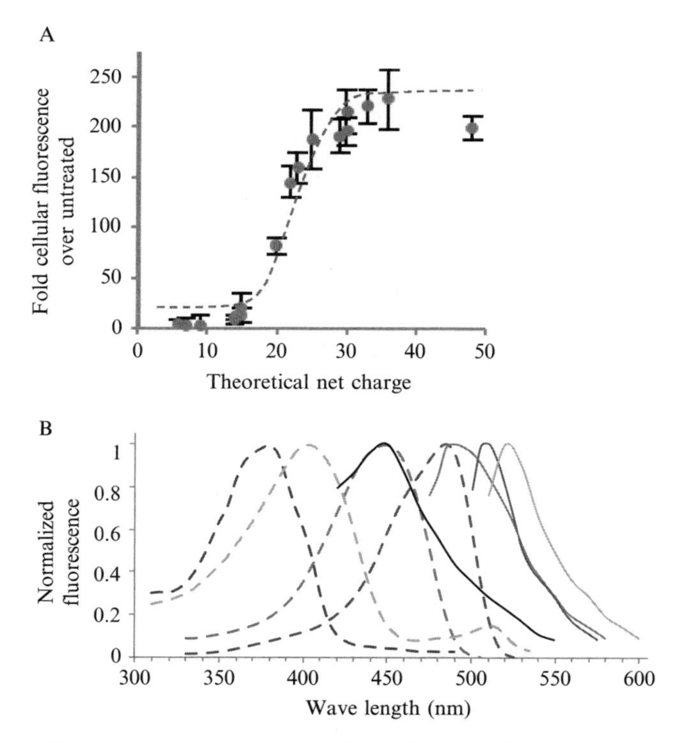

Figure 12.4 Properties of supercharged GFP variants. (A) The charge dependence of supercharged GFP uptake in cultured HeLa cells treated with 200 n*M* protein for 4 h at 37° C. (B) The excitation (dashed line) and emission spectra (solid line) of blue, cyan, green, and yellow fluorescent variants of +36 GFP. The yellow variant is notable, as it has a large stokes shift with a 400 nm absorption and a 520 nm emission maxima. (See Color Insert.)

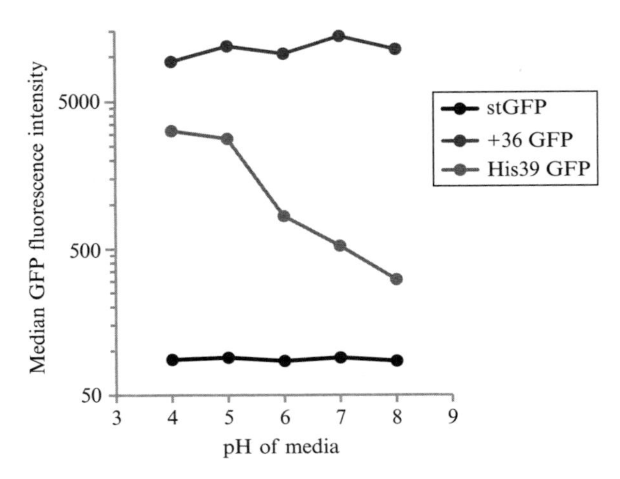

Figure 12.5 His_{39} GFP penetrates mammalian cells in a pH-dependent manner consistent with the protonation state of its histidine side chains. (For color version of this figure, the reader is referred to the Web version of this chapter.)

8 exhibit a pH-dependent internalization whereas the uptake of +36 GFP and nonsupercharged starting GFP (stGFP) are unaffected by changes in pH over this range (Fig. 12.5).

This sigmoidal charge-cell penetration relationship has not been observed for oligoarginine peptide reagents, where uptake efficiency generally *decreases* past ~+15 charge magnitude (Thompson *et al.*, submitted). It therefore appears that supercharged protein surfaces can attain higher net charges than cationic peptide reagents without eroding cell penetration capabilities. Even when comparing the cell-penetration potency of supercharged proteins with that of synthetic oligo-Lys/Arg peptides of similar or identical theoretical net charge, we observed that the cationic peptides were consistently outperformed by supercharged GFPs (Thompson *et al.*, submitted). The mechanism underlying this difference remains the subject of active investigation in our laboratory. It is tempting to speculate that the structure of the surface of a folded, globular protein may engage in cooperative binding or cross-linking of anionic cell-surface receptors more effectively than similarly charged unstructured peptides. Indeed, we have observed certain forms of endocytic stimulation from supercharged proteins that are not observed upon treatment with unstructured cationic peptides of similar theoretical net charge (Thompson *et al.*, submitted).

1.4. Macromolecule delivery by supercharged proteins

Supercharged proteins are general and potent vehicles for delivery of macromolecules into mammalian cells. As summarized above, simple mixing of siRNA or plasmid DNA with +36 GFP results in the formation of electrostatic complexes. Incubation of these complexes with mammalian cells leads to delivery of the associated nucleic acid even into cell lines known to be resistant to lipid-mediated transfection (McNaughton *et al.*, 2009).

Supercharged proteins have also been shown to deliver a variety of functional protein molecules directly into cells by translational fusion to supercharged proteins (Cronican *et al.*, 2010). We initially demonstrated delivery of mCherry, ubiquitin, and Cre recombinase into multiple cell types using +36 GFP (Cronican *et al.*, 2010). These three proteins provide separate complementary measures of delivery in cultured mammalian cells. Delivery of mCherry provides a quantitative measure of simple protein uptake into target cells. Ubiquitin delivery enables a measure of endosomal escape due to the cytosol-specific cleavage of ubiquitin by deubiquitinating enzymes (Cronican *et al.*, 2010). Cre recombinase functions a general measure of extraendosomal delivery of functional proteins, as a positive delivery phenotype is only observed upon nuclear localization and enzymatic activity of the Cre protein in appropriate reporter cell lines. We observed effective protein delivery into mammalian cells in all three cases using supercharged proteins, including some examples *in vivo* (Cronican *et al.*, 2010).

Most recently, we have demonstrated that these delivery properties are not unique to engineered cationic supercharged proteins but are instead present in a diverse class of naturally occurring supercharged human proteins (Cronican, 2011). We analyzed the human proteome for proteins with unusually high net positive charge and found a large, diverse class of proteins (possibly >2% of the human proteome) that potently delivers protein in functional form into mammalian cells both *in vitro* and in retinal, pancreatic, and white adipose tissues *in vivo* (Fig. 12.6; Cronican, 2011). These findings reveal a diverse set of macromolecule delivery agents for *in vivo* applications and also raise the possibility that some human proteins may penetrate cells as part of their native biological functions.

2. Methods

2.1. Theory underlying protein supercharging

Our initial motivation for generating supercharged proteins was the desire to test the hypothesis that highly charged proteins were less likely to aggregate than lesser charged proteins. Supercharged proteins enabled the limits of this hypothesis to be tested. In theory, protein surfaces are the more likely region of a protein to tolerate substitution compared to the protein core because protein folding is driven primarily by the loss of solvation of hydrophobic residues that make up the core of the protein (Dill *et al.*, 2008).

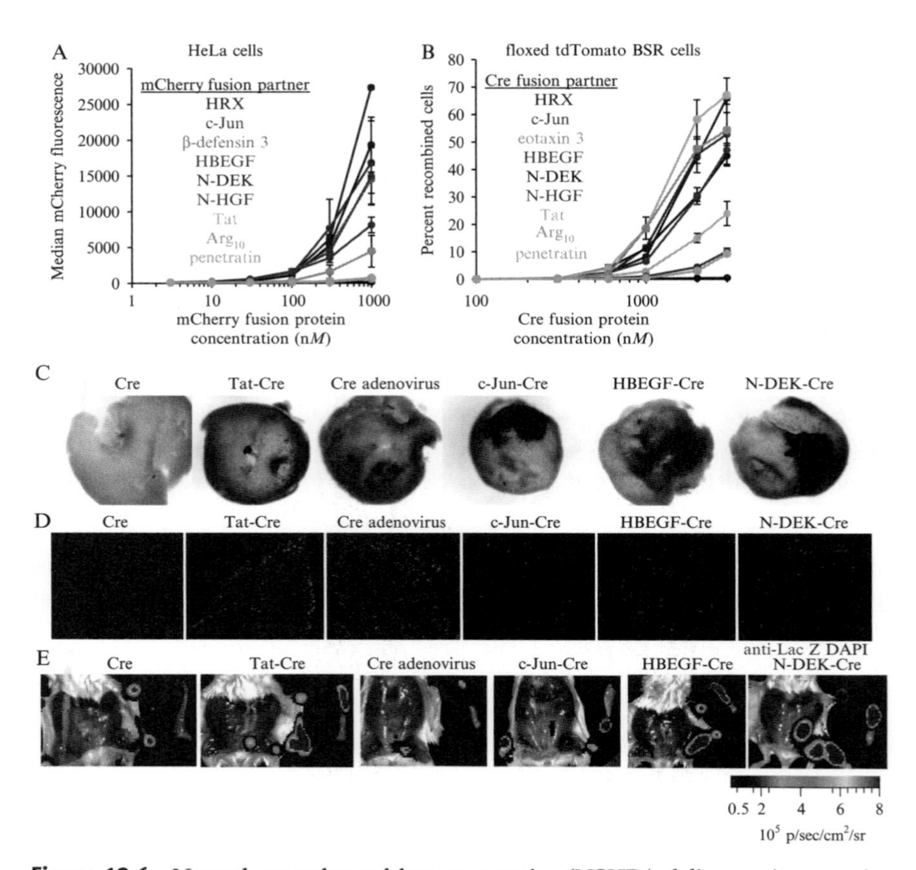

Figure 12.6 Natural supercharged human proteins (NSHPs) deliver active proteins *in vitro* and *in vivo*. (A) Median mCherry fluorescence of HeLa cells incubated with NSHP-mCherry fusions as measured by flow cytometry. (B) Percent recombined cells among floxed tdTomato BSR cells incubated with NSHP-Cre fusions as measured by flow cytometry. (C) Adult floxed LacZ mice were injected subretinally with Cre fusion proteins. Recombination results in LacZ activity, which was visualized with X-gal stain (blue) 3 days after injection. (D) Adult floxed LacZ mice injected in the pancreas with Cre fusion proteins exhibit recombination in the exocrine tissues as indicated by LacZ immunostaining (red) 5 days after injection. (E) Adult floxed luciferase mice injected subcutaneously with Cre fusion proteins exhibit recombination in the white adipose tissue as visualized by luminescence 3 days after injection. White adipose tissue was extracted and placed to the right of each mouse. (See Color Insert.)

Water molecules, which possess limited freedom when interacting with hydrophobic amino acid side chains, gain entropy as the hydrophobic residues collapse into the interior of the protein. Amino acid side chains that do not become buried upon protein folding remain solvent exposed and are in a very similar environment in the folded or unfolded state. Due to the similar environments in the folded and unfolded states, solvent-exposed amino acids are thought to contribute less energy toward stabilizing the folded state compared with buried residues (Tokuriki *et al.*, 2007). In idealized cases in which the side chain of an amino acid is completely solvent exposed and makes no interactions with the rest of the protein, the identity of the side chain should not contribute at all to the thermodynamics of protein folding. This simple model was supported by our experiments in which up to 36, 32, and 30 mutations were installed in GFP, in the streptavidin tetramer, and in the GST dimer, respectively, without abolishing the fluorescence, biotin-binding, or catalytic function of these three proteins. Because three to five random mutations in a 50kDa protein typically eliminates protein function (Tokuriki *et al.*, 2007), it is highly unlikely that such a large number of mutations (up to 15% of the total number of residues) could be made to these proteins without destroying their fold and function had they not been restricted to the most solvent-exposed amino acids.

2.2. Engineering supercharged proteins

Engineering a supercharged protein first requires that the structure of the parent protein has been determined and is available for analysis. PDB files containing three-dimensional proteins structures are widely available on public databases including the Research Collaboratory for Structural Bioinformatics (http://www.rcsb.org). To supercharge GFP, for example, the structure of an optimized GFP was downloaded as the file 2B3P.pdb. In the case of GFP, the solvent-exposed residues were selected by manually inspecting the crystal structure using a PDB imaging software such as PYMOL. Engineering of supercharged streptavidin and GST proteins was performed computationally by ranking residues by their average number of neighboring atoms (within 10Å) per side-chain atom (AvNAPSA). Charged or highly polar solvent-exposed residues (Asp, Glu, Arg, Lys, Asn, and Gln) were mutated either to Glu (unless the original residue was Asn, in which case to Asp) for negative supercharging or to Lys for positive supercharging. The AvNAPSA Perl script is provided upon request.

As an example, supercharging streptavidin proceeds as follows:

(1) Download the pdb file of interest. In this case, 1stp.pdb can be downloaded from http://www.pdb.org/pdb/explore/explore.do?structureId=1STP
(2) Install Perl (available at http://www.perl.org/get.html)
(3) Save the AvNAPSA file to the same folder as the 1stp.pdb file. For example, save both files to C:\Temp.
(4) In the command prompt, enter:

```
C:\Temp>perl avnapsa 1STP.pdb
```

(5) The AvNAPSA program will output a list of values associated with each amino acid in the protein.
(6) Copy the AvNAPSA amino acid list to a spreadsheet program and rank the list by AvNAPSA values, starting from the residues with the lowest AvNAPSAs (the highest degree of solvent exposure).
(7) To positively supercharge the protein, change the polar nonpositive amino acids (D, E, N, or Q) to lysine (K), starting from the top of the list. To negatively supercharge the protein, change the polar nonnegative amino acids (R, K, or Q) to glutamate (E), except for asparagine (N) which should be changed to aspartate (D). The mutations should be made in order of increasing AvNAPSA values. The total number of mutations to supercharge a protein is dependent upon the size of the protein being analyzed and the desired degree of supercharging.
(8) Optional: if a family of protein sequences related to the protein of interest is available, perform a sequence homology alignment and avoid mutation of evolutionarily conserved residues, regardless of AvNAPSA.
(9) Rank the sequence-altered amino acid list by residue number to reform the correct amino acid sequence.

If a PDB file is not available for a desired protein, a related protein PDB file may be identified by searching the PDB database for sequence homology to the desire protein. BLASTP is a powerful application for this purpose (http://blast.ncbi.nlm.nih.gov/). A family of related structures can be used as a template to generate a PDB file for the desired protein using a modeling program such as MODELLER (http://salilab.org/modeller/). The generated PDB file can be used as the input for the supercharging algorithm or to model the finished supercharged protein.

The supercharging method is compatible with other means of identifying mutable residues. For example, Biopython (biopython.org) has built-in modules for calculating amino acid exposure by the number of neighboring amino acid alpha carbons within a specified radius of each amino acid alpha carbon or by the number of alpha carbons in the half-sphere defined by the alpha carbon to beta carbon vector. Other programs such as DSSP (http://swift.cmbi.ru.nl/gv/dssp/) are also able to assign accessible surface area values to each amino acid to provide another estimation of relative solvent exposure. Mutable amino acids can also be identified without any necessary dependence on solvent exposure by using protein design software such as ROSETTADesign. Each implementation for the purposes of supercharging will likely have different advantages and disadvantages. Incomplete knowledge of a protein's folding and functional requirements can result in nonfunctional supercharged variants. It is therefore useful to combine the above method with techniques for generating and screening libraries of supercharged proteins for variants that function in the desired context.

2.3. Screening for functional supercharged proteins

Incomplete knowledge of a protein's folding and functional requirements can result in nonfunctional supercharged variants for a variety of reasons presented below. It is therefore useful to combine the above method with techniques for generating and screening libraries of supercharged proteins for variants that function in the desired context.

While solvent-exposed amino acids should be more easily mutated without impairing protein structure or function, the relationship between solvent exposure and mutability is imperfect. For example, the folding energy of most proteins is 5–15 kcal/mol while the strength of a hydrogen bond is ~0–9 kcal/mol depending on its environment (Dill *et al.*, 2008). If a solvent-exposed side-chain contributes one or more intramolecular hydrogen bonds or other electrostatic interactions, then it may not be mutable. Indeed, there is evidence that some naturally thermostable proteins are stabilized by the presence of multiple surface salt bridges (Karshikoff and Ladenstein, 2001). Similarly, since the native protein structure is in equilibrium with an ensemble of nonnative folded states, a mutated side chain may stabilize a nonnative structure in which it is no longer as solvent exposed or in which it makes new interactions, reducing the relative stability of the native protein. Moreover, side chains regardless of their degree of solvent exposure may play important roles in the expression and folding of some proteins (Dill *et al.*, 2008). Finally, some residues in X-ray or even NMR structures of proteins may adopt nonnative conformations that alter their apparent AvNAPSA values.

Given the challenges associated with the correct prediction of many mutable residues in a protein, it is prudent to generate a panel of candidate supercharged variants of a protein and screen or select those with desired properties. For example, using the AvNAPSA method, one can generate a hypothetical protein that contains many more supercharging mutations than the final protein is likely to withstand, then fragment and shuffle the gene encoding the "overcharged" protein with the gene encoding the wild-type protein, and screen or select to isolate variants that possess the desired function. For example, we designed by manual inspection (prior to development of the AvNAPSA method) a GFP with a net theoretical charge of +36 containing 29 mutations. When the gene corresponding to this protein was overexpressed, +36 GFP expressed well and was fluorescent. However, when we designed using the same principles a GFP with a net theoretical charge of −39, the resulting protein did not express in *E. coli*. The following is a summary of the method we used to generate negatively supercharged GFP variants:

(1) The starting GFP and −39 GFP genes were ordered as two sets of overlapping 40-bp oligonucleotides and resuspended in water to 100 μ*M* (Stemmer *et al.*, 1995).

(2) The oligonucleotides for starting GFP and −39 GFP were mixed in a 1:20 molar ratio and 6 μL of the mixture was added to 28 μL water, 4 μL 10× T4 Ligase Buffer and 2 μL T4 polynucleotide kinase (10 U/μL, New England Biolabs). The kinase reaction was incubated for 20 min at 37 °C, 2 min at 94 °C, and then cooled at 0.1 °C/s to 70 °C.
(3) While the oligonucleotides were being phosphorylated, a mixture of 10 μL 10× T4 Ligase Buffer and 50 μL water was warmed to 70 °C and added to the kinase reaction once it reached 70 °C. The reaction was incubated for another 30 min at 70 °C, then cooled at 0.1 °C/s to 16 °C. A 5 μL aliquot was removed as the sample before ligation.
(4) To the reaction was added 2 μL of T4 DNA ligase (5 U/μL, Invitrogen). The reaction was incubated for 1 h at 16 °C. Another 5 μL aliquot was removed as the sample after ligation.
(5) The ligation product (1 μL) was used as the template for a PCR reaction using a terminal 5′ forward oligo and the 3′ reverse oligo. The integrity of the library was confirmed by running out the "before ligation," "after ligation," and PCR samples on a 1.2% agarose gel containing ethidium bromide. Ligation should result in formation of a broad smear upward that is present in the sample after ligation but not before ligation. An ideal PCR reaction will predominantly contain a single band at the expected gene length.
(6) The resulting shuffled library of GFP genes was then digested with restriction enzymes, purified by agarose gel electrophoresis, ligated into a protein expression plasmid, and transformed into a cloning strain of *E. coli*.
(7) The resulting colonies from transformation were scraped from the agar plate, grown in liquid culture, and harvested for plasmid purification.
(8) The plasmid library was transformed into a protein expression host such as BL21(DE3) cells and plated on an IPTG-containing agar plate.
(9) The next morning, the plate was illuminated with ultraviolet light to reveal colonies that contain functional fluorescent GFP clones. These colonies were picked for sequencing of the plasmid and larger scale protein expression and purification.

In this manner, −30 GFP, which contained 15 of the 20 planned mutations, and −25 GFP, which contained 12 mutations, were identified. As the costs of gene synthesis decrease, libraries can also be generated by *in vitro* recombination of full-length genes using methods such as DNA shuffling (Zhao and Arnold, 1997), StEP (Zhao *et al.*, 1998), and NEBNext (New England Biolabs). While most proteins are not fluorescent, a wide variety of screens that use chromogenic (Huimin and Frances, 1997) or fluorogenic assays (Yanisch-Perron *et al.*, 1985) or that couple the function of a protein to transcription of a reporter gene (Karimova *et al.*, 1998) can also be used to identify functional protein variants.

2.4. Identifying naturally supercharged proteins

One concern of using engineered proteins such as +36 GFP for *in vivo* applications is that the protein may provoke an immune system response. To identify naturally occurring supercharged human proteins that may be less immunogenic than engineered supercharged proteins, we sorted the human proteome to identify the most positively charged proteins (Fig. 12.7). We found that these proteins are also able to penetrate mammalian cells and functionally deliver associated macromolecules *in vitro* and *in vivo*. The naturally supercharged human protein (NSHP) Python script that we use to identify NSHPs from the PDB and Swissprot databases will be provided upon request. The following is a protocol for identifying supercharged proteins from collection of protein sequences:

(1) Supercharged proteins can be identified by first collecting all of the primary sequences for the proteins you are interested in. A curated collection of all known natural proteins can be downloaded from the Uniprot Web site (http://www.uniprot.org/downloads) as uniprot_sprot.fasta.gz. The uniprot_sprot.fasta file can be extracted and each protein entry will contain the name of the protein, the source organism, whether the protein is experimentally confirmed or predicted, and the protein primary sequence in FASTA format. Another important source of proteomic data is from the Research Collaboratory for Structural Bioinformatics Web site (www.rcsb.org) which contains only proteins with X-ray crystallography or NMR structural information. The advanced search function on the RCSB Web site allows for filtering for proteins that are from the source organism, such as *Homo sapiens*, and expression host, such as *Escherichia coli*. The resulting list will provide proteins that will likely express in sufficient yield for macromolecule delivery experiments and will also likely have accompanying biochemical or biological information.

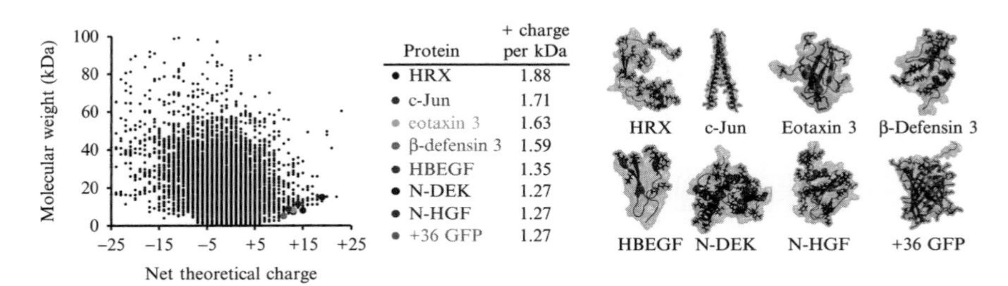

Figure 12.7 Plot of human proteins expressed from *E. coli* within the Protein Data Bank. The blue dots represent proteins with positive charge:molecular weight ratios exceeding +0.75/kDa. (See Color Insert.)

(2) A scripting program such as Python or Perl is used to generate a calculated molecular weight and net theoretical charge from the primary sequences of the protein entries. It is helpful to create a graph of the proteome with net theoretical charge on the *x*-axis and molecular weight on the *y*-axis. Supercharged proteins can be identified from this dot plot by manual inspection, or the proteins can be ranked by net theoretical charge or by the ratio of net theoretical charge to molecular weight. Our experience working with naturally supercharged proteins and with various charged GFPs suggests that a positive charge to molecular weight (in kDa) ratio greater than 0.75 represents a good starting point to identify proteins that can penetrate cells.

(3) Genes corresponding to the supercharged protein can be ordered from various companies. Genes for natural proteins, for example, can be ordered from companies with cDNA libraries such as Open Biosystems (www.openbiosystems.com). Genes for proteins identified through the PDB database can be requested from the lab that produced the structural information. Genes can also be ordered from gene synthesis companies with codon optimization for recombinant expression.

This approach is distinct from and complementary to approaches that identify cell-penetrating peptides such as QSAR, which is based on algorithms developed using an empirical training set of known cell-penetrating and nonpenetrating examples (Jones *et al.*, 2010).

2.5. Protein expression

In our experience, supercharged proteins generally express well in standard *E. coli* protein expression strains such as BL21(DE3). Protein-specific variations in protocol may be necessary, and parameters including induction time and temperature can significantly impact the yield of the expressed supercharged protein on a case-by-case basis.

Small-scale expression cultures should be tested to determine the optimal expression conditions and the solubility of the expressed protein before performing large-scale expression experiments. The following procedure can be used as a starting point for protein-specific protocol optimization:

Materials

LB broth, 2× YT broth and agar plates containing appropriate antibiotic.
pET expression plasmid.
Competent BL21(DE3) cells.
Resuspension buffer: PBS with 2*M* NaCl, 20m*M* imidazole, pH 7.5, with one tablet of EDTA-free Complete Protease Inhibitor (Roche) per 50mL buffer.
Elution buffer: PBS with 2*M* NaCl, 500m*M* imidazole, pH 7.5.
Ni-NTA agarose resin (Qiagen).

(1) Transform the supercharged protein pET expression plasmid into BL21 (DE3) and plate on 2× YT media with appropriate antibiotic. Incubate overnight at 30 °C to minimize the risk of colony overgrowth, auto-induction, and toxicity from leaky protein expression.
(2) The following day, pick a medium-sized, isolated colony and inoculate a 5 mL overnight seed culture of 2× YT media. Incubate overnight at 30 °C.
(3) The following day, pellet the seed culture by centrifugation at 4000×*g* for 5 min and resuspend the pellet in LB media.
(4) Inoculate a 1 L expression culture of LB media with the resuspended overnight seed culture. Grow the culture at 37 °C to an OD_{600} of ~0.6 and induce expression with 1 m*M* IPTG.
(5) Following induction, incubate the expression culture for 4 h at 30 °C.
(6) Harvest the culture by centrifugation at 6000×*g* for 10 min. Pellets can be stored at −80 °C or processed immediately.

2.6. Protein purification

The following protocol is for standard His-tagged proteins; other purification protocols are also suitable for supercharged proteins. The protein samples should be kept on ice at all times.

(1) Thaw frozen pellets and resuspend in 40 mL resuspension buffer.
(2) Split the resuspended pellets into two 20 mL fractions and lyse each by sonication in a Sonic Dismembrator 550 (Fisher Scientific) for 3 min with a cycle of 1 s on and 1 s off at power level of 4.
(3) Pellet cell debris by centrifugation at 4000×*g* for 10 min. For soluble protein preparations, transfer the supernatant to a new 50 mL conical tube. For insoluble protein preparations, resuspend and homogenize the pellet in resuspension buffer containing 1% Tween 20 to remove the lipid contents of the pellet. Pellet the still insoluble inclusion bodies by centrifugation at 4000×*g* for 10 min. Resuspend and homogenize in resuspension buffer containing 8 *M* urea, and incubate with agitation for 30 min at room temperature.
(4) Add 1 mL of settled Ni-NTA resin to the lysate and incubate at 4 °C for 30–45 min with gentle agitation on a rocker or rotary.
(5) Pellet the Ni-NTA resin by centrifugation at 2500×*g* for 1 min. Discard the supernatant.
(6) Transfer the Ni-NTA resin to a column.
(7) Wash the resin first with 20 mL PBS + 2 *M* NaCl, then 15 mL of PBS +2 *M* NaCl +20 m*M* imidazole, and finally elute with 4 mL PBS + 2 *M* NaCl + 500 m*M* imidazole.
(8) Dialyze the protein against 1 L of PBS or an appropriate buffer at 4 °C for 1 h, replace with 2 L of fresh buffer, and dialyze overnight. Concentrate as necessary by ultrafiltration.

Upon desalting or dialysis into buffers with lower salt concentration, a small amount of precipitate may form, giving the solution a cloudy appearance. This is likely due to complexation of the supercharged protein with copurifying charged cell components such as phospholipids and nucleic acids. Pellet these precipitates to recover the remaining soluble fraction. Most supercharged proteins and their fusions in our experience are soluble in PBS. Certain fusion proteins, including fusions to Cre recombinase, may require additional salt (up to 500 m*M*) or other additives to promote protein solubility and stability.

The protein preparation may either be used as is or subjected it to further purification steps including ion exchange chromatography and endotoxin removal (described below). Quantitate proteins either spectrophotometrically or via bicinchoninic acid protein assay. Protein expression and size should be confirmed by PAGE and/or by MALDI-TOF mass spectrometry. Following quantitation of protein, multiple aliquots should be prepared at volumes sufficient for single sets of assays to minimize freeze–thaw cycles. Aliquots should be snap-frozen in liquid nitrogen and stored at −80 °C.

2.7. Cation exchange and endotoxin removal

Copurified anionic contaminants may still be present following Ni-NTA-based purification. The presence of anionic contaminants may alter the interaction of supercharged proteins with desired anionic molecules in downstream applications. For example, high levels of endotoxin (>0.5 EU/mL) can alter phenotypes in cellular assays. As such, we recommend further purification steps for most applications, though the Ni-NTA purified material may be sufficient for some experiments. The following is a representative FPLC protocol for purifying superpositive proteins.

(1) The cation exchange column is first washed with deionized water and equilibrated with five column volumes of PBS. We use an Akta FPLC and a 1 mL HiTrap Capto SP XL column.
(2) Dialyze or desalt protein into PBS prior to injection onto FPLC cation exchange column. Many of the supercharged protein-Cre fusion proteins are not soluble in PBS and were dialyzed against PBS + 0.5 *M* NaCl. To cation exchange these proteins, they were diluted 1:4 into PBS immediately prior to injection onto the column.
(3) After injection, five column volumes are passed through the column to remove contaminant proteins and degradation products.
(4) We recommend an elution protocol utilizing a linear gradient from PBS to PBS + 1 *M* NaCL over 10 column volumes.
(5) Following completion of the elution cycle, pool fractions, concentrate the sample to ~1.5 mg/mL and dialyze into storage buffers as needed. To avoid introducing molecules that could influence cell penetration or

delivery, we dialyze proteins against PBS for storage except for Cre proteins which are dialyzed against PBS + 0.5 *M* NaCl.

We have found that Ni-NTA purification, cation exchange and anion exchange often do not fully remove endotoxin from protein samples. If the protein is to be used in endotoxin-sensitive applications, we use an endotoxin removal column and often repeat this process until endotoxin levels are acceptably low (e.g., <0.5 EU/mL).

(1) Endotoxin levels are measured using an endpoint chromogenic limulus amoebocyte lysate endotoxin assay kit (Lonza). Standard endotoxin is diluted to create a standard curve from 0 to 1 EU/mL according to the manufacturer's protocol. Protein is known to inhibit the endotoxin reaction and we have found that +36 GFP concentrations greater than 1.5 μM are inhibitory.
(2) The protein sample is then diluted in endotoxin-free water at non-inhibitory concentrations such as 0%, 0.25%, 0.5%, and 1% of the endotoxin sample.
(3) To remove endotoxin, we use a polymixin B column (Pierce catalog number 20344) according to the manufacturer's instructions. We equilibrate the column with five volumes of freshly prepared PBS.
(4) Protein is eluted with five additional volumes of PBS, concentrated, and retested for endotoxin levels. We have found some protein preparations to initially contain >30 EU/mL endotoxin. The polymixin B columns can remove as much as >90% of endotoxin but may need to be repeated to lower endotoxin concentrations to acceptable levels.

2.8. Construction of fusion proteins

Unfused supercharged proteins are typically purified by affinity chromatography with either N- or C-terminal His_6-tagging. For fusion of the supercharged protein to other protein cargoes, the orientation of the fusion partners and affinity tags can be optimized to maximize the yield of the full-length fusion. We have tested and optimized the constructs in multiple cases and have arrived at a general-purpose starting architecture shown in Fig. 12.8.

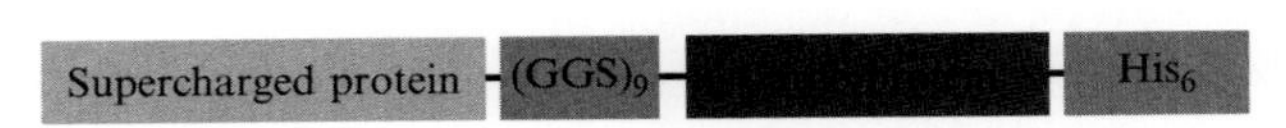

Figure 12.8 Optimized supercharged protein fusion architecture for expression, purification, and protein delivery. (For color version of this figure, the reader is referred to the Web version of this chapter.)

We recommend that the supercharged protein domain be located at the N-terminus for initial fusion attempts. We have found that this orientation generally results in higher expression levels and provides advantages in purification described below. A long flexible glycine-serine linker, $(GGS)_9$, ensures sufficient spacing between folded domains, stable linkage of the purified fusion, and efficient intracellular cleavage of the fusion partners. The fusion partner of interest is located at the C-terminus, followed by a C-terminal His_6 tag. This orientation allows for efficient isolation of full-length fusion protein without additional size-exclusion or gel-filtration chromatography steps. The fusion is first pulled down by the C-terminal tag using Ni-NTA and then further purified via cation exchange to isolate proteins possessing an intact N-terminal supercharged protein domain. This construction and purification strategy ensures that both ends of the fusion are present in the final collected material.

2.9. Nucleic acid delivery by supercharged proteins *in vitro*

The nonviral delivery of siRNA and plasmid DNA into mammalian cells are valuable both for research and therapeutic applications (Akinc *et al.*, 2010). Purified +36 GFP protein (or other superpositively charged protein) is mixed with siRNAs in the appropriate serum-free media and allowed to complex prior addition to cells. Inclusion of serum at this stage inhibits formation of the supercharged protein-siRNA complexes and reduces the effectiveness of the treatment. The following protocol has been found to be effective for a variety of cell lines (McNaughton *et al.*, 2009). However, pilot experiments varying the dose of protein and siRNA should be performed to optimize the procedure for specific cell lines.

(1) One day before treatment, plate 1×10^5 cells per well in a 48-well plate.
(2) On the day of treatment, dilute purified +36 GFP protein in serum-free media to a final concentration 200 n*M*. Add siRNA to a final concentration of 50 n*M*. Vortex to mix and incubate at room temperature for 10 min.
(3) During incubation, aspirate media from cells and wash once with PBS.
(4) Following incubation of +36 GFP and siRNA, add the protein-siRNA complexes to cells.
(5) Incubate cells with complexes at 37 °C for 4 h.
(6) Following incubation, aspirate the media and wash three times with 20 U/mL heparin PBS. Incubate cells with serum-containing media for a further 48 h or longer depending upon the assay for knockdown.
(7) Analyze cells by immunoblot, qPCR, phenotypic assay, or other appropriate method.

We have further found +36 GFP to be an effective plasmid delivery reagent in a range of cells. As plasmid DNA is a larger cargo than siRNA, proportionately more +36 GFP protein is required to effectively complex plasmids. For effective plasmid delivery we have developed a variant of +36 GFP bearing a C-terminal HA2 peptide tag, a known endosome-disrupting peptide derived from the influenza virus hemagglutinin protein. The following protocol has been effective in a variety of cells, but as above it is advised that plasmid DNA and supercharged protein doses be optimized for specific cell lines and delivery applications.

(1) One day before treatment, plate 1×10^5 per well in a 48-well plate.
(2) On the day of treatment, dilute purified +36 GFP protein in serum-free media to a final concentration 2 μ*M*. Add 1 μg of plasmid DNA. Vortex to mix and incubate at room temperature for 10 min.
(3) During incubation, aspirate media from cells and wash once with PBS.
(4) Following incubation of +36 GFP and plasmid DNA, gently add the protein-DNA complexes to cells.
(5) Incubate cells with complexes at 37 °C for 4 h.
(6) Following incubation, aspirate the media and wash with PBS. Incubate cells in serum-containing media and incubate for a further 24–48 h.
(7) Analyze plasmid delivery (e.g., by plasmid-driven gene expression) as appropriate.

2.10. Protein delivery by supercharged proteins *in vitro*

Protein delivery is a valuable strategy for experimental biological research and therapeutics. The direct delivery of proteins without exogenous nucleic acids minimizes the risk of oncogenic mutation or undesired genome alteration. Further, protein delivery is particularly well suited for applications in which prolonged activity of the delivered molecule is not necessary or potentially harmful. Finally, extracellular nucleic acids is known to induce strong antiviral responses (Diebold *et al.*, 2004). Here, we describe a representative protocol for the delivery of proteins fused to supercharged proteins that has been effective across several cell lines:

(1) One day before treatment, plate 1×10^5 per well in a 48-well plate.
(2) On the day of treatment, dilute purified +36 GFP fusion protein into serum-free media.
    (a) If delivering fluorescent reagents such as +36 GFP-mCherry to monitor uptake, significant levels of uptake are observed dose-dependently in treatments ranging from 25 n*M* to 2 μ*M*. Internalization can be visualized within 15 min of treatment, rapidly increasing over the course of 2 h before reaching a plateau after 4 h of incubation. These parameters should be varied as needed for the assay of interest.
    (b) If using ubiquitin-+36GFP as a means to monitor cytosolic localization of internalized proteins, cells should be treated with no more

than 100 n*M* protein in serum-free media for 1 h. The use of higher doses of protein in this assay is not recommended, as removal of extracellular proteins by washing can be increasingly incomplete or variable, confounding downstream analysis. Treatment with a noncleavable ubiquitin-+36GFP, such as the G76V ubiquitin mutant protein, is recommended as a control for nonspecific cleavage by extracellular proteases prior to internalization and within endosomes postinternalization.

(c) If delivering +36 GFP-Cre protein (or other functional protein cargo of interest), we recommend treatment with ~10 n*M* to 2 μ*M*. Incubate cells with proteins for at least 4 h prior to removal and further incubation in serum-containing media.

(3) If desired, treatment with endosome-disrupting reagents such as chloroquine can be included during the incubation with proteins. Cell-line-specific optimization of these agents should be performed, as they are generally toxic. For chloroquine, we recommend a starting dose of ~100 μ*M* chloroquine during the protein treatment, with an optional continued treatment for up to 12 h following removal of the protein-containing media. Optimization of dose and duration of chloroquine (or other endosome disrupting agent) treatment should be performed to maximize protein response while minimizing cyotoxicity.

(4) Following incubation with protein, wash cells at least three times with PBS containing 20 U/mL heparin to remove cell-surface-bound protein. This washing is required for measurement of internalized protein signal, as any remaining extracellular protein can confound interpretation of results.

(5) Analyze cells as appropriate.

(a) For fluorescent protein fusion uptake assays: immediately trypsinize cells in wells to detach from plate. This treatment also works to further remove any remaining uninternalized protein signal, as surface proteins are extensively cleaved during trypsin treatment. Following quenching of trypsin with serum-containing media, we recommend analysis by flow cytometry to determine the extent of protein internalization.

(b) For ubiquitin fusion analysis: immediately following incubation and washing, lyse cells on ice directly in wells with cold LDS sample buffer containing 1 m*M* PMSF. Scrape cells to ensure consistent and complete removal. Analyze by Western blot to observe the fraction of ubiquitin-+36GFP fusion protein that has been specifically cleaved by cytosolic deubiquitinating enzymes. The cleaved product will run ~8 kDa smaller on a 12% SDS-PAGE gel and can be readily quantified by standard densitometry methods.

(c) For Cre protein fusion delivery analysis: incubate cells for 24–48 h posttreatment to allow reporter signal to maximize. Cells should be analyzed for reporter protein expression as appropriate. We have previously used a fluorescent Cre reporter, floxed tdTomato BSR cells, which carry a genomically integrated floxed STOP cassette followed by the tdTomato fluorescent protein. Following successful Cre recombination, the STOP is deleted to allow high expression of tdTomato from the upstream CAG enhancer and CMV promoter. To analyze recombination driven by +36 GFP-delivered Cre protein, trypsinize cells and analyze by flow cytometry. Recombinant cells will display strong fluorescence at 581 nm corresponding to expression of the tdTomato protein. Treatment with 1 to 2 μ*M* +36 GFP-Cre as described above will yield 20–30% recombination in this cell line, with levels as high as 70% being achieved upon cotreatment with 100 μ*M* chloroquine at doses as low as 200 n*M* +36 GFP-Cre.

2.11. Protein delivery by supercharged proteins *in vivo*

Supercharged proteins are also able to deliver protein *in vivo*. We have found that supercharged proteins can deliver functional Cre recombinase to many cell types in vivo including cells of the retina, pancreas, or white adipose tissue (Cronican, 2011). Before performing experiments with animals, consult your local Experimental Review Board or Institutional Animal Care and Use Committee to receive approval for your experimental protocol.

The following is a standard protocol for injection of protein into the subretinal space of mice (Matsuda and Cepko, 2008):

(1) Anesthetize adult mice and keep on a 37 °C pad during and after surgery.
(2) Expose the eyeball pulling down the skin around the eye. Make a small incision in the sclera near the lens using a 30-gauge needle.
(3) Insert an injection needle such as a Hamilton syringe with a 32- or 33-gauge blunt-ended needle into the eyeball through the incision until you feel resistance.
(4) Inject 1 μL of protein sample.
(5) For visualization of protein delivery with a β-galactosidase reporter, harvest retinae 3 days postinjection, fix with 0.5% glutaraldehyde for 30 min, stain with X-Gal overnight, and image.
(6) For sectioning, embed retinae in 50% OCT, 50% of 30% sucrose in PBS and store at −80 °C.
(7) Cut retinae into 30 μm sections and image for X-Gal on a brightfield microscope.

The following is a standard protocol for pancreatic injections:

(1) Spike in adenovirus GFP (1.5×10^8 pfu) or an injection dye to each 100 μL of protein samples that are not fluorescent to identify the sections of the pancreas that have been injected.
(2) Anesthetize adult mice and keep on a 37 °C pad during and after surgery.
(3) Shear hair from the abdomen and sterilize three times with betadine and ethanol.
(4) Make an incision in the abdominal wall, identify and expose the pancreas. Using a 27-gauge needle, directly inject 100 μL of the protein sample into 2–3 foci of the dorsal lobe with a 3/10 cc insulin syringe.
(5) Replace the pancreas into the abdominal cavity and close the incision with a suture and surgical staple. Provide mice with 48 h of analgesia to facilitate recovery.
(6) After 5 days, remove the pancreases and isolate the GFP-positive (or other delivery reporter-positive) regions of the pancreata with a fluorescent dissecting microscope.
(7) For visualization of protein delivery, for example, with a β-galactosidase reporter, fix the pancreas tissues in 20 mL of 4% paraformaldehyde by rocking at 4 °C for 1.5 h and wash three times with 20 mL of PBS by rocking at 4 °C for 5 min. Equilibrate the pancreas in 25 mL of 30% sucrose PBS by rocking at 4 °C overnight. Place the pancreas into a mold and cover with OCT. Incubate at room temperature for 30 min, freeze on dry ice, and store at −80 °C. Cut the frozen molds into 12 μm sections. Immunostain the sections with anti-beta galactosidase antibody and image with a fluorescence microscope.
(8) For quantification of protein delivery with a β-galactosidase reporter, suspend the pancreas in 200 μL of β-galactosidase lysis buffer (2.5 m*M* EDTA, 0.25% NP-40, 250 m*M* Tris, pH 4) plus one Complete Protease Inhibitor tablet (Roche) per 10 mL. Homogenize pancreas tissues by electric pestle (2×15 s) in a 1.5 mL tube and incubate on a rotating drum at 4 °C for 30 min. Centrifuge lysates at 13,000×*g* for 5 min and use 20 μL of supernatant for the β-galactosidase assay (Stratagene). Incubate the β-galactosidase assay reactions at 37 °C for ~30 min and assay for absorbance at 575 nm. The β-galactosidase assay can be normalized to a standard curve using commercially available β-galactosidase enzyme (Abcam). The pancreas supernatants (5 μL) can also be added to a bicinchoninic acid protein assay (Pierce) to quantify total protein concentration against a BSA standard.

The following is a standard protocol for subcutaneous injection near white adipose tissue:

(1) Inject adult mice (anesthesia optional) subcutaneously above the pelvic bone on either side of the mouse abdomen with 100 μL of protein sample. Insert a 3/10 cc insulin syringe underneath the mouse skin and prior to injection the tip has to be angled to confirm that the syringe is underneath the skin but has not penetrated the peritoneum.
(2) For visualization of protein delivery, for example, with a luciferase reporter, after 3 days, inject the mice with 400 μL of 7.5 μg/mL luciferin. Sacrifice the mice after 5 min, extract white adipose tissue, and image for luminescence using the IVIS molecular imaging system (Caliper).
(3) For quantification, suspend the white adipose tissue in 2 mL of 0.1% Triton in PBS plus one Complete protease inhibitor tablet (Roche) per 10 mL buffer. Homogenize the white adipose tissues using an Omni tissue homogenizer (2×30 s) in a 15 mL tube and then incubate on a rotating drum at 4 °C for 30 min. Centrifuge lysates at 13,000×*g* for 5 min and add 40 μL of supernatant to 200 μL of the luciferase reaction buffer (Stratagene). The white adipose tissue supernatant was also added (5 μL) to a bicinchoninic acid protein assay (Pierce) to quantify total protein concentration against a BSA standard.

Supercharged proteins have been demonstrated to deliver functional protein to the retina, pancreas, and white adipose tissue in mouse *in vivo*.

3. Conclusion

Supercharged proteins are a class of engineered and naturally existing proteins with highly positive or negative net theoretical charge (typically >1 net charge unit per kDa of molecular weight). Both negatively and positively supercharged proteins display remarkable chemical and biological properties, including resistance to chemically or thermally induced aggregation. Superpositively charged proteins can bind and potently penetrate mammalian cells and can effectively deliver both nucleic acid and protein cargoes into cells in functional form. Protein delivery by supercharged proteins can also function *in vivo* in multiple tissue types of therapeutic interest, including the retina, the pancreas, and white adipose tissue. The potency of functional delivery effected by supercharged protein reagents can, in some tissues, exceed the current best methods for macromolecule delivery including the Tat peptide and adenoviral vectors. As such, supercharged proteins represent a powerful tool to deliver macromolecules into cells and tissues of interest, in both basic research and prospective therapeutic applications.

REFERENCES

Akdim, F., *et al.* (2010). Effect of mipomersen, an apolipoprotein B synthesis inhibitor, on low-density lipoprotein cholesterol in patients with familial hypercholesterolemia. *Am. J. Cardiol.* **105,** 1413–1419.

Akinc, A., *et al.* (2010). Targeted delivery of RNAi therapeutics with endogenous and exogenous ligand-based mechanisms. *Mol. Ther.* **18,** 1357–1364.

Akinc, A., *et al.* (2008). A combinatorial library of lipid-like materials for delivery of RNAi therapeutics. *Nat. Biotech.* **26,** 561–569.

Arkin, M. R., and Wells, J. A. (2004). Small-molecule inhibitors of protein-protein interactions: progressing towards the dream. *Nat. Rev. Drug. Discov.* **3,** 301–317.

Cronican, J. J., *et al.* (2010). Potent delivery of functional proteins into Mammalian cells in vitro and in vivo using a supercharged protein. *ACS Chemical Biology* **5,** 747–752.

Cronican, J. J., *et al.* (2011). A class of human proteins that deliver functional proteins into mammalian cells in vitro and in vivo. *Chem. Biol.* **18,** 833–838.

Diebold, S. S., Kaisho, T., Hemmi, H., Akira, S., and Reis e Sousa, C. (2004). Innate antiviral responses by means of TLR7-mediated recognition of single-stranded RNA. *Science.* **303,** 1529–1531.

Dill, K. A., Ozkan, S. B., Shell, M. S., and Weikl, T. R. (2008). The protein folding problem. *Annu. Rev. Biophys.* **37,** 289–316.

Durcan, N., Murphy, C., and Cryan, S. A. (2008). Inhalable siRNA: potential as a therapeutic agent in the lungs. *Mol. Pharm.* **5,** 559–566.

Fuchs, S. M., and Raines, R. T. (2007). Arginine grafting to endow cell permeability. *ACS Chem. Biol.* **2,** 167–170.

Gump, J. M., June, R. K., and Dowdy, S. F. (2010). Revised role of glycosaminoglycans in TAT PTD-mediated cellular transduction. *J. Biol. Chem.* **285,** 1500–1507.

Harding, C., Heuser, J., and Stahl, P. (1983). Receptor-mediated endocytosis of transferrin and recycling of the transferrin receptor in rat reticulocytes. *J. Cell. Biol.* **97,** 329–339.

Jantsch, J., *et al.* (2008). Small interfering RNA (siRNA) delivery into murine bone marrow-derived dendritic cells by electroporation. *J. Immunol. Methods* **337,** 71–77.

Jo, D., *et al.* (2001). Epigenetic regulation of gene structure and function with a cell-permeable Cre recombinase. *Nat. Biotech.* **19,** 929–933.

Jones, S., Holm, T., Mäger, I., Langel, U., and Howl, J. (2010). Characterization of bioactive cell penetrating peptides from human cytochrome c: protein mimicry and the development of a novel apoptogenic agent. *Chem. Biol.* **17,** 735–744.

Kam, N. W., Liu, Z., and Dai, H. (2005). Functionalization of carbon nanotubes via cleavable disulfide bonds for efficient intracellular delivery of siRNA and potent gene silencing. *J. Am. Chem. Soc.* **127,** 12492–12493.

Karimova, G., Pidoux, J., Ullmann, A., and Ladant, D. (1998). A bacterial two-hybrid system based on a reconstituted signal transduction pathway. *Proc. Natl. Acad. Sci.* **95,** 5752–5756.

Karshikoff, A., and Ladenstein, R. (2001). Ion pairs and the thermotolerance of proteins from hyperthermophiles: a "traffic rule" for hot roads. *Biochem. Sci.* **26,** 550–556.

Lawrence, M. S., Phillips, K. J., and Liu, D. R. (2007). Supercharging proteins can impart unusual resilience. *Journal of the American Chemical Society* **129,** 10110–10112.

Lee, S. K., Han, M. S., Asokan, S., and Tung, C. H. (2011). Effective Gene Silencing by Multilayered siRNA-Coated Gold Nanoparticles. *Small* **7,** 364–370.

Matsuda, T., and Cepko, C. L. (2007). Controlled expression of transgenes introduced by in vivo electroporation. *Proc. Natl. Acad. Sci. USA* **104,** 1027–1032.

McNaughton, B. R., Cronican, J. J., Thompson, D. B., and Liu, D. R. (2009). Mammalian cell penetration, siRNA transfection, and DNA transfection by supercharged proteins. *Proc. Natl. Acad. Sci. USA* **106,** 6111–6116.

Nishina, K., *et al.* (2008). Efficient in vivo delivery of siRNA to the liver by conjugation of alpha-tocopherol. *Mol. Ther.* **16,** 734–740.

Overington, J. P., Al-Lazikani, B., and Hopkins, A. L. (2006). How many drug targets are there? *Nat. Rev. Drug. Discov.* **5,** 993–996.

Patil, L. M., *et al.* (2009). Internally Cationic Polyamidoamine PAMAM-OH Dendrimers for siRNA Delivery: Effect of the Degree of Quaternization and Cancer Targeting. *Biomacromol.* **10,** 258–266.

Shen, Z. P., Brayman, A. A., Chen, L., and Miao, C. H. (2008). Ultrasound with microbubbles enhances gene expression of plasmid DNA in the liver via intraportal delivery. *Gene. Ther.* **15,** 1147–1155.

Soutschek, J., *et al.* (2004). Therapeutic silencing of an endogenous gene by systemic administration of modified siRNAs. *Nature (London)* **432,** 173–178.

Stemmer, W. P., Crameri, A., Ha, K. D., Brennan, T. M., and Heyneker, H. L. (1995). Single-step assembly of a gene and entire plasmid from large numbers of oligodeoxyribonucleotides. *Gene* **164,** 49–53.

Thompson, D. B., Villasenor, R., Zerial, M., and Liu, D. R. Probing the Mechanism of Cellular Uptake and Trafficking of Supercharged Proteins. *Submitted.*

Tokuriki, N., Stricher, F., Schymkowitz, J., Serrano, L., and Tawfik, D. S. (2007). The stability effects of protein mutations appear to be universally distributed. *J. Mol. Biol.* **369,** 1318–1332.

Turcotte, R. F., Lavis, L. D., and Raines, R. T. (2009). Onconase cytotoxicity relies on the distribution of its positive charge. *FEBS J.* **276,** 3846–3857.

Veldhoen, S., Laufer, S. D., Trampe, A., and Restle, T. (2006). Cellular delivery of small interfering RNA by a non-covalently attached cell-penetrating peptide: quantitative analysis of uptake and biological effect. *Nucleic Acids Res.* **34,** 6561–6573.

Wadia, J. S., Stan, R. V., and Dowdy, S. F. (2004). Transducible TAT-HA fusogenic peptide enhances escape of TAT-fusion proteins after lipid raft macropinocytosis. *Nature Medicine* **10,** 310–315.

Watanabe, K., *et al.* (2009). In vivo siRNA delivery with dendritic poly(l-lysine) for the treatment of hypercholesterolemia. *Mol. Biosyst.* **5,** 1306–1310.

Yanisch-Perron, C., Vieira, J., and Messing, J. (1985). Improved M13 phage cloning vectors and host strains: nucleotide sequences of the M13mp18 and pUC19 vectors. *Gene* **33,** 103–119.

Zhao, H., and Arnold, F. H. (1997). Functional and Nonfunctional Mutations Distinguished by Random Recombination of Homologous Genes. *A* **94,** 7997–8000.

Zhao, H., Giver, L., Shao, Z., Affholter, J. A., and Arnold, F. H. (1998). Molecular evolution by staggered extension process (StEP) in vitro recombination. *Nat. Biotech.* **16,** 258–261.

Zhou, Q., Brown, J., Kanarek, A., Rajagopal, J., and Melton, D. A. (2008). In vivo reprogramming of adult pancreatic exocrine cells to beta-cells. *Nature* **455,** 627–632.

Author Index

Note: Page numbers followed by "*f*" indicate figures, and "*t*" indicate tables.

A

Abbatiello, S., 152–153
Abdeljabbar, D. M., 79
Abreu, L. C., 245
Achilefu, S., 36–37
Ackerman, M. E., 140*t*, 147, 148, 229, 234
Adams, D. J., 58–60, 61*f*, 65*f*, 67*f*
Adams, G. P., 108, 109, 262, 263
Adams, J. J., 190
Adamson, G. N., 260
Adamson, S., 161–162
Affholter, J. A., 306
Agemy, L., 38*t*, 39–41, 45, 51
Aggen, D. H., 191, 192–193, 195, 201, 206, 207–208, 209, 213–214
Aha, P. M., 137, 140*t*, 144–145, 148
Aird, D., 148
Akcan, M., 58–59
Akdim, F., 294–295
Akerman, M. E., 36–37
Åkerström, B., 159
Akinc, A., 294–295, 312
Akira, S., 313–315
Alamed, J., 277–278, 286–287
Albericio, F., 239
Albert, A., 58
Aleksic, M., 190–191
Alewood, D., 60
Alewood, P. F., 58–60, 61*f*, 65*f*, 66–67, 238–239
Al-Lazikani, B., 294
Allen, P. M., 192–193, 195, 198–200, 207–208, 209
Allen, T. M., 36, 109–110, 259–261, 260*f*
Allison, J. P., 164–166
Alpaugh, R. K., 109, 262
al-Ramadi, B. K., 198–200
Alsina, J., 239
Altman, J. D., 198–200
Altmann, A., 78*t*
Alvarez-Salas, L. M., 5
Alvarmo, C., 58, 59–60, 61*f*
Amalfitano, A., 111
Amemiya, C. T., 103
Ametamey, S. M., 281–282
Amstutz, P., 102, 103–106, 104*f*, 107, 112–113, 116, 136
Anbazhagan, R., 260
Anderson, C. J., 260*f*
Anderson, J. M., 40
Anderson, M. A., 58, 67–68
Andrusiak, R., 37–39
Angell, Y. M., 239
Anton van der Merwe, P., 207–208
Arai, M., 59–60, 61*f*
Arap, W., 36–39, 38*t*, 41
Ardelt, P. U., 36–37
Arden, B., 190–191, 207–208
Arendash, G. W., 286–287
Arkin, M. R., 294
Armishaw, C. J., 58, 59–60, 61*f*
Armstrong, K. M., 190
Arndt, K. M., 106–107
Arnold, B., 37–39
Arnold, F. H., 306
Arnold, S., 163
Arora, P. S., 5–6
Artyomov, M. N., 191
Ashikaga, T., 36–37
Asokan, S., 294–295
Aspmo, S. I., 5–6
Attard, P., 59–60, 61*f*
Atwal, J. K., 4–5
Auckenthaler, A., 163
Avrutina, O., 226, 238
Axmann, M., 190–191
Azimzadeh, A., 181–182

B

Bachmann, B. J., 167
Baggio, R., 144–145, 227
Bahrami, S. B., 58–59, 225
Bai, A., 191, 200–201, 209
Baici, A., 106
Bailey, C. W., 136–137
Baker, A. H., 111
Baker, B. M., 190–191, 192–193, 207–208, 209, 213–214
Baker, J. R. Jr., 260
Bakshi, S., 260*f*
Balatoni, J., 111
Ballmaier, M., 226
Balogh, L. P., 260
Bandeiras, T. M., 112
Bander, N. H., 260*f*
Bankovich, A. J., 192, 207–208, 209
Banks, D., 213–214

Bao, A., 226–227
Barany, G., 239
Barbour, R., 277–278
Bargou, R., 191
Barisic, I., 287
Barker, J. J., 58
Barkinge, J., 138, 140*t*, 142, 144, 152–153
Barnathan, E. S., 36
Barouch, D. H., 198–200
Barsh, G. S., 225–226, 238
Bart, J., 260*f*
Bartlett, D. W., 259–261, 260*f*
Bastidas, D., 260*f*
Baston, E., 190–191, 192, 207–208
Basu, S., 231
Baumann, M. J., 112
Baum, C., 201
Baynes, B. M., 148, 231
Becker, H., 285–286
Becker, P. M., 40
Becker, S., 225, 238, 241
Beck, J. L., 59–60, 61*f*
Becskei, A., 286–287
Behnke, S., 107–108
Belayev, L., 285–286
Bell, J. I., 198–200, 207–208
Bellows, M. L., 79
Belmont, H., 207–208, 209
Belousova, N., 106, 111
Beltzer, J., 90
Bendotti, C., 286
Bennett, A. D., 191
Bennett, K. L., 138, 140*t*, 142, 144, 152–153
Benz, C. C., 259–261, 260*f*
Bergers, G., 36–37
Berk, D. A., 259
Bernal, F., 8*t*, 9–10, 11–12, 17–18, 20–25, 22*f*, 24*f*, 26, 27–28, 29*t*
Bernard, M. A., 38*t*
Bernasconi, M., 36–37, 38*t*, 41
Bernkop-Schnurch, A., 225
Bertozzi, C. R., 117–118
Bessette, P. H., 76–77, 79, 93
Beste, G., 159, 161, 163, 164, 168*f*, 170, 174–175, 178, 180
Betzi, S., 4
Bhatia, S. N., 36–37, 38*t*, 39–41
Bhatt, R. R., 106
Biancalana, M., 153
Bibi, E., 106–107
Bickel, U., 271–272, 281–282
Biddison, W. E., 207–208, 213–214
Biddle, W. C., 260*f*
Bieler, J. G., 198–200
Biggers, K., 109–110
Binder, U., 76–77, 78*t*, 164, 165*f*, 168*f*, 176–177
Binz, H. K., 102, 103–107, 104*f*, 108, 109, 112–113, 116, 124, 136, 257–258, 262
Bird, G. H., 8*t*, 9–10, 12, 17–18, 21, 23–25
Birtalan, S., 144
Bjugstad, K. B., 286–287
Black, M., 37–39, 40–41
Black, R. S., 280–282, 286–287
Blackwell, H. E., 6–7
Blangy, S., 106
Blankenstein, T., 201
Bloom, L., 136, 137, 140*t*, 144–145
Bluhm, M. E., 162, 166–167
Blumenthal, R. D., 263
Boado, R. J., 271–272, 274–282, 275*t*, 276*f*, 277*f*, 280*f*, 281*f*, 282*t*, 283, 283*f*, 284–288, 284*t*
Boczek, E., 106, 111
Boder, E. T., 89–90, 192, 227, 228–229, 234, 236, 237
Boerman, O. C., 263
Boersma, Y. L., 106–107
Bohn, M., 285–286
Boigegrain, R. A., 224–225
Boldicke, T., 19
Bolin, K. A., 225–226
Bombardieri, E., 260*f*
Bond, T., 58, 225
Bongarzone, I., 260*f*
Boniface, J. J., 190–191, 207–208
Bookman, M. A., 108
Boo, L. M., 58
Borgna, L., 37–39
Bork, P., 103
Borregaard, N., 159, 162, 166–167
Bosley, A. D., 236
Bosshard, H. R., 102, 106, 229
Bothwell, A. L., 198–200
Boulter, J. M., 192, 207–208
Boulware, K. T., 76–77, 79, 93
Bourel, L., 19
Bourette, R. P., 93
Bowerman, N. A., 190–191, 207–208, 213–214
Bowers, W. J., 270–271
Bowley, D. R., 229
Bradbury, A. R., 164, 172–173
Brady, J., 77
Brady, R. L., 58
Braeckmans, K., 121
Brameshuber, M., 190–191
Bramlage, B., 77
Braun, P. G., 106–107
Brayman, A. A., 294–295
Bredesen, D. E., 37–39, 38*t*
Brennan, T. M., 305
Brenner, S. E., 49
Breustedt, D. A., 159, 162, 167, 170
Brewer, J. E., 213–214
Briand, C., 104*f*, 106, 107, 112–113, 116
Brid, C., 58, 59–60, 61*f*, 67*f*
Bristol, G., 58
Brodin, P., 4–5

Brophy, S. E., 192, 207–208
Brown, D. M., 38*t*, 39–40
Brown, J., 153, 294–295
Brown, N., 159, 160*f*, 162–163, 164–166, 170, 184
Brugman, M. H., 201
Bruhlmeier, M., 281–282
Buchholz, C. J., 111
Buciak, J. B., 271–273, 284
Buciak, J. L., 272, 274, 281–282
Buck, C. A., 36
Bueno, M. M., 58
Bu, G., 273
Bugaj, J. E., 36–37
Builes, J., 207–208, 209
Buining, R. J., 106–107
Bullock, W. O., 171–172
Burdette-Radoux, S., 36–37
Burgstaller, P., 144–145, 227
Burton, D. R., 106, 229
Buschor, P., 112
Busto, R., 285–286

C

Caflisch, A., 103–105
Cai, W., 40–41
Calabro, V., 136, 137
Calvaresi, N., 286
Camarero, J. A., 59–60, 61*f*
Cambillau, C., 106, 136–137
Cameron, B. J., 213–214
Cam, H., 201
Campanacci, V., 106
Campbell, A., 77
Campbell, I. D., 136–137
Cancasci, V. J., 144
Candido, K. A., 260
Canevari, S., 260*f*
Cantley, L. C., 40
Cao, C., 153
Cao, Z., 260
Capitani, G., 106
Capretto, D. A., 260*f*
Card, K. F., 207–208, 209
Cardo-Vila, M., 36–37
Carion, O., 19
Carlsson, J., 257–258
Carmeliet, P., 36
Carr, L. M., 195
Carroll, R. G., 192
Carugo, O., 224–225
Carver, L. A., 36–37
Castro, B., 224–226, 238–239
Ceitlin, J., 103
Cemazar, M., 58
Cepko, C. L., 315–317
Cerundolo, V., 190–191
Chabner, B. A., 37–39
Chai, B. X., 225–226, 238
Chakrabarti, P., 5–6
Chakrabarty, T., 113
Chakraborty, A. K., 191
Chan, A. C., 40
Chan, C. M., 209
Chandonia, J. M., 49
Chan, L. A., 91
Chanson, A. H., 58
Chan, T. R., 117–118
Chan, W. C., 238–239
Chao, G., 147, 148, 153, 228–229, 232, 233, 234, 235–236, 237
Chapman, A. P., 108, 116
Chatwell, L., 159, 160*f*, 162–163, 164–166, 170, 184
Chaudhary, V. K., 127
Chen, C. L., 76–77
Chen, D., 12
Chene, P., 4
Cheng, P., 140*t*, 147
Cheng, Y., 243–244
Cheng, Z., 58–59, 225, 226–227, 230–231, 241, 243*f*, 244–246
Chen, J. L., 36–39, 190–191, 200–201, 209
Chen, K., 40–41
Chen, L., 36–37, 294–295
Chen, X., 40–41
Chen, Y., 4–5, 37–39, 140*t*, 144–145, 225–226
Chen, Y. H., 141
Chen, Y. N., 225, 238–239
Chervin, A. S., 191, 192–193, 198–201, 206, 207–208, 209, 213–214
Chiche, L., 67–68, 224–225
Chien, Y., 190–191, 207–208
Chilkoti, A., 256, 259
Chiorean, E. G., 136, 153
Chishti, A. H., 40
Chlewicki, L. K., 190–191, 198–200
Choi, J. Y., 36–37
Chothia, C., 143
Chow, K. C., 171–172
Cho, Y. K., 227
Christian, B. T., 281–282
Christian, S., 36–37
Christinger, H. W., 90
Christmann, A., 76–77, 226, 227, 230
Christopher Garcia, K., 190–191
Chung, S., 207–208, 209
Chu, Q., 25
Cichutek, K., 111
Cines, D. B., 36
Clarke, A. R., 58
Clarke, N., 79
Clark-Lewis, I., 60
Clark, R. J., 58–60, 61*f*, 65*f*, 67–68, 67*f*, 225–226
Claus, R., 226

Cochran, A. G., 90
Cochran, F. V., 226–227, 228–229, 230–231, 232, 235*f*, 237, 238, 240, 244–246
Cochran, J. R., 58–59, 225, 226–227, 228–229, 228*f*, 230–231, 232, 233, 235*f*, 236, 237, 238, 240, 241, 243*f*, 244–246
Cohen, F. E., 106
Colby, D. W., 228–229, 232, 233, 234, 237, 244
Cole, A. M., 58
Coleman, J. R., 229
Coletti-Previero, M. A., 224–226
Colf, L. A., 192, 207–208, 209
Colgrave, M. L., 58, 59–60, 61*f*, 225–226
Coliva, A., 260*f*
Converse, A. K., 281–282
Cooke, R., 113
Coombs, D., 190–191
Cooper, M. D., 103
Cornejo, M., 6, 8*t*, 9–12, 17–18, 21–25, 26, 27–28
Corti, A., 37–39
Cotton, J., 225
Couillin, I., 161–162
Covell, D. G., 263
Cowburn, D., 59–60, 61*f*
Craik, D. J., 58–60, 61*f*, 65*f*, 66–68, 67*f*, 224–226, 238–239
Crameri, A., 305
Creswell, K., 37–39
Crofts, T. S., 190–191
Crompton, A., 40
Cronican, J. J., 4–5, 298–299, 300–301, 312, 315–317
Crooks, G. E., 49
Cropp, T. A., 117–118
Crusius, K., 112
Cryan, S. A., 294–295
Cullis, P. R., 36
Culter, C. S., 226–227
Cunningham, B. C., 90, 142, 143–144
Curnis, F., 37–39
Curreli, F., 17–18
Cuthbertson, A., 238–239

D

Dai, H., 40–41, 294–295
Daly, N. D., 65*f*
Daly, N. L., 58–60, 61*f*, 66, 67–68, 67*f*, 224–226, 238–239
Dane, K. Y., 78*t*, 79, 91
Daugherty, P. S., 76–77, 78*t*, 79, 86–87, 88, 90, 91, 93
Davidson, R. J., 281–282
Davis, M. E., 259–261, 260*f*
Davis, M. M., 190–191, 198–200, 207–208
Dawson, P. E., 60–61
Day, M., 225
Dean, R. A., 286–287
Debets, M. F., 117–118, 120
De Caterina, R., 36
Deechongkit, S., 59–60, 61*f*
De Groot, A. S., 288
Deichmann, A., 201
Deiters, A., 117–118
de Jong, S., 260*f*
de Leij, L. F., 109–110
de Leij, L. M., 109–110
Delic, M., 90
Dellian, M., 259
Del Rosario Aleman, M., 77
Demattos, R. B., 286–287
Demeester, J., 121
DeMong, P., 12
Dempster, L., 58, 59–60, 61*f*, 67*f*
DeNardo, G. L., 260
DeNardo, S. J., 260
Dennis, M. S., 230
Denzin, L. K., 207–208
de Pasquale, C., 108, 109, 124, 257–258, 262
Derfus, A. M., 36–37, 38*t*, 39–40
Derossi, D., 4–5
Deshpande, N., 226–227
De Smedt, S. C., 121
Deuber, S. A., 112
Devadas, S., 191
Devin, F., 112
Devlin, M., 90
de Vries, E. G., 260*f*
Dewhirst, M. W., 256, 259
de Wolf, F. A., 117–118
Dickson, P., 287–288
Di Cresce, C., 5
Diebold, S. S., 313–315
Diederichsen, U., 226, 238
Dill, K. A., 301–303, 305
Dinner, A. R., 191
Di Paolo, C., 109–110, 127
Diril, H., 259
Dixon, N. E., 59–60, 61*f*
Dobson, C. M., 136–137
Doherty, P. C., 190
Dohn, K., 128
Domoki, F., 286–287
Donermeyer, D. L., 192–193, 195, 198–200, 207–208, 209
Donzeau, M., 128
Do, P., 190–191
Dossett, M. L., 192–193, 195
Douglas, N. R., 238
Dowdy, S. F., 4–5, 297–299
Dower, W. J., 106
Dreher, M. L., 177–178
Dreher, M. R., 256, 259
Dreier, B., 106, 107–108, 109, 111, 112–113, 124, 257–258, 262

Driessen, A. J., 106–107
Drinkwater, R. D., 58–59
Dröge, M. J., 106–107
Drummond, D. C., 259–261, 260*f*
Duan, W. M., 286
Dubel, S., 76–77, 79, 164, 172–173
Duffy, K. R., 284
Dunn, S. M., 190–191, 192, 207–208
Durcan, N., 294–295
Durr, E., 36–37
Dushek, O., 190–191
Dutton, J. L., 65*f*
Dux, M., 286–287
Duza, T., 36–37, 38*t*, 39–40

E

Eaton, C., 136, 153
Eberspaecher, U., 112
Edwards, L. J., 190–191
Eggel, A., 112
Ehrenreich, H., 285–286
Ehrhardt, G. R., 103
Eichinger, A., 159, 160*f*, 162–163, 167, 170, 184
Eigenbrot, C., 230
Eisenberg, J., 274
Eisen, H. N., 191, 200–201, 209
Eisenhut, M., 78*t*
El-Gamil, M., 191
Elias, J. A., 40
Ellerby, H. M., 37–39, 38*t*
Ellerby, L. M., 37–39
Ellgaard, L., 227
Elliot, S., 286
Elliott, A. G., 58
Ellison, E. C., 36–37
Ellman, G. L., 128
Ely, L. K., 190
Emanuel, S. L., 153
Engelhardt, B., 272
Engels, B., 200–201
Engle, L. J., 153
Eriksson, T. L., 257–258
Erion, J. L., 36–37
Esaki, K., 138, 141, 143, 144, 147, 152–153
Essler, M., 36–37, 38*t*, 39–40
Evavold, B. D., 190–191
Ewald, C., 103–105

F

Fabri, L. J., 225–226
Fahnert, B., 209
Fairbrother, W. J., 90
Famm, K., 141
Fand, I., 108, 263
Fanghänel, J., 112
Fanning, A. S., 40
Fan, X., 38*t*
Farlow, M. R., 286–287
Farokhzad, O. C., 259–260, 260*f*
Fasman, G. D., 20–21
Favel, A., 224–226
Fedosov, S. N., 77
Fee, L., 241, 243
Fekkes, P., 106–107
Feldhaus, J. M., 229
Feldhaus, M. J., 229
Feldman, S. A., 191
Feller, S. M., 93
Fellouse, F. A., 144
Felsher, D., 58–59, 226–227
Feng, D., 190
Fernandez, J. M., 171–172
Ferrara, N., 166
Fey, R., 38*t*
Figini, M., 260*f*
Finn, M. G., 117–118
Fire, E., 8*t*, 9–10, 17–18, 26
Fischer, H., 58, 59–60, 61*f*, 67*f*
Fischer, J., 286–287
Fischer, M., 102, 163
FitzGerald, D. J., 127
Fitzpatrick, E., 153
Flamm, A., 36–37
Fleig, J., 78*t*
Fleming, T. J., 90
Fletcher, J. E., 263
Fletterick, R. J., 106
Fling, S. P., 169–170
Flippen-Anderson, J. L., 5–6
Flores, L. II., 111
Floudas, C. A., 79
Flower, D. A., 159
Flower, D. R., 159
Fogal, V., 36–37, 39–40, 51, 52
Fokin, V. V., 117–118
Foley, F. M., 225–226
Folkman, J., 36
Foon, K. A., 260*f*
Foreman, A. R., 260
Forrer, P., 102, 103–107, 104*f*, 111, 112–113, 116, 142
Fox, A. S., 281–282
Fraaije, M. W., 79
Frances, H. A., 306
Fraser, C. C., 201
Freymuller, E., 245
Friden, P. M., 272
Friedman, S. M., 207–208, 209
Friedrich, K., 226
Friedrich, S., 286–287
Fritsch, E. F., 167–169
Fritzberg, A. R., 260*f*
Fry, D. C., 4
Fu, A., 284–287, 284*t*
Fu, C., 40

Fuchs, S. M., 297–298
Fuh, C., 90
Fuh, G., 230
Fuhrmann, M., 163
Fujimori, K., 263
Fukumura, D., 259
Fuller, S., 36–37
Fumic, K., 287
Furfine, E., 136, 153
Fushman, D., 59–60, 61*f*
Fu, Y. X., 207–208, 213–214

G

Gabikian, P., 58–59, 225
Gabizon, A., 259–260, 260*f*
Gabrijelcic-Geiger, D., 226, 238
Gagnon, J., 67–68
Gahmberg, C. G., 38*t*
Gai, S. A., 200, 227
Galvez, A., 58
Gambhir, S. S., 58–59, 225, 226–227, 241, 243*f*, 244–246
Ganss, R., 37–39
Gantz, I., 225–226, 238
Gao, G. F., 207–208
Gao, Y. G., 209
Garboczi, D. N., 207–208, 209
Garcia, K. C., 190, 192, 207–208, 209
Garcia, M. L., 58
Garcia, R. S., 58
Garrigue, A., 201
Gartland, G. L., 103
Gasparri, A., 37–39
Gavathiotis, E., 8*t*, 26
Gay, D. A., 36–37
Gebauer, M., 158–159, 162–163
Geisbert, J. B., 111
Geisbert, T. W., 111
Ge, L., 102
Gelly, J. C., 224–225
Georgiou, G., 76–77
Geretti, E., 40
Gerlowski, L. E., 36
Getmanova, E. V., 140*t*, 144–145
Getz, E. B., 113
Getz, J. A., 77
Ghosh, P., 207–208
Gibson, T. J., 49
Gietz, R. D., 194
Gilbert, R. J. C., 93
Gilbreth, R. N., 138, 139–141, 143, 147, 152–153
Gill, D., 277–278
Gille, H., 159, 160*f*, 162–163, 164–166, 170, 184
Gill, S. C., 169–170
Ginsberg, M. D., 285–286
Giordano, R. J., 36–37
Girard, O. M., 38*t*, 39–40, 45, 51
Giraudo, E., 36–37, 38*t*
Giriat, I., 59–60, 61*f*
Giussani, A., 260*f*
Giver, L., 306
Gladow, M., 201
Glass, D., 259–260
Glick, M., 192, 207–208
Glimm, H., 201
Glockshuber, R., 102
Goebeler, M., 191
Goetz, D. H., 162, 166–167
Gokemeijer, J., 136, 140*t*, 144–145, 153
Goldenberg, D. M., 259, 263
Goldrosen, M. H., 260*f*
Goldsmith, S. J., 260*f*
Gonzales, C. R., 286–287
Gonzalez, C., 58
Gonzalez-Lepera, C., 111
Gordon, M. N., 277–278, 286–287
Goren, D., 259–260, 260*f*
Gorina, S., 28
Gottschall, P. E., 277–278, 286–287
Gottstein, C., 78*t*, 79, 91
Goulder, P. J. R., 198–200
Grabski, A. C., 210
Grabstein, K., 118, 120
Gracy, J., 224–225
Graff, C. P., 228–229, 232, 233, 234, 237, 244, 263
Gray, T., 241, 243
Grebien, F., 138, 140*t*, 142, 144, 152–153
Greenberg, P. D., 192–193, 195, 200–201
Greenfield, N., 20–21
Greenwald, D. R., 40–41
Gregerson, D. S., 169–170
Gregson, M. W., 140*t*, 147
Griffin, J. H., 60–61
Griffiths, G. L., 259
Grignaschi, G., 286
Grimm, J., 277–278, 286–287
Grimsley, G., 241, 243
Grishin, A. A., 58, 59–60, 61*f*
Groner, B., 127
Groom, C. R., 4
Grossmann, T. N., 12
Grubbs, R. H., 6–7
Gruettner, C., 260
Grundman, M., 280–282, 286–287
Grütter, M. G., 103–105, 104*f*, 106, 107, 112–113, 116
Guerlesquin, F., 4
Gu, F., 259–260, 260*f*
Guharoy, M., 5–6
Gu, K., 137, 140*t*, 144–145
Gulotti-Georgieva, M., 106, 112
Gump, J. M., 298–299
Gunasekera, S., 58, 59–60, 61*f*, 225–226

Gunn, M. D., 141
Guo, W. P., 286
Guo, Z., 259–260, 260*f*
Gu, X., 224–225
Gyapay, G., 201

H

Haberkorn, U., 78*t*
Hacein-Bey-Abina, S., 201
Hackel, B. J., 140*t*, 147, 148, 201–202, 228–229, 230, 232, 233, 234, 235–236, 237
Hackeng, T. M., 60–61
Hackman, R. C., 58–59, 225
Haedicke, W., 38*t*
Hafler, D. A., 207–208, 209
Hait, W. N., 37–39
Ha, K. D., 305
Hale, S. P., 144–145, 227
Hall, S. S., 79, 87
Hambley, T. W., 37–39
Hammerling, G. J., 37–39
Hampl, J., 190–191, 207–208
Hamzah, J., 37–40
Hanahan, D., 36–37, 38*t*, 39–40, 45, 51
Hanes, J., 102, 106, 229
Hanick, N. A., 192, 207–208
Han, M. S., 294–295
Hansen, R. E., 128–129
Hansen, S. J., 58–59, 225
Hantschel, O., 138, 140*t*, 142, 144, 152–153
Han, Z., 152–153
Harding, C., 298
Hardman, K. D., 207–208
Harkiolaki, M., 93
Harmsen, M. C., 109–110
Harrington, K. J., 259–260
Harwerth, I. M., 127
Hasadsri, L., 4–5
Healy, K. E., 91
Heavner, J. E., 225
Heckman, K. L., 170
Heikkila, P., 38*t*
Heinzelman, P., 229
Heitz, A., 67–68, 224–225
Heitz, F., 4–5
He, L., 40–41
Helenius, A., 227
Hemmi, H., 313–315
Henchey, L. K., 5–6
Hengen, P. N., 171–172
Hernandez, A., 17–18
Hernandez, J. F., 67–68, 207–208, 209
Herrman, M., 285–286
Hershberger, K. A., 226–227, 230–231, 236
Herve, M., 225
Hess, G., 191
Heuser, J., 298
Heya, T., 263
Heyneker, H. L., 305
Higgins, D. G., 49
Hilbert, D. M., 90
Hildebrandt, I. J., 259–261, 260*f*
Hilgraf, R., 117–118
Hilinski, G. J., 29–30
Hillig, R. C., 112
Hill, M. E., 213–214
Hill, T., 36–37
Hilpert, K., 226
Hinkle, G., 36–37
Hinni, K., 36–37
Hippenstiel, S., 226
Hjelm, B., 78*t*
Hobbs, C. A., 127
Hocke, A. C., 226
Hoess, A., 102
Hoffman, J. A., 36–37, 38*t*, 39–40, 41
Hoffman, R. M., 36–37, 38*t*, 39–40
Hohlbaum, A. M., 159, 160*f*, 162–163, 164–166, 170, 184
Höhne, W., 106–107, 226
Hoiden-Guthenberg, I., 257–258
Holler, P. D., 191, 192, 195, 198–201, 207–208, 209
Holman, P. O., 192, 198–200, 207–208
Holmes, M. A., 162, 166–167
Holm, T., 308
Honegger, A., 102
Honer, H., 281–282
Hon, G., 49
Hong, H. Y., 36–37
Hong, K., 259–261, 260*f*
Hong, S., 37–39, 271–272
Hong, T. T., 58, 67–68
Hooijberg, E., 191, 201
Hopkins, A. L., 4, 294
Horak, E. M., 109
Horn, G., 163
Horowitz, A. T., 259–260, 260*f*
Hosbach, J., 153
Houston, L. L., 108
Howland, S. W., 140*t*, 147, 148
Howl, J., 308
Hruby, V. J., 58
Huang, H. L., 225
Huang, J., 153, 190–191
Huang, R. H., 209
Huang, S. K., 259
Huang, X., 136–137
Huber, T., 106
Hui, E. K. H., 275–278, 275*t*, 281–282, 282*t*, 283, 283*f*, 284–288, 284*t*
Hui, E. K. W., 275–277, 275*t*, 279–281, 281*f*, 284*t*, 286, 287–288
Huimin, Z., 306
Hu, J., 201

Hülsmeyer, M., 159, 160*f*, 162–163, 164–166, 170, 184
Hung, D. T., 209
Hunte, C., 166–167
Huppa, J. B., 190–191
Huse, M., 190
Hussain, S., 109–110, 259–260, 260*f*
Huston, J. S., 108
Huzjak, N., 287
Hynes, N. E., 127

I

Iacono, L., 136, 153
Im, M. N., 12
Indraccolo, S., 201
Indrevoll, B., 238–239
Ingold, F., 107–108
Inman, J. K., 116
Inoue, M., 36–37, 38*t*
Insaidoo, F. K., 192–193, 207–208, 209, 213–214
Interlandi, G., 103–105
Irvine, D. J., 191
Ivkov, R., 260
Iwai, H., 59–60, 61*f*
Iwakura, M., 59–60, 61*f*
Iwatsuki, S., 58

J

Jabagat, C., 287–288
Jabaiah, A., 93
Jackson, P. J., 225–226, 238
Jacob, J., 192
Jacobsen, J. S., 277–278
Jacobs, M., 161–162
Jacobson, S., 213–214
Jager, E., 213–214
Jager, P. L., 260*f*
Jahnig, P., 285–286
Jahrling, P. B., 111
Jain, R. K., 36, 259
Jakobsen, B. K., 191, 192, 207–208
Janin, J., 143
Jansco, G., 286–287
Jantsch, J., 294–295
Järvinen, T. A., 38*t*
Jelesarov, I., 106, 111
Jennings, C. V., 67–68
Jensen, J., 58, 59–60, 61*f*, 67*f*
Jenssen, H., 5–6
Jermutus, L., 102, 106, 229
Jespers, L., 141
Jhurani, P., 144
Jiang, L., 226–227, 230–231, 241, 243*f*, 244–246
Jiang, N., 190–191
Jiao, J. A., 207–208, 209
Jimenez, H. N., 36–37
Jochim, A. L., 5–6
Jo, D., 294–295
Johansson, B. B., 270
Johnson, L. A., 191
Johnson, R. S., 118, 120
Johnson-Wood, K., 277–278
Joliot, A. H., 4–5
Jones, A. T., 58, 60, 121
Jones, D. S., 226–227, 230–231, 236
Jones, E. Y., 93
Jones, L. L., 192, 207–208, 209
Jones, M. D., 200
Jones, R. B., 138, 140*t*, 142, 144, 152–153
Jones, R. M., 106
Jones, S., 4, 308
Jonsson, A., 76–77, 78*t*
Joos, T. O., 106
Jose, J., 176
Josephson, L., 40–41
Jost, C., 111
Jostock, T., 76–77, 79
Joyce, J. A., 36–37
Jullian, M., 17–18
Julsing, M. K., 106–107
June, R. K., 298–299
Jurt, S., 103–105

K

Kaas, Q., 224–225
Kain, R., 37–39, 38*t*
Kaisho, T., 313–315
Kajava, A. V., 103
Kakkis, E. D., 287–288
Kalaria, R. N., 286–287
Kalimo, H., 270
Kalin, N. H., 281–282
Kallab, V., 259–260, 260*f*
Kam, N. W., 294–295
Kanarek, A., 294–295
Kang, Y. S., 271–272, 274, 281–282
Kantor, C., 38*t*
Kanyo, Z., 106
Kapila, A., 140*t*, 147, 229, 230
Karatan, E., 152–153
Karbach, J., 213–214
Karimova, G., 306
Kariolis, M. S., 226–227, 230–231, 236, 241, 243*f*, 245
Karle, I. L., 5–6
Karlsson, R., 183–184
Karmali, P. P., 37–41, 38*t*, 45, 51
Karshikoff, A., 305
Kashi, R., 263
Kastantin, M., 37–39, 40–41
Kaupe, I., 138, 140*t*, 142, 144, 152–153
Kawai, Y., 58
Kawe, M., 108, 109, 124, 257–258, 262
Kay, B. K., 152–153

Kaye, P. M., 159, 160*f*, 162–163, 164–166, 170, 184
Kayushin, A. L., 163
Keenan, C. J., 90
Kellogg, B. A., 228–229, 230–231, 232, 233, 234, 237, 244
Kelly, J. W., 59–60, 61*f*
Kelly, K., 40–41, 260
Kemperman, R., 58
Kenan, D. J., 141
Kenig, M., 105, 106–107
Kenrick, S. A., 77, 78*t*, 79, 86–87, 88, 90
Kent, S. B. H., 60
Kerbel, R. S., 166
Kerenga, B., 58
Kerns, J. A., 225–226
Kessels, H. W., 191, 201
Khan, K., 277–278
Khan, M. K., 260
Kholodenko, D., 277–278
Kieke, M. C., 192, 195, 198
Kiick, K. L., 117–118
Kim, H. J., 159, 160*f*, 162–163, 167, 170, 184
Kim, I. S., 36–37
Kim, J., 207–208
Kimura, R. H., 58–59, 225, 226–227, 229, 230–231, 232, 235*f*, 237, 238, 240, 241, 243*f*, 244–246
Kimura, T., 225, 238–239
Kim, Y. W., 7–9, 12, 16, 20–21
Kinuya, S., 263
Kirpotin, D. B., 259–261, 260*f*
Kirschfink, M., 78*t*
Kitamura, T., 203–204
Kjaer, A., 58–59, 226–227
Klagsbrun, M., 40
Klauser, T., 176
Klein, L. O., 190–191
Klein-Szanto, A. J., 109, 262, 263
Klinger, M., 191
Kling, K., 277–278
Klintmalm, G. B., 58
Knappik, A., 102, 106
Kneissl, S., 111
Knop, S., 191
Kobe, B., 103
Koch, H., 112
Kohl, A., 104*f*, 105, 106, 107, 112–113, 116
Koide, A., 136–137, 138, 139–141, 140*t*, 142, 143, 144, 147, 152–153
Koide, S., 136–137, 138, 139–141, 140*t*, 142, 143, 144, 147, 152–153
Koivunen, E., 36–37, 38*t*
Kojima, T., 203–204
Kok, J., 58
Kolmar, H., 58–59, 76–77, 79, 224–225, 226, 227, 230, 238, 241
Kolonin, M. G., 36–37
Komatsu, M., 38*t*
Konarev, A. V., 58
Konkle, B. A., 36
Konttinen, Y. T., 38*t*
Korman, A. J., 164–166
Korndörfer, I. P., 161, 162
Koropatnick, J., 5
Korosteleva, M. D., 163
Kosch, W., 163
Kossiakoff, A. A., 144
Kosterink, J. G., 260*f*
Kotamraju, V. R., 37–41, 38*t*, 45, 46, 48, 51–52
Koup, R. A., 111
Krag, D. N., 36–37
Krajewski, S., 36–40, 38*t*, 51, 52
Kramer, J., 176
Kramer, M. A., 103–105
Kramer, S., 78*t*
Kranz, D. M., 190–191, 192–193, 195, 198–201, 206, 207–208, 209, 213–214
Krasinska, K. M., 36–37
Krasnykh, V., 106, 111
Kratzner, R., 76–77, 226, 227, 230
Krause, S., 226
Kreuter, J., 4–5
Krieger, A., 287–288
Krogsgaard, M., 190
Kronqvist, N., 76–77, 78*t*
Kubetzko, S., 108, 109–110, 115, 116, 127
Kubo, S., 225, 238–239
Kuimelis, R., 137, 140*t*, 144–145
Kukowska-Latallo, J. F., 260
Kumagaye, K. Y., 225, 238–239
Kung, A. L., 6, 8*t*, 9–10, 11–12, 17–18, 21–25, 27–28
Kunkel, T. A., 143
Kurz, M., 137, 140*t*, 144–145
Kussie, P. H., 28
Kusuma, Y., 58–59, 225
Kutchukian, P. S., 7–9, 16, 20–21
Kwok, D., 58–59, 225
Kwon, M. K., 36–37
Kwon, T. G., 36–37
Kwon, T. H., 36–37

L

Laakkonen, P., 36–37, 38*t*, 39–40, 41
Labrijn, A. F., 229
Lacy, M. J., 207–208
Ladant, D., 306
Ladbury, J. E., 207–208
Ladenstein, R., 305
Ladner, R. C., 90
Lahdenranta, J., 36–37
Lahti, J. L., 225, 226–227, 228*f*, 230–231, 232, 233, 237, 241
Lampel, S., 4

Lam, T., 137, 140*t*, 144–145
Lane, M., 144–145, 227
Langel, U., 308
Langer, R., 259–260, 260*f*
Langmann, T., 163
Larsson, B., 257–258
Laufer, S. D., 298
Lauffer, I., 109–110, 127
Laugel, B., 192
Lau, W. L., 147, 148, 228–229, 232, 233, 234, 235–236, 237
LaVallie, E. R., 76–77
Lavis, L. D., 299–300
Lawrence, M. C., 106
Lawrence, M. S., 4–5, 295–296, 297
Lazarides, E., 38*t*
Lazari, M. F., 245
Lebowitz, M. S., 198–200
Lechler, R. I., 200
Lee, C. G., 40
Lee, C. V., 230
Lee, E. J., 36–37
Lee, H. J., 272, 273
Lees, A., 116
Lee, S. K., 294–295
Lee, S. M., 36–37
Lefort, J., 161–162
Lehrer, R. I., 58
Lei, P., 78*t*
Le, N., 263
Le-Nguyen, D., 67–68, 224–225, 238–239
Leo, E., 191
Leroy, S., 225
Le, S., 287–288
Lesley, J., 272
Lester, T., 287–288
Letourneur, F., 201, 203
Leunig, M., 259
Levary, D., 229, 234
Levin, A. M., 90, 225, 226–227, 229, 230–231, 232, 235*f*, 237, 238, 240, 241, 244–246
Lewis, R. J., 58–59, 65*f*, 66–67
Lewitzky, M., 93
Liang, S., 224–225
Liang, W. C., 230
Libby, P., 36
Lichière, J., 106
Liddy, N., 190–191, 192, 207–208
Liepinsh, E., 59–60, 61*f*
Li, J. Y., 103, 271–272
Lilie, H., 209
Lillemeier, B. F., 190–191
Lim, A., 137, 140*t*, 144–145
Lingel, A., 59–60, 61*f*
Ling, V., 140*t*, 144–145
Link, A. J., 79
Lin, Z., 153
Li, P., 226–227, 241, 243*f*, 244–246
Lipes, B. D., 141
Lipovšek, D., 136, 137, 138, 140*t*, 144–145, 147, 148, 153–154, 227, 231
Lippow, S. M., 140*t*, 147, 148, 228–229, 231, 232, 233, 234, 235–236, 237
Li, Q. J., 190, 191
Lis, M., 191
Lissina, O., 161–162
Lissin, N., 213–214
Little, L. E., 91
Liu, B., 190–191, 207–208, 209
Liu, D. R., 295–296, 297, 298–301, 312
Liu, H., 137, 140*t*, 144–145, 226–227, 241, 243*f*, 244–246
Liu, N., 37–39
Liu, R., 144–145, 227
Liu, W., 256, 259
Liu, Z., 40–41, 294–295
Li, W., 78*t*
Li, Y. F., 36–37, 191, 192
Lo Conte, L., 143
Lofblom, J., 76–77, 78*t*
Lohka, M. J., 76–77
Lohse, P., 137, 140*t*, 144–145
Lombardo, C. R., 37–39
Lopes de Menezes, D. E., 259–260, 260*f*
Loscalzo, J., 36
Losche, W., 226
Loughnan, M. L., 65*f*
Lou, J., 109, 262, 263
Lovelace, E. S., 58, 59–60, 61*f*
Love, S., 238–239
Lowman, H. B., 142, 143–144
Lub-de Hooge, M. N., 260*f*
Luchniak, A., 4–5
Luckett, S., 58
Luginbühl, B., 106
Luison, E., 260*f*
Lu, J. Z., 275–278, 275*t*, 279–282, 281*f*, 282*t*, 283, 283*f*, 284–288, 284*t*
Luo, J., 224–225
Lu, S., 224–225
Lutz, A. M., 226–227
Lu, Z., 76–77
Lyons, D., 190–191, 207–208
Lyons, S. A., 225

M

Macdougall, I. C., 286
Mack, H., 168*f*
Madani, N., 8*t*, 9–10
Maecke, H. R., 36–37
Mäger, I., 308
Magni, F., 37–39
Magnusson, M., 257–258
Ma, H., 141
Mahajan, A., 103

Mahmut, M., 78*t*
Mahon, T. M., 190–191, 192, 207–208, 213–214
Maillard, I., 10
Maillere, B., 225
Maillet, I., 161–162
Majoros, I. J., 260
Makabe, K., 153
Malia, M., 102
Malissen, B., 201, 203
Mamot, C., 259–260, 260*f*
Maniatis, T., 167–169
Mann, N., 259–260, 260*f*
Manuel, H., 287–288
Maqueda, M., 58
Marfatia, S. M., 40
Marks, J. D., 109, 259–261, 260*f*, 262, 263
Martin, F., 260*f*
Martin-Killias, P., 110, 127, 128
Martin, W., 288
Marx, U. C., 66
Matias, P. M., 112
Matschiner, G., 76–77, 78*t*, 159, 160*f*, 162–163, 164–166, 165*f*, 168*f*, 170, 176–177, 184
Matsuda, T., 315–317
Mattes, M. J., 259
Mattheakis, L. C., 106
Matthews, J. M., 59–60, 61*f*
Mattras, H., 225–226
Mattrey, R. F., 38*t*, 39–40, 45, 51
Mattsson, L., 183–184
Matysiak, S., 19
Mauget-Faysse, M., 112
Mayo, S. L., 139
McCafferty, J., 164, 172–173
McCall, A. M., 109
McCallum, E. J., 58
McCartney, J. E., 108
McCoy, J. M., 76–77
McCrae, K. R., 36
McEntee, M., 287–288
McEver, R. P., 36
McHeyzer-Williams, M. G., 198–200
McKern, N. M., 106
McLaughlin, P. M., 109–110
McMicheal, A. J., 198–200
McMurry, J. A., 288
McNaughton, B. R., 4–5, 298–299, 300–301, 312
McNulty, J. C., 225–226, 238
Medina, O. P., 38*t*
Meeuwissen, S. A., 117–118
Melchers, L. J., 109–110
Melton, D. A., 294–295
Mena, M., 231
Menez, A., 225
Merguerian, M., 152–153
Mescalchin, A., 5
Messing, J., 167, 306
Meunier, F. A., 58, 59–60, 61*f*, 67*f*
Meyer, L. A., 260*f*
Meyer, T. F., 176
Miao, C. H., 294–295
Miao, H. Q., 40
Miao, Z., 58–59, 226–227, 230–231, 241, 243*f*, 244–246
Michaelsson, A., 183–184
Michelich, C. R., 256, 259
Midelfort, K. S., 229, 234, 236, 237
Miers, L. A., 260
Mier, W., 78*t*
Mikheeva, G., 106, 111
Milicic, A., 192
Miljanich, G. P., 58–59
Millard, E. L., 65*f*
Miller, C. J., 58
Miller, K. D., 229
Millhauser, G. L., 225–226, 238
Minard, P., 106–107
Minari, Y., 77
Mintz, P. J., 36–37
Miranda, S., 112
Miroshnikov, A. I., 163
Mirzapoiazova, T., 40
Misselwitz, R., 226
Missimer, J., 281–282
Missirlis, D., 37–39, 40–41
Mita, A. C., 136, 153
Mitchell, D. J., 4–5
Mittl, P. R., 103–105
Moarefi, I., 93
Moch, H., 107–108
Moellering, R. E., 6, 8*t*, 9–12, 17–18, 21–25, 26, 27–28
Moffatt, B. A., 167
Mohammadtaghi, S., 259–260
Moineau, S., 106
Moingeon, P., 288
Moise, L., 288
Moldenhauer, G., 37–39
Molloy, P. E., 190–191, 192, 207–208
Momiyama, A., 225, 238–239
Moore, S. J., 228–229
Morgan, A. C. Jr., 260*f*
Morgan, D., 277–278, 286–287
Morgan, R. A., 191
Morin, P., 153
Morita, S., 203–204
Moroney, S. E., 102
Morris, M. C., 4–5
Mortelmaier, M. A., 190–191
Moser, R., 161–162
Moss, P. A. H., 198–200
Motter, R. N., 277–278, 280–282, 286–287
Mouchiroud, G., 93
Mourier, G., 225
Moysey, R., 192

Mueller, M., 127
Mühlebach, M. D., 111
Muir, T. W., 59–60, 61*f*
Mukhapadhyay, U., 111
Mukherji, M., 117–118
Mullan, M., 286–287
Müller, K. M., 106–107
Mullings, R. E., 213–214
Münch, R. C., 111
Munoz, N. M., 58–59, 225
Murali, D., 281–282
Murphy, C., 294–295
Murray, K. S., 76–77
Muyldermans, S., 136–137
Mylne, J. S., 58
Myszka, D. G., 182–184

N

Nabel, G. J., 111
Nagai, K., 209
Nagaya, M., 271–272
Nagy-Davidescu, G., 107–108, 109, 124, 257–258, 262
Najjar, A. M., 111
Nakayama, N., 40–41
Nalis, D., 225, 238–239
Nam, Y., 10
Nangola, S., 106–107
Naranjo, D., 66–67
Natarajan, A., 260
Navarro, V., 260*f*
Nazer, B., 271–272
Nelson, D., 37–39
Nemerow, G. R., 111
Neubauer, P., 209
Neumann, R. D., 263
Nevin, S. T., 58, 59–60, 61*f*, 65*f*, 67*f*
Newell, E. W., 190–191
Nguyen, T. M., 58, 67–68
Nicke, A., 65*f*
Nickles, R. J., 281–282
Nicklin, S. A., 111
Nicols, A., 280–282, 286–287
Nielsen, C. H., 58–59, 226–227
Nielsen, K. J., 224–225
Nielsen, U. B., 259–261, 260*f*
Nieves, E., 207–208, 209
Nigavekar, S. S., 260
Nijnuis, H., 260*f*
Nilsson, F. Y., 257–258
Nilsson, M., 257–258
Nishina, K., 294–295
Nishio, H., 225, 238–239
Nixon, A. E., 90
Noble, C. O., 259–260, 260*f*
Noppeney, R., 191
Norgaard, P., 128–129
Norman, D. G., 67–68
Norton, R. S., 224–225
Nunn, M. A., 161–162
Nuttall, P. A., 161–162

O

Oakes, T. R., 281–282
Oberle, V., 226
O'Connell, D., 163
O'Connell, T., 163
O'Dorisio, M. S., 36–37
O'Dorisio, T. M., 36–37
Ogawa, K., 111
O'Herrin, S. M., 192, 198–200, 207–208
Ohno, S., 111
Oh, P., 36–37
Olivera, B. M., 58–59
Ollmann, M. M., 225–226
Olsen, J. O., 36–37
Olsen, M. J., 228–229
Olson, J. M., 58–59
Olsson, Y., 270
O'Neal, J., 225
Ong, G. L., 259
Oppermann, H., 108
Opresko, L. K., 229
Orcutt, K. D., 229, 234
Orlova, A., 257–258
Orning, L., 77
Orosz, K., 286–287
Orr, B. A., 195
Ösapay, G., 58
Ösapay, K., 58
Ostergaard, H., 128–129
Ostermeier, M., 236
Otting, G., 59–60, 61*f*
Ouellette, A. J., 58
Overington, J. P., 294
Ozkan, S. B., 301–303, 305

P

Pace, C. N., 241, 243
Pack, P., 102
Paesen, G. C., 161–162
Paik, C. H., 263
Pal, A., 111
Pallaghy, P. K., 224–225
Pallay, A., 280–282, 286–287
Pancer, Z., 103
Pancook, J., 260*f*
Papadopoulos, K., 136, 153
Papahadjopoulos, D., 259
Pardridge, W. M., 270, 271–273, 274–282, 275*t*, 276*f*, 277*f*, 280*f*, 281*f*, 282*t*, 283, 283*f*, 284–288, 284*t*
Parizek, P., 106, 111, 112
Parke, E., 198

Park, J. H., 36–37, 38*t*, 39–40
Park, J. W., 259–261, 260*f*
Park, K. M., 270–271
Park, R. W., 36–37
Parks, R. J., 111
Paschke, E., 287
Paschke, M., 106–107
Pasqualini, R., 36–39, 38*t*, 41
Passage, M., 287–288
Pastan, I., 127
Patel, M., 112
Patil, M. L., 294–295
Patnaik, A., 136, 153
Paul, S. M., 286–287
Pease, L. R., 170
Pécorari, F., 106
Peggs, K. S., 164–166
Peinovich, M., 287–288
Pellequer, J. L., 181–182
Peng, J., 78*t*
Penke, B., 286–287
Pepper, L. R., 227
Perik, P. J., 260*f*
Pero, S., 36–37
Perry, M. C., 37–39
Peters, A. M., 259–260
Peters, D., 38*t*
Petit Frère, C., 128
Petkovic, G., 287
Petrenko, V. A., 36–37, 77
Pfitzinger, I., 102
Pham, E., 286
Pham, T. T., 67–68
Philip, R., 58
Phillips, K. J., 4–5, 295–296, 297
Picarelli, Z. P., 245
Pidoux, J., 306
Piehler, J., 139–141
Piel, N., 163
Piepenbrink, K. H., 190, 192–193, 207–208, 209, 213–214
Pilarski, L. M., 259–260, 260*f*
Pilch, J., 36–37, 38*t*, 39–40
Ping Li, W., 260*f*
Pitter, K., 8*t*, 17–18, 23–25, 26
Plan, M. R., 67–68
Plaxco, K. W., 136–137
Plückthun, A., 59–60, 61*f*, 102, 103–108, 104*f*, 109–110, 111, 112–113, 115, 116, 124, 127, 128, 136, 139, 142, 229, 257–258, 259–260, 260*f*, 262
Pober, J. S., 36
Pohlner, J., 176
Po, J., 6, 7–9, 12, 20–21, 23
Pollak, E. S., 36
Pongor, S., 224–225
Pons, J., 277–278, 286–287
Popp, A., 128
Porkka, K., 36–37, 38*t*, 39–40, 41
Porter, K. A., 58
Porto, C. S., 245
Poth, A. G., 58
Pozderac, R. V., 36–37
Prange, H., 285–286
Prenosil, E., 106
Price-Schiavi, S. A., 207–208, 209
Prosselkov, P., 59–60, 61*f*
Prum, B., 201
Prusiner, S. B., 106
Prusoff, W. H., 243–244
Pupo, A., 77
Purbhoo, M. A., 191, 213–214
Putney, A. R., 144–145, 227
Pütter, V., 112

Q

Qi, S., 191
Qiu, W., 12
Quadros, E. V., 77
Quan, C., 90
Quax, W. J., 106–107
Quezada, S. A., 164–166

R

Raba, M., 128
Radovic-Moreno, A. F., 259–260, 260*f*
Ragnarsson, G. B., 200–201
Raines, R. T., 297–298, 299–300
Rainisalo, A., 38*t*
Rajagopal, J., 294–295
Rajotte, D., 38*t*
Rakestraw, J. A., 148
Raleigh, D. P., 59–60, 61*f*
Rao, R., 37–39
Raptis, D., 144
Raseman, J. M., 191, 192, 201, 206
Ravanpay, A. C., 58–59, 225
Raymond, K. N., 162, 166–167
Redl, B., 162
Reichel, A., 139–141
Reichelt, P., 128
Reich, Z., 190–191, 207–208
ReiseSousa, C., 313–315
Reiss, S., 226
Rejman, J., 121
Ren, G., 226–227, 241, 243*f*, 244–246
Restle, T., 5, 298
Reubi, J. C., 36–37
Rhoden, J. J., 255–268
Rian, A., 77
Rice, J. J., 76–77, 79, 86–87, 91, 93
Richards, F. M., 66
Richey, C. W., 120–121
Richman, S. A., 192–193, 195, 200
Rico, M., 58

Rio, G. D., 37–39
Rizkallah, P., 192, 207–208
Rizk, S. S., 4–5
Robbins, P. F., 191
Roberts, J. D., 143
Roberts, R. W., 144–145, 227
Roberts, T. G. Jr., 37–39
Robinson, E. A., 225–226
Robinson, M. K., 262
Rockberg, J., 78*t*
Roederer, M., 111
Rojas, G., 77
Ronan, V., 277–278, 286–287
Rood, H. L., 238
Roschitzki-Voser, H., 106
Roschke, V., 90
Rosenberg, S. A., 200–201
Rosengren, K. J., 58, 59–60, 61*f*, 67–68, 67*f*, 225–226
Rosenthal, A., 277–278, 286–287
Ross, R., 260
Rothgery, L., 152–153
Röthlisberger, D., 106
Rothschild, S., 110, 127, 128
Rots, M. G., 109–110
Rowley, D. A., 207–208, 213–214
Roy, E. J., 195
Rubin, M. A., 260
Rudnick, S. I., 109, 262, 263
Rudolph, M. G., 190, 198–200
Ruiters, M. H., 109–110
Ruoslahti, E., 36–41, 38*t*, 45, 46, 48, 51–52
Russ, A. P., 4
Rutjes, F. P., 117–118, 120
Ryffel, B., 161–162

S

Sacchi, A., 37–39
Sachdeva, M., 90
Safirstein, B., 280–282, 286–287
Saga, T., 263
Sailor, M. J., 36–37, 38*t*, 39–41
Saito, T., 58
Sakakibara, S., 225, 238–239
Salahuddin, T. S., 270
Salama, J. K., 207–208, 213–214
Salier, J.-S., 159
Salo, T., 38*t*
Sambrook, J., 167–169
Sami, M., 192, 207–208
Sanchez-Barrena, M. J., 58
Sanders, N. N., 121
Sando, L., 225–226
Sankhala, K., 136, 153
Santha, P., 286–287
Sapra, P., 260–261
Sarkar, C. A., 106, 108, 115, 116
Sato, J., 263
Sato, S., 59–60, 61*f*
Satzger, M., 128
Saxon, E., 117–118
Sazinsky, S. L., 147, 148, 228–229, 232, 233, 234, 235–236, 237
Scanlon, M. J., 66–68
Schaffer, D. V., 91
Schaffitzel, C., 106
Schafmeister, C. E., 6, 7–9, 12, 20–21, 23
Schaser, T., 111
Schatz, O., 163
Scheinfeld, N., 109–110
Schellinger, P. D., 285–286
Schenk, D., 277–278
Schibli, R., 108, 109, 124, 257–258, 262
Schier, R., 109
Schietinger, A., 207–208, 213–214
Schirmer, W. J., 36–37
Schlachetzki, F., 277–278, 284
Schlapschy, M., 168*f*
Schlehuber, S., 159–161, 160*f*, 162–163, 164–166, 168*f*, 170, 174–175, 178, 180, 184
Schmidt, F. S., 159, 161, 163, 170
Schmidt, H., 207–208, 213–214
Schmidt, M., 201
Schmidt, M. A., 36–37
Schmidt, M. M., 109, 256, 257*f*, 258–259, 260*f*, 261*f*, 263, 264, 265*f*
Schmidt, T. G., 166–167, 168*f*, 172–173, 179–180, 184–185
Schmitt, T. M., 200–201
Schmitz, T., 225
Schmoldt, H. U., 225, 226, 238, 241
Schneck, J. P., 198–200
Schneider, D., 285–286
Schneider-Mergener, J., 226
Schnitzer, J. E., 36–37
Schnölzer, M., 60
Schoep, T. D., 75–97
Schoffelen, S., 117–118, 120
Schohn, A., 76–77, 79, 93
Scholle, M. D., 152–153
Schönfeld, D. L., 159, 160*f*, 162–163, 164–166, 167, 170, 184
Schon, O., 141
Schottelius, M., 168*f*
Schroeter, S., 277–278
Schroter, G. P., 58
Schubiger, A. P., 281–282
Schubiger, P. A., 108
Schuler, M., 191
Schuler, T., 201
Schultz, P. G., 117–118
Schumacher, T. N., 191, 201
Schutz, G. J., 190–191
Schwartz, A. L., 273
Schwartz, B. S., 36

Schwarz, C., 128
Schweizer, A., 106
Schwenk, J. M., 106
Schwer, H., 163
Schwill, M., 101–134
Schymkowitz, J., 301–303
Scott, D. W., 288
Sedgwick, S. G., 103
Seely, J. E., 120–121
Seifert, B., 107–108
Seifert, C., 166–167
Seki, F., 111
Selkoe, D. J., 271–272
Selles, K. G., 106–107
Selsted, M. E., 58
Selvin, P. R., 113
Seregni, E., 260*f*
Serrano, L., 301–303
Seth, A., 207–208
Setlur, S. R., 260
Settanni, G., 103–105, 106–107
Seubert, P., 277–278
Sexton, D. J., 90
Shafer, D. E., 116
Shalaby, M. R., 259–261, 260*f*
Shaller, C. C., 109, 262, 263
Shamah, S., 140*t*, 144–145
Sham, J. S., 40
Shao, Y., 259–261, 260*f*
Shao, Z., 306
Sharkey, R. M., 263
Sharma, A., 161–162
Sharpless, K. B., 117–118
Shavrin, A., 111
Shell, M. S., 301–303, 305
Shelton, S. E., 281–282
Shenderov, E., 190–191
Shen, G. P., 36–37, 78*t*
Shen, Z. P., 294–295
Sherman, C. D., 260*f*
Shewry, P. R., 58
Shih, L. B., 259
Shmeeda, H., 259–260, 260*f*
Short, J. M., 171–172
Shriver, S. K., 90
Shtatland, T., 40–41
Shukla, G. S., 36–37
Shusta, E. V., 192, 195, 198–200, 207–208, 227
Siani, M. A., 238
Sidhu, S. S., 138, 141, 142, 143–144, 147, 152–153, 164, 172–173, 230
Sieber, M., 226
Siegall, C. B., 127
Siegel, R. W., 229
Siemers, E. R., 286–287
Silver, B. A., 136, 153
Silverman, A. P., 225, 226–227, 230–231, 232, 233, 237, 241, 243*f*, 244–246
Simberg, D., 36–37, 38*t*, 39–40
Simmons, H. H., 109
Simon, M., 118
Simonsen, S. M., 225–226
Singh, M., 36–37, 38*t*
Skarlatos, S., 284
Skerra, A., 76–77, 78*t*, 102, 136, 158–161, 160*f*, 162–163, 164–167, 165*f*, 168*f*, 170, 172–173, 174–175, 176–178, 179–180, 182, 184–185
Sliz, P., 10
Smerdon, S. J., 103
Smith, G. P., 36–37, 77, 90
Smith-Jones, P. M., 260*f*
Soghomonyan, S., 111
Soloshonok, V. A., 12
Sommerhoff, C. P., 226, 238
Sondermann, H., 93
Songyang, Z., 40
Sontheimer, H., 225
Soo Hoo, W. F., 207–208
Sorsa, T., 38*t*
Sota, H., 59–60, 61*f*
Souhami, R. L., 37–39
Souied, E. H., 112
Soutschek, J., 294–295
Spangenberg, P., 226
Speck, J., 106–107
Sperling, R. A., 280–282, 286–287
Spinelli, S., 106
Spiotto, M. T., 207–208, 213–214
Spits, H., 191, 201
Spitzfaden, C., 136–137
Srinivasan, A., 36–37
Staats, H. F., 141
Stadler, B. M., 112
Stafford, W. F. III., 108
Stahel, R. A., 109–110, 127
Stahl, M. L., 76–77
Stahl, P., 298
Stahl, S., 76–77, 78*t*, 257–258
Stanfield, R. L., 190, 198–200
Stan, R. V., 297–298
Starzl, T. E., 58
Stefan, N., 110, 127, 128
Steiner, D., 105–107, 112–113, 142
Stemmer, W. P., 237, 305
Stevens, J., 67–68
Stewart, J. S., 259–260
Stewart, M. L., 8*t*, 9–10, 17–18, 26
Stibora, T., 159, 161, 163, 170
Stone, J. D., 191, 195, 198–201, 209
Strauch, A., 76–77
Straumann, N., 106
Stricher, F., 301–303
Strobl, S., 224–225
Strominger, J. L., 207–208, 209
Strong, R. K., 162, 166–167

Stroud, M. R., 58–59, 225
Studier, F. W., 167
Stumpp, M. T., 102, 103–107, 104*f*, 108, 109, 112–113, 116, 124, 142, 257–258, 262
Sugahara, K. N., 38*t*, 39–41, 45, 46, 48, 51–52
Su, G. C., 286–287
Su, H., 259–261, 260*f*
Suhoski, M. M., 192
Sullivan, N. J., 111
Summerer, D., 117–118
Sun, E. Y., 40–41
Sun, J. J., 79
Sun, L., 137, 140*t*, 144–145
Sun, X., 40–41
Superti-Furga, G., 138, 140*t*, 142, 144, 152–153
Sutton, D. H., 192, 213–214
Suurmeijer, A. J., 260*f*
Suzuki, M., 8*t*, 26
Sweeney, C. J., 136, 153
Swers, J. S., 228–229, 230–231, 232, 233, 234, 237, 244
Sykes, B. D., 66
Symmonds, K., 277–278, 286–287
Szostak, J. W., 144–145, 227
Szymbowski, D. E., 270–271

T

Tai, M. S., 108
Takahashi, H., 59–60, 61*f*
Takahashi, T., 225, 238–239
Takeda, M., 111
Takenawa, T., 59–60, 61*f*
Tamaskovic, R., 108, 109, 124, 257–258, 262
Tanaka, Y., 77
Tang, P., 277–278
Tang, Y.-Q., 58, 109, 262, 263
Tansey, M. G., 270–271
Tartoff, K. D., 127
Tawfik, D. S., 301–303
Tayapiwatana, C., 106–107
Teesalu, T., 38*t*, 39–41, 45, 46, 48, 51–52
Teeuwen, R. L., 117–118
Tegoni, M., 106
Teply, B. A., 259–260, 260*f*
Tereshko, V., 138, 141, 143, 147, 152–153
Terlau, H., 58–59
Testa, J. E., 36–37
Teyton, L., 192
Theobald, I., 76–77, 78*t*, 164, 165*f*, 168*f*, 176–177
Theurillat, J. P., 107–108
Thiele, E. W., 264
Thogersen, H. C., 209
Thomas, C., 36–37
Thomas, J. M., 78*t*, 79
Thomas, L., 66–67
Thomas, T. P., 260
Thomaz, M., 112
Thompson, D. A., 225–226, 238
Thompson, D. B., 4–5, 295–296, 298–301, 312
Thompson, J. D., 49
Thompson, K., 77
Thomson, E., 207–208, 209
Thornton, J. M., 4
Thurber, G. M., 109, 257, 259, 263, 264, 265*f*
Tirrell, D. A., 117–118, 120
Tirrell, M., 37–39, 40–41
Tjia, W. M., 76–77
Tobias, J. S., 37–39
Tobon, G., 229, 234
Todorov, P. T., 192, 207–208
Tokic, V., 287
Tokui, N., 77
Tokuriki, N., 301–303
Tolcher, A. W., 136, 153
Tolmachev, V., 257–258
Tomassetti, A., 260*f*
Tom, J. Y. K., 90
Torchilin, V. P., 259
Tortarolo, M., 286
Toth, L., 286–287
Trabi, M., 58
Trampe, A., 298
Tran, D., 58
Tran, L., 260*f*
Tran, P. T., 58–59, 226–227
Tremblay, D., 106
Trentmann, S., 159, 160*f*, 162–163, 164–166, 170, 184
Trepel, M., 36–37, 41
Triguero, D., 271–273, 284
Trzpis, M., 109–110
Tsai, M. D., 103
Tschudi, D., 109–110, 127
Tung, C. H., 294–295
Tung, W. L., 171–172
Turcotte, R. F., 299–300
Tyler, A. F., 8*t*, 9–10, 11–12, 17–18, 20–25, 22*f*, 24*f*, 26, 28, 29*t*
Tzemach, D., 259–260, 260*f*

U

Uchiyama, F., 77
Uckert, W., 200–201
Uhlen, M., 78*t*
Ullmann, A., 306
Underhill, C. B., 37–39
Uster, P. S., 259–260
Utz, U., 198–200, 207–208, 213–214

V

Vajdos, F., 241, 243
Valdivia, E., 58
Vallabhajosula, S., 260*f*

Valtanen, H., 38*t*
VanAntwerp, J. J., 228–229
van Berkel, S. S., 117–118, 120
van Bloois, E., 79
Van Cleave, V., 76–77
Vandehey, N. T., 281–282
van Delft, F. L., 117–118, 120
van den Berg, A., 4–5
van Den Boom, M. D., 191, 201
Vandenbroucke, R. E., 121
Van den Brulle, J., 163
van der Gun, B. T., 109–110
van der Heyde, H., 38*t*
Vanderheyden, J. L., 260*f*
van der Merwe, P. A., 190–191
van Dijl, J. M., 106–107
van Dulmen, T. H., 117–118
van Hest, J. C., 117–118, 120
Van Overtvelt, L., 288
Van Regenmortel, M. H., 181–182
VanTienhoven, E. A., 207–208
Varela-Rohena, A., 192
Vargaftig, B. B., 161–162
Varga, J., 286–287
Vasser, M., 144
Vassilev, L. T., 4
Veesler, D., 106
Veiseh, M., 58–59, 225
Veiseh, O., 58–59, 225
Veldhoen, S., 298
Venegas, R., 38*t*
Venturi, M., 166–167
Vercauteren, D., 121
Verdine, G. L., 4, 7–9, 20–21, 25, 29–30
Verin, A. D., 40
Viardot, A., 191
Vidal, C. I., 36–37
Vieira, J., 167, 306
Vile, R. G., 259–260
Virnekäs, B., 102
Vispo, N. S., 77
Vives, E., 4–5
Vogel, M., 112
Volgin, A. Y., 111
von Hippel, P. H., 169–170
Voss, E. W. J., 207–208
Voss, S., 182
Vuidepot, A. L., 190–191, 192, 207–208

W

Wade, M., 8*t*, 26, 27–28
Wadia, J. S., 297–298
Wagner, R. W., 137, 140*t*, 144–145, 227
Wahlstrom, M. E., 58, 59–60, 61*f*
Waibel, R., 108, 109–110, 124, 127, 257–258, 262
Waine, C., 58, 67–68, 225
Waldherr, C., 36–37
Waldmann, T., 163
Walensky, L. D., 4, 6, 8*t*, 9–10, 11–12, 17–18, 21–25, 26, 27–28
Walker, R. G., 103
Waltenberger, J., 40
Walter, K., 226, 227
Wambre, E., 288
Wang, A., 118, 120, 259–260, 260*f*
Wang, B., 38*t*
Wang, C. K., 225–226
Wang, C. L., 76–77
Wang, J., 140*t*, 144–145
Wang, Q. J., 117–118, 191
Wang, Y., 274, 275*t*, 278–279, 280*f*
Ward, C. W., 106
Wargo, J. A., 191
Warikoo, V., 140*t*, 144–145
Waring, A. J., 58
Warrens, A. N., 200
Warshawsky, I., 273
Wartenberg, K., 285–286
Waser, B., 36–37
Watanabe, K., 294–295
Waterman, P., 260
Watt, S. J., 59–60, 61*f*
Weaver, D. L., 36–37
Weber-Bornhauser, S., 102, 106, 229
Weber, K. S., 192–193, 195, 198–200, 207–208, 209
Weber, W. A., 259–261, 260*f*
Wegrzyn, M., 285–286
Weichselbaum, R. R., 207–208, 213–214
Weidanz, J., 207–208, 209
Weikl, T. R., 301–303, 305
Weimar, C., 285–286
Weiner, L. M., 108, 109, 262, 263
Weinstein, J. N., 263
Weisbach, M., 128
Weissenborn, K., 285–286
Weiss, G. A., 90
Weissleder, R., 40–41, 260
Welfle, H., 226
Welfle, K., 226
Wellnhofer, G., 102
Wells, J. A., 90, 142, 143–144, 294
Wels, W., 127
Wender, P. A., 4–5
Weng, A. P., 10
Weng, S., 137, 140*t*, 144–145
Wentzel, A., 76–77, 225, 226, 227, 230, 238, 241
Werle, M., 225
Wernerus, H., 76–77, 78*t*
Wescott, C. R., 90
Wessner, H., 226
Wester, H. J., 168*f*
Weston-Davies, W., 161–162
Wetzel, S. K., 103–105, 106–107

White, P. D., 238–239
Wick, T. M., 36
Widstrom, C., 257–258
Wiesmann, C., 144
Wilcock, D. M., 277–278, 286–287
Wiley, D. C., 207–208, 209
Wiley, H. S., 229
Wilken, J., 225–226
Willcox, B. E., 207–208
Williams, J. A., 225
Williams, N. K., 59–60, 61*f*
Williamson, R. A., 106
Williams, R. M., 12
Williams, R. S., 12
Willingham, A. T., 103
Willis, A. C., 161–162
Willmann, J. K., 58–59, 226–227
Willuda, J., 108, 109–110, 127
Wilson, B. D., 225–226
Wilson, I. A., 190, 198–200
Winblade Nairn, N., 118, 120
Winkler, J., 110
Winter, G., 141, 177–178
Winter, R. T., 79
Winther, J. R., 128–129
Wishart, D. S., 66
Withofs, N., 58–59, 226–227
Wittrup, K. D., 89–90, 109, 140*t*, 147, 148, 149, 192, 195, 198–200, 201–202, 207–208, 227, 228–229, 230–231, 232, 233, 234, 235–236, 237, 244, 256, 257, 257*f*, 258–259, 260*f*, 261*f*, 263, 264, 265*f*
Wojcik, J., 138, 139–141, 140*t*, 142, 144, 152–153
Wolf, S., 112
Wolf-Schnurrbusch, U. E., 112
Wölle, J., 102
Wong, S. L., 76–77
Woods, C. M., 38*t*
Woods, R. A., 194
Wörn, A., 102
Wright, M. C., 144–145, 227
Wucherpfenning, K. A., 207–208, 209
Wu, D., 281–282
Wu, E., 111
Wu, S. C., 76–77, 78*t*
Wyer, J. R., 207–208
Wyler, E., 106
Wyns, L., 136–137

X

Xia, C. F., 275*t*, 277–278, 277*f*
Xiao, M., 113
Xie, Q. H., 59–60, 61*f*
Xue, Y. Q., 286
Xu, H., 191
Xu, J., 40
Xu, L., 111, 137, 140*t*, 144–145
Xu, X., 37–39

Y

Yachechko, R., 40
Yamniuk, A. P., 153
Yanagi, Y., 111
Yang, J., 272, 274, 281–282
Yang, M., 36–37, 38*t*, 39–40
Yang, O. O., 58
Yang, Y. K., 225–226, 238
Yang, Z. Y., 111
Yanisch-Perron, C., 167, 306
Yao, V. J., 36–37
Yeh, H. H., 111
Yeung, Y. A., 149, 228–229, 232, 233, 234, 237, 244
Yoo, E. S., 36–37
Yoon, G. S., 36–37
Yoshikawa, T., 271–272, 284
Young, D., 111
Yuan, F., 256, 259
Yuan, J., 58
Yu, B., 78*t*
Yu, J. N., 36–37, 90
Yu, L. Y., 191
Yu, P., 207–208, 213–214
Yu, Y. J., 4–5

Z

Zacchetti, A., 260*f*
Zack, J. A., 58
Zahnd, C., 106, 108, 109, 112–113, 124, 257–258, 262
Zajic, S. C., 257, 263, 264
Zakour, R. A., 143
Zalipsky, S., 259–260, 260*f*
Zangemeister-Wittke, U., 108, 109–110, 127, 128, 259–260, 260*f*
Zarnitsyna, V. I., 190–191
Zennaro, E., 286
Zerbe, O., 103–105
Zhai, D., 12
Zhang, B., 207–208, 213–214
Zhang, C., 271–272
Zhang, H., 8*t*, 9–10, 12, 17–18, 21–23, 36–37, 190–191
Zhang, L., 36–40, 38*t*, 51, 52, 259–260, 260*f*
Zhang, M., 58–59, 225
Zhang, Y. F., 4–5, 271–272, 274, 275*t*, 277–279, 277*f*, 280*f*, 284
Zhao, C., 58
Zhao, H., 306
Zhao, L. R., 286
Zhao, Q., 8*t*, 9–10, 12, 17–18, 21–23
Zhao, W., 285–286
Zhao, X., 78*t*

Zhao, Y., 191
Zheng, Z., 191
Zhou, Q. H., 275–278, 281–282, 282*t*, 283, 283*f*, 284*t*, 286–288, 294–295
Zhu, C., 190–191, 277–278
Zhu, H., 78*t*
Zhu, R. R., 153
Zimmerman, G. A., 36
Zimmerman, T., 128
Zinkernagel, R. M., 190
Zitzmann, S., 78*t*
Zubov, D., 163
Zugmaier, G., 191
Zuilhof, H., 117–118
Zuker, C. S., 103
Zurrer-Hardi, U., 107–108
Zwick, M. B., 229
Zwilling, R., 225–226

Subject Index

Note: Page numbers followed by "*f*" indicate figures, and "*t*" indicate tables.

A

AAA. *See* Antiamyloid antibody
ADCC. *See* Antibody directed cellular cytotoxicity
Affinity and selection, TCR
  CDR loops, 198–200
  high-affinity variants, 200
  mutagenesis, CDR loops, 200
  scTv ligand binding, 199*f*
  T cell display
    design and construction, libraries, 201–203
    isolation, sequences, 206–207
    packaging, transduction and characterization, libraries, 203–205
    receptors, high affinity, 205–206
Affinity–uptake relationship, 261*f*, 262
Antiamyloid antibody (AAA), 277–278
Antibodies
  anti-drug, 162
  cognate, 164–166
  fragments, 257
  humanized/human recombinant, 158–159
Antibody directed cellular cytotoxicity (ADCC), 262
Anticalins
  bacterial surface display library
    DY634 and FITC fluorophores, 177
    FACS, 176
    incubation, 176–177
    mutagenized gene cassette, 176
  biochemical research and drug development, 184–186
  colony screening
    hydrophobic PVDF membrane, 178
    immobilized anticalins, 179
    induction, 177–178
    mutagenized gene cassette, 178
  CTLA-4, 164–166
  genetic library construction, 170–172
  lipocalins, cloning and expression, 166–170
  phage display library
    phagemid selection system, 172–173
    procedure, 173–175
    solid-phase panning, 172–173
  screening, target-binding activity, 179–181
  selection strategies, 165*f*
  structural basis, libraries, 160*f*
  target affinity measurement
    ELISA, 181–182
    surface plasmon resonance, 182–184
Antigen
  cross-reactivity, 180
  cytotoxic T lymphocyte antigen 4 (CTLA-4), 162–163
  specificity, 159–161

B

Bacterial display and flow cytometry
  binding affinity, displayed peptides
    dissociation constants, 89
    individual clones ranking, 90
    kinetics and affinity, 89*f*
    mean cell fluorescence, 89–90
  cell-binding peptides identification, 91–93
  construction, peptide libraries
    cell competency measurement, 86
    electrocompetent cells, 85–86
    insert DNA, 81–82
    protocols, 84–85
    test ligations, 82–84
    vector DNA, 79–81
  display scaffolds
    gram-negative based systems, 76–77
    structurally constrained peptide libraries, 77
  display systems, 79
  peptide affinity maturation, 90
  peptide isolation
    identification, target-binding peptides, 77*f*
    libraries, 78*t*
    MACS and FACS, 76
  protease substrate identification
    CLiPSs, 93
    incubation time and protease concentration, 93–94
    screening, 94*f*
    steps, 93
  sorting, libraries
    avidity and rebinding effects, 88
    positive control bacteria, 88
    protocols, MACS and FACS, 86–87
    streptavidin binders, 88
BBB. *See* Blood–brain barrier
Binding site barrier, tumor-targeting agents
  clearance modulus, 263–264
  dimensionless ratio, inertial, 263
  IgGs, 264

Binding site barrier, tumor-targeting agents (*cont.*)
  rigidity and permanence, 263
  scFv-sized agent, 263–264
  Thiele modulus, 264
  xenografted tumor model, 263
  xenograft experiment, 264, 265*f*
Biopharmaceuticals
  AD and PD, 270–271
  BBB RMT and MTH, 271–273
  genetic engineering, expression plasmid DNA
    DHFR, 278–279
    PCR reaction, 278–279, 280*f*
    tandem vector (TV), 278–279, 279*f*
  IgG fusion proteins
    brain targeting, 283
    brain uptake, 281–282
    CNS pharmacological effects, 284–287
    immune response, 287–288
    pharmacokinetics, 279–281
  recombinant proteins
    AAAs, 277–278
    antibody fusion, functional domains, 277–278, 277*f*
    decoy receptor, 275–277
    HIRMAb, 274
    HIRMAb-GUSB fusion proteins structures, 274–275, 276*f*
    IgG fusion proteins, 274–275, 275*t*
    in vitro model, 274
    neurotrophins, 275–277
    reverse transcytosis, 277–278
    single chain Fv (ScFv) antibody, 277–278
  TNFI and TNFR, 270–271
Blood–brain barrier (BBB)
  disruption, 286–287
  PS product, 281–282
  RMT and MTH, 271–273

C

CD. *See* Circular dichroism
CDRs. *See* Complementarity-determining regions
Cell-binding peptides
  bacterial display libraries, 91
  fluorescence microscopy, 91–93
  screening, 91*f*
Cell-penetrating peptides (CPPs)
  description, 4–5
  negative charge, 28
Cell permeability, stapled peptides
  confocal fluorescence microscopy, 24*f*, 25
  flow cytometry, 23–25, 24*f*
  SAH-p53, 23–25, 24*f*
Cellular libraries of peptide substrates (CLiPSs)
  pBAD33-eCPX-CLiPS, 81*f*
  procedure, 93, 94*f*
  protease substrates identification, 80*f*
CendR peptides
  features, 40
  peptide-mediated systemic delivery, 40–41
Chlorotoxin, 58–59
Circular dichroism (CD), 20–21
"Clickable" DARPins
  analysis
    nonnatural amino acids, 119
    vapor phase hydrolysis, 119
  azido-DARPins, 120
  expression, 118
  IMAC purification, 119
"Click chemistry"
  azide-alkyne Huisgen cycloaddition, 117–118, 117*f*
  methionine analogs, 118
  site-specific PEGylation, DARPins, 120
CLiPSs. *See* Cellular libraries of peptide substrates
Complementarity-determining regions (CDRs)
  antibody, 297
  diversity, 202
  ^{10}Fn3 scaffold, 136–137, 137*f*
  loops
    interactions, 190
    mutagenesis, 198–200
  site-directed libraries, 198–200
Conotoxins
  description, 58–59
  generic α-conotoxin framework, 65*f*
  neuropathic pain treatment, 58–59
  three-dimensional structure, 67*f*
CPPs. *See* Cell-penetrating peptides
Cyclic peptide toxins
  combinatorial synthetic strategy, 71*f*
  conotoxins, 58–59
  cyclization, 58
  design
    backbone cyclization, 59, 60*f*
    distance, N- and C-termini, 59, 61*f*
    linker sequences, 59–60
  resin-splitting strategy, 70
  stability assays
    proteolytic, 68
    serum, 69
    SGF, 68–69
    SIF, 69
  structural analysis
    chemical shift analysis, 66
    distance and angle restraints, 67–68
    NMR, 66, 67*f*
    three-dimensional, 66–67
    two-dimensional TOCSY and NOESY spectra, 66
  synthesis, 60–65
  venomous creatures, 58
Cyclosporin A, 58
Cyclotides, 58
Cystine-knot miniproteins, 224–225

D

DARPins. *See* Designed ankyrin repeat proteins
Designed ankyrin repeat proteins (DARPins)
  biomedical applications
    binding affinity, 121
    "clickable" DARPins (*see* "Clickable" DARPins)
    "click chemistry" (*see* "Click chemistry")
    dissociation constant determination, 122–124
    kinetic parameters determination, 124–127
    purification, "click" PEGylated, 120–121
    quantitative PEGylation, 115–117
    site-specific PEGylation, 120
    stoichiometric cysteine labeling, 113–114
    stoichiometric N-terminal labeling, 114–115
    toxin fusion proteins, 127–129
  clinic, 112
  diagnostics, 107–108
  motivation, 102
  potential therapeutic significance, 112
  properties
    consensus design, 103
    3D structure, 104*f*
    "full-consensus", 105
    hydrophobic interface, 103–105
    self-compatible, 105–106
  pure proteins, 107
  repeat proteins properties
    binding interface, 103
    modules, 103
  selection technologies, libraries
    phage display, 106–107
    phagemids, 106–107
    ribosome display, 106
  targeted tumor therapy
    bivalent binders, 110
    EpCAM, 109–110
    EpCAM-positive tumors, 110
    small-interfering RNA (siRNA), 110
  tumor targeting
    high affinity and specificity, 108, 109
    parameter regions, 108
    PEGylated DARPins, 108–109
  viral retargeting, 111
DHFR. *See* Dihydrofolate reductase
Dihydrofolate reductase (DHFR), 278–279

E

Ecballium elaterium trypsin inhibitor-II (EETI-II)
  AgRP knottins, 226–227
  functional loop, 226
  mutants against trypsin, 227
  structural loop, 226
EETI-II. *See Ecballium elaterium* trypsin inhibitor-II
Enhanced permeability and retention (EPR), 259–260
Enzyme replacement therapy (ERT), 287
EPO. *See* Erythropoietin
EPR. *See* Enhanced permeability and retention
Equilibrium titration
  cellular receptors, labeled ligands, 122*f*
  high-affinity binders, 123–124
  mean fluorescence intensity, 123
  receptor concentration, 124
ERT. *See* Enzyme replacement therapy
Erythropoietin (EPO), 286
Escherichia coli. (E. coli)
  lipocalin and anticalin cloning and expression, 166–170
  microtiter plate expression, 179–181
  XL1-Blue, 174–175

F

FACS. *See* Fluorescence-activated cell sorting
10th Fibronectin type III domain (^{10}Fn3)
  E. coli
    high-throughput production, 152
    individual variants production, 151–152
  fusion proteins
    affinity clamps, 153
    modularity, 153
    partners, 152–153
Fluorescence-activated cell sorting (FACS)
  knottin libraries, 234–235
  library screening protocols, 86–87
  preparation and analysis, 229
  real-time analysis, peptide properties, 76, 77*f*
  scTv fusions isolation, 197–198
  single-color, 235–236
Fluorescence polarization (FP), 21–23
^{10}Fn3. *See* 10th Fibronectin type III domain
Fusion proteins, 152–153

G

Genetic anticalin library
  BstXI restriction site, 170
  cloning, DNA fragment, 171
  electrocompetent bacterial cells, *E. coli*, 171–172
  electroporation, 172
  Lcn2 scaffold, 170
  ligation reaction, 171
  one-pot PCR assembly reaction, 171
  PCR steps, 170–171
GFP. *See* Green fluorescent protein
Green fluorescent protein (GFP)
  aggregation resistance, 296
  excitation and emission maxima, 295–296
  and folds, 297
  and plasmid DNA, 313
  serum-free media, 312

Green fluorescent protein (GFP) (*cont.*)
siRNA, 312
stGFP, 299–300
supercharge, 303
variants, 305–306

H

Homing peptides
bioactive, 52
in vivo studies, 51
in vivo targeting specificity, 38*t*
play-off display, 50
receptors identification, 39–40, 51–52
screen design, 46
uses, 52
vascular, 37–39

I

IgG fusion proteins
brain targeting
molecular Trojan horse, 283, 283*f*
permeability–surface area (PS) product, 283
TNRF, 283
brain uptake
HIRMAb and cTfRMAb, 281–282
mouse ranges, 281–282, 282*t*
PET, 281–282
CNS pharmacological effects
capillary depletion method, 284
cTfRMAb-EPO, 285–286
cTfRMAb-ScFv, 286–287
EPO, 286
HIRMAb-EPO, 284–285
Hurler's syndrome, 287
in vivo, brain, 284–285, 284*t*
MCAO, 284–285
MPSI mice, 287
vs. immune response
chronic treatment study, 287–288
C-region, 288
cTfRMAb-GDNF, 287–288
T-cell immune tolerance, 288
pharmacokinetics (PKs), 279–281

K

Knottins
cell binding assays
measuring high-affinity interactions, 244
receptors expressed, mammalian cells, 243–244, 243*f*
clinical use, ω-conotoxin MVIIa, 225
competition binding assay
factors, 246
incubate cells, 246
use detection method, 246
direct binding assay
cell culture media, 244
FlowJo/CellQuest, 245
incubate cells, 245
PBSA, 245
single concentration, 244, 245*f*
library, yeast-displayed, 227–230
production, chemical synthesis, 237
properties, 225
recombinant expression, *Pichia pastoris*, 241–243
scaffolds, molecular recognition
AgRP, 225–226
combinatorial methods, 226–227
disintegrin proteins, 226
radiotherapy applications, 226–227
structural loop, EETI-II, 226
screening, 234–237
synthetic production
analysis and purification, crude peptide, 240*t*
automated solid phase, 239*t*
epitope tags, 238
folding and analysis, 240*t*
peptide cleavage and deprotection, 239*t*
RP-HPLC purification, 238–239, 238*f*
tag-free format, 238
three-dimensional structures, 224–225, 224*f*
topology, 224–225
yeast surface display
biochemical and biophysical properties, 227
combinatorial technology, 227
FACS, 228–229
genotype–phenotype linking, 229–230
library sizes, 229
MACS, 229
pCT yeast, construction, 228–229, 228*f*
Saccharomyces cerevesiae, 228–229
Kunkel mutagenesis, 142–143

L

Library construction
mRNA display, 145
phage display, 143–144
yeast-surface display, 148
Lipocalins
periplasmic secretion, 166–167
plasmid vectors, 166–167
shake flask scale expression, 167–169
spheroplasts, 167–169
unbound protein removal, 169–170
vectors and libraries, 168*f*

M

MACS. *See* Magnetic-activated cell sorting
Magnetic-activated cell sorting (MACS), 76, 229
Major histocompatibility complex (MHC)
CDR loops, 190
pepMHC

detection, 212–214, 213*f*
endogenous, 213–214
TCR:pepMHC interactions, 190–191
Mammalian cells. *See* Supercharged proteins, mammalian cells
Molecular Trojan horses (MTH)
antibody-based, species specific, 272
conjugation/fusion, pharmaceutical, 272–273
fusion protein, 277–278
MAb against, 272
plasma AUC, 272–273
properties, BBB, 272–273, 272*t*
technology, 288
MTH. *See* Molecular Trojan horses

N

Nanoparticle accumulation, tumor
biodistribution, targeted and untargeted, 259–260, 260*f*
cell internalization, 260–261
dendrimers, 260
EPR effect, 259–260
iron oxide, 260
Natural supercharged human proteins (NSHPs), 302*f*, 307–308
NSHPs. *See* Natural supercharged human proteins

O

Olefin metathesis, 16

P

PCR reaction. *See* Polymerase chain reaction reaction
Peptide
discovery (*see* Bacterial display and flow cytometry)
synthesis
amino acids, 61
anhydrous hydrofluoric acid (HF), 00015:p0700
aqueous buffer, 64–65
chemical ligation reaction, 60–61
% coupling, 61–63
cyclization, backbone, 64, 65*f*
disulfide bonds, 64
double coupling, 63
HBTU/DMF/DIPEA mixture, 63
linear precursors, 62*f*
purification, 64
PET. *See* Positron emission tomography
Phage display. *See also*Target-binding proteins, ^{10}Fn3
anticalins, s0040
vascular ZIP code (*see* Phage display and vascular ZIP code)
Phage display and vascular ZIP code
ex vivo
binding and washes, 47
blocking, 47
cell suspension preparation, 46–47
selections, 48
in vivo
anesthesia, perfusion and dissection, 48–49
display, washes, 49
dosing, 48
spread sheet, 49
molecular basis and exploration
cytotoxic anticancer drugs, 37–39
endothelial cells, 36
homing peptides, 38*t*
in vivo biopanning, 39–40
in vivo screening, 36–37
structure-activity studies, 36–37
T7 phage display (*see* T7 phage display)
Pichia pastoris, 241–243
Polymerase chain reaction (PCR) reaction, 278–279, 280*f*
Positron emission tomography (PET), 281–282

R

Receptor-mediated transport (RMT) systems
AAA, 277–278
BBB, 272
in vivo CNS pharmacologic effects, 271–272
insulin receptor/TfR, 271–272
ligands, 272
scavenger receptor (SR), 271–272
Rink Amide MBHA resin, 12–15
RMT systems. *See* Receptor-mediated transport systems

S

SGF assay. *See* Simulated gastric fluid assay
SIF assay. *See* Simulated intestinal fluid assay
Sigmoidal curve fit, 258–259
Simulated gastric fluid (SGF) assay, 68–69
Simulated intestinal fluid (SIF) assay, 69
siRNA. *See* Small interfering RNA
Small interfering RNA (siRNA)
plasmid DNA, 312–313
supercharged protein, 312
SOE. *See* Splicing by overlap extension
Solid-phase peptide synthesis (SPPS)
Fmoc-based, 15*f*, 19
ruthenium-mediated olefin metathesis, 6–7
Soluble scTv proteins
E. coli
design and cloning, 209
high-affinity, 208
growth and induction, 209–210
isolation, inclusion bodies

Soluble scTv proteins (*cont.*)
microfluidization, 210
pellet, 210–211
monitoring, ELISA
biotinylated scTv A6, 215
human Vα2 domain, 214
quantitation, 215*f*
pepMHC epitopes, 212–214
purification
His_6 sequence, 211–212
size exclusion chromatography, 212, 212*f*
solubilization and refolding, 211
Splicing by overlap extension (SOE), 197
SPPS. *See* Solid-phase peptide synthesis
Stapled peptides
biophysical characterization
CD, 20–21
FP and SPR, 21–23
p53 α-helix, 22*f*
pH modulation, 21
proteolytic susceptibility, 23
cell permeability (*see* Cell permeability, stapled peptides)
cleavage, solid support, 19–20
design
MAML peptide, 10
negative control, 11–12
NOTCH-CSL binary complex, 10, 11*f*
NOTCH-CSL-MAML ternary complex, 10–11
protein-protein interactions, 9–10
α-helical
bioactive, 8*t*
C-C bond-forming reaction, 6–7
description, 6
multiple staples incorporation, 9
optimized versions, 7–9
types, 7*f*
in vitro target interaction and activity assays
antibody, 26
BID $SAHB_A$ peptide, 27
immunoprecipitation/pull-down assays, 26
NOTCH-dependent transcriptional activation, 27
survival and proliferation, cancer cells, 26–27
in vivo efficacy, 27–28
N-terminal and internal modifications
biotinylation, 18
categories, 16–17
charged moiety, 16
fluorescein and rhodamine, 17–18
lysine/cysteine, 18–19
N-α-Fmoc-Lysine(Mmt)-OH, 19
structure, 17*t*
olefin metathesis, 16
purification, 20
strategies
aqueous solubility, 29–30
hydrophobic patch, 28–29
mutations, 30
optimization, 29*t*
SAH-p53, 28
synthesis
acid-labile side chain protecting groups, 13*t*
deprotection and coupling, 14–15
dry and shrink, resin, 16
Fmoc-based SPPS, 15*f*
Rink Amide MBHA resin, 12–15
SPPS, 12
Stoichiometric cysteine labeling
Alexa Fluor-488-C5-maleimide, 113, 114
DARPins, molar absorbance, 114
elution behavior, 113–114
labeling efficiency, 114
site-directed mutagenesis, 113
Streptavidin
binders, 88
identification, target-binding peptides, 77*f*
SAPE solution, 87
streptavidin–phycoerythrin, 88, 90
Streptavidin-R-phycoerythrin (SAPE) solution, 87
Supercharged proteins, mammalian cells
aggregation resistance
centrifugation, 297
GFP variants, samples, 296, 296*f*
His_{39} GFP, samples, 297, 297*f*
pathways, 296
thermal and chemical induced, 296
alpha carbon to beta carbon vector, 304
cation exchange and endotoxin removal, 310–311
cationic, 294–295
cell penetration
cell-surface, 298–299
fluorescence microscopy, 298
GFP variants, 299–300, 299*f*
His39 GFP, 299–300, 300*f*
HIV Tat, peptide, 297–298
internalization assays, 298
membrane recycling, 297–298
synthetic oligo-Lys/Arg peptides, 300
trafficking kinetics, 298
discovery and properties
aggregation-resistant and cell-penetrating, 295–296
electrostatic surface potentials, GFP, 295–296, 295*f*
extensive mutagenesis, 295–296
DSSP, 304
expression
E. coli strains, 308
materials, 308*t*
small-scale cultures, 308–309
functional, screening

AvNAPSA method, 305–306
DNA shuffling, 306
E. coli., 305–306
and generating, 305
negative, GFP variants method, 305–306
nonnative structure, 305
solvent exposure *vs.* mutability, 305
X-ray/NMR structure, 305
fusion proteins, construction
architecture, 311, 311*f*
glycine-serine linker, 312
GFP, 303
identification
human proteins expression, *E. coli*, 307–308, 307*f*
immune system response, 307–308
protocol, 307, 308
QSAR, 308
macromolecule
general and potent vehicles, 300–301
lipid-mediated transfection, 300–301
mCherry providers, 301
NSHPs deliver, *in vitro* and *in vivo*, 301, 302*f*
translational fusion, 301
macromolecule therapeutics, 294
nucleic acid delivery, *in vitro*, 312–313
PDB file, 304
protein delivery
in vitro, 313–315
in vivo, 315–317
purification
desalting/dialysis, buffer, 310
His-tagged proteins, 309
PAGE, MALDI-TOF mass spectrometry size, 310
phospholipids and nucleic acids, 310
streptavidin, 303–304
theory
amino acid side chains, 301–303
streptavidin, 301–303
thermodynamice, protien folding, 301–303
three-dimensional structures, 303
tissue types, therapeutic, 317–318
Synthetic peptides, *in vivo* homing, 50–51

T

Target-binding proteins, ^{10}Fn3
combination display, 151
crystal structures, antigen-binding domain, 137*f*
immunoglobulins, 136
library design
diversity, 138
principles, 138
"monobodies", 137
mRNA display
affinity maturation, 147
clone isolation and characterization, 147
feature, 145
library construction, 145
selection, 145–146
phage display
affinity maturation, 144
clone isolation and characterization, 144
library construction, 143–144
sorting, library, 144
vector design and display optimization, 142–143
production
expression, inside cells, 153
fusion proteins, 152–153
high-throughput, *E. coli*, 152
individual variants, *E. coli*, 151–152
selection platform
aggregated/misfolded species, 139
biotinylation, 139–141
directed-evolution methods, 140*t*
in vitro selection, 139
solution conditions, 141
surface loops, 136–137
yeast-surface display, 147–151
T cell receptor (TCR)
affinity
and selection (*see* Affinity and selection, TCR)
threshold, 191
coreceptors CD4 and CD8, 191
design and cloning
analysis, flow cytometry, 194–195
detection, folded scTv variable domain, 194*f*
LiOAc-mediated transformation, 194
"negative" cells, 194–195
pCT302 vector, 192–193
scTv gene cloning, 192–193
T cell clones, 192–193
yeast display, 192, 193*f*
intracellular epitopes, 191
MHC ligands, 190
recipes, media and buffers, 215–218
soluble scTv proteins
design and cloning, *E. coli*, 209
growth and induction, 209–210
high-affinity, *E. coli*, 208
isolation, inclusion bodies, 210–211
monitoring, ELISA, 214–215
pepMHC epitopes, 212–214
purification, 211–212
solubilization and refolding, 211
stability and selection
electrocompetent EBY100 cells, 195–197
plasmid DNA, 198
random mutagenesis, 195
scTv library, 195–197, 196*f*
stabilized scTv mutants, 197–198
temperature-based selections, 198

TCR. *See* T cell receptor
T7 phage display
  amplification and purification
    density gradient purification, 44
    stocks preparation, 44
  cross contamination, 41
  detection and amplification, noninfectious phage, 45–46
  libraries construction and individual peptide phage
    cloning, 42
    oligonucleotides design, 42, 43*t*
  media, amplification, 41
  titration and sequencing, 44–45
  vectors and *Escherichia coli* strains, 42
Tumor-targeting agents
  affinity and biodistribution
    ADCC, 262
    nonresidualizing radioisotopes, 262
    pharmacodynamic effect, 262
    plasma clearance kinetics, 262
    series, anti-HER2 Darpins, 262
    topographical plot, size effect, 261, 261*f*
  binding site barrier, 263–264
  molecular size
    advantage, 257–258
    compartmental model, relationship, 258–259
    DARPin uptake, 257*f*, 258
    vs. death valley, 257
    genetic and materials engineers, 256
    hydrodynamic diameters, 257–258
    measurement, metabolic half-life, 259
    relationship, uptake, 256, 257*f*
    sigmoidal curve fit, 258–259
    time trajectory, 259
    vascular extravasation rate drops, 256
    vascular permeability coefficient, 259
  nanoparticle accumulation (*see* Nanoparticle accumulation, tumor)
  peptidic scaffolds, 256

V

Vascular ZIP code mapping
  CendR peptides, 40–41
  in vivo homing, synthetic peptides
    configuration, 51
    fluorescent labels, 51
  peptide receptors identification
    affinity chromatography, 51–52
    colocalization studies, 52
    importance, 51
  phage biopanning, 46
  phage display
    ex vivo, 46–48
    in vivo, 48–49
    molecular basis and exploration, 36–40
    T7 phage display, 41–46
  postscreening phage auditions
    titration-based, 49–50
    titration-independent, 50
Viral retargeting, 111

Y

Yeast-displayed knottin library
  construction
    assembly and amplification methods, 231
    DMSO, 230–231
    generation, 230–231, 231*f*
    gene synthesis costs, 231–232
    NNS/NNK codons, 230–231
    protein and peptide loops, 230
    RGD tripeptide, 230
  DNA
    amplification reaction, 232*t*
    assembly reaction, 232*t*
    transformation, 233*t*
  screening
    c-myc antibodies, 235–236
    FACS, 234
    glycerol stocks, 235–236
    isolate mutants, 236
    kinetic off-rates, 236
    kinetic sorts, 236
    negative sorting, 236
    recombinant/synthetic approach, 237
    second-generation, 237
    single-cell resolution, 234–235
    sort progression, 234–235, 235*f*
    streptavidin/antibodies, 236
Yeast-surface display
  affinity maturation, 151
  clone isolation and characterization, 150
  ^{10}Fn3 variant, 148
  library
    construction, 148
    sorting, 149–150
  selection unit, 148

A

N-terminal capping AR

α1 α2

MRGSHHHHHHGSDLGKKLLEAARAGQDDEVRILMANGADVNAx

Designed AR module

βt α1 α2

DxxGxTPLHLAAxxGHLEIVEVLLKzGADVNAx

C-terminal capping AR

βt α1 α2

QDKFGKTAFDISIDNGNEDLAEILQ

B

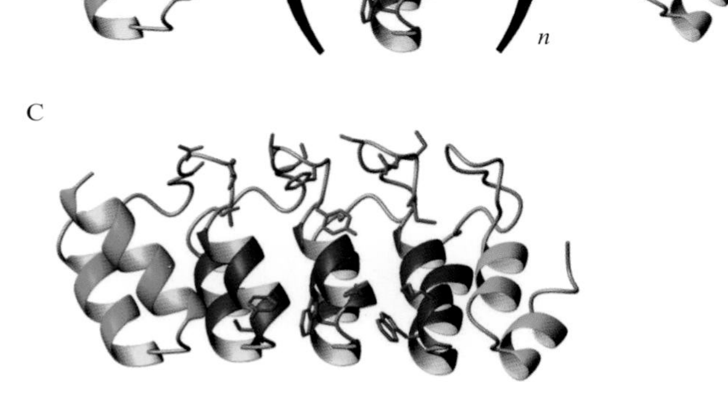

C

Rastislav Tamaskovic, *et al.*, Figure 5.1 DARPin modules and typical DARPin 3D Structure. (A) Sequences of the N-terminal capping ankyrin repeat (AR), the designed AR module, and the C-terminal capping AR. The secondary structure elements are indicated above the sequences. The designed AR module consists of 26 defined framework residues, 6 randomized potential interaction residues (red x, any of the 20 natural amino acids except cysteine, glycine, or proline), and 1 randomized framework residue (z, any of the amino acids asparagine, histidine, or tyrosine). The designed AR module was derived via sequence and structure consensus analyses. (B) Schematic representation of the library generation of designed AR proteins (DARPins). Note that this assembly is represented on the protein level, whereas the real library assembly is on the DNA level. By assembling an N-terminal capping AR (green, left), varying numbers of the designed AR module (blue, middle), and a C-terminal capping AR (cyan, right), combinatorial libraries of DARPins of different repeat numbers were generated (side chains of the randomized potential interaction residues are shown in stick-mode in red). (C) Ribbon representation of the selected MBP binding DARPin off7 (colors as in B). This binder is derived from a library consisting of an N-terminal capping AR, three designed AR modules, and a C-terminal capping AR. Figure reproduced from Binz *et al.* (2004).

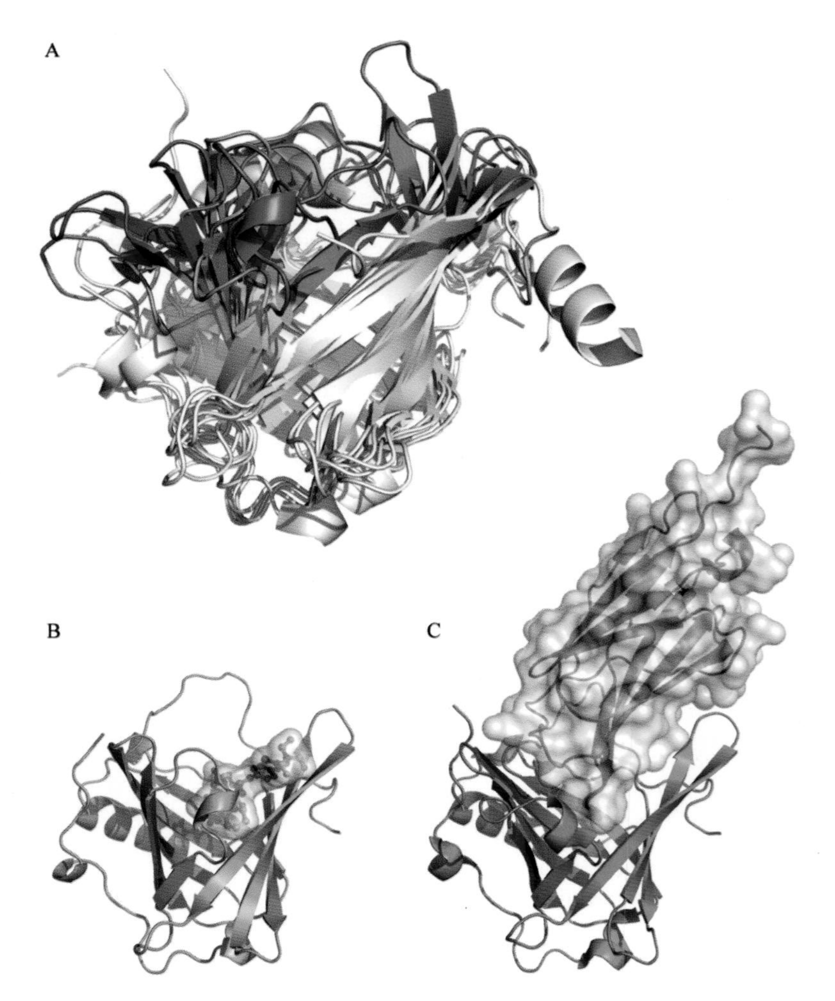

Michaela Gebauer and Arne Skerra, Figure 7.1 Structural basis for the design of Anticalin libraries. (A) Conformational plasticity in the loop region of natural lipocalins. The crystal structures of six human members of the lipocalin family—RBP, Tlc, ApoD, AGP, NGAL (Lcn2), c8γ (PDB entries 1RBP, 3EYC, 2HZQ, 3KQ0, 1L6M, 2QOS, respectively)—were superimposed via their β-barrel Cα positions (Skerra, 2000b). While the β-barrel itself as well as the loops at its bottom show a conserved three-dimensional structure, the four loops at the open end (top, colored) exhibit high conformational variability, not unlike the hypervariable region of immunoglobulins. (B) Crystal structure of an Anticalin based on Lcn2 recognizing a small molecule (PDB entry 3DSZ). Ribbon diagram of the Anticalin selected against an Y(III)–DTPA derivative, which is depicted as ball and sticks with a translucent surface (Kim *et al.*, 2009). Charged with suitable radionuclides, this Anticalin offers applications in radio-immunotherapy (RIT) and diagnostics (RID). (C) Crystal structure of an Lcn2-based Anticalin recognizing a macromolecular target (PDB entry 3BX7). Ribbon diagram of the Anticalin selected against the extracellular domain of the deactivating T-cell coreceptor CTLA-4, which is depicted in darker color with a translucent surface (Schönfeld *et al.*, 2009). This Anticalin efficiently blocks the negative regulatory activity of CTLA-4 and thus shows potential for stimulating the cellular immune response to treat infectious diseases or for the immunotherapy of cancer.